BASIC TECHNICAL DRAWING

7th Edition

HENRY CECIL SPENCER

JOHN THOMAS DYGDON

JAMES E. NOVAK

Glencoe
McGraw-Hill

New York, New York   Columbus, Ohio   Woodland Hills, California   Peoria, Illinois

ACKNOWLEDGMENTS

The Publisher gratefully acknowledges the cooperation and assistance received from the many individuals and organizations who have contributed to this book. Special recognition is given to the following reviewers:

Gerald Bradshaw
Technical Drafting Instructor
Crown Point High School
Crown Point, IN
Purdue University

John Clark
Ben Davis High School
Indianapolis, IN

Terry Cobb
Hendersonville High School
Hendersonville, TN

Leonard Groce
Starmount High School
Booneville, NC

Bill Hyde
Caprock High School
Amarillo, TX

Warren Jennison
Seminole High School
Sanford, FL

Carole Kersey
Kendrick High School
Columbus, GA

Ronald F. Logan
Drafting/Design Technology
   Instructor
Vocational Dep't. Chair
Austin-East Magnet High School
Knoxville, TN

Roger McSween
Ryan High School
Denton, TX

Bill Ross
Wynne High School
Wynne, AR

The American National Standards Institute (ANSI) tables in the Appendix have been reproduced from the indicated publications with the permission of the publisher, The American National Standards Institute, 11 West 42nd Street, New York, NY 10036.

NOTICE
The Internet listings in this book are a source for extended information related to our text. We have made every effort to recommend sites that are informative and accurate. However, these sites are not under the control of Glencoe/McGraw-Hill, and, therefore, Glencoe/McGraw-Hill makes no representation concerning the content of these sites. We strongly encourage teachers to preview Internet sites before students use them. Many sites may eventually contain "hot links" to other sites that could lead to exposure to inappropriate material. Internet sites are sometimes "under construction" and may not always be available. Sites may also move or have been discontinued completely by the time you or your students attempt to access them.

Glencoe/McGraw-Hill
A Division of The McGraw-Hill Companies

Send all inquiries to:
Glencoe/McGraw-Hill
3008 W. Willow Knolls Drive
Peoria, IL 61614-1083

ISBN 0-02-682553-8 (Student text)
ISBN 0-02-682554-6 (Student Workbook)
ISBN 0-02-682555-4 (Teacher's Resource Binder)

Printed in the United States of America.

2 3 4 5 6 7 8 9 10   027   03 02 01 00

CONTENTS IN BRIEF

About the Authors

HENRY CECIL SPENCER was Professor Emeritus and Director of the Department of Engineering Graphics, Illinois Institute of Technology. He previously taught engineering drawing at Texas A & M University and mechanical drawing at Ballinger High School, Ballinger, Texas. He authored or coauthored twelve books in the field of technical drawing and engineering graphics. He served as national chairman of the Division of Engineering Graphics of the American Society for Engineering Education and received its Distinguished Service Award. Professor Spencer passed away in June 1972. This book is a tribute to him and will serve to pass on his knowledge.

JOHN THOMAS DYGDON is Professor Emeritus of Engineering Graphics, formerly Chairman of the Department and Director of the Division of Academic Services at Illinois Institute of Technology. Prior to joining the IIT faculty in 1952, he worked in industry as an engineer and designer and continues to serve as an engineering and management consultant. He has authored or coauthored fifteen books in the field of technical drawing, engineering graphics, and management. He is a member of many engineering and management societies, including the Division of Engineering Design Graphics of the American Society for Engineering Education.

JAMES E. NOVAK joined the faculty of the Department of Engineering Graphics at Illinois Institute of Technology in 1967 and currently teaches in the Department of Civil and Architectural Engineering. He has coauthored ten other texts and workbooks, has served as a consultant to industry, and is a member of several professional societies, including the Division of Engineering Design Graphics of the American Society for Engineering Education.

CONTENTS

PREFACE

Basic Technical Drawing is a classroom textbook and reference book for the beginning student of technical drawing. It provides essential information as clearly and completely as possible in straightforward language and with ample illustrations.

This new edition provides updated information on the basics of technical drawing; yet it retains the practical focus and open format that have made it popular with thousands of students.

RATIONALE

Anyone can use this book. No prior training or skill in technical drawing is assumed. *Basic Technical Drawing* is designed primarily for the beginner, whether in high school, community college, a technical institute, or a four-year college. It is for those interested in technical drawing because they wish to be a drafter, an engineer, an industrial designer, or an architect. It is designed also for those with an interest in one of the many other professions that require a knowledge of technical drawing. Even the student of fine arts will find much useful information about freehand sketching, pictorial drawings, and mechanical drafting techniques. For everyone, this book will provide some insight into modern industry and technology.

It is the authors' firm conviction that technical drawing is so interesting a subject, and its value in a technological civilization so immense, that the primary objective of *Basic Technical Drawing* should be to teach the fundamentals of its subject matter so that the student can achieve real skill through solid accomplishment and thorough understanding.

This book has been designed to be as completely self-teaching as possible. The teacher is relieved of many tedious explanations of minor details. He or she may devote time and effort to teaching the more important principles in an effective manner.

ORGANIZATION

The organization of *Basic Technical Drawing* was dictated by the subject matter. Although each chapter deals with a different topic, the sequence of chapters progresses from the elementary to the advanced. This sequence follows the topic sequence of most drawing courses. Teachers, however, can easily adapt this book to their own course requirements and preferences in organization. For example, if the subject of dimensioning is not discussed in class until the chapter on dimensioning (Chapter 10) is reached, no drawings need be dimensioned until then. If, however, instructors wish to teach dimensioning gradually and require students to dimension drawings from the beginning, they may refer to Chapter 10 as necessary.

Complete explanations and frequent cross-references in each chapter allow the teacher to use this book effectively, regardless of any special emphases of the course, the order of topics, or the individual requirements of students.

To make the subject matter more logical and understandable, all important terms are defined, explained, and illustrated in detail. The technical terms include not only those concerned with drafting, but also those associated with the related manufacturing processes. A special feature of this revision is an expanded glossary of key terms in technical drawing. Thus, *Basic Technical Drawing* can be a valuable reference for the trained craft worker as well as for the professional drafter, technician, or engineer.

DRAWING PROBLEMS

Many practical problems have been provided. A drafting text would not be complete without them. The problems, arranged from the easy to the more difficult, are given after each chapter. Every problem is one that is encountered in industry. All have been thoroughly tested in class use in all parts of the United States.

The early problems in each group are given in sheet-layout form so that the beginner can get started without difficulty. Most are designed to fit a standard 8.5" x 11.0" sheet. The use of large sheets partitioned for individual problems is avoided. Beginners can work more neatly and easily by using one smaller sheet for each problem. In addition, they derive a greater sense of accomplishment as they complete each problem and give it to the instructor for evaluation rather than using a larger sheet and completing additional problems before the first problems are graded.

In addition to the problems in this text, a complete workbook has been prepared especially for use with this text: *Basic Technical Drawing Student Workbook,* by H.C. Spencer, J.T. Dygdon, and J.E. Novak.

Text material, illustrations, and problems have been updated in accordance with the standards of the American National Standards Institute (ANSI), particularly those dealing with drafting standards. The authors provide introductory information about metric dimensioning at the end of Chapter 10.

AUTHORS' ACKNOWLEDGMENTS

The authors wish to thank the many companies, schools, and individuals who have contributed material or provided suggestions for this book. We especially appreciate the assistance of Mr. Robert Krawczyk, Department of Architecture, Illinois Institute of Technology; Professor Jamshid Mohammadi, Chair, Department of Civil and Architectural Engineering, Illinois Institute of Technology; and Mr. Stephen A. Smith, Director of Product Development, Plastofilm Division, IVEX Packaging Corporation.

Special thanks are due to Michael J. Eleder, President, Instructional Design Systems, who wrote Chapters 3 and 9 and contributed other material related to computer-aided drawing (CAD). We are also most grateful to our editor, Ms. Trudy Muller, for her insightful assistance and encouragement.

The authors hope that both teachers and students will continue to feel free to write with their suggestions and criticisms.

John Thomas Dygdon

James E. Novak

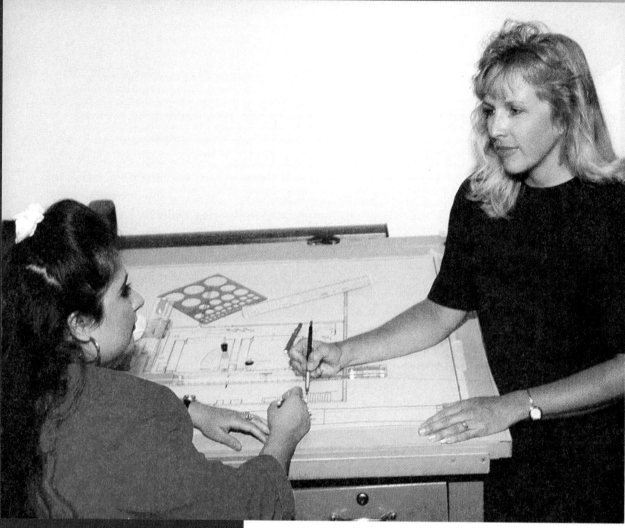

Arnold & Brown

IMPORTANT TERMS

working drawing

architectural drawing

machine drawing

structural drawing

sheet-metal drawing

electrical drawing

aeronautical drafting

marine drawing

entrepreneurs

OBJECTIVES

After studying this chapter, you should be able to:

- Outline the procedure for identifying a career for which you would be suited.

- Identify the main stages in the development of a drawing.

- Identify the main branches of technical drawing.

- Identify the six specific functions in each branch of engineering.

- Describe the steps of the problem-solving process.

CHAPTER 1

The Graphic Language

This chapter describes the skills and behaviors needed for a successful career in the world of work. While several careers relating to technical drawing are discussed, the chapter also presents general advice regarding career growth.

Technical drawing is the oldest type of written expression. This chapter describes the types of technical drawings and the historical uses of drafting. While the techniques may change, drafting will continue to be essential in the future.

Now and in the future, successful design requires people who can think creatively and work together. Special emphasis is given to the need for teamwork. The skills needed by each member of a design team are identified and discussed.

CAREER LINK

Skills USA-VICA is a national organization with nearly 250,000 members, including high school and college students as well as teachers and school administrators. Many corporations, trade associations, and labor unions work with Skills USA in support of its goal to provide quality education in leadership, teamwork, citizenship, and character development. To learn more about this organization, including the Skills USA Championships, write to National VICA, P.O. Box 3000, Leesburg, VA 20177 or visit their web site at *www.skillsusa.org*

The American Design Drafting Association, established in 1959, is a professional society that brings together industrial design drafters, educators, and students. To promote recognition of the drafter as a professional, ADDA has established a Drafter Certification Program. For drafting students, ADDA sponsors a national design drafting contest. There are ADDA student chapters at schools throughout the United States. For more information about ADDA, write to ADDA, P.O. Box 11937, Columbia, SC 29211 or visit their web site at *www.adda.org*

1.1 YOUR LIFE'S WORK. Have you decided yet what your life's work will be? If not, you should start thinking very seriously now so that you can intelligently prepare for it.

The first rule in selecting a vocation is to find the activities in which you are deeply *interested.* For example, you might be interested in commercial work such as building construction. Your friends might like scientific work such as engineering or research.

The most successful people are not those who set out to make a lot of money. The most successful are those who choose work they enjoy. When you like your job, you perform so well that people are willing to pay well for your performance. When you do a good job, the money will take care of itself.

Do not make the mistake of assuming that you will necessarily be successful in a given field just because your parents, relatives, or friends have been successful in that line of work. The kind of work that is best for you is the work *you* like to do.

The second rule is to select a type of work that you can *do well.* Everyone has some special abilities. There are some things that you can do better than the next person. Your challenge is to learn as much as possible about a wide variety of jobs or professions. You then need to select the one that you like best and can do best. How do you select such a job? Your library has some good books on vocational guidance. Read some of them. See your school counselor who can give you expert advice. In most larger cities there are vocational testing agencies where you can take scientific tests to find out what your abilities and interests are.

Seize every opportunity to learn by firsthand observation any time you are around a business office, a construction site, or a manufacturing plant. Talk to people. Ask questions. Learn all you can about the jobs that interest you. Continue until you find a job you like and can do well. Avoid those types of work that clash with your known weaknesses. For example, if you are consistently poor in mathematics, do not plan to be an engineer.

Don't forget for a moment that regardless of your interests and abilities, your success will depend largely upon your *personality* and *character.* You must be able to get along with other people. You must be absolutely honest. If you are a "square shooter," everyone will know it. If you are not, they will know that, too.

You must be dependable at all times. No one will have confidence in you or give you any important responsibility if you cannot be depended on to do what you are supposed to do.

You must be a hard worker. Thomas A. Edison once said, "Success is one-tenth inspiration and nine-tenths perspiration."

Think on the job. Do things that ought to be done *before* the boss reminds you. Be cooperative. Be a team worker. Industry has no place for people who cannot work with others.

In this wonderful age of science and technology, opportunities are greater than ever before. Each new discovery, invention, or improvement opens up new horizons. These discoveries allow scientists, engineers, designers, drafters, mechanics, technicians, businesspeople, teachers, lawyers, manufacturers, and many others to make further advancements. There are jobs for qualified people, regardless of their particular abilities, if they will only find them. Several of these jobs are discussed towards the end of this chapter.

1.2 THE LANGUAGE OF DRAWING. In a technical civilization such as ours, there are many good opportunities in technical or mechanical fields or in allied fields, Fig. 1-1. If you are interested in planning or building things, you may find your life's work somewhere in these areas. To help you "find yourself," many schools offer a wide variety of technical courses such as technical drawing, manufacturing, construction, and electronics. Technical drawing is considered basic to all of these courses. It is the principal means of expressing ideas in a technical world. Drafting is a *graphic language*. It has its own alphabet and grammar.

The industrial history of the United States has been written in terms of the graphic language. If you do not understand this language, you will be illiterate. Even if you find yourself in a field only indirectly associated with industry, a knowledge of the graphic language is essential in order for you to read blueprints. Regardless of your future vocation, you will have many uses for your knowledge of technical drawing.

Technical drawing is the oldest type of written expression. It is understood the world over. A word is an abstract symbol representing a thing or an idea, but a picture represents an object the way it appears in real life. Confucius said that "one picture is worth a thousand words." To understand the truth of this statement, try to tell in words how to build a house, a model, or a mechanical device. Imagine how hard it would be to tell someone how to build a jet plane. You cannot accurately and completely describe in words how to make simple objects like a wood screw or gear. However, no object is so complicated that it cannot be drawn. In fact, if it cannot be drawn, it cannot be made.

ComputerVision Corp.

Fig. 1-1. Modern tools of the graphic language.

Technical drawing even has great value for nontechnical people. It allows you to communicate ideas clearly and effectively. Usually this is done by means of a freehand sketch on the back of an envelope or on a piece of scratch paper. How many times have you heard someone say, when words have failed, "Oh, I guess I'll have to draw a picture"? If you have good ideas, but cannot communicate them to other people, your ideas will not be used.

1.3 INDUSTRIAL DRAFTING. Aside from general idea-generating sketches, there are two main classes of drawings: artistic and technical. The artist expresses philosophic or aesthetic ideas or emotions. The objects created by an artist are drawn in the artist's own personal style. An artist's drawing looks the way the artist sees the object, not necessarily the way the object really appears. Personal style is one of the main objectives of artistic drawings. The technical person is concerned with how objects really are. Technical drawings show objects as true to life as possible. *Accuracy* is one of the main objectives of a technical drawing.

Every new invention or development starts with an idea in the mind of the originator. Engineers, inventors, or designers usually create the first drawings of their design themselves. They are the only people who can express exactly what their design will look like. These people are usually well trained in technical drawing. They find it easy to communicate their new ideas this way.

Refining a new idea usually takes several stages of development. The first stage of development is often a *freehand sketch*. This sketch is followed by several more accurate *mechanical drawings*. Each mechanical drawing is accurately created with precision drawing instruments or on a computer.

When the designer is satisfied with the development of the idea, *working drawings* are created for the manufacturing shop. The designer may create these drawings, or he or she may turn them over to a *drafter* to create. When creating working drawings, many of the construction details are finalized.

The drafter may also be required to assist in the design process. Salaries of drafters depend upon how much designing they are able to perform. The beginning drafter is in a favorable position to learn all about the products of a company and may advance into key positions if his or her ability merits promotion.

Let us do one thing at a time. Your job right now is to learn how to make drawings skillfully, correctly, and quickly.

1.4 WORKING DRAWINGS. A *working drawing* is a complete drawing or set of drawings prepared so that the object represented can be produced without additional information. The working drawing is a complete size and shape *description* of the object. The drawing is composed of two parts, the *views* and the *dimensions*.

Just like the written language, the graphic language has its own grammar and symbols. Examples of these are the *alphabet of lines* and the rules of presentation.

By using *views,* the people reading the working drawing can visualize the objects that must be produced. If the working drawing is correctly drawn, you can visualize exactly what the designer had in mind. This ability to visualize, or "think in three dimensions," is essential to the designer, the drafter, and the people who must build the object. Visualization is one of the principal skills you will learn in a technical drawing course. It is also one of the best means of developing the "constructive imagination" that is essential in all original design work.

To build an object, you need to know the exact size of each part. The *dimensions* tell you the size and shape of each part of the design. Dimensions can describe large sizes measured in hundreds of feet as well as extremely tiny measurements as small as ten-thousandths of an inch.

The working drawings tell a complete story. Reproduction *prints* made from these drawings can be sent to manufacturing plants in any country in the world to be built. Also, all the separate parts described in the drawing can be made exactly as designed. They can then be shipped to a central assembly plant where the parts can be assembled as originally designed.

1.5 THE BRANCHES OF TECHNICAL DRAWING. The term *mechanical drawing* is often used to describe industrial drawings. However, the term does not include *freehand sketching,* which is an important part of this subject. Therefore, *technical drawing* has become the accepted term today. It more accurately describes the broad scope of drawing for industry.

Technical drawing is composed of many specialized types of drawing applied to various fields. *Architectural drawing* is used in the building industry. *Machine drawing* is used in the manufacturing industries. *Structural drawing* is used in the construction industries where structural steel is used for large buildings and bridges. *Sheet-metal drawing* is used in the heating, ventilating, and air-conditioning industries. *Electrical drawing* is used in the electronics industries. *Aeronautical drafting* is used in aircraft manufacturing. *Marine drawing* is used in ship construction. Some industries use the term *drafting* instead of *technical drawing*.

The terms *engineering graphics, graphic science,* and *graphics* are generally used to describe drawing courses at the college level. These terms accurately characterize the scope of work in drawing at this level. Such courses include graphic solutions to engineering and mathematical problems, as well as the description of objects for manufacturing.

Computer graphics is a term applied to the use of computers to produce drawings, designs, graphs, charts, and other graphical data.

1.6 THE GOALS OF TECHNICAL DRAWING. First, a technical drawing must be *accurate.* The requirements of industry are exacting. Drafters must be accurate in everything they do. A drawing that is not accurate may be worthless. Such a drawing may lead to costly errors by those who depend on the information described in the drawing. The success of industry depends on accurate technical drawings.

Second, a technical drawing must be drawn with the proper *technique,* or good workmanship. This means that the lines must have "sparkle" or "snap" and exhibit good contrast. A sloppy drawing does not show good tech-

nique. It is likely to be incorrect or unclear.

Third, a technical drawing must be *neat.* Neatness is a habit that can be acquired. It is gained by observing orderly arrangement and handling of the equipment and keeping the drawing clean.

Fourth, technical drawings must be made with *speed,* for "time is money." Slow drafters will soon find themselves looking for other jobs. Speed in drafting comes from mental and physical alertness. It comes naturally as a result of concentration on the job and intelligent planning. Slowness is the inevitable product of a dull or disinterested mind. A slow drafter can become faster by working continuously and not wasting valuable time talking, daydreaming, or trying to work out solutions by trial and error without first learning the fundamental principles.

The following quotation from a drafter's handbook illustrates the attitude of industry toward a drafter's work:

"Our drawings are considered by the management, the buyers, the outside sources, the pattern and die makers, the inspectors, and the shop supervisor as the last and only word in specifications. Our drawings must stand alone in conveying the ideas of the Chief Engineer to the thousands of people who use them. They must tell all that needs to be known about the parts they represent. They must be so clear and complete that every one of the thousands of users arrives at exactly the same interpretation."

1.7 MONUMENTS OF HISTORY. Technical drawing has played an important part in human progress. It has been used from the time of the ancient Egyptian pyramids and the classical Greek Parthenon to the geodesic domes of Buckminster Fuller and NASA's

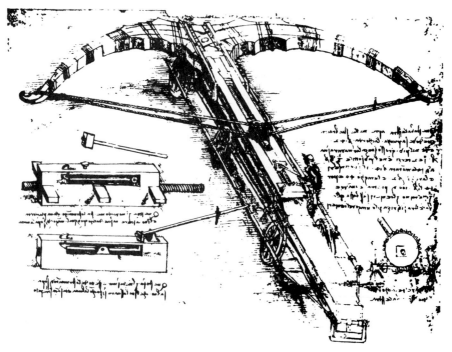

Lieb Museum

Fig. 1-2. Fifteenth-century design by Leonardo da Vinci for a great crossbow.

Massachusetts Historical Society

Fig. 1-3. One of Thomas Jefferson's plans for Monticello, his home in Virginia.

space stations. Drafting has made it possible to develop and record a vast amount of knowledge. Without drafting, technological progress would have been much slower.

Drafting is considered a universal language that can be understood around the world. The methods used in drafting did not develop suddenly. Men and women struggled for centuries with the problem of representing three-dimensional objects on a flat drawing surface with only two dimensions. A giant step forward was made by Leonardo da Vinci (1452-1519). He was a master of art and technical drawing. His three-dimensional sketches were easy to understand. His studies were published in 1651. There are 7,000 pages of his notes and sketches that reveal his genius as a designer, architect, and inventor, Fig. 1-2.

In the eighteenth century, the Industrial Revolution ushered in the machine age. This was dependent on sketches and detailed drawings for new machinery and industrial plants. It was during this time that drafting was labeled the "Language of Industry."

1.8 AMERICAN KNOW-HOW. Americans have always made use of the graphic language. Many prominent leaders used instruments for preparing accurate technical drawings. George Washington was a surveyor. His military officers used drafting skills for mapping new territories and preparing drawings of military structures. Thomas Jefferson prepared house plans for Monticello, his beautiful mansion in Virginia, Fig. 1-3. He also prepared plans for the University of Virginia.

Cadets from West Point, the United States military academy, led the way in preparing plans for structures needed by the military and government. They learned how to prepare technical drawings for the construction of waterways, dams, bridges, lighthouses, and buildings. They were perhaps the first academically trained engineers in the United States, Fig. 1-4.

Technical drawing is the most precise graphic method of recording and communicating our ideas. This is why applications for a patent often include technical drawings. (A *patent* gives an inventor exclusive rights to his

United States Military Academy

Fig. 1-4. Minot's Ledge lighthouse, which was designed by West Point cadets.

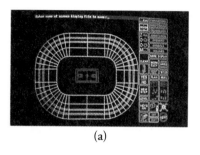
(a)

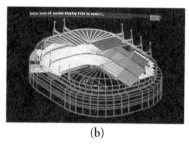
(b)

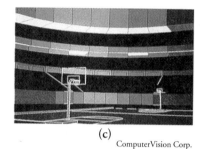
(c)

ComputerVision Corp.

Fig. 1-5. Computer-generated plans for an athletic complex: (a) the general plan, (b) cutaway pictorial, (c) interior pictorial.

or her invention. For a specified number of years, no one else can make, use, or sell the invention without the inventor's permission.)

Today, drawings made with the use of computers expand our thinking and present us with new spatial experiences, Fig. 1-5. In the future, we will use even more remarkable tools. Designers will need the imagination to think of new ideas, the knowledge to turn the ideas into plans, and the ability to communicate those plans to others.

1.9 DESIGNING THE FUTURE. Imagine the opportunities that will emerge during the next twenty years. Can you imagine yourself working on a futuristic design team in the year 2020? The high-tech corporation that

Courtesy of NASA

Fig. 1-6. Space station designs will be adapted to meet changing needs.

employs you might assign you to work as a design technician on an engineering team for their new space station. They might ask your team to take a flight into space, or ride on a NASA commercial shuttle. You might need to visit an older space station to review and develop plans for designing a new space station, Fig. 1-6.

Your team will have to think about how the space station will be used. What kinds of work will be done at the station? How many people will work there? Since they can't go home to Earth every evening, will their families live on the station with them?

Your team will need knowledge of how people and materials react to conditions in space. You will need to gather information from many fields, ranging from materials science to human psychology.

Will you be prepared to work with design tools and computers for this type of professional job and lifestyle? Could you prepare the necessary detailed design drawings? Will you understand the graphic language used by the design team? Can you begin to learn the basic skills that will be necessary if you are to be successful years from now? There are standards for drafting that control the production of drawings as legal documents. These standards are changing with the new technology.

Courtesy of NEC Technologies, Inc.

Fig. 1-7. **Computers are used to help plan the technologies of the future.**

Yet, there are tougher questions. How do you become prepared to live and work in the twenty-first century? Have you decided what kind of job you would like to have? How can you build your power of imagination? Will you be involved in design? A course in technical drawing could prepare you for a job relating to design, Fig 1-7.

1.10 SOME THINGS DO NOT CHANGE. As you prepare yourself to serve on a design team, you must realize that you are in the middle of major changes in the high-tech world of design. These changes may be the largest since cave painters discovered how to use charcoals to create picture stories on their cave walls. Most of these new changes involve the use of computers by the design team. Yet with all the changes in computer software and hardware, one important thing has stayed the same when generating graphics — the designer has to make the decisions.

The computer may provide graphic formats, layout data, and generate art forms, but it cannot make design decisions. Successful design detailers who contribute to award-winning designs develop solutions using the computer-aided drafting (CAD) system as a design tool. They have not yet found a computer that can make their decisions, Fig. 1-8.

While computers can help the design team get its job done faster and more accurately, computers cannot do your original designing and layout work for you. The key to successful designs will be original creativity. To be successful you must keep up with the constantly changing tools, techniques, and computer software. These will make you a productive member of the design team.

1.11 TECHNICAL DRAWING LEADS TO DESIGN. In every branch of engineering there are six specific functions. Simply stated, engineering starts with the research engineer. *Research* is

ComputerVision Corp.

Fig. 1-8. **Design teams use computer graphics to visualize solutions.**

the first function. Research engineers create experiments that employ theories of science, mathematics, and technology in an attempt to create new processes. Technical drawing assists in creating graphic communication that defines procedures and clarifies new ideas in all six functions of engineering.

Development engineering is the second function. The first two functions are always identified together as research and development (R&D). The development engineer uses research data creatively to try out new ingenious products. The knowledge base for new drawings is refining ideas into probable materials and processes that can be used by the designer.

The *design engineers* work through the third function of engineering. They assume the role of one who creates products by determining construction methods, materials, and physical shapes that use the technology available. They assume economic restraints, marketing needs, and aesthetic requirements.

The fourth function of engineering is fulfilled by the *production engineer.* These engineers select equipment and plan plant facilities to work out human and economic factors that allow efficient flow of materials for new products, testing, and inspection.

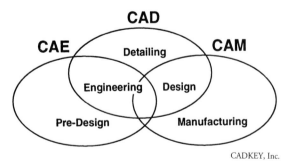

CADKEY, Inc.

Fig. 1-9. Computer-aided-engineering and manufacturing is called concurrent engineering.

Construction engineers serve in the fifth function as supervisors of the building processes. They prepare the substructure and superstructure (building framework) for new building environments. The materials assembled form high-quality spaces for commercial businesses, institutions, and residential requirements.

The *operating engineer* supervises the normal functions essential for control of machines and facilities. This sixth function of engineering ensures that power, communication, and transportation functions of all the basics required within the physical plant are always available.

The specific work of the engineer may vary at each level. The drafting technician and CAD operator are always essential members of the engineering team. Computer-integrated manufacturing (CIM) is the result of all six functions of engineering, Fig. 1-9.

1.12 TECHNOLOGY IS PLANNING YOUR FUTURE. Technology is changing every day, and the designer is the key to that change. The designer and drafting technician work well together in preparing graphic communications that describe the changes that are constantly evolving. The designer has to think about the complete design process and how it will affect lifestyles in the twenty-first century. When you decide that you would like to be creating images for tomorrow, just remember how technology might change in the next twenty years.

In this chapter we have outlined over twenty types of jobs for you to review and consider. Technical drawing skills are related to all of them. In some jobs, computer graphics skills are essential.

Traditional drafting and computer-aided drafting (CAD) are courses available in most high schools and colleges. Computer-aided drafting became available to the whole design team with the development of the personal computer (PC). The PC has emerged as a powerful tool for the designer and drafting technician as new software programs became available for creating design geometry. Computers have zoomed in on the importance of drafting techniques and learning basic drafting skills. The basics for learning and understanding the graphic language are the same, whether the designs are created on the drawing board or on the monitor (screen) of a CAD system.

1.13 PLANS FOR JOINING THE TEAM. Careers generally begin with a high school program that includes drafting course work with CAD experiences. Science, mathematics, and English classes are essential for those beginning a career on the design team. Many engineering firms are interested in offering young people opportunities if they have drafting skills and aptitude for solving problems. Technology positions generally require a drafting technician to be a mechanically oriented person who likes to know how things work. You must be able to prepare drawings that describe this interaction with details, Fig. 1-10. The design team may have entry level positions for the following:

1. *Beginning level drafters* who have good lettering skills for revising notations on

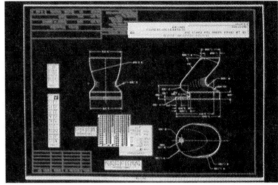

ComputerVision Corp.

Fig. 1-10. Layout is only one part of the total design.

existing drawings. Additional skills and techniques for drawing changes would be essential to maintain company drawings.

2. *Drafting – CAD technicians* with one year of college experience after at least one year of high school course work in technical drawing. A strong mechanical aptitude with CAD skills in a specific area of machine part or product layout would be required.

3. *Design – CAD technicians* with two years of college coursework, and knowledge of a specialized field of layout and design. For example, one might have die design, mold design, piping layout, or architectural and landscape experience.

The technician may be registered by an organization to qualify for specific titles. New technologies are emerging. No one person can know all areas. You must specialize in an area to be qualified.

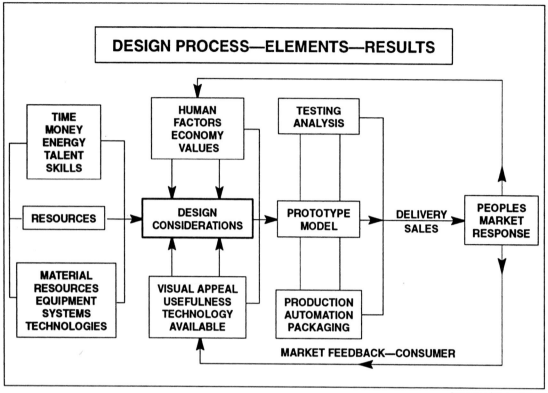

Instructional Design Systems

Fig. 1-11. The design process is a complex communication system.

1.14 THE DESIGN TEAM'S SPECIAL SKILLS. If the corporation is large, the design staff may be well defined and include all the positions shown below. Smaller departments may include only some of these jobs.

- Researchers
- Development staff
- Project coordinator
- Engineers (e.g., mechanical, electrical)
- Illustrators (solids modeling)
- Designers
- Senior detailers
- CAD technicians
- Computer graphics programmers
- Design drafting technicians
- Junior drafting technicians
- Clerical engineering staff
- Specification writers
- Contract specialists

The design staff is oriented around graphic communication and the legal documents prepared for production and construction. The designer's job is to solve problems using creative solutions. Solving problems requires knowledge, skills, resources, and the ability to communicate, Fig. 1-11. Few people work by themselves. Work teams solve problems as a group. Successful members of the team will be able to present their ideas in a clear and concise manner. They will have good writing skills. They will be able to create reports that clearly describe problems and outline solutions.

Most important of all, however, every member of the design team must be able to communicate using graphic descriptions or drawings. Much of the communication between members will relate to the creation, analysis, modification, and revision of drawings. If you do not understand the graphic language, you cannot contribute to the team's success.

1.15 COMPUTER-AIDED DRAFTING (CAD). This textbook will introduce you to CAD hardware and software. You will learn how to take advantage of your knowledge of the graphic language to make this high-tech tool assist you in solving graphic problems.

CAD software does not teach the basic principles of drawing or the graphic language any more than a table saw teaches carpentry. Therefore, it is important to learn the basics of both CAD and traditional drawing.

The ability to coordinate keyboard skills with the use of a mouse on a CAD system requires practice. Quick eye/hand coordination between the computer monitor and the keyboard/mouse is a skill that can be mastered with practice.

Once the skills of using the CAD tool are learned, the power of CAD becomes evident. However, learning the CAD software can take a lot of practice. Software companies continue to make CAD easier to use and more powerful. Most CAD programs contain an on-line tutorial (series of lessons) that demonstrates basic drawing procedures. Remember, these programs do not teach drafting principles. To use CAD effectively, you must have a firm understanding of technical drawing basics.

While there are many definitions for technical drawing, they all mean graphic communication. Technical drawing involves more than simply drawing lines, geometric shapes,

orthographic views, and pictorial drawings. The successful student will learn how to visualize (or see) three dimensions, Fig. 1-12.

1.16 BACK TO BASICS. What does it take to become a designer or other member of the design team? The principles of design drafting are common to both traditional and computer-aided drafting. The standards for each are basically the same.

In traditional drafting, the pencil skills for lettering and drawing various line weights are essential for success. Good techniques for using hand-held tools such as pencils, compasses, and templates are essential for good results.

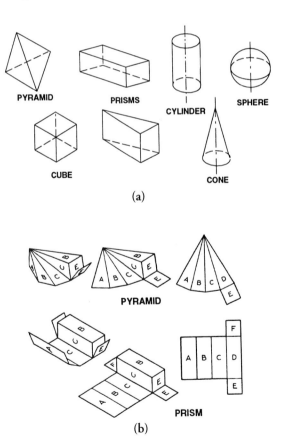

Fig. 1-12. (a) Designers understand two-dimensional forms. (b) Can you create a 3-D image?

In CAD drafting, the basic skills for generating geometric forms, dimensioning, and lettering must also be mastered to be successful. CAD has replaced some traditional design work. It has replaced the work that was formerly done on drafting tables with hand-held tools. CAD has also replaced manual filing systems since the ability to store and recall drawings electronically is available with CAD software programs.

1.17 DESIGNER'S TRAITS. The main job of a designer is to solve problems by generating new ideas, Fig. 1-13. The designer should strongly believe in his or her own ideas that are based in formal training and on-the-job experience.

The designer, however, does not work independently. A successful designer must be willing to accommodate the ideas of clients and other members of the design team. Enthusiasm for reviewing, refining, and revising an idea over and over are desirable traits for a designer. A vivid imagination and being able to think of alternate solutions are also essential.

A skilled designer depends on graphic communication as a tool for selling ideas. To sell their ideas, they must be proficient in the presentation and detailing of drawings. Technical drawing skills can be the basis for all successful graphic presentations and technical documents. Sometimes the best creative thoughts are developed in solitude and presented in drawings.

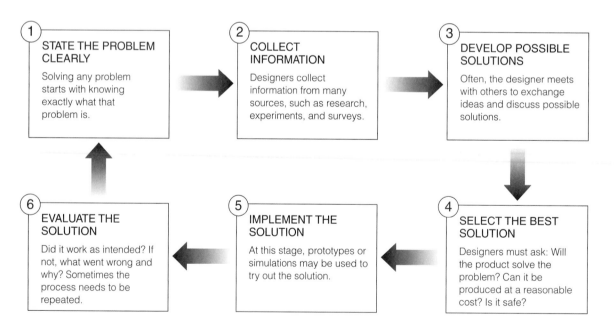

Fig. 1-13. Designers use the problem-solving process to create new products.

INTEGRATED DESIGN TEAM

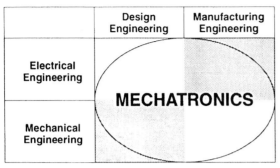

ComputerVision Corp.

Fig. 1-14. Mechatronics combines electrical engineering, mechanical engineering, design engineering, and manufacturing engineering.

Courtesy of Sigma Design International

Fig. 1-15. Architects use CAD for designing buildings.

There are a few personal traits that many creative professionals possess. Ask yourself how many of these traits you possess.

- Do you like to take things apart to see how they work?
- Do you like to make things with your hands?
- Do you notice the importance of shape in objects with good design?
- Do you ever think that you can improve the design of an object?
- Do you find yourself concerned about other people's needs?
- Do you like to sketch or make drawings?
- Do you like to study products that have unusual shapes?
- Are you open to new ideas?
- Do you like to do research?
- Would you like to investigate new ways of working with computer systems?

Successful people on the design team think clearly. They are organized and resourceful, curious, and imaginative. They are good communicators and like to sketch. Are you qualified to be a member of a design team? If not, what can you do to become qualified?

1.18 A LOOK AT SOME PROFESSIONS. The design-related professions listed in Table 1-A (page 16) include various types of engineers and architects. Their design activities require them to work closely with other design professionals. In Fig. 1-14, you can see an example of how one company combines several branches of engineering.

During the design process some people are involved in project planning. Other members of the team must carry out the plan in the proper sequence to reach the desired solution.

For example, in the design of a building, the team members include architects, interior designers, landscape architects, mechanical engineers, drafters, contractors, and material suppliers. All the team members work together to implement the project plan, Fig. 1-15.

The technical drawings and accompanying written specifications prepared by the team become legal contracts for the construction of the building. The finished building is the result of a creative idea, a sound plan, and well-executed activities performed by the team members.

TABLE 1-A. PROFESSIONS USING TECHNICAL DRAWING AND DESIGN SKILLS

Profession	Designs and Work Performed	Knowledge and Skills
Aeronautical Engineer	Aircraft designs for NASA, public transportation, and military applications.	Aerodynamics, structures, analysis, powerful engines, materials, technology.
Aerospace Engineer	Creates space shuttles, plans satellites and power systems with living environments.	Aerodynamics, power structures, interplanetary studies, analysis, orbits, experimental projections.
Architect	Provide aesthetically pleasing, functional designs for buildings and structures.	Human factors, building technology, three-dimension visual powers, prepare drawings, supervise construction.
Automotive Stylist/Designer	Cars shaped to look appealing and be comfortable to drive.	Human factors, materials, production technology.
Cartographer (mapping)	Maps for describing and locating natural and man-made features of cities, states, and countries.	Measuring, surveying, photography, symbolic forms, changing environments.
Civil Engineer	Structures, environmental systems, construction projects, bridges, surveying.	Structures and construction material methods, mathematics, analysis, laser technology.
Computer Engineer	Input and output systems with memory banks, networking of information.	Mathematics, hi-tech materials, electronic devices, programming skills, how information flows.
Construction Engineer	Supervision, structural construction, material assembly, methods of contracting trades.	Interpretation of specification and construction documents. Logistic control of construction.
Design Educator	Teaches the fundamentals of design and a specialized area of design.	Teaching skills with basic tools of design and knowledge of materials, professional designer.
Electrical Engineer	Electric power devices, controls, mechanisms, systems development.	Electrical principles and symbols, energy systems, utility controls.
Electronic Engineer	Electronic systems and power transmission, analog and digital circuits, communications.	Electronic systems analysis, controls technology, audio, video production methods.
Fashion Designer	Clothes, accessories, model presentation.	Human factors, fabrics, pattern drafting, color, design balance.

Profession	Designs and Work Performed	Knowledge and Skills
Graphic Designer	Visual communications, corporate images, advertising.	Layout, typography, art production, colors, media presentation styles.
Human Factors Bio-engineer	Ergonomics studies, research forms, physiological shapes.	Anatomy, physical forms, psychology, nervous systems.
Industrial Designer	Products, commercial systems, displays, layout, packaging.	Human factors, three-dimensional visual skills, materials, production systems.
Industrial Engineer	Industrial manufacturing processes, management, assembly, materials.	Machines, assembly lines and plant design, robotics, production management.
Interior Designer	Interior spaces, functional unity, human needs, specifying materials, colors, furnishings.	Human factors planning, lighting, furniture, room layout.
Landscape Architect	Gardens, parks, recreation facilities, landscapes, traffic patterns.	Nature, horticulture, methods of landscaping, planning, layout.
Mechanical Engineer	Machines, equipment, power plants, heating, ventilation, AC systems.	Machine design, analysis materials, mechanics technology.
Naval Architect	Military and civilian ships, boats, marine structures, aerofloat, sailing.	Human factors, naval technology, power plants, planning, structural materials.
Product Designer	Unique products of glass, textiles, leather, pottery, plastic, furniture, jewelry.	Aesthetics, tools and skills to shape and form materials.
Structural Engineer	Building shapes, steel forms, analysis, material technology.	Planning with architects, analysis, building materials.
Systems Engineer	Build a series of interactive productions, combine tooling, assembly, automation.	Machines, control systems, computers, manufacturing processes.
Theater Designer	Stage sets, props, costumes, platforms, backdrops, sound systems.	Imagination for creating environments in 3-D, colors, lighting, art-drama settings.
Urban Designer City Planner	Planning of communities, villages, cities, environmental development.	Geography, demography, plan layout of total environment.

As you can see, there are many promising careers on the design team. The keys to success in design are the creative solutions, Fig. 1-16. These take shape on the technical drawing board or CAD system.

The automotive stylist is one type of designer. He or she works with models and transforms technical drawings into three-dimensional forms in the design process, Fig. 1-17. Can you imagine the size of the design team needed by a major automobile or jet plane manufacturer? At Boeing, it took a design team of over 5,000 people to design the Boeing 777 jetliner. Can you imagine how important every technical drawing is to the total design and manufacturing process?

1.19 ENTREPRENEURSHIP. Have you ever had an idea for a new product? Learning to make technical drawings can help you turn your idea into reality. As you read earlier, every new invention starts with an idea. Being able to communicate that idea with drawings is an important step.

Some people take the process further by becoming the maker and seller of the product. They become *entrepreneurs,* people who start a business.

Entrepreneurs must have a clear idea of what they want to offer to customers. They must be able to answer the question: "How will this product or service benefit the person who buys it?" They must be creative and able to solve problems. They must be willing to take risks. Most people who start a business put much of their own money and time into the venture.

Most entrepreneurs need additional money to start their company or to help it grow. They may borrow the money. They may also sell shares of stock in their company. Each person who buys one or more shares owns a portion of

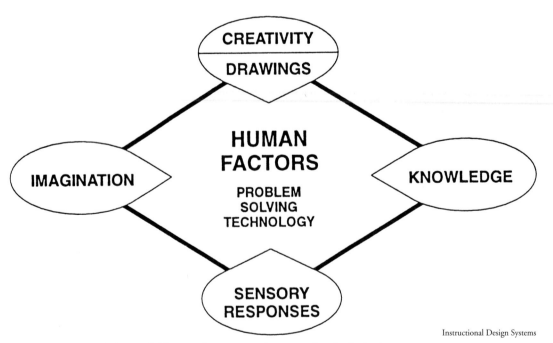

Instructional Design Systems

Fig. 1-16. Human factors are a key to technological advancement.

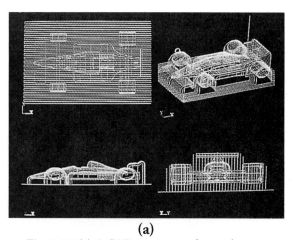

ComputerVision Corp.

Fig. 1-17. (a) A CAD system can be used to create automotive designs. (b) A CAD-designed car in action.

the company. Entrepreneurs may also make an arrangement with a venture capitalist. Venture capitalists are people who provide money and advice to entrepreneurs. In turn, they own a portion of the entrepreneur's company.

Entrepreneurs are important to our country's economy. Each year, entrepreneurs start millions of new businesses. Their companies create most of the new jobs in this country.

The companies pay taxes and otherwise benefit the community.

People who are creative, ambitious, and hard-working often make good entrepreneurs. If you have these qualities, you may want to study business and marketing in school. Many schools now offer courses in entrepreneurship to help students prepare for owning and running their own business.

PROBLEMS

PROBLEM 1-A. Drafting skills are used in a variety of careers. Using the resources available at your school or public library, research those careers that require mastery of computer-aided drafting (CAD) skills. The Department of Labor of the federal government publishes the *Dictionary of Occupational Titles*. This comprehensive resource, which is updated frequently, lists and describes many careers requiring drafting skills. Check the listed careers for those that might interest you. Identify the education and training needed to pursue each of the careers you have identified. Identify those schools in your area that offer such training and education.

PROBLEM 1-B. The development of modern drafting techniques has been prompted by technological changes over the years. The importance of drafting increased greatly with the advent of the Industrial Revolution. Research the Industrial Revolution to identify those ways by which the development of standardized drafting practices aided the development of modern manufacturing methods.

Hewlett Packard

IMPORTANT TERMS

proportions   major axis

ellipse   minor axis

OBJECTIVES

After studying this chapter, you should be able to:

- Explain why sketching is important.
- Identify the thicknesses and kinds of lines used in sketching.
- Demonstrate the sketching of a straight line.
- Demonstrate the estimating of proportions.
- Demonstrate the sketching of an arc, a circle, and an ellipse.
- List the steps in sketching a view.

CHAPTER 2
Freehand Sketching

Freehand sketching is a valuable skill in both technical and non-technical work. For example, graphic designers use freehand sketches as they develop ideas.

A graphic designer may be involved with any area of visual communications concerning an individual product or an entire company's corporate image. This may involve signage, logos, advertising, or appearance of an Internet web site. The graphic designer must have a good sense of color, style, typography, layout, and media presentation styles to address the appropriate market for a product.

Working from freehand sketches, the designer will use computer graphics systems that have high-resolution, photo-quality color printers to produce finished renderings that once required extremely tedious and time-consuming artwork drawn by hand.

CAREER LINK

To learn more about careers in graphic design, contact the American Institute of Graphic Arts, 164 Fifth Avenue, New York, NY 10010.

2.1 THE IMPORTANCE OF SKETCHING. The importance of freehand sketching to the drafter, engineer, architect, and others in technical or nontechnical work cannot be overemphasized. It is a valuable means of expression for anyone. It is an effective way to get an idea across when words fail. In this way, graphic language becomes an important aid to verbal language.

Most original mechanical ideas or inventions are recorded for the first time in the form of a sketch, Fig. 2-1. Sketches help designers organize their ideas and recall from day to day what was figured out before. The sketch is also used to show others what the designer has in mind. In case of a lawsuit over a patent, the original sketch may connect the idea to the inventor.

Sketches are often used instead of complete mechanical drawings where changes of design must be made in a hurry. They may also be used when time is not available for a finished mechanical drawing to be made. The greatest use of sketches, however, is in formulating, expressing, and recording new ideas.

Arnold & Brown

Fig. 2-1. A technical drawing begins with a sketch.

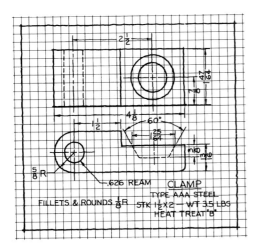

Fig. 2-2. Sketch on cross-section paper.

2.2 WHAT SKETCHING MATERIALS TO USE. You need only three objects to make a freehand sketch: a pencil, an eraser, and a piece of paper. Sketches are often made on the backs of envelopes or on scraps of paper. However, the drafter or engineer usually has a sketching pad or several sheets fastened to a clipboard. Cross-section paper (Fig. 2-2) is often used in sketching. The ruled lines help keep the lines straight. The squares (usually $\frac{1}{8}''$ or $\frac{1}{4}''$) can be used to sketch approximately to scale, if desired. However, most sketches need not be drawn to scale, but only in proportion, Sec. 2.7. The beginner should learn to sketch without the aid of cross-section paper as soon as possible.

For sketching, whether you use a mechani-

cal or wood pencil, always use a soft lead, such as F or HB. Two erasers are desirable, a white plastic drafting eraser and an ordinary pink pencil eraser, Fig. 4-24(a) and (b).

2.3 SHARPENING THE WOODEN PENCIL. Sharpen the pencil to a conical point, Fig. 2-3. Three thicknesses of lines are used in sketching: *thin, medium,* and *thick,* as shown at (a), (b), and (c). To make these, the pencil should be *sharp, nearly sharp,* or *slightly dull,* respectively. All three types of lines should be clean-cut and *dark.* Avoid fuzzy, gray, or sloppy lines. Only construction lines (d) should be *very light* and *gray.* They should have medium thickness.

2.4 FREEHAND LINES. A good freehand line should not be rigid and stiff like a mechanical line, Fig. 2-4(a). The effectiveness of a mechanical line lies in its *exacting uniformity.* In sketching, no attempt should be made to imitate mechanical lines. The important thing is to sketch the line in the right direction, and not as shown at Fig. 2-4(b).

A good freehand line has the qualities of *freedom* and *variety.* It continues in the correct path, as shown at Fig. 2-4(c) and (d). A long line should not be drawn in a single stroke. It should be drawn in several strokes end-to-end, the hand being shifted after each

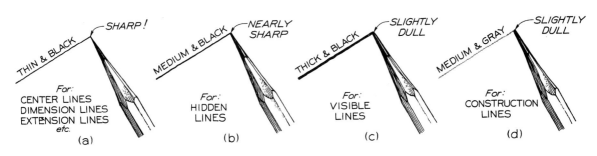

Fig. 2-3. Pencil points.

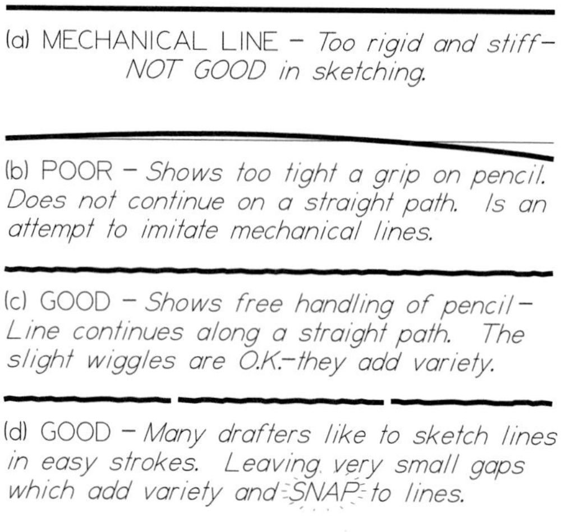

(a) MECHANICAL LINE — *Too rigid and stiff— NOT GOOD in sketching.*

(b) POOR — *Shows too tight a grip on pencil. Does not continue on a straight path. Is an attempt to imitate mechanical lines.*

(c) GOOD — *Shows free handling of pencil— Line continues along a straight path. The slight wiggles are O.K.—they add variety.*

(d) GOOD — *Many drafters like to sketch lines in easy strokes. Leaving very small gaps which add variety and SNAP to lines.*

Fig. 2-4. Character of lines.

stroke. Small gaps (if *very small*) may be left between strokes if desired, (d). All final lines should be clean-cut and dark. They should never overlap or be sloppy and indefinite.

Avoid uncertain intermingled lines, appropriately referred to as "hen-scratch lines."

2.5 SKETCHING STRAIGHT LINES. Hold the pencil naturally, about $1\frac{1}{2}$ inches from the point. To draw horizontal lines, Fig. 2-5(a), first spot your beginning and end points. Then swing the pencil back and forth between the points, barely touching the paper, until the proper direction is clearly established. Finally, draw the line firmly with a free and easy wrist-and-arm motion. *Keep your eye on the point toward which you are drawing*, not on the pencil point. This is something like golf, where you watch the ball, not the club.

Draw vertical lines downward, or toward your stomach, with a free finger-and-wrist motion, as shown in Fig. 2-5(b). If an inclined line is to be nearly vertical, draw it

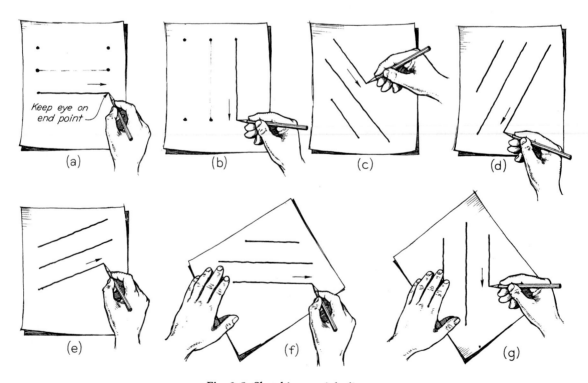

Fig. 2-5. Sketching straight lines.

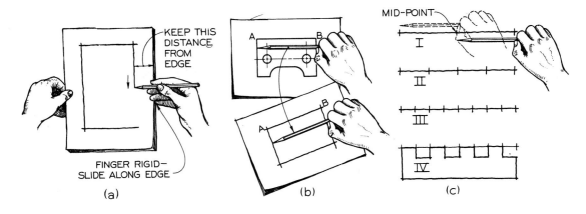

Fig. 2-6. Aids in sketching.

downward, as shown at Fig. 2-5(c) and (d). If it is to be nearly horizontal, draw it to the right, Fig. 2-5(e). You can draw an inclined line as a horizontal or a vertical line merely by turning the paper to the desired position, Fig. 2-5(f) and (g).

2.6 AIDS IN SKETCHING. There are many "tricks" you can use as aids in sketching. Figure 2-6(a) shows a useful method of blocking in horizontal or vertical lines by drawing lines parallel to the edges of the tablet or pad.

At (b) is shown a method of transferring a distance by using a pencil as a measuring stick. At (c) is shown a method of dividing a line into a number of equal parts. At I, the pencil is used to estimate half the distance by trying an estimated distance on the left and then on the right. At II and III, the divisions are further subdivided by eye. The final drawing is shown at IV.

2.7 ESTIMATING PROPORTIONS. You may have heard the advice, "Be the labor great or small,

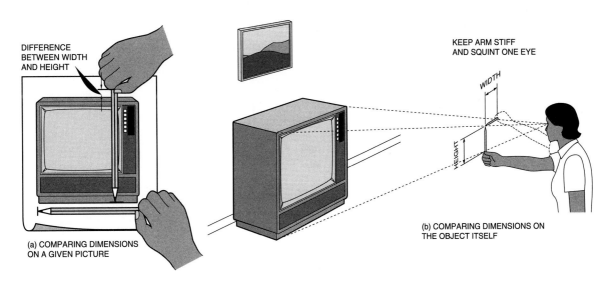

Fig. 2-7. Estimating proportions of an object.

do it well or not at all." This rule particularly applies to *proportions* in sketching. The proportions "make or break" a sketch! A drawing is *in proportion* if all its parts are the correct sizes as compared to all other parts. The larger the drawing, the more important it is to draw it in correct proportion.

For example, if you want to sketch the front view of an object, you must first get its width and height correctly proportioned. If you are working from a given picture, Fig. 2-7(a), you can compare measurements with your pencil, as shown. If you are sketching directly from the object, you can compare measurements by sighting dimensions on the object and noting how long they appear on your pencil, as shown at (b). For these comparisons, always hold your pencil at arm's length.

Another method, if you are working from a picture, is to mark off by eye convenient units on the picture, as shown in Fig. 2-8. You might also use a strip of paper on whose edge you have marked off one unit. See how many units high and how many units wide the drawing is. Assume that you are drawing a computer. In this case, it is convenient to

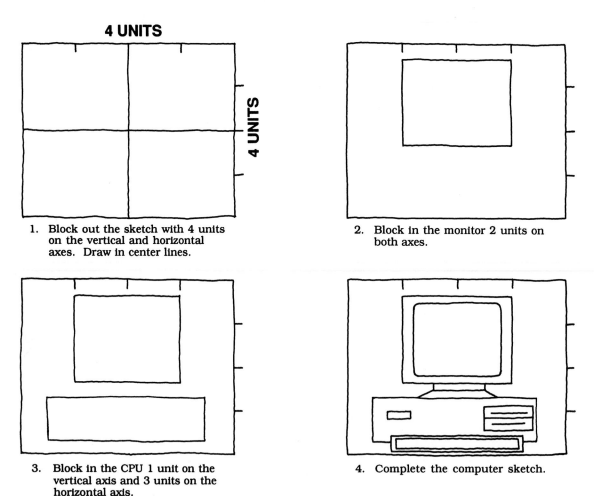

4 UNITS

4 UNITS

1. Block out the sketch with 4 units on the vertical and horizontal axes. Draw in center lines.

2. Block in the monitor 2 units on both axes.

3. Block in the CPU 1 unit on the vertical axis and 3 units on the horizontal axis.

4. Complete the computer sketch.

Fig. 2-8. Steps in sketching a computer.

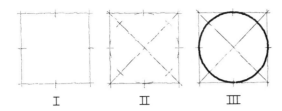

Fig. 2-9. Starting with a square.

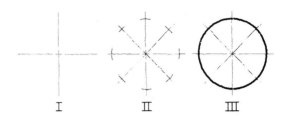

Fig. 2-10. Starting with center lines.

mark off four units for the width. About four units are also required for the height. Start the sketch, Fig. 2-8, 1, by sketching a large rectangle four units high and four units wide. Carefully compare each unit by eye with every other unit. Then draw the monitor as shown in 2. Next, as shown in 3, block in the CPU. Then, as shown in 4, complete the sketch, adding the remaining details as shown. All lines should be very light up to this point.

Finally, dim all the lines with the eraser until they can hardly be seen. Then "punch in" the final lines, making them clean-cut and dark. The main outline should be slightly heavier than the interior lines.

Remember, *the secret of sketching is to first draw the large areas in correct proportion. You* *should then add the smaller features in their correct relative sizes.*

2.8 ARCS AND CIRCLES. You can easily sketch a small circle, Fig. 2-9(I). First, lightly sketch an enclosing square and mark the midpoints of the sides. Then, II, draw light diagonals and mark off the estimated radius-distance on each. Finally, draw the circle through the eight points, III.

Another method is to start with the center lines, Fig. 2-10(I). Add light radial lines, or "spokes," between these as at II. Sketch small arcs at the radius-distance from the center on each. Finally, III, sketch the full circle.

In both methods, keep the construction lines very light. If necessary, dim all lines with an eraser before heavying-in the final circle.

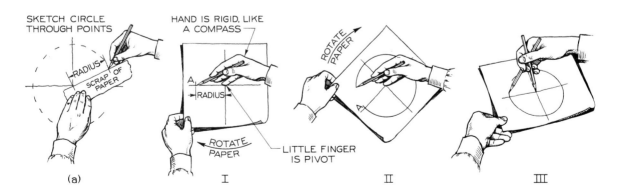

Fig. 2-11. Sketching a large circle.

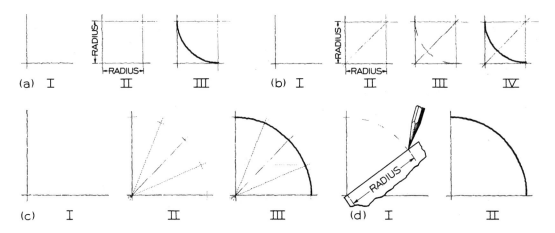

Fig. 2-12. Sketching arcs.

Another excellent method, Fig. 2-11(a), especially useful for large circles or arcs, is to set off the radius on a scrap of paper. Use this radius to set off from the center as many points on the circle as desired. You can then sketch in the circle through these points.

After a little practice, you can make excellent large circles by using your pencil and hand as a compass. Place your little finger at the center to serve as a pivot. Set the pencil point at the radius-distance from the center, Fig. 2-11(I). While your hand is held rigidly in this position, rotate the paper slowly, as shown at II, while the pencil marks the circle. You can use the same procedure while holding two pencils rigidly in position, III.

In sketching arcs, Fig. 2-12, use the same general methods as in sketching circles. Where the construction lines are too noticeable, dim the lines with the eraser before heavying-in the final arcs.

2.9 SKETCHING ELLIPSES. If you look straight at circles, they appear as true circles, Fig. 2-13(a). If you view circles at an angle, they appear as *ellipses,* (b). If you view them edge-

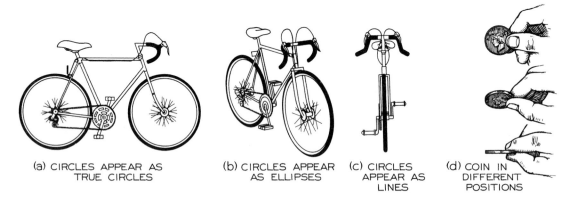

(a) CIRCLES APPEAR AS TRUE CIRCLES

(b) CIRCLES APPEAR AS ELLIPSES

(c) CIRCLES APPEAR AS LINES

(d) COIN IN DIFFERENT POSITIONS

Fig. 2-13. Circles and ellipses.

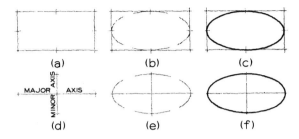

Fig. 2-14. Sketching ellipses.

wise, they appear as lines, (c). You can demonstrate this principle another way by viewing a coin in different positions, as shown at (d). The long axis of an ellipse is called the *major axis.* The short axis is called the *minor axis.*

To sketch an ellipse, Fig. 2-14(a), lightly sketch the enclosing rectangle and mark the approximate mid-points of the sides. Then, (b), sketch light tangent arcs at the mid-points, and complete the ellipse, (c). Before heavying-in the final ellipse, dim all construction lines with an eraser.

A second method is to start with the major and minor axes and sketch the ellipse as shown at (d) to (f).

An excellent way to sketch a large ellipse is to use the *trammel method,* Fig. 6-21. You first sketch the two axes at right angles to each other. You then prepare a "trammel" on the edge of a piece of scrap paper. As the trammel is moved to different positions, points on the ellipse can be marked. The final ellipse can be sketched through these points.

2.10 STEPS IN SKETCHING A VIEW. If we look at a Lock Plate, Fig. 2-15(a), in the direction of the arrow, we see a "view" of the object. The steps in sketching this view are shown in the figure. The sketch is not made to scale, but is carefully proportioned.

I. Sketch in the large main areas lightly. This is the most important part of the sketch. No sketch can be satisfactory if these main areas are incorrectly proportioned.

II. Lightly block in construction for arcs and circles.

III. Dim all lines with the eraser. Then heavy-in all final lines, making them clean-cut and dark.

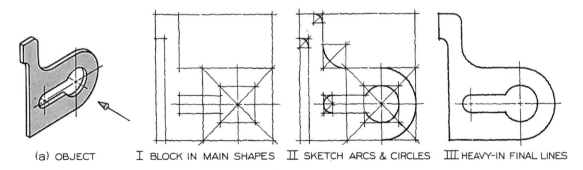

(a) OBJECT   I BLOCK IN MAIN SHAPES   II SKETCH ARCS & CIRCLES   III HEAVY-IN FINAL LINES

Fig. 2-15. Steps in sketching a lock plate.

PROBLEMS

Several one-view sketching problems are given in Figs. 2-16 to 2-31. For each problem, use an $8\frac{1}{2}'' \times 11''$ sheet of cross-section paper, plain bond typewriter paper, or drawing paper. Sketch a border, leaving a margin of about $\frac{1}{2}''$ on all sides. Then sketch the assigned problem carefully in proportion, fitting it on the sheet approximately as shown. Write the date and your name at the bottom of each sheet. For problems 2-24 and 2-25, you may sketch the plans as shown. With your instructor's approval, you may sketch a plan of your own home.

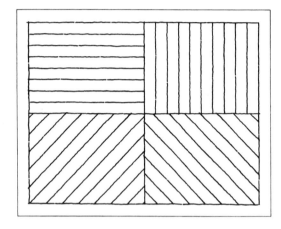

Fig. 2-16. Straight lines.

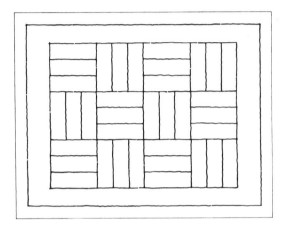

Fig. 2-17. Parquet floor.

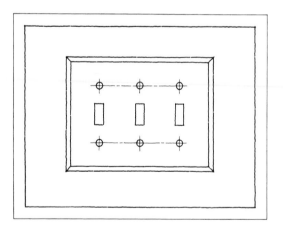

Fig. 2-18. Switch cover.

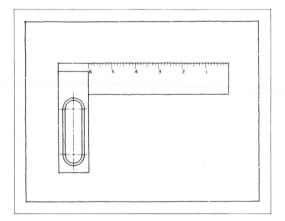

Fig. 2-19. Try square.

Additional one-view sketching problems are available in Figs. 4-44 to 4-65. Sketching from actual objects similar to those shown below will be most helpful. Select an object. Obtain your instructor's approval before you start work.

If assigned by the instructor, construct drawings using decimal-inch or metric scales by converting given fractional dimensions to decimal-inch or metric equivalents. Refer to Table 20 in the Appendix.

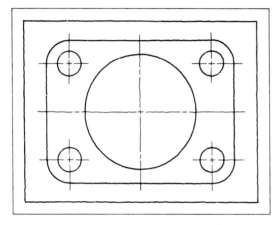

Fig. 2-20. Cover gasket.

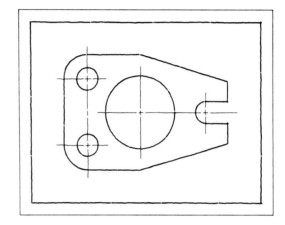

Fig. 2-21. Stamping.

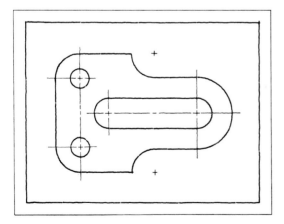

Fig. 2-22. Stamping.

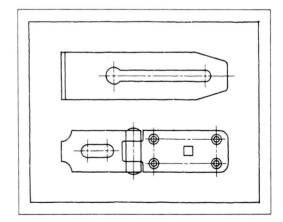

Fig. 2-23. Plane blade and hasp.

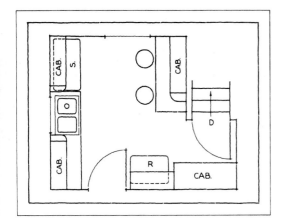

Fig. 2-24. Kitchen plan.

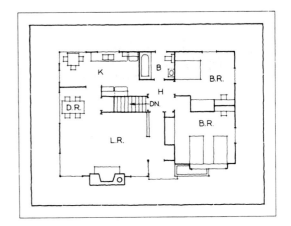

Fig. 2-25. House plan.

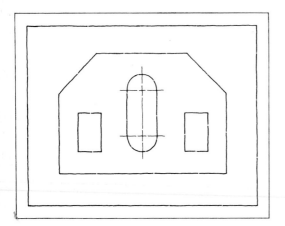

Fig. 2-26. Cover plate.

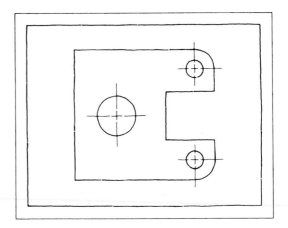

Fig. 2-27. Rail stop.

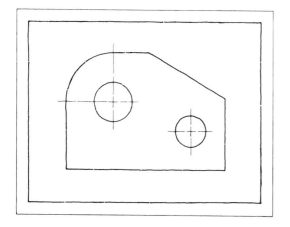

Fig. 2-28. Hinge plate.

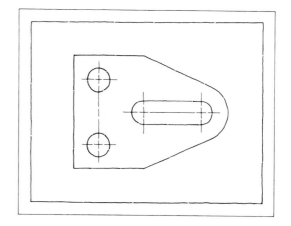

Fig. 2-29. Slide.

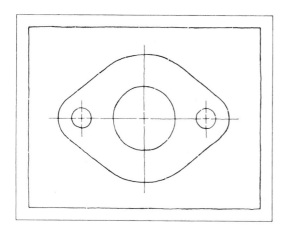

Fig. 2-30. Gasket.

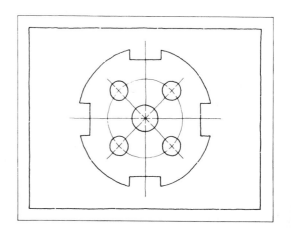

Fig. 2-31. Index head.

CHAPTER 3

OBJECTIVES

After studying this chapter, you should be able to:

- Identify the two main parts of a CAD system.
- Name three important qualities of a CAD program.
- Identify the four common input devices.
- Identify five types of storage devices.
- Describe the two main methods of transferring a CAD drawing to paper.

CHAPTER 3

Introduction to Computer-Aided Drafting (CAD)

Many products made today are designed and drawn using a computer-aided drafting (CAD) system. The Nintendo® game in your living room and the cellular telephone in your hand were created by a design team using a CAD system.

Architects are designing "smart homes" today using CAD. These homes, which were drawn on a computer, have computer-controlled heating, lighting, security, and home entertainment systems. Structural engineers are using CAD to create steel shapes for tomorrow's skyscrapers and sports arenas. Manufacturers are designing next year's products with CAD.

This chapter will introduce you to the basic operations of a computer-aided drafting (CAD) system. It discusses the CAD hardware and software that are used to create, store, and print CAD drawings.

CAREER LINK

Computer engineers, software engineers, and software developers design new computer hardware and software for CAD and other applications. You can learn more about these careers from the Occupational Outlook Handbook. Look for this publication in your school library or on the Internet at
http://stats.bls.gov/ocohome.htm

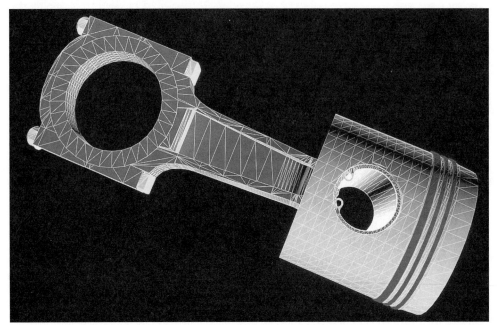

Alfred Pasieka/Science Photo Library/Photo Researchers

Fig. 3-1. Today's designers use CAD to turn their ideas into 3D images, such as this one.

3.1 USING CAD TO DESIGN THE FUTURE. Only a few years ago, all designers and drafters created and refined their ideas on paper. They used drawing boards, pencils, ink, and a variety of mechanical drawing instruments. Today, most designers and manufacturers are using CAD systems. With CAD systems, companies can produce quality products quickly and at lower cost.

Without CAD, most companies could not compete in the world marketplace. Without the benefits of CAD systems, VCRs would be too expensive. New car models would take longer to get into the showroom. Camcorders would be too big to carry around. Cellular phones would be only a designer's dream. Without CAD, the technological gadgets you use every day would not exist.

3.2 THE DESIGN TEAM USES CAD. Today a typical CAD system consists of two parts: a fast computer workstation and CAD software. Together, they transform a designer's ideas into accurate, three-dimensional computer-generated geometric shapes, Fig. 3-1.

If a CAD system were a taxicab, the driver would be the software program commanding the vehicle to turn left or right. The taxicab's engine and body would be the hardware that translates the driver's commands into action. Neither the taxicab nor the CAD system will go anywhere until YOU tell them where you want to go. CAD systems do not replace people. They enhance the value of the design team members who have the skills needed to drive these high-tech tools, Fig. 3-2.

CAD software programs contain hundreds of commands. These make designing and drawing easier, faster, and more accurate. One CAD command might be to DRAW a circle of a given diameter at a certain location. Another command might be to CALCU-

36

LATE the area of a floor plan. The designer or drafter "tells" the CAD software program what command to perform. The software program then performs the mathematical calculations needed to complete the command.

Some CAD programs present the user with a command line where he or she can type the various commands. Newer programs present the user with a graphical user interface (GUI). The user clicks on icons and menu options to issue commands.

Speed, accuracy, and ease of use are three important qualities of CAD software programs. When running on a fast computer workstation, a CAD software program can draw a circle tangent to (touching) two lines in microseconds. CAD software programs can redraw the entire floor plan of a house in seconds. Large networked CAD systems can generate the three-dimensional (3D) images of a college campus. Multiple users can share drawings from one computer to another across a network. Advanced CAD software can even calculate the angles of lights and shadows on buildings, Fig. 3-3.

3.3 THE BENEFITS OF A CAD SYSTEM.
CAD systems today are being used by most members of the design team. To be an effective member of the design team, you must be trained in both technical drawing skills and the operation of CAD systems.

Designing and drawing with a CAD system is faster than traditional drawing techniques. For example, an object can be drawn once, saved in an electronic library, and reused later as part of another drawing. By reusing these items, a designer can prepare new drawings in a fraction of the time it would take using pencil and paper.

Changing and refining drawings on CAD

is also faster than manual techniques. Instead of erasing and painstakingly redrawing lines on paper, members of the design team can open the computer file. They can make changes and generate a new set of drawings in minutes.

CAD-generated drawings are also more accurate than traditional paper drawings. Lines are drawn to accuracies of millionths of an inch, compared with sixty-fourths of an inch with traditional drawing methods. Many CAD systems can also calculate areas, volumes, stresses, forces, loads, sizes, intersections, and shadows. They do this with great accuracy and speed.

3.4 TYPICAL CAD SYSTEMS.
A modern CAD system uses a fast computer to run the CAD software program. The calculations made by the computer every time it displays the drawing on the monitor require a powerful processor and large amounts of memory.

John Coletti/Stock Boston

Fig. 3-2. CAD has greatly reduced the time needed for product design.

Courtesy of Sigma Design International

Fig. 3-3. CAD software calculates the angles of lights and shadows.

The *central processing unit* (CPU) is the brain of the computer. Also called a microprocessor, the CPU is the main chip in the computer. This chip receives instructions, analyzes them, performs the required operations, and sends information to storage.

The speed of a microprocessor chip is usually measured in millions of processing cycles per second or megahertz (MHz). A 500 MHz processor is five times faster than a 100 MHz processor. Some workstation computers have more than one microprocessor chip. On workstation computers, the processor's speed is measured in millions of instructions per second (MIPS). For example, a processor rated at 100 MIPS can process 100 million instructions per second. A CAD system on a desktop computer used for designing houses might run at only 10 MIPS. A CAD system connected to other workstations on a network used for designing skyscrapers might run at more than 150 MIPS.

The microprocessor performs most of the calculations. The information it processes is stored in the computer's temporary memory. This memory, called *random access memory* (RAM), is located on modules about the size of a stick of gum. The RAM on a computer is similar to the memory in your pocket calculator. When you turn off the computer, the information stored in RAM disappears.

To keep information permanently, a computer needs storage media. Typical storage media include the computer's *hard disk* and 3.5″ diskettes. Various other media are also available. These are discussed in Section 3.9.

The CAD software program and the individual drawings you create are typically stored on the hard disk. When you turn off the computer, the information remains on the hard disk. This memory, however, is not really permanent. You can always erase files from a hard disk, but the CAD system will not erase the information on the hard disk until you tell it to do so. If you never erase old files from the hard disk, it will eventually fill up.

The rest of the CAD system consists of input and output devices such as a mouse, a keyboard, the display monitor, speakers and microphone, a modem, printers, and plotters.

Some CAD computers are stand-alone units. This means they are one-person computers that contain everything a designer needs in one computer, Fig. 3-4. Other CAD computers are networked together with wires, phone lines, or optical cables. Such a system allows the design team to share ideas, files, printers, and software programs.

Whether the computers are stand-alone units or networked, care should be taken to avoid injury to the user or damage to the equipment. See Table 3-A for tips on using computer systems.

3.5 CAD SOFTWARE PROGRAMS. CAD software programs contain the hundreds of commands needed to create and draw objects. Some CAD software programs use a command language of verbs and nouns. For

TABLE 3-A. TIPS FOR PROPER COMPUTER USE AND CARE

Personal Health and Safety

- If you have impaired sight, hearing, or motion, check the user's manual to learn about the computer's accessibility features.
- Take breaks every 20 to 30 minutes.

Avoid Eye Strain
- Adjust room lighting and screen brightness to reduce glare.
- Look away from the screen often and refocus your eyes on distant objects.

Avoid Strain on Your Neck, Back, Arms, or Wrists
- Monitor should be 20 to 30 inches away, directly in front of you, at a height that does not require tilting the head up or down.
- Back of chair should make full contact with your back. Most people are more comfortable if they recline slightly rather than sit upright.
- Height of chair should allow your feet to be flat on the floor.
- Chair should be deep enough to support your thighs.
- For keyboarding, the angle between upper and lower arm should be 100 to 110 degrees.
- Wrists should be neutral, not bent up or down.
- Keep mice and other pointing devices within easy reach.

Proper Care of Equipment

- The computer system should be on a sturdy, level surface.
- The computer and monitor should have at least 4 inches of space around the sides and back for heat to escape. Do not block any of the ventilation slots.
- The computer system should not be in direct sunlight or next to a heating device.
- Avoid dust, high humidity, and extreme heat or cold.
- Make sure there is surge protection.
- Keep liquids away from the computer and keyboard.

- Store diskettes, CDs, etc., in a clean, dry place. Avoid high humidity and extreme heat or cold.
- Keep diskettes away from magnets.
- Before moving any computer component, check the user's manual.
- Turn off and unplug the system before cleaning it. Check the user's manual for approved cleaning products and methods. Never spray cleaning products directly on the computer or other components. Spray a clean, lint-free cloth and then use the cloth to wipe the components.

example, you might tell a CAD software program to DRAW a LINE. The verb (DRAW) is the action command—drawing. The noun (LINE) identifies the object the verb will act on—a line. If the CAD program recognizes your command, it executes the command.

If you tell the software program to DRAW a MOTORCYCLE, it will probably respond with an error message. Drawing a motorcycle is not in its command vocabulary. You would draw the motorcycle by directing the system to draw lines, circles, arcs, etc. You would determine the proper size and position of these items.

Newer CAD programs use graphical icons—buttons that you can click with the mouse to issue commands. Pictures on the

Courtesy of IBM

Fig. 3-4. A CAD system may consist of a single, stand-alone computer, or it could be many computers that are connected in a network.

icon buttons help the user remember the function or command of the icon button.

To draw with a computer, you must learn the CAD command language. Learning the CAD language will take time and practice. However, once you can communicate with the program, you can use it to draw almost anything.

There are many CAD software programs on the market today. Some cost less than one hundred dollars and run on a small home computer. Others cost hundreds of thousands of dollars and require expensive workstation networks. Companies select a CAD system according to their needs and budgets. When you learn the basic skills needed to operate a CAD software program, you can apply those skills to any CAD system. You can then become a valuable member of the design team.

By developing skills that employers need, you can assure your place in the high-tech work force of the twenty-first century. Even the boss needs to know CAD to be an effective member of the design team.

3.6 OPERATING SYSTEMS. All software programs, including CAD programs, require an *operating system.* The operating system translates the commands from the CAD software program into language the computer hardware can understand. Windows® is a widely used operating system.

Operating systems also contain many utility programs that are used by the system administrator to maintain the hardware and software loaded on the computer. On many large design teams, the system administrator's only job is to manage the computer facility so that work proceeds smoothly. Do you think you could be a system administrator?

3.7 USING INPUT DEVICES. An input device translates commands from you into commands the CAD software program can understand. The most common types of input devices are the *keyboard,* the *digitizer pad,* the *mouse,* and *voice recognition.*

The *keyboard* consists of typewriter keys, navigation keys, function keys, and a number keypad, Fig. 3-5. The typewriter keys work just like a typewriter for typing letters, numbers, and punctuation. When entering command names, file names, or dimensions and notes, you will probably use the typewriter keys of the keyboard.

The navigation keys move you around the monitor screen. The main navigation keys are the four cursor keys. These keys have arrows on them. They are used to move the cursor up, down, left, or right.

The function keys, labeled F1 through F12, are shortcut keys. They issue commands with the push of a button. For example, the F1 key often provides help information. The function keys are usually in a row across the top of the keyboard.

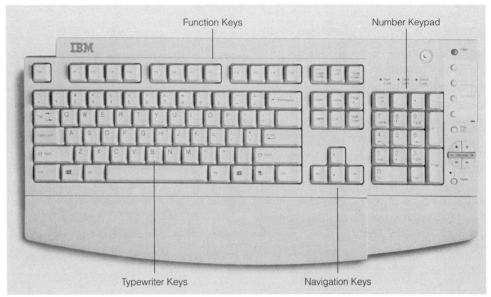

Courtesy of IBM

Fig. 3-5. Typical computer keyboard.

The number keypad on the right side of the keyboard duplicates the number keys above the typewriter keys. When typing a lengthy series of numbers, it is more efficient to use the number keypad.

CAD students must learn a few special keys. One important key is ESC, the escape key. This key usually releases you from the current command sequence and returns you to the main menu. Two other special keys are ALT, or the alternate key, and CTRL, or the control key. These keys, found on both sides of the space bar, are used in combination with another key to perform special functions. For example, holding the CTRL key down while you press the letter T might turn a CAD function on or off.

The ENTER or RETURN key is at the right side of the typewriter keys, above the right SHIFT key. A duplicate ENTER key is located to the right of the number keypad. Pressing this key communicates a message to the computer. You may be telling the computer, "Execute the command I just gave you" or "Return to the command prompt."

The BACKSPACE key will delete the letters of a command one letter at a time. It deletes letters to the left of the screen cursor. For example, if you type SAVR instead of SAVE, you can press the BACKSPACE key to delete the incorrect letter R and then type the correct letter E.

Many CAD software programs come with a keyboard template that fits on the keyboard above the function keys. The template lists the most commonly used key functions. Thus, you save time because you do not have to refer to the operator's manual to look up every command.

The *digitizer pad* is another input device, though it is not used as often as the keyboard or mouse. It consists of (1) a tablet that is a little larger than a notebook, (2) a plastic sheet that fits over the tablet, and (3) a hand-held puck with several buttons. See Fig. 3-6. The tablet contains invisible spots that sense the position of the puck. Each spot is programmed with a different CAD command. For example, when you move the puck over the spot for SAVE and press the appropriate button on the puck, the tablet sends the save command to the computer. Each command is printed on the plastic sheet so you know what command each spot represents. Experienced CAD users can usually draw much faster when they use a digitizer tablet and puck.

In the middle of the digitizer tablet is a blank drawing area, just like the drawing area on the display screen. By moving the puck around this area you can draw and edit objects.

The *mouse* is a simplified version of a puck. It has only one, two, or three buttons and does not use a tablet. Instead it has a rubber ball on the bottom that rolls on the desktop. When you move the mouse up or down, the cursor on the screen moves up or down. When you move the mouse left or right, the cursor mark moves left or right. Some mice use a light beam and a special mouse tablet instead of a rubber ball to sense movement.

Voice recognition is the newest type of input device. Voice recognition consists of a microphone and special software that recognizes your unique voice patterns. Most voice recognition software requires you to read several pages of text into the microphone so that the software can learn your speech patterns. After that, you can speak the command name into the microphone and the computer will perform the command.

Most CAD software programs display a list of commonly used commands on the screen as words or icon buttons. If you move the mouse so the cursor points to one of these commands

Fig. 3-6. A tablet digitizer.

or icon buttons on the screen and press the left mouse button, the command will be executed. When these commands are words, they are called menus. When the commands are graphical pictures, they are called icons. Using the mouse to choose menu commands or icons on the screen is called selecting.

You can also select parts of a drawing with the mouse or puck. For example, if you select a circle with the mouse or puck, you can perform a CAD command on the circle, such as COPY, DELETE, or MOVE.

Some software programs have so many commands they cannot be displayed on the screen at one time. Most of the CAD commands are hidden or nested behind visible menu headings. If you select the command FILE from the screen menu, for example, a listing of all the file commands is displayed. There might be FILE OPEN, FILE SAVE, FILE DELETE, or FILE PRINT. Learning where these nested commands are located is an important part of becoming an effective user of a CAD program.

Some students have difficulty developing the eye-to-hand coordination needed to manipulate the mouse or puck. The more you practice, the easier it becomes. Also remember that the sensitivity (speed) of the mouse or puck can be changed to suit you.

3.8 DISPLAY MONITORS. One of the most important parts of a CAD system is the display monitor. The display monitor is the primary visual method of feedback from the computer. The display monitor will show you what the computer is doing in response to your instructions.

When you type a command into the keyboard, the computer lets you know that it received the command by displaying a response on the screen. If you issue an incorrect command, the computer will display an error message on the screen. This lets you know that something is not correct. If you tell the computer to draw a circle, the quickest way to know if it performed correctly is to see the circle on the screen. If you see a rectangle instead of a circle, you know that something is wrong.

There are several types of display monitors. The most common is the *super video graphics array* (SVGA) color display monitor. The *resolution* of a display monitor is determined by the number of dots, or *pixels,* displayed on the screen. Regular VGA displays 640 pixels horizontally and 480 pixels vertically. Since CAD drawings need to show a lot of detail, many CAD systems use super-VGA or XGA (extended graphics array) monitors that display 1280 or more pixels horizontally and 1024 or more pixels vertically.

The higher the resolution, the more information there is on the screen. However, if too much information is displayed on a small monitor, it becomes hard to read. While simple CAD computers might use a 17″ monitor, powerful CAD systems use 21″ high-resolution monitors, Fig. 3-7.

Today's monitors can display many different colors. This feature is useful in CAD, especially when objects are on different levels, or layers. A CAD layer is like a transparent drawing medium. Layers make it easy to keep track of a complicated drawing. In architectural drawing, the walls of a floor plan might be on layer 1. The dimensions might be on layer 2. The electrical wiring might be on layer 3. A CAD system can maintain numerous drawing layers in one drawing file.

One of the main advantages of using layers is that each layer can be turned on or off to make the drawing easier to read and edit. To clarify which objects are on what layers, each layer may have a different color. The walls on layer 1, for example, might be green. The dimensions on layer 2 might be yellow.

3.9 STORAGE MEDIA. As mentioned earlier, various storage media are used to permanently store drawings and software programs. Without storage media, there would be no way to save a drawing. This section will discuss hard disks, diskettes, CD-ROMs, DVDs, and tapes, Fig. 3-8.

The *hard disk* is the primary storage medi-

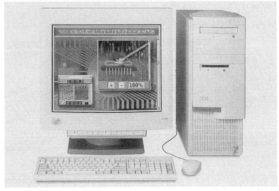

Courtesy of IBM

Fig. 3-7. The display monitor shows the results of commands.

um. It stores the CAD software program and your drawing files. Hard disk drives are measured by their size in bytes and by their speed in milliseconds. For example, a hard drive may have a capacity of 2 gigabytes. One gigabyte equals about one billion bytes. One byte is equal to the space needed to store one letter or number. A page of double-spaced typewritten copy contains about 2,000 bytes (2 kilobytes) of information.

The speed of a drive is a measure of how fast you can read or write information to the drive. For example, a 10-millisecond hard drive is three times as fast as a 30-millisecond hard drive.

The hard drive is generally located inside the computer. It cannot easily be removed. However, a computer can have more than one hard drive. Each drive is given a letter name. The first hard drive is called drive C. If a second hard drive is installed, it is usually called drive D. The designation "Drive A" is reserved for the diskette drive. "Drive B" is

seldom used in computers today. It is reserved for a second diskette drive.

Diskettes (sometimes called floppy diskettes) are transportable storage media. The 3.5″ diskette is the standard today. However, the disk cartridges for a high-density removable drive such as the Zip™ drive can hold 100 megabytes of data. Each cartridge is only slightly thicker than a diskette.

CD-ROMs and *DVDs* are similar to audio CDs and movie DVDs. Instead of holding digital music and movie information, they hold digital computer information. CD-ROM stands for compact-disk read-only memory. DVD stands for digital versatile disk. (The spelling *disc* is also used.)

There are three types of CD-ROMs: read only, write-once read-many (WORM), and writable. The advantages of CD-ROMs and DVDs are convenience, low cost, and huge storage capacity. Many software companies now supply their programs on CD-ROM or DVD instead of on diskettes.

Streaming tape is another type of storage medium. Like audio cassette tape, this is a magnetic tape. Streaming tapes come in various sizes and formats. Their advantage over diskettes is convenience and speed. One streaming tape can hold several gigabytes of data. A diskette typically holds 1.4 megabytes.

The various types of storage media have different uses. Diskettes are used mainly to store files that users have created. Most CD-ROMs and DVDs store commercially made files and programs. Such products are read-only. However, there are some types of CDs and DVDs that are erasable and rewritable. Streaming tape is used mainly for backing up data on a regular basis. The hard disk is used to store the programs and files that are used most often.

GTS Graphics

Fig. 3-8. Storage media commonly used by computers.

Courtesy of CalComp Inc.

Fig. 3-9. This plotter uses pens to produce a CAD drawing.

It is important to have a method of restoring the data on a hard disk in case its contents become damaged and unreadable. The slightest piece of dirt on the surface of a hard disk can ruin the whole disk drive. If you had all your drawings on one disk, you would lose all your work if the disk were to become unreadable.

To safeguard your drawings, a backup of the hard disk should be made on another medium, usually every day or week. That way, if the hard disk becomes ruined, you can restore the drawings onto a new hard disk. You will not lose any of your drawings. This type of copying is called a normal backup. It uses diskettes, tape, or a high-capacity disk such as a Zip™ cartridge.

Another type of backup is an archival backup. This type of backup is performed on old files that are no longer needed for current projects. For example, when an engineering office completes a design project, it could transfer the files from the hard disk onto streaming tape. This frees space on the hard disk for new work. If the client comes back for changes, the original drawings can be restored from the archival backups.

Hard disk maintenance and administration are an important part of using a CAD system. Electronic file storage is fast and convenient. However, if you do not back up the hard disk on a regular basis, the work could be lost if the disk becomes damaged.

Larger design teams have a system administrator whose job is to maintain and administer the CAD network. Good system administrators need solid technical drawing backgrounds, CAD experience, and computer interest.

3.10 OUTPUT DEVICES. After the design team completes a set of drawings on CAD, they are usually printed on paper. The paper copies are sent to the construction or manufacturing team for review and approval. (Sometimes files are sent electronically rather than on paper.)

There are two basic methods of transferring an electronic CAD drawing to paper: plotting and printing. Plotting is a process of drawing every line, curve, and letter with an ink pen or pencil point mounted in a mechanical device called a *plotter*, Fig. 3-9. The plotter moves the pen horizontally and the paper vertically. (Some plotters, such as electrostatic and ink jet plotters, do not use pens.)

Plotters come in various sizes. They can hold as many as eight different pens. Each pen can be a different color and line width. The CAD software program tells the plotter which pen to use and where to draw each line. Line quality depends on the type of pen and the amount of ink left in the pen.

Printing uses a different method of transferring the CAD drawing to paper. Pens are not used. Instead, the drawing is reproduced by means of various image transfer techniques, such as laser printing on large paper sheets.

Many companies have eliminated printing on paper by using integrated manufacturing software programs. These programs translate electronic CAD drawings directly into manufacturing commands that production equipment (such as mills and lathes) can understand.

3.11 INTEGRATED DESIGN AND MANUFACTURING. To be competitive in a world market, companies must reduce costs and decrease the time it takes to produce a product. By combining CAD drawings with manufacturing processes, companies can cut costs and reduce production time. This is called *integrated manufacturing*. The software programs that integrate design and manufacturing are called computer-aided drafting and computer-aided manufacturing (CAD/CAM).

CAD/CAM software saves time and money by streamlining the transfer of information from the design team to the manufacturing team. A company using CAD/CAM software can design a product on the computer. It can then use the computer to control the production process.

3.12 CAD AND THE WORK TEAM. Communication between the design team and manufacturing team is essential if a company is to be competitive. That is why communication skills are so important if you are to be a successful member of the design or manufacturing team. The more skills you have, the more successful you will become.

The interaction between team members is a key to a company's success. Everyone must understand the graphic language. They must know how to interact with the CAD/CAM software programs. They must be able to communicate with other members of the team.

New communication tools such as e-mail and portable document image files (PDF) allow all members of a design team to share files no matter where in the world they are located. E-mail is an electronic memo or letter. The electronic letter is sent across a network or the Internet. It reaches its destination in seconds. Electronic images and files can be attached to e-mail, so everyone on the design team knows what is happening at every moment.

In many companies, the design and production teams work together to develop new products. The production team no longer has to wait until a design is finished before they see it. By seeing the design in its early stages, production people can help find ways to make the product easier to manufacture. This process of working together is called *concurrent engineering*.

CAD/CAM software programs also ensure that the high standard of accuracy of the original CAD drawing is transferred to the manufacturing process. You can see how CAD/CAM helps companies reduce costs and cut the time it takes to design and manufacture products.

PROBLEMS

Even if you are doing primarily mechanical drawings at this point, it is a good idea to become familiar with the CAD systems available to you. As you work with these problems, record your responses in your drafting notebook, where they will be readily accessible.

PROBLEM 3-A. Go to your school's CAD lab. Find out more about the following:
- Which CAD program or programs does your school use?
- What operating system is installed on the CAD computers?
- What choices are available for input? (These might include keyboard, mouse, digitizer, and voice recognition.)

Write these and any other significant specifications in your drafting notebook for future reference.

PROBLEM 3-B. Find out what type of backup system is used by the CAD lab. Find out how it is used. Summarize its operation in one or two paragraphs.

PROBLEM 3-C. Find out what type of plotting or printing devices are set up for use in the CAD lab. Describe, in writing, the general method by which drawings are printed or plotted.

PROBLEM 3-D. Suppose that you were going to use a CAD program to create the drawing shown in Fig. 3-10. List the layers (levels) you would set up. Explain why each layer might be needed.

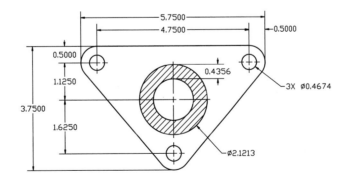

Fig. 3-10

Arnold & Brown

CHAPTER 4

OBJECTIVES

After studying this chapter, you should be able to:

- Distinguish between mechanical drawing and freehand sketching.

- Demonstrate the drawing of a horizontal line.

- Demonstrate the drawing of a line at various angles to the horizontal.

- Demonstrate the drawing of parallel lines.

- Describe the techniques for drawing to scale.

- Demonstrate the use of an irregular curve.

Mechanical Drawing

This chapter explains how to use drafting instruments. You will learn the basic skills that form the foundation of all mechanical drawing.

Skillful use of instruments is essential to success in a technical drawing career. Thus, the people who teach these skills to others can make a real difference in their lives.

A successful design educator understands the theory of drawing well and knows how to use the various tools required to prepare a drawing, either by hand or with a CAD program. He or she must also be able to communicate well with all members of the class, all the while making the subject interesting and meaningful. It is not enough that the design educator's students are able to produce good-looking, technically correct drawings. They must also learn how and why a drawing is made in a certain way.

Teaching a technical subject such as drawing requires that the design educator always be on the lookout for new techniques, methods, materials, equipment, computer programs, and industrial standards. These must first be learned by the teacher and then passed on to the students.

■ C A R E E R L I N K ■

You can learn more about careers in teaching from the *Occupational Outlook Handbook*. Look for this publication in your school library or on the Internet at *http://stats.bls.gov/ocohome.htm*

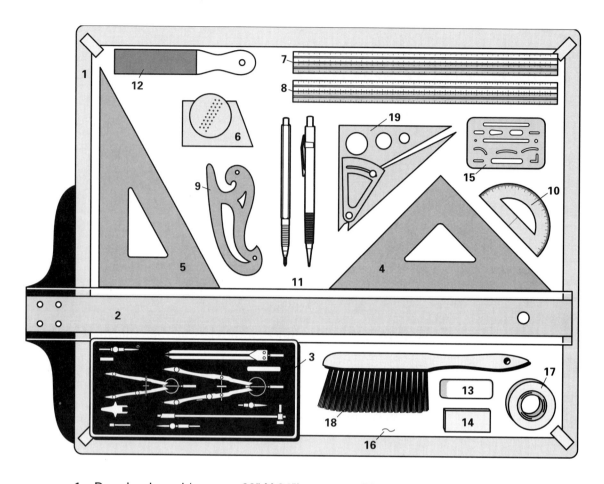

1. Drawing board (approx. 20" X 24")
2. T-square (24" transparent edge)
3. Set of drawing instruments
4. 45° triangle (8" sides)
5. 30° X 60° triangle (10" long side)
6. Ames lettering instrument
   (or lettering triangle)
7. Engineers scale
8. Architects scale
9. Irregular curve
10. Protractor
11. Mechanical drawing pencils
12. Sandpaper pad (or file)
13. Pencil eraser

14. Plastic eraser
15. Erasing shield
16. Drawing paper
17. Drafting tape
18. Dusting brush (or dust cloth)
19. Adjustable triangle

Optional equipment (not illustrated above):

Cleaning powder (or pad)

Fig. 4-1. Equipment used in technical drawing.

4.1 WHAT IS MECHANICAL DRAWING? Most people who think they could never learn to draw just "think" they couldn't. Their usual apology is: "I never could learn to draw. I can't even draw a straight line." This is in a sense true. No one can draw a really straight line without a guiding edge.

A clear distinction should be made between *mechanical drawing* and *freehand drawing* or *sketching,* both of which are used to express the graphic language. As a matter of fact, Fig. 2-4, a good freehand sketch should not be drawn with rigidly straight lines. The lines should have a certain freedom and variety, unlike mechanically drawn lines. Mechanical drawings, however, with which we will concern ourselves in this chapter, are made with precision drawing instruments. Any intelligent person can learn to execute good mechanical drawings rapidly and skillfully.

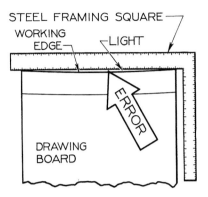

Fig. 4-2. **Testing working edge of board.**

4.2 DRAWING EQUIPMENT. The basic drawing equipment used by drafters is shown in Fig. 4-1. Their uses are described on the following pages. Good drawing instruments are fairly expensive. Since it is difficult for beginners to tell inferior instruments from those of high quality, you should consult your instructor before purchasing.

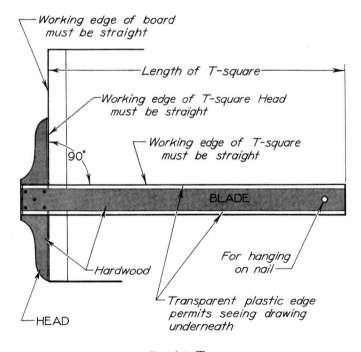

Fig. 4-3. T-square.

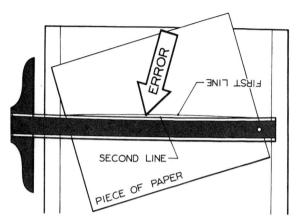

ERROR

FIRST LINE

SECOND LINE

PIECE OF PAPER

Fig. 4-4. Testing T-square blade.

4.3 DRAWING BOARDS. At least one end of the drawing board must be true. Test both ends of the board with a framing square, Fig. 4-2, or with a *T-square* whose blade is known to be true. If neither edge of the board is true, use a hand plane or a jointer to produce at least one straight edge. Mark this edge "working edge." Always use your T-square head against this edge.

4.4 T-SQUARES. The T-square has two parts, the *head* and the *blade,* Fig. 4-3. They must be rigidly fastened at right angles to each other. Their *working edges* must be straight. Test the working edge of the head with a framing square, a triangle, or any true straightedge. You can easily test the working edge of the blade, Fig. 4-4. Do this by drawing an accurate sharp line along the working edge on a piece of paper. Then turn the paper around until the other side of the line is against the T-square. Draw a second line. If the two lines do not coincide, the space between them represents *double the error* of the blade.

A new but faulty T-square should be returned to the dealer. Slight errors in the working edges can be corrected with a fine sandpaper block. However, to avoid ruining the T-square, you should obtain your instructor's assistance.

Do not use the T-square to drive tacks into the board or for any rough purpose. Never cut paper along its working edge. This will produce nicks that will ruin the T-square.

Although the T-square is used in the following illustrations, many drafting stations will use parallel edges in place of the T-square. (See Sec. 4.31.) The procedures will be the same, but will be more accurate and more easily carried out.

4.5 DRAWING PAPER. White, cream, and light green drawing papers are used. The greatest preference is for cream or "buff" paper. White paper is used for display drawings. It is seldom used for working drawings because it soils so easily. Light green paper is used by some to lessen eyestrain.

Industrial drawings are made in pencil directly on tracing paper, on vellum, or on tracing film. Thus, blueprints can be made from them immediately without doing "tracing."

Ink is seldom used today. When ink is used, it is generally applied to tracing film. For more information on inking, see Sec. 8.21.

Two systems of sheet sizes are listed by the American National Standards Institute (ANSI). One (architectural) is based on 9.0" × 12.0" (trimmed size), followed by multiples 12.0" × 18.0", 18.0" × 24.0", etc. The other (engineering) is based on 8.5" × 11.0", 11.0" × 17.0", 17.0" × 22.0", and 22.0" × 34.0". These sheets when folded fit into standard letter files. Sheet sizes are also available from suppliers in fractional inch or metric equivalents.

For sheet layouts for problems, see the Appendix.

4.6 FASTENING PAPER TO THE DRAWING BOARD. Drafting tape is used to fasten the paper to the drawing board. It is available in rolls, Fig. 4-5(b). It is used in a variety of ways, as shown at (c), (d), and (e). The tape may also be used on hard surfaces such as Masonite, steel, or glass. Transparent cellulose tape is more difficult to remove. It may damage the paper if not handled carefully.

Place the paper fairly close to the working edge of the board, Fig. 4-5(a). This will decrease the error from the slight swing or "give" of the T-square blade. Place the paper about equally far from the top and bottom of the board.

Hold the paper steady with one hand. Slide the T-square to the middle of the paper. Smooth the paper from the center and fasten the upper-left corner and lower-right corner.

Then fasten the remaining corners. Large sheets may require fastening also at the middle of each edge. To remove the tape, pull it slowly outward toward the edge of the paper.

Left-handers: Place the working edge of the drawing board and the head of the T-square on your right. Tape first the upper-right corner.

4.7 DRAWING PENCILS. High-quality *drawing pencils* should be used in industrial drafting. Never use ordinary writing pencils. Your drawing pencils are your most important tools. The leads are made of graphite with clay added in varying amounts to make eighteen grades from 9H (the hardest) down to 7B (the softest). The general uses of the different grades are explained in Fig. 4-6.

Pencils of less than 3" should not be used. They are too short to be handled properly.

Many makes of mechanical pencils are also available, together with refill drafting leads in all grades, Fig. 4-7. These are preferred by the

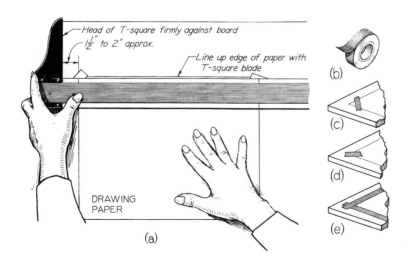

Fig. 4-5. Fastening paper to drawing board.

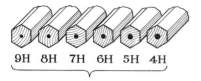

9H 8H 7H 6H 5H 4H

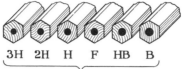

3H 2H H F HB B

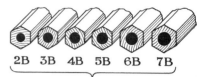

2B 3B 4B 5B 6B 7B

HARD

The harder pencils in this group (left) are used where extreme accuracy is required, as on graphical computations, charts and diagrams. The softer pencils in this group (right) are used by some for line work on engineering drawings, but their use is restricted because the lines are apt to be too light.

MEDIUM

These grades are for general-purpose work in technical drawing. The softer grades (right) are used for technical sketching, for lettering, arrowheads, and other freehand work on mechanical drawings. The harder pencils (left) are used for line work on machine and architectural drawings. The H and 2H pencils are used on pencil tracings for blueprinting.

SOFT

These pencils are too soft to be useful in mechanical drafting. Their use for such work results in smudged, rough lines which are hard to erase, and the pencil must be sharpened continually. These grades are used for art work of various kinds, and for full-size details in architectural drawing.

Fig. 4-6. Pencil grade chart.

professional drafter. Thin lead pencils (0.3, 0.7, etc.) are also in common use.

4.8 CHOICE OF PENCIL. The grade mark, Fig. 4-8(b), is supposed to indicate the exact hardness of lead. Actually it is only approximate. Since you cannot depend entirely on the grade mark, you must learn to use judgment in selecting pencils for the kind of lines required.

For light *construction lines,* Sec. 4.10, guide lines for lettering, and for constructions where great accuracy is necessary, use a hard pencil, such as 4H, 5H, or 6H. For general line work and lettering, all lines should be dark, so they will show up clearly. Select a

pencil that is soft enough to produce a black line without smudging too easily or requiring too-frequent sharpening, such as an F or H. The texture of the paper must be considered. For example, a hard paper will take a harder pencil than will a soft paper. Also, the weather must be considered. For example, in humid weather all papers tend to soften and require softer pencils. The drafter must learn to select the pencil that produces the quality of line he or she needs.

4.9 SHARPENING THE WOODEN PENCIL. *Keep your pencil sharp!* This is certainly the instruction most frequently needed by the beginning student. A dull pencil produces fuzzy, indefinite, sloppy lines. Only a *sharp* pencil can produce accurate, clean-cut, dark lines.

First, use a pencil sharpener with special drafter's cutters to cut away the wood from the *unmarked end* of the pencil a full $1\frac{1}{2}''$ from the point. Leave about $\frac{3}{8}''$ of lead extending uncut beyond the wood. See Fig. 4-8(a) and (b).

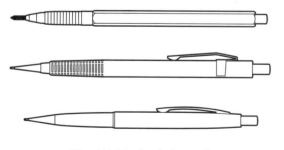

Fig. 4-7. Mechanical pencil.

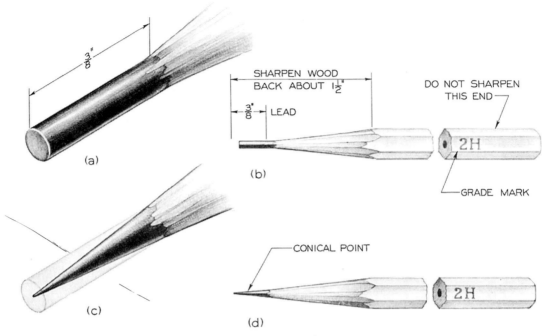

Fig. 4-8. Pencil points.

Second, Fig. 4-8(c) and (d), to produce the conical point (for all line work and lettering), dress the lead down to a long, sharp, symmetrical cone. Do this by rubbing it while rotating it on a sandpaper pad or file, Fig. 4-9(a). *Keep the pencil almost flat on the file or pad.* Many drafters prefer to finish the point by "burnishing" on a scrap of rough paper, such as drawing paper, (b).

Be careful not to get loose graphite on your drawing, your tools, or your hands. Never sharpen your pencil over your drawing or over any of your equipment. Instead, sharpen to one side, where the loose graphite will fall on the floor, Fig. 4-9(c). After sharpening, wipe the point on a clean cloth. Keep your

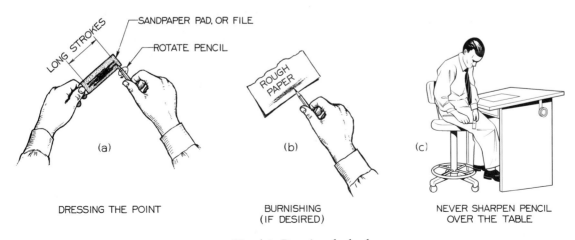

Fig. 4-9. Dressing the lead.

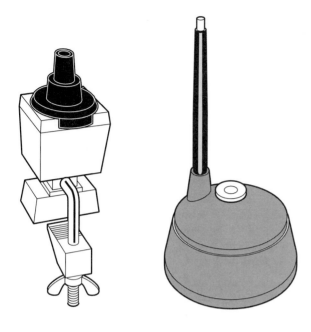

Fig. 4-10. Pencil lead pointers are designed for pointing leads in standard drafting mechanical pencils and in wooden pencils after the wood has been cut back.

sandpaper pad or file in an envelope to prevent it from soiling your equipment. See Fig. 4-25.

If a mechanical pencil is used, a conical point may be produced with the sandpaper pad or file as described before or by using a lead pointer, Fig. 4-10. The pencil lead is inserted in the receptacle in the top of the pointer and rotated to shape the point.

Thin-lead mechanical pencils do not require sharpening. They are becoming very popular with drafters.

4.10 ALPHABET OF LINES. The lines used in drafting are often referred to as the *Alphabet of Lines*, Fig. 4-11. They have the same relation to drawings that letters do to words. Pencil lines are shown on the left, and ink lines on the right. The heavier types of ink lines are thicker than the corresponding pen-

cil lines. The thinner lines are about the same in pencil or in ink.

There are three distinct thicknesses of lines: (1) *thick* (border lines, visible lines, cutting-plane lines, and short-break lines); (2) *medium* (hidden lines); and (3) *thin* (long-break lines, section lines, center lines, dimension lines, extension lines, and phantom lines).

For pencil drawing, all lines except the construction line should be dense black — never gray, fuzzy, or indefinite — so that they will reproduce clearly when copies are made, Secs. 7.3 to 7.6. The thin lines, such as the center line, should be as dark as the visible or hidden lines. The contrast should be in the thickness and not in the degree of blackness. A fairly soft pencil, such as F or H, should be used for the thicker lines. A slightly harder pencil, such as the 2H, should be used for the thinner lines.

Construction lines should always be extremely light. They should be so light that they can be barely seen when viewed at arm's length. They should be made with a hard pencil, such as the 4H. Construction lines are used for "blocking in," or constructing a drawing before the lines are made heavy, Fig. 4-45(V). They are also used for guide lines for lettering, Sec. 5.7.

Applications of the various lines are illustrated in the center column of Fig. 4-11.

On a drawing the visible lines should be the outstanding feature.

4.11 HORIZONTAL LINES. To draw a horizontal line, Fig. 4-12(a): With the left hand press the T-square head firmly against the working edge of the board. With the same hand smooth the blade to the right, pressing the blade firmly against the paper. Then draw the line from left to right, with the little finger

56

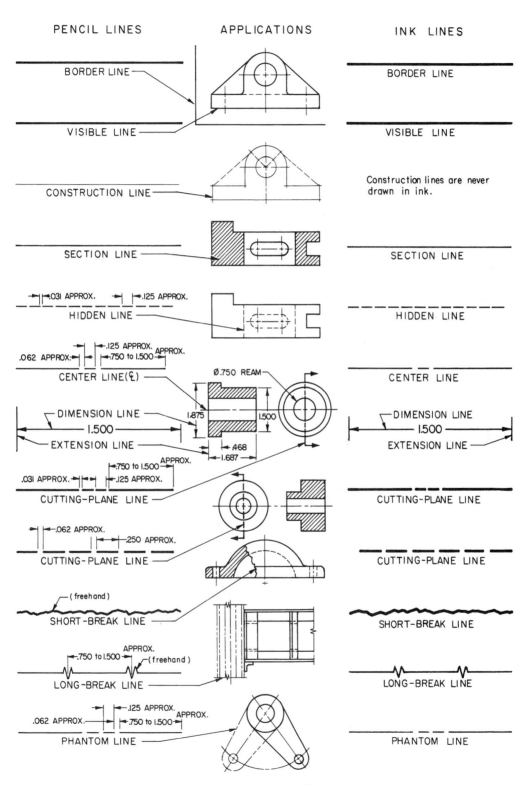

Fig. 4-11. Alphabet of lines.

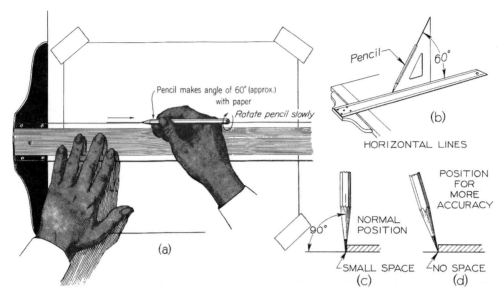

Fig. 4-12. Drawing a horizontal line.

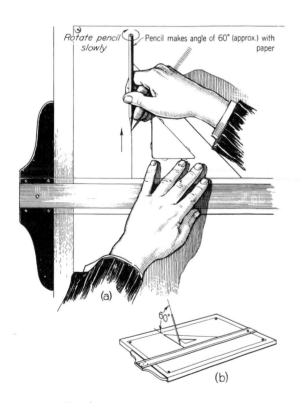

Fig. 4-13. Drawing a vertical line.

gliding lightly along the T-square blade. At the same time, to produce lines of uniform width, rotate the pencil slowly by pressing the thumb forward so as to roll the pencil. This prevents the pencil from wearing down in one place, which would increase the thickness of the line and make the point lopsided.

Lean the pencil at an angle of about 60° with the paper in the direction of the line, as shown by the triangle in Fig. 4-12(b), while keeping the pencil in a vertical plane. In this position, the point will be slightly away from the T-square, (c). The line will not be straight if the pencil is tilted from this position while a line is being drawn. Where great accuracy is required, the pencil may be "toed in," (d), to assure the straightness of the line.

Never draw lines along the lower edge of the T-square.

Left-handers: If you are left-handed, press the T-square head against the right edge of the board. Draw the line from right to left.

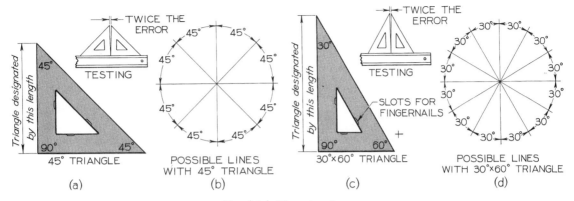

Fig. 4-14. The triangles.

4.12 VERTICAL LINES. Either of the two triangles may be used to draw vertical lines, Fig. 4-13. Notice that the *vertical side of the triangle is on the left,* from which direction the light comes. Press the head of the T-square against the board. Then slide the left hand to the position shown in Fig. 4-13(a), where it holds both the T-square and the triangle together firmly in position. With the pencil leaning at 60° to the paper, (b), *draw the line upward.* Rotate the pencil slowly throughout the progress of the line.

Left-handers: Since the T-square head will be on your right, the triangle will have its vertical edge on the right. Lean the pencil as shown, and draw the line *upward.*

4.13 TRIANGLES. All vertical lines and most inclined lines are drawn with the 45° triangle or the 30° × 60° triangle, Fig. 4-14. As shown at (b), the complete 360° can be divided into eight 45° sectors with the 45° triangle. As shown at (d), twelve 30° sectors can be made with the 30° × 60° triangle.

4.14 INCLINED LINES. The method of drawing lines at 45° with horizontal is shown in Fig. 4-15. When the inclined lines to be drawn are long, use the positions at (a) or (b). The position at (c) is more rigid and accurate. It is recommended for general use. Arrows indicate the directions in which lines are drawn.

The method of drawing lines at 30° with horizontal is shown in Fig. 4-16(a), and at 60° with horizontal at (b). These are recommended where the lines to be drawn are long. However, the method shown at (c) is best for general use. It offers greater stability and accuracy.

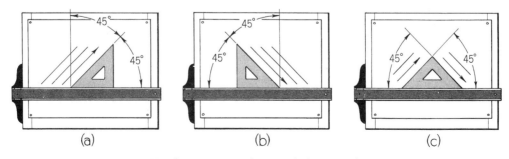

Fig. 4-15. Drawing lines with 45° triangle.

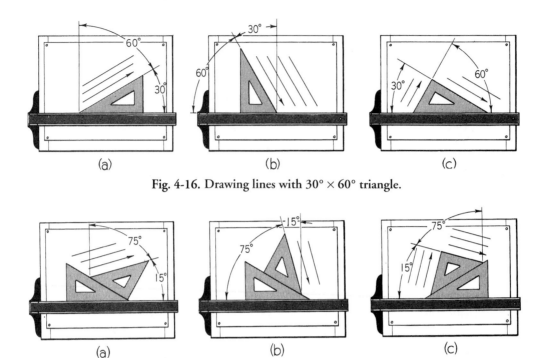

Fig. 4-16. Drawing lines with 30° × 60° triangle.

Fig. 4-17. Drawing lines with triangles in combination.

If the two triangles are combined, lines may be drawn at 15° with horizontal, Fig. 4-17(a), or 75° with horizontal, (b). If the top triangle is arranged to rest upon its hypotenuse, more accurate lines at either 15° or 75° with horizontal may be drawn, (c).

The entire 360° can be divided into 24 sectors of 15° each with the T-square and the tri-

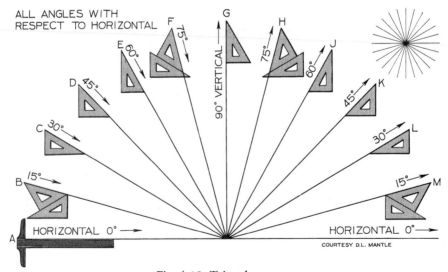

Fig. 4-18. Triangle summary.

angles, either singly or in combination, Fig. 4-18. The arrows indicate the directions in which lines should be drawn.

Angles other than those shown in Fig. 4-18 are measured with the protractor, Fig. 4-19. The drafting machine, Fig. 4-42, and adjustable triangle could also be used.

4.15 PARALLEL LINES. You can easily draw parallel lines at any of the standard angles by sliding the triangles and repeating the lines, as shown in Figs. 4-15, 4-16, and 4-17.

To draw a line parallel to *any* line, Fig. 4-20(a), with the underside of the T-square up, move the T-square and triangle together until the hypotenuse of the triangle lines up with the given line. Then, with the T-square held firmly in position, slide the triangle along the T-square away from the line, (b). Draw the required line, (c). The 30° × 60° triangle could be used equally well for this. Any side of either triangle could be used in contact with the T-square. Instead of the T-square as the supporting member, another triangle may be used.

4.16 PERPENDICULAR LINES. In Figs. 4-15, 4-16, and 4-17, the lines at (b) are perpendicular to those shown at (a). At (c) the lines are perpendicular to each other. To draw a line

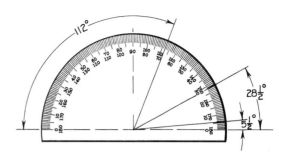

Fig. 4-19. Protractor.

perpendicular to *any* given line, Fig. 4-21(a), move the T-square and triangle (either triangle, resting upon its hypotenuse), until a side of the triangle lines up with the given line. Then, with the T-square held firmly in position, slide the triangle across the line, (b). Draw the perpendicular, (c). Instead of the T-square as the supporting member, another triangle may be used.

A perpendicular to a given line may also be drawn by the *revolved-triangle method*, Figs. 4-22 and 4-23. This method is used when the relatively long hypotenuse is needed for the required perpendicular.

Again, instead of the T-square as the supporting member, another triangle may be used.

4.17 ERASING. Erasers are made because we all make mistakes. You can avoid most mis-

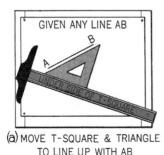
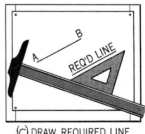

(a) MOVE T-SQUARE & TRIANGLE TO LINE UP WITH AB

(b) SLIDE TRIANGLE ALONG T-SQUARE

(c) DRAW REQUIRED LINE PARALLEL TO AB

Fig. 4-20. Drawing parallel lines.

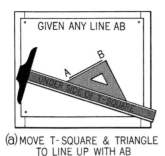

GIVEN ANY LINE AB

(a) MOVE T-SQUARE & TRIANGLE
TO LINE UP WITH AB

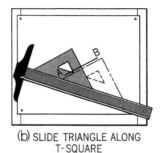

(b) SLIDE TRIANGLE ALONG
T-SQUARE

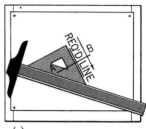

(c) DRAW REQUIRED LINE
PERPENDICULAR TO AB

Fig. 4-21. Drawing perpendicular lines—adjacent-sides method.

takes by following the rule: *Do not draw a line until you know what you are doing!* When you do have to erase, take care to avoid spoiling the drawing.

The soft plastic eraser, Fig. 4-24(a), is used for cleaning large areas, not for general erasing. Avoid the practice of making a careless, dirty pencil drawing and then scrubbing it with the eraser before heavying-in the final lines.

For general use, a pink eraser, (b), is recommended for both pencil and ink erasing. Never use gritty ink erasers. Never use a razor blade or a knife for erasing purposes. They will ruin the drawing.

The erasing shield, (c) to (e), is used to protect lines near the line being erased. Electric *erasing machines,* (e), are available in industrial drafting rooms to save valuable time.

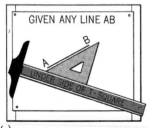

GIVEN ANY LINE AB

(a) MOVE T-SQUARE & TRIANGLE
TO LINE UP WITH AB

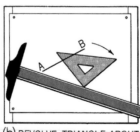

(b) REVOLVE TRIANGLE ABOUT
90° CORNER

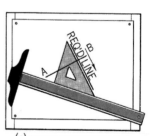

(c) DRAW REQUIRED LINE
PERPENDICULAR TO AB

Fig. 4-22. Drawing perpendicular lines—revolved-triangle method (45° triangle).

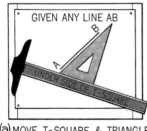

GIVEN ANY LINE AB

(a) MOVE T-SQUARE & TRIANGLE
TO LINE UP WITH AB

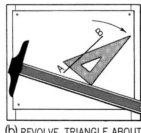

(b) REVOLVE TRIANGLE ABOUT
90° CORNER

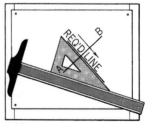

(c) DRAW REQUIRED LINE
PERPENDICULAR TO AB

Fig. 4-23. Drawing perpendicular lines—revolved-triangle method (30° × 60° triangle).

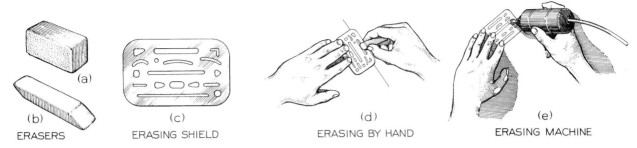

ERASERS | ERASING SHIELD | ERASING BY HAND | ERASING MACHINE
(a) (b) | (c) | (d) | (e)

Fig. 4-24. Erasing.

4.18 NEATNESS. Neatness is essential. There is no reason for sloppy, dirty drawings. Most of the "dirt" on drawings is really graphite from the pencil lead. It is the result of carelessness in letting particles of graphite drop on the drawing or in smearing the graphite of the lines already drawn. When a line is drawn, loose graphite particles are left along the line. These should be blown off with the breath at frequent intervals. Do not allow the drawing to accumulate dust or graphite, *but keep the drawing clean from the beginning.* Keep your hands and your equipment clean. Do everything you can to protect your drawing from being smeared. For perspiring hands, wash frequently with soap and water, and apply talcum powder. Some of the precautions are illustrated in Fig. 4-25.

4.19 DRAWING TO SCALE. Drawings are made on certain standard sizes of paper, Sec. 4.5. Objects to be drawn may be relatively small (such as a machine part) or large (such as a building). Thus, it is necessary to consider what size to make the drawing. Whenever possible, an object should be drawn full size.

If necessary it may be drawn $\frac{1}{2}$ size, $\frac{1}{4}$ size, or smaller. A building may have to be drawn $\frac{1}{48}$ size or a map $\frac{1}{1200}$ size.

The conventional unit of measurement used on engineering drawings in the United States is the inch. However, many large industries have converted to the metric system. In some industries, all dimensions on engineering drawings are expressed in inches. In others, both feet and inches are used. The standard symbols are " for inch and ' for foot.

Small objects such as machine parts, sheet-metal parts, or furniture are drawn in terms of decimal or fractional inches. Since this is understood, *all inch marks are omitted* from dimensions on such drawings. Thus, 2.375 by itself means 2.375 inches. Likewise, $2\frac{1}{2}$ by itself means $2\frac{1}{2}$ inches.

Large structures, such as buildings, bridges, ships, dams, and the terrain (as on maps), are drawn in feet and inches. It is common practice on such drawings to show the symbol for feet and to omit the symbol for inches. Dimensions in feet and inches together would therefore be given as: 5'-0, 3'-2$\frac{1}{2}$, 23'-0$\frac{1}{4}$, etc.

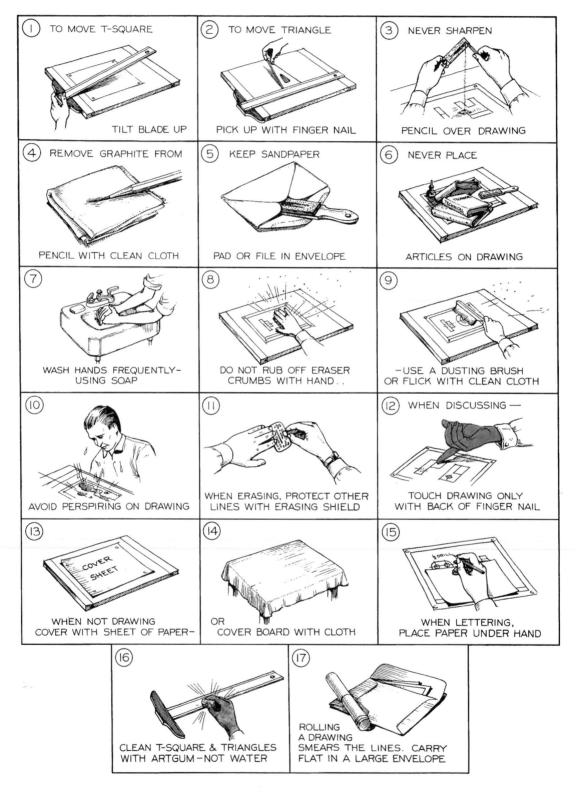

Fig. 4-25. Neatness is not accidental.

The scale to which a drawing is made should be indicated on the drawing, either in the title box or in another appropriate space. For example, if a machine drawing is made using the architects scale, the scale would be indicated as:

SCALE: FULL SIZE or 1.00 = 1.00, 1:1, 1 = 1, or $\frac{1}{1}$

SCALE: HALF SIZE or .50 = 1.00, 1:2, 1 = 2, or $\frac{1}{2}$

The scale of drawings made quarter or eighth size would be indicated in a similar manner. If an enlarged drawing is made, the scale would be shown as:

SCALE: TWICE SIZE or 2.00 = 1.00 or 2 = 1 or $\frac{2}{1}$

For architectural and construction drawings in which the dimensions are expressed in terms of feet and inches, the scale would be indicated as 3″ = 1′-0, 1½″ = 1′-0, ¼″ = 1′-0, etc. For drawings consisting of maps, diagrams, or other graphical constructions for which the engineers scale is used, the scale of the drawing might be indicated to scales such as 1″ = 400′, 1″ = 200 lbs., 1″ = 30° Fahrenheit, etc.

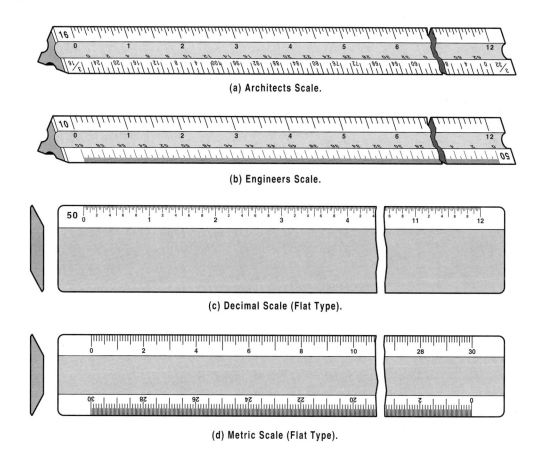

(a) Architects Scale.

(b) Engineers Scale.

(c) Decimal Scale (Flat Type).

(d) Metric Scale (Flat Type).

Fig. 4-26. Types of scales.

4.20 TYPES OF SCALES. Never call a scale a "rule" or a "ruler."

The *architects scale,* Fig. 4-26(a), is an all-round scale for many uses. It has a full-size scale of inches divided into sixteenths. It also has a number of reduced-size scales in which inches or fractions of an inch represent *feet*.

Drawings can be made full size, $\frac{1}{2}$ size, $\frac{1}{4}$ size, and so on down to $\frac{1}{128}$ size.

The *engineers scale,* Fig. 4-26(b), has a series of scales in which inches are divided into 10, 20, 30, 40, 50, 60, or 80 parts. These scales are used in many ways. For example, on a map the "50-scale" can be used so that 1″ = 50′, or

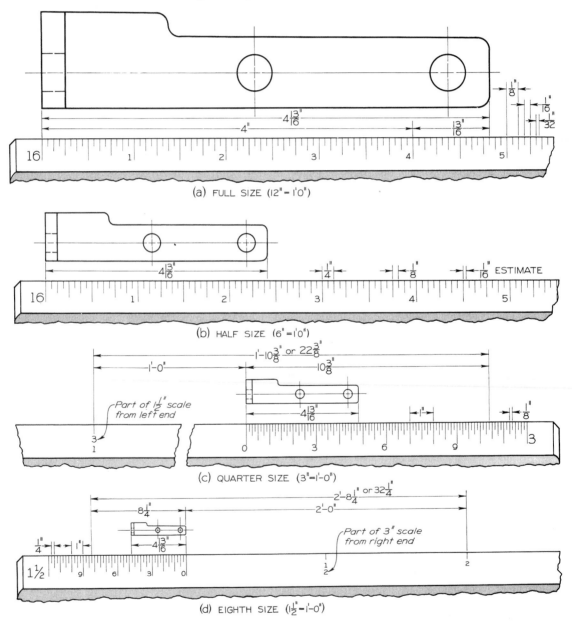

(a) FULL SIZE (12″ = 1′0″)

(b) HALF SIZE (6″ = 1′0″)

(c) QUARTER SIZE (3″ = 1′-0″)

(d) EIGHTH SIZE (1½″ = 1′-0″)

Fig. 4-27. Use of architects scale.

1″ = 500′, or 1″ = 5 miles. Using the "10-scale" on a machine drawing, you can lay off a decimal dimension of, say, 3.652 by measuring 3″ and then adding six-tenths and then slightly more than half a tenth.

The engineers scale is often called the civil engineers scale. It was originally used primarily in civil engineering.

The *decimal scale* is shown in Fig. 4-26(c). On this scale, the inches are divided into 10, 30, 40, and 50 parts. The decimal scale is used in the same way as the engineers scale.

The examples given so far show scales using the customary-inch measurement system. There also are scales that use the metric system. The modern form of the metric system, SI, was established in 1960 by international agreement. The SI stands for *Systeme International d'Unites*. It is now the world standard for measurement. Many large corporations in the United States have adopted this standard. They have converted their products to a metric design.

In the metric system, all larger and smaller units are based on multiples of ten, with no fractions. The common units of linear measurement are the millimeter, centimeter, meter, and kilometer.

1 millimeter (1 mm) is about 0.039 inch
1 centimeter (1 cm) = 10 mm
1 meter (1 m) = 100 cm = 1000 mm
1 kilometer (1 km) = 1000 m = 100,000 cm = 1,000,000 mm

The millimeter (mm) is the primary unit of measurement for engineering drawing and design. The meter (m) and kilometer (km) are used in architectural and surveying.

Metric scales, Fig. 4-26(d), are available in both flat and triangular scales. The full-size and half-size scales are common.

Full-size metric scale. The 1:1 scale is full size, with each division being 1 mm in width and the numbering of the calibrating being at 10-mm intervals. This scale can also be used for scale ratios of 1:10, 1:100, and 1:1000.

Half-size metric scale. The 1:2 scale is one-half size. Each division is at 2-mm intervals. This scale can also be used for scale ratios of 1:20, 1:200, and 1:2000.

Most metric scales have four remaining scales in the scale ratios of 1:5, 1:25, 1:33, and 1:75. Each of these scales can be enlarged or reduced by multiplying or dividing by 10.

Scales are made either in triangular form or in flat form. The triangular form is the more economical. It shows many scales on one stick, whereas several flat scales would be required to show the same number. However, the flat-type scale shown in Fig. 4-26 is very popular with professional drafters because of its simplicity.

Fully divided scales have all of the main units subdivided throughout the entire length of the scale. Fully divided scales have the advantage that several values can be read from the same origin without resetting, Fig. 4-26(a) and (b). Engineers fully divided scales are often called chain scales. Open divided scales have only the main units graduated. However, an extra unit is fully subdivided in the opposite direction from the zero point.

4.21 USE OF ARCHITECTS SCALE. The standard scales for machine drawing are the full size, half size, quarter size, and eighth size. Figure 4-27 shows an object $4\frac{13}{16}″$ long drawn to each of the following four scales:

(a) *Full Size* (12 ″ = 1′-0″). Simply set off 4″ from zero. Add to this $\frac{13}{16}″$, as shown. Note that to set off $\frac{1}{32}″$ it would be necessary to estimate half of $\frac{1}{16}″$.

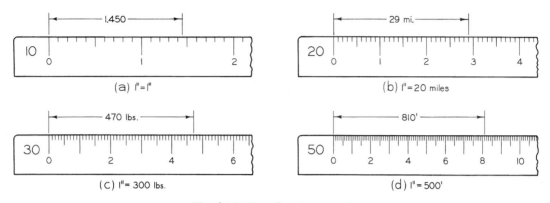

Fig. 4-28. Use of engineers scale.

(b) *Half Size* (6 " = 1 ´-0"). Use the full-size scale, but regard each $\frac{1}{2}$ " on the scale as 1" on the drawing. (Do not use the $\frac{1}{2}$ " scale for half size because this is an architectural scale in which $\frac{1}{2}$ " = 1´-0", or $\frac{1}{24}$ " size.) To set off $4\frac{13}{16}$ " to half size, set off 2" (half of 4"), and add $6\frac{1}{2}$ sixteenths (half of $\frac{13}{16}$).

(c) *Quarter Size* (3 " = 1´-0"). To the right of zero on the scale is a *foot* reduced to 3", or quarter size. This foot is subdivided into inches and fractions of an inch. To set off $4\frac{13}{16}$ ", set off 4" toward the right from zero and then add to this $6\frac{1}{2}$ small divisions. (Each division represents $\frac{1}{8}$ " or $\frac{2}{16}$ ".) To set off 1´-10$\frac{3}{8}$ ", take the foot from the first main division to the left of zero, and the 10$\frac{3}{8}$ " from the right of zero.

(d) *Eighth Size* (1$\frac{1}{2}$ " = 1´-0"). To the left of zero on the scale is a foot reduced to 1$\frac{1}{2}$ ", or eighth size. This foot is subdivided into inches and fractions of an inch. To set off $4\frac{13}{16}$ ", set off 4 " toward the left from zero. Then add to this 3$\frac{1}{4}$ small divisions. (Each small division = $\frac{1}{4}$ " or $\frac{4}{16}$ ".) To set off 2´-8$\frac{1}{4}$ ", take the 2´ from the second main division to the right of zero, and the 8$\frac{1}{4}$ " from the left of zero.

All the remaining scales on the architects scale are reduced scales *in which inches or fractions of an inch stand for feet*. The reduced scales marked 1", $\frac{3}{4}$ ", $\frac{1}{2}$ ", $\frac{3}{8}$ ", $\frac{1}{4}$ ", $\frac{3}{16}$ ", $\frac{1}{8}$ ", and $\frac{3}{32}$ " each have main divisions representing feet. One foot at the end is divided into inches and fractions. Architects frequently use the $\frac{3}{4}$ " scale for large details of doors, windows, etc. They use the $\frac{1}{4}$ " and $\frac{1}{8}$ " scales for plans and elevations of buildings.

If a machine part is very small, it is drawn oversize, usually *double size*. The full-size scale is used, 1" = $\frac{1}{2}$ " on the object.

4.22 USE OF ENGINEERS SCALE. The engineers scale is used in drawing machine parts, maps, charts, and diagrams and whenever decimal dimensions are encountered graphically. A linear measurement of 1" on any of the scales may be used to represent distances such as inches, feet, miles, etc., or to represent other quantities such as time, weight, force, etc.

For example, to set off a distance of 1.450" full size, Fig. 4-28(a), the 10-scale would be used by simply measuring one main division and adding 4$\frac{1}{2}$ " subdivisions. At (b) a distance of 29 miles is set off by using the 20-scale (1 " = 20 miles) and measuring two main divisions plus nine subdivisions. Other examples of use of the engineers scale are shown at (c) and (d).

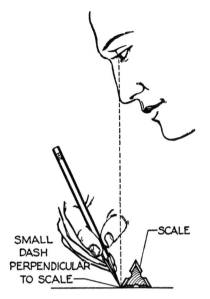

Fig. 4-29. Accuracy in use of scale.

4.23 MEASURING WITH THE SCALE. The habit of accuracy is important in drafting. It is essential to success in any technical work. Accurate drafting depends largely upon the correct use of the scale. Measurements should not be taken directly off the scale with a compass or dividers. This will mar the subdivisions and ruin the scale. Always place the scale along the line to be measured. (If no line is present, draw a construction line.) With your eye directly above the graduation mark on the scale, Fig. 4-29, make a short dash (use a sharp pencil) at right angles to the scale.

After setting off a dimension, always dou-

ble-check with the scale to make sure the distance is accurate. For horizontal measurements, place the scale on the paper so that the scale in use is at the top. For vertical measurements, place the scale so that the scale in use is on the left.

Avoid cumulative errors in the use of the scale. If a series of measurements is to be set off end to end, such as $\frac{9}{16}$, $\frac{3}{4}$, $\frac{1}{2}$, and $\frac{1}{8}$, do not set off the first dimension and then move the scale to start the next dimension from zero, etc. All of these should be set off *without moving the scale*, as shown in Fig. 4-30(a). The scale is thus a simple adding machine. The scale may also be used for subtraction and multiplication, as shown at (b) and (c).

4.24 DRAWING INSTRUMENTS. A set of drawing instruments suitable for school or professional use is illustrated in Fig. 4-31. This set contains a center-wheel bow compass, dividers, and various accessories.

To do highly skilled work you must have good tools. This is particularly true in mechanical drawing. First-class work is impossible with cheap or defective instruments. However, the qualities of high-grade instruments are generally not recognized by the beginner. Therefore, you should obtain the advice of your instructor, an experienced drafter, or a reliable dealer.

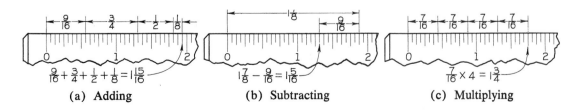

(a) Adding   (b) Subtracting   (c) Multiplying

Fig. 4-30. Scale arithmetic.

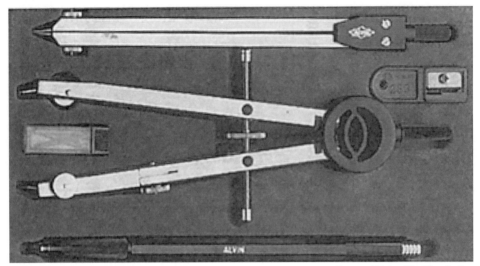

Courtesy of Modern School Supplies, Inc.

Fig. 4-31. Basic bow compass and divider set.

4.25 THE GIANT BOW SET. In modern drafting, reproductions are often made directly from pencil drawings or tracings. Thus, the lines must be dense black. A very rigid compass is needed so that the drafter can "bear down" in drawing arcs and circles. The *giant bow compass* is the best device for this. This is simply a large bow compass with a maximum radius about equal to that of the conventional large compass, but with the rigidity of the small bow instrument. A typical giant bow set is shown in Fig. 4-32. This set contains a large bow compass, a small bow compass, dividers, a pencil, and various accessories.

4.26 THE COMPASS. The compass, with attachments, is used to draw circles from about 1″ radius to about 6″ radius, Fig. 4-33.

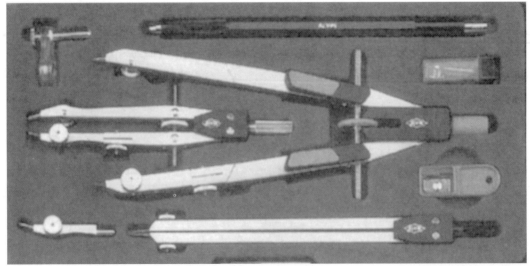

Courtesy of Modern School Supplies, Inc.

Fig. 4-32. Combination rapid acting bow compass and divider set.

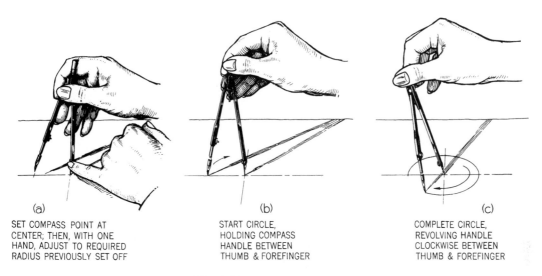

(a)
SET COMPASS POINT AT
CENTER; THEN, WITH ONE
HAND, ADJUST TO REQUIRED
RADIUS PREVIOUSLY SET OFF

(b)
START CIRCLE,
HOLDING COMPASS
HANDLE BETWEEN
THUMB & FOREFINGER

(c)
COMPLETE CIRCLE,
REVOLVING HANDLE
CLOCKWISE BETWEEN
THUMB & FOREFINGER

Fig. 4-33. Using the compass.

If the lengthening bar is used, circles up to about 12″ radius can be drawn. A special *beam compass* must be used for larger sizes.

When the compass is lifted from the case, the legs are squeezed apart with the thumb and forefinger. The needle point is set at the center of the circle. The compass is adjusted to the required radius previously set off with a scale on the center line as in Fig. 4-33(a). Start drawing the circle on the left-hand side, (b). Lean the compass slightly forward. Draw the circle in a clockwise direction with the handle rotating between the thumb and forefinger, (c).

Left-handers: Hold the compass with the left hand and draw circles *counterclockwise*.

As shown in Fig. 4-34, the scale should be used to set off the radius along a center line. Do not place the compass directly on the scale to obtain the radius. This will eventually damage the subdivisions on the scale. After the circle is drawn, III, check the diameter with the scale, IV, to make sure an error in radius has not resulted in a double error in diameter. The circle should be drawn lightly and not made heavy until the correctness is established. A good method is to draw a trial circle on a piece of scrap paper and check it

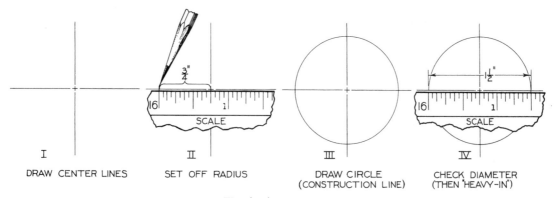

I
DRAW CENTER LINES

II
SET OFF RADIUS

III
DRAW CIRCLE
(CONSTRUCTION LINE)

IV
CHECK DIAMETER
(THEN "HEAVY-IN")

Fig. 4-34. Use of scale.

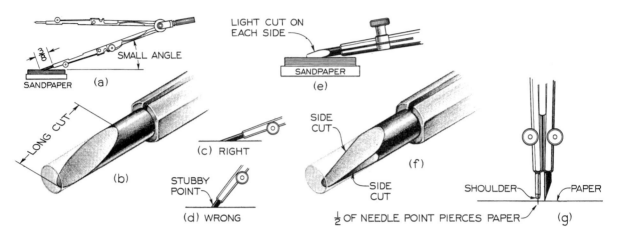

Fig. 4-35. Sharpening compass lead.

$1\frac{1}{2}''$

with the scale before drawing the circle on the drawing.

In drawing circles, you usually know the diameter. You must determine the radius by dividing the diameter in half. If a diameter of $3\frac{1}{2}''$ is given, the radius is figured as follows: Half of 3" = $1\frac{1}{2}''$. Half of $\frac{1}{2}'' = \frac{1}{4}''$. The radius = $1\frac{1}{2}'' + \frac{1}{4}'' = 1\frac{3}{4}''$. If the drawing is to half scale, it is necessary to divide $1\frac{3}{4}''$ again in the same manner.

4.27 SHARPENING THE COMPASS LEAD. For construction arcs or circles, use a hard lead,

such as 4H, 5H, or 6H. For general work, use a softer lead that will produce dark lines without smudging too easily, such as an F or H. You cannot exert as much pressure on a compass as on a drawing pencil. Thus, you may need to use a compass lead about one grade softer than is used on the straight-line work.

Compass leads come with your drawing set. They can also be purchased separately in all grades. If you do not have the right lead, cut off a piece of your drawing pencil, remove the wood, and use the lead. Used pencil stubs should be saved for this purpose.

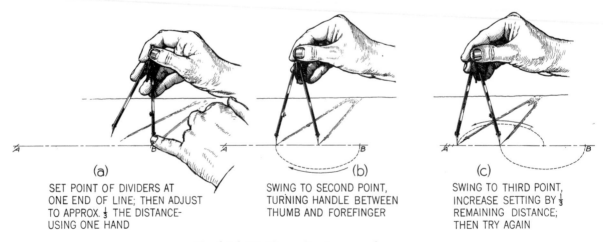

Fig. 4-36. Dividing a line into equal parts.

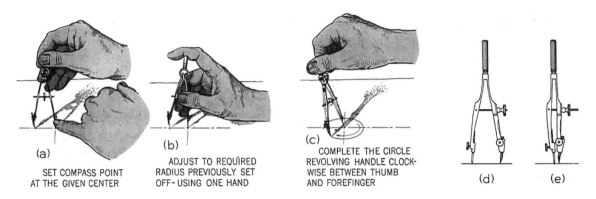

(a)
SET COMPASS POINT
AT THE GIVEN CENTER

(b)
ADJUST TO REQUIRED
RADIUS PREVIOUSLY SET
OFF – USING ONE HAND

(c)
COMPLETE THE CIRCLE
REVOLVING HANDLE CLOCK-
WISE BETWEEN THUMB
AND FOREFINGER

(d)   (e)

Fig. 4-37. Using the bow pencil.

Adjust the lead so that it extends about $\frac{3}{8}''$ from the compass. Rub the lead on the sandpaper pad or file to produce a long inclined cut on the lead, Fig. 4-35(a) to (c). Always keep a liberal length of lead in the compass so that a long cut can be made. Avoid a stubby point as shown at (d).

Some drafters prefer to dress the point still further by making two additional very light side cuts, as shown at (e) and (f). Adjust the needle point so that it extends slightly longer than the lead, as shown at (g). Use the "shoulder" end of the needle point, not the plain end.

4.28 DIVIDERS. The *dividers*, Fig. 4-36, are used for subdividing distances into equal spaces or for transferring distances in which the spacing between the points is approximately $1''$ or over. To divide a line AB into equal parts, say three, lift the dividers from the case. Squeeze the legs apart between the thumb and forefinger until the points are separated an estimated one-third of the total distance. Set the point at one end of the line as in Fig. 4-36(a). Then turn the dividers between the thumb and forefinger to the second point, as shown at (b), and finally to the third point, as at (c). Here the spacing of the divider points is too short. It is necessary to

increase the space by an estimated one-third of the remaining distance. Try again. If this is carefully done, the correct spacing can be obtained in two or three trials.

4.29 BOW INSTRUMENTS. The *bow pencil, bow pen,* and *bow dividers* are known collectively as the *bow instruments,* Fig. 4-31. They are used in a manner similar to the larger instruments. However, they are smaller and more rigid. They should always be used when spacings or radii are under $1''$, which is the approximate capacity of these instruments. The proper use of the bow pencil is shown in Fig. 4-37. Some drafters prefer the "center-wheel" instruments shown in this figure, (a) to (c). Others prefer the "side-wheel" instruments, (d) and (e). One is as good as the other. Notice that bow instruments, like all drawing instruments, are operated with one hand.

Sharpen the compass lead as described in Sec. 4.27. For average use, turn the slanted side of the lead to the outside, Fig. 4-37(d). For small radii, turn the slanted side to the inside, as shown above at (e).

4.30 IRREGULAR CURVES. *Irregular,* or *French, curves* are made of amber or clear plastic, Fig. 4-38(a) to (f). They are used to draw irregular curves. The irregular curve is not used to establish the original curve, but only to make the final curve smooth, Fig. 4-39. First, plot enough points to establish the curve accurate-

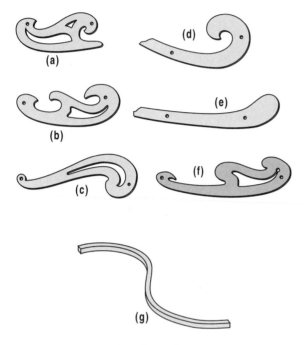

Fig. 4-38. Irregular curves.

ly. Second, sketch *very lightly,* by eye, a smooth curve through the points. Third, match the irregular curve to the sketched curve, the sketched line determining the direction or flow of the curve.

The student usually tries to make the curve fit too many points in one setting in order to reduce the number of settings. Watch the original curve carefully. Make sure that each setting *flows smoothly* from the previous one. You do this by never drawing the curve the full length of the portion that appears to match the sketched curve, but by overlapping each successive setting, Fig. 4-39.

For ink work, the drawing pen should be held nearly vertical. The point should be kept approximately parallel to the edge of the irregular curve. A triangle placed under the curve will keep ink from running under. Leave small gaps between the inked segments. Fill these in by hand. See Sec. 8-21.

An *adjustable curve* is shown in Fig. 4-38(g).

If the curve is symmetrical, such as an ellipse, it is well to use the same segment of

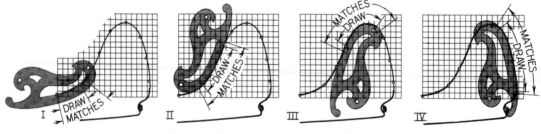

Fig. 4-39. Settings to draw a smooth curve.

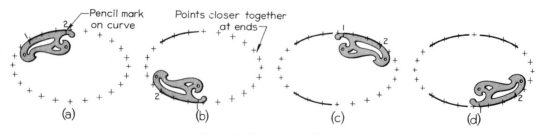

Fig. 4-40. Symmetrical figures.

Fig. 4-41. Parallel ruling straightedge.

David Young-Wolf/PhotoEdit

Fig. 4-42. Drafting machine.

the irregular curve in two or more opposite places. For example, Fig. 4-40(a), the irregular curve has been fitted and the line drawn from 1 to 2. Light pencil dashes are drawn directly on the irregular curve at these points. (The curve will take pencil marks better if it is "frosted" by light sandpapering with #00 sandpaper.) At (b) the irregular curve is turned over and matched so that the line may be drawn from 2 to 1. In similar manner, the same segment is used again at (c) and (d). To complete the ellipse, the gaps at the ends may be filled in by means of the irregular curve or, if desired, with the compass.

4.31 PARALLEL RULING STRAIGHTEDGE. For large drawings, the T-square becomes unwieldy because of the length of the blade. Considerable inaccuracy may result from the "give," or swing, of the blade. In such cases the *parallel ruling straightedge,* Fig. 4-41, is recommended. The ends of the straightedge are controlled by a system of cords and pulleys. These permit it to be moved up or down on the board, always maintaining a true horizontal position.

4.32 DRAFTING MACHINE. The *drafting machine,* Fig. 4-42, has long been regarded as a practical and effective time-saver in drafting. The removable scales are attached securely to the head. A knob is provided to turn the scales to any desired angle as indicated on the protractor around the knob. The scales may then be moved anywhere on the board without changing the angle. The scales are used both for measuring and for drawing lines. The drafting machine thus combines in convenient form all the functions of the T-square, triangles, scales, and protractor.

4.33 SHEET LAYOUTS. Standard sheet layouts for all problems are given in the Appendix of this book. These conform to ANSI Standard sheet sizes, based on 8.5″ × 11.0″ and multiples thereof, Sec. 4.5. The 8.5″ × 11.0″ basic size has been adopted by the majority of industries largely because the 8.5″ × 11.0″ size and the folded multiples, will fit in standard envelopes and can be filed in standard drawing files. Three sizes of sheets are shown — 8.5″ × 11.0″, 11.0″ × 17.0″, and 17.0″ × 22.0″, with title blocks or strips for each layout.

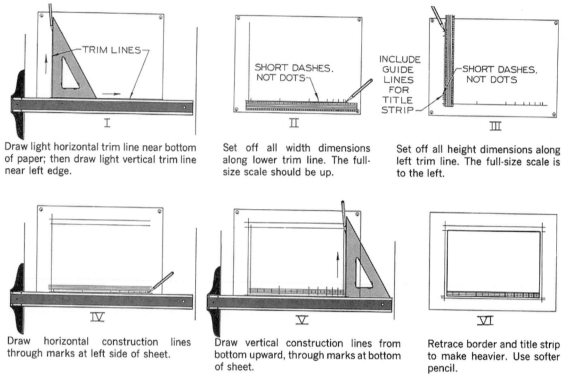

I

Draw light horizontal trim line near bottom of paper; then draw light vertical trim line near left edge.

II

Set off all width dimensions along lower trim line. The full-size scale should be up.

III

Set off all height dimensions along left trim line. The full-size scale is to the left.

IV

Draw horizontal construction lines through marks at left side of sheet.

V

Draw vertical construction lines from bottom upward, through marks at bottom of sheet.

VI

Retrace border and title strip to make heavier. Use softer pencil.

Fig. 4-43. Laying out sheet. Use Layout A in Appendix.

4.34 TO LAY OUT A SHEET. After the sheet has been attached to the board as described in Sec. 4.6, lay out the sheet as shown in Fig. 4-43. Construction lines, drawn lightly with a sharp hard pencil, are used in the first five steps. Clean black border lines, Fig. 4-11, are used in step VI. Note that in steps II and III all measurements are set off at once for the title strip, including guide lines for lettering.

PROBLEMS

The problems that follow for this chapter are all drawn on Layout A, as shown in the Appendix. If the instructor wishes to use 9.0 " × 12.0 " sheets, this can be easily done by lengthening the two left-hand spaces in the title strip of Layout A. The problems may be dimensioned, if desired by the instructor. In such a case, the student should study Sec. 5.11 or 5.13, on the lettering of numerals, and Secs. 10.1 to 10.12, on dimensioning conventions.

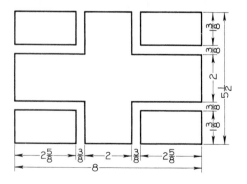

Fig. 4-44. Inlaid linoleum design.

The problems for this chapter are all one-view problems. These afford practice in using the T-square, triangles, drawing instruments, and other equipment.

In Fig. 4-45, the steps in making a one-view mechanical drawing of Fig. 4-44, involving only horizontal and vertical lines, are given. It is suggested that this problem be drawn first. Then assignments may be made from Figs. 4-46 to 4-51.

Next, the student should draw the problem in Figs. 4-52 and 4-53 involving inclined lines. Further assignments can be made from Figs. 4-54 and 4-55.

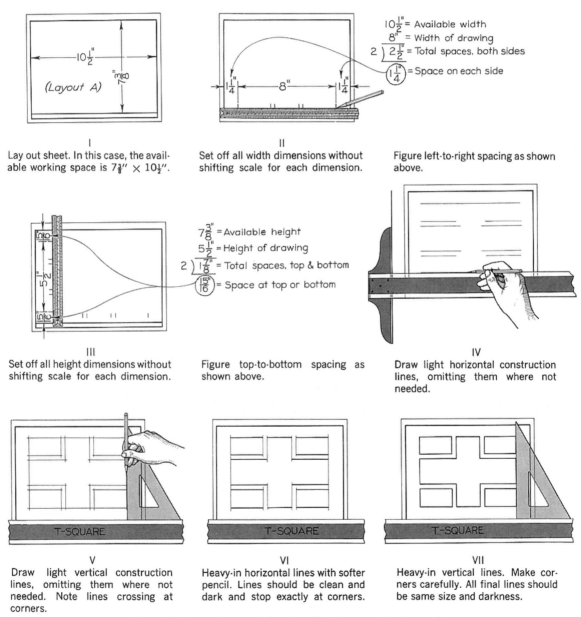

I
Lay out sheet. In this case, the available working space is $7\frac{3}{8}'' \times 10\frac{1}{2}''$.

II
Set off all width dimensions without shifting scale for each dimension.

III
Set off all height dimensions without shifting scale for each dimension.

IV
Draw light horizontal construction lines, omitting them where not needed.

V
Draw light vertical construction lines, omitting them where not needed. Note lines crossing at corners.

VI
Heavy-in horizontal lines with softer pencil. Lines should be clean and dark and stop exactly at corners.

VII
Heavy-in vertical lines. Make corners carefully. All final lines should be same size and darkness.

Fig. 4-45. Steps in pencil drawing. Use Layout A in Appendix.

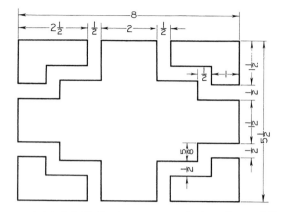

Fig. 4-46. Inlaid linoleum center design. Draw full size. Use Layout A in Appendix.

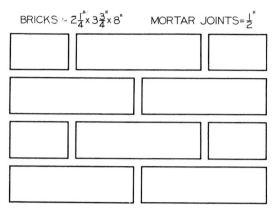

Fig. 4-47. Brick wall. Draw half size. Use Layout A in Appendix. (Small rectangles are ends of bricks.)

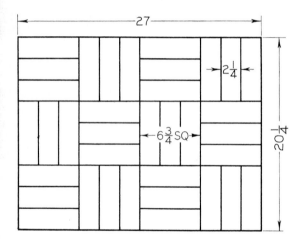

Fig. 4-48. Oak floor pattern. Draw quarter size. Use Layout A in Appendix.

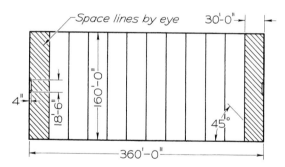

Fig. 4-49. Football gridiron. Draw to scale 1″ = 40′-0″. Use Layout A in Appendix. Use architects or engineers scale.

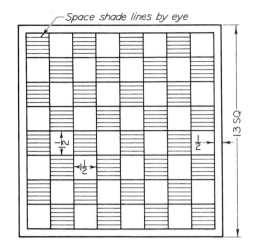

Fig. 4-50. Inlaid wood checker board. Draw half size. Use Layout A in Appendix.

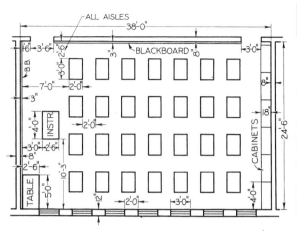

Fig. 4-51. Drafting room layout. Draw to scale $\frac{1}{4}'' = 1'\text{-}0''$. Use Layout A in Appendix.

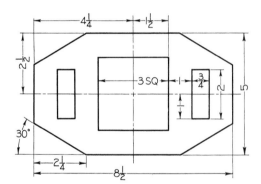

Fig. 4-52. Base plate.

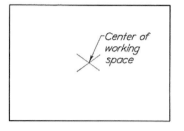

I. Locate center by drawing light diagonals.

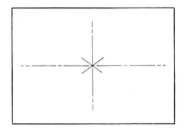

II. Draw center lines.

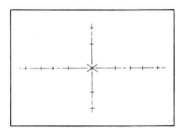

III. Set off vertical and horizontal distances.

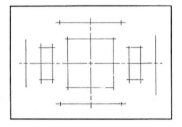

IV. Draw horizontal and vertical construction lines.

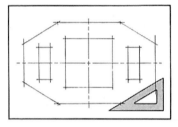

V. Draw angles (note lines crossing at corners).

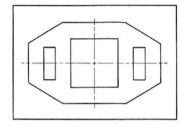

VI. Heavy-in final lines.

Fig. 4-53. Steps in drawing base place. Use Layout A in Appendix.

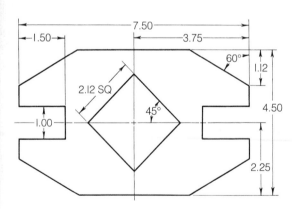

Fig. 4-54. Shim (Layout A).

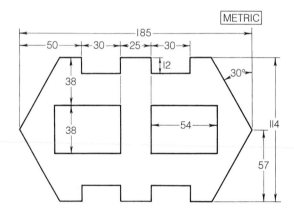

Fig. 4-55. Sheet metal stamping (Layout A).

A problem involving circular arcs and circles is given in Figs. 4-56 and 4-57. Further assignments can be made from Figs. 4-58 to 4-65.

The steps in drawing the Inlaid Linoleum Design, Fig. 4-44, are shown in Fig. 4-45. First, lay out the sheet, as shown in Fig. 4-43 (Layout A, Appendix). The working space inside the border is $10\frac{1}{2}''$ wide and $7\frac{3}{8}''$ high. The drawing is to be centered in this space. Make spacing calculations on a scrap of paper or on your sheet outside the trim line. Follow the steps below. Apply them to all of the problems that follow.

The steps in drawing the Base Plate, Fig 4-52, are shown in Fig. 4-53. First, lay out the sheet as shown in Fig. 4-43 (Layout A, Appendix). The working space inside the border is $10\frac{1}{2}''$ wide × $7\frac{3}{8}''$ high. The drawing is to be centered in this space. One method of spacing is

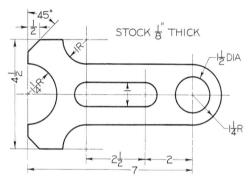

STOCK $\frac{1}{8}''$ THICK

Fig. 4-56. Adjusting arm.

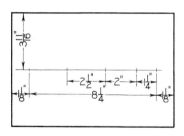

I. Draw horizontal center line and set off horizontal distances with scale.

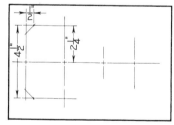

II. Draw vertical lines and construction as shown.

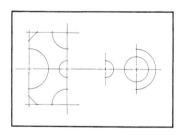

III. Draw construction arcs.

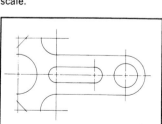

IV. Draw connecting straight lines.

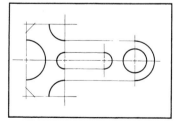

V. Heavy-in arcs and circles.

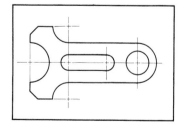

VI. Heavy-in straight lines to complete drawing.

Fig. 4-57. Steps in drawing adjusting arm. Use Layout A in Appendix.

shown in Fig. 4-45. Another is shown in Fig. 4-53, in which you locate the center of the working space by drawing diagonals and then making all measurements from this point.

The steps in drawing the Adjusting Arm, Fig. 4-56, are shown in Fig. 4-57. First, lay out the sheet, using Layout A in the Appendix. The working space inside the border is $10\frac{1}{2}''$ wide $\times 7\frac{3}{8}''$ high. The drawing is to be centered in this space. As shown in step I, draw the horizontal center line at mid-height on the sheet ($7\frac{3}{8}'' \div 2 = 3\frac{11}{16}''$). Since the object is $8\frac{1}{4}''$ long overall, and the space is $10\frac{1}{2}''$ wide, the space on each side is $1\frac{1}{8}''$, as shown.

If assigned by the instructor, construct drawings using decimal-inch or metric scales by converting given fractional dimensions to decimal-inch or metric equivalents. Refer to Table 20 in the Appendix.

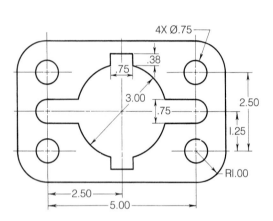

Fig. 4-58. Key plate (Layout A).

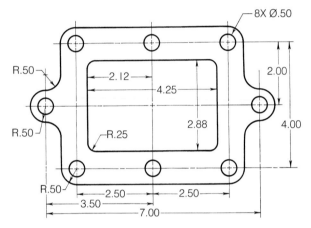

Fig. 4-59. Gasket (Layout A).

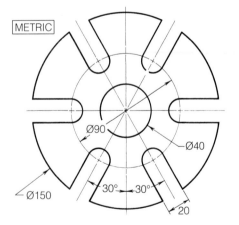

Fig. 4-60. Slotted cam. Use Layout A in Appendix.

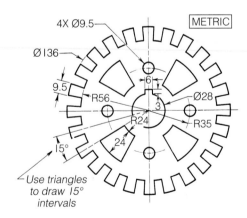

Fig. 4-61. Armature lamination. Use Layout A in Appendix. Use triangles to draw 15° sectors, Fig. 4-18.

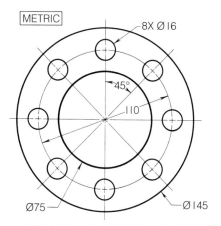

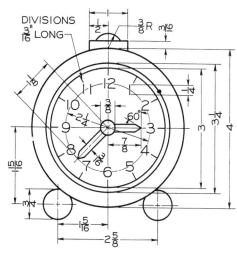

Fig. 4-62. Gasket. Use Layout A in Appendix. Use 45° triangle to draw 45° intervals, Fig. 4-14(b).

Fig. 4-63. Clock. Use Layout A in Appendix. Use 30° × 60° triangle to draw 30° intervals, Fig. 4-14(d).

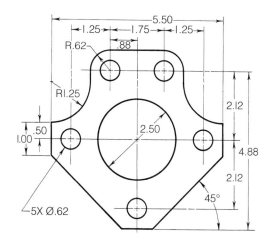

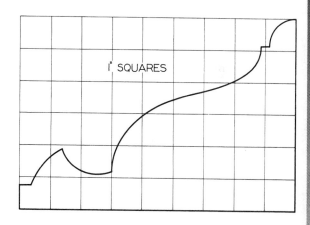

Fig. 4-64. Template. Use Layout A in Appendix.

Fig. 4-65. Table leaf support wing. Draw full size. Use Layout A in Appendix. Use French curve.

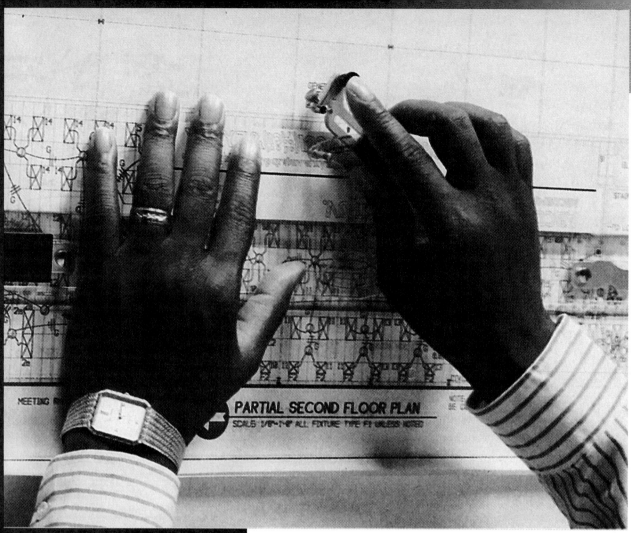

Arnold & Brown

CHAPTER 5

IMPORTANT TERMS

pictograph

lowercase
letters

vertical letters

inclined
letters

OBJECTIVES

After studying this chapter, you should be able to:

• Briefly describe the way letters evolved.

• Identify the three modern letter forms.

• List the techniques that will lead to uniformity in lettering.

• Identify the width of each vertical capital letter.

• Space words and letters correctly.

• Letter correctly, using the correct strokes.

84

CHAPTER 5

Lettering

Technical drawings must be correctly lettered in order to communicate information to others. In this chapter, you will learn the standards of good lettering. Such knowledge is important to all fields of technical drawing.

In some careers, it can be valuable to know not only the current lettering practices but also those of the past. A city planner, for example, may need to work with drawings that were made long ago.

The city planner, or urban designer, is concerned with the way communities are developed or redeveloped. Typical responsibilities include determining best use of the land, such as whether an area will be zoned for residential, business, or industrial use. Density of use—how closely the buildings are placed—is an important issue, as are adequate parking, lighting, streets, water and utilities, etc.

Urban planners must have a good working knowledge of rules that would have an impact upon the city in which they work. Among these are local, county, state, and federal standards for zoning, environmental issues, historical districts and landmark buildings, and other such designations. They must be able to read and interpret maps and architectural drawings, and they should have good sketching and visualization abilities.

CAREER LINK

For more information, contact the American Planning Association, Education Division, 122 South Michigan Avenue, Suite 1600, Chicago, IL 60630.

5.1 ORIGIN OF LETTERS. Printing is done from type. Letters are made by hand. Children print; drafters letter. The letters of our alphabet are *symbols* for sounds. The word *alphabet* comes from the Greek letters alpha and beta. When letters are placed in combinations, words result that have meanings. All writing or lettering started in some form of *pictograph* (picture writing). However, it took humans thousands of years to develop letters from the earliest cave drawings, Fig. 5-1, then from pictographs. Among the earliest forms of pictographs were the ancient Egyptian hieroglyphics (pronounced *high-a-row-glif 'iks*), Fig. 5-2. The characters in the Chinese alphabet are a form of picture writing. However, one must go back to earlier forms of picture writing to see the resemblance to real things, Fig. 5-3. To understand how ideas can be expressed by pictures, try drawing a pictograph yourself, such as the one in Fig. 5-4.

The story of the alphabet is an interesting subject that would require a book in itself. It is enough for our purposes to know that the early Phoenicians developed an alphabet of

Fig. 5-1. Cave drawings. (From *The Story of Writing*, American Council on Education.)

Fig. 5-2. Egyptian hieroglyphics. (This hieroglyphic means "Cleopatra".)

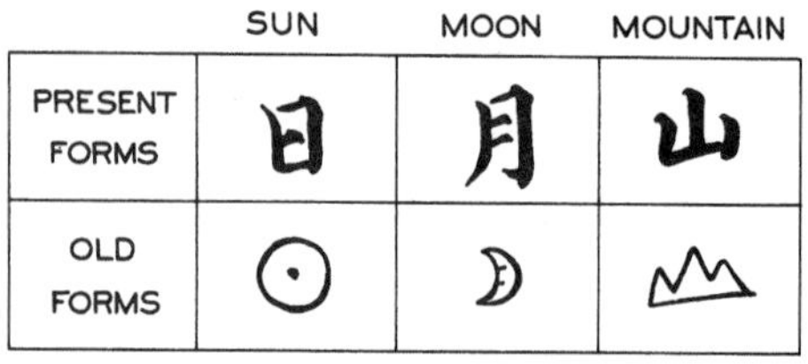

	SUN	MOON	MOUNTAIN
PRESENT FORMS	日	月	山
OLD FORMS	☉	☽	⌒⌒

Fig. 5-3. Characters in the Chinese alphabet. (From *The Story of Writing*, American Council on Education.)

twenty-two characters from earlier picture writing. These characters were later further developed by the Greeks and then the Romans. As a result of Roman conquests, the Latin alphabet was adopted throughout most of the then-civilized world. The Old Roman alphabet is the direct parent of all our present letter forms.

5.2 MODERN LETTER FORMS. The three basic types of letters used in the western world are the Roman, Gothic, and Text letters, Fig. 5-5. These may be uppercase letters or lowercase letters. (They are called *lowercase* because they were kept by compositors in the lower cases of type.) If the letters stand upright, they are *vertical letters*. If they are inclined, they are called *inclined letters* or *italic letters*. Various styles of letters are shown in Figs. 5-29 to 5-31.

Fig. 5-4. "I Saw Many People Walking." (From *The Story of Writing*, American Council on Education.)

Fig. 5-5. The three basic groups of letters. (Courtesy Ross F. George, *Speedball Text Book,* published by the Hunt Pen Co., Camden, N. J.)

5.3 SINGLE-STROKE GOTHIC LETTERS. Early industrial drawings were lettered with what we would regard as "fancy" letters. These letters conformed to the historical styles, usually Roman. It was then the fashion for houses, furniture, and other manufactured products to be decorated with "curlicues." As industry advanced, everything became more streamlined and functional. Fancy lettering frills were abandoned. In the 1890s, C. W. Reinhardt developed alphabets based upon the Gothic letters. These letters could be easily made with single strokes of an ordinary pencil or pen. These letters were called single-stroke Gothic letters. They are very similar to our present letters.

5.4 ANSI STANDARD LETTERS. In 1935 the letters were further standardized when the American National Standards Institute adopted standard alphabets of vertical and inclined letters. With slight revisions, these alphabets have become generally accepted as the "last word" for use in drafting, Figs. 5-6 and 5-7. These letters are used on all drawings in this book. These standards have been published in

ABCDEFGHIJKLMNOP
QRSTUVWXYZ&
1234567890

abcdefghijklmnopqrstuvwxyz

Fig. 5-6. Standard vertical letters. (Adapted from ASME Y14.2M.)

ABCDEFGHIJKLMNOP
QRSTUVWXYZ&
1234567890

abcdefghijklmnopqrstuvwxyz

Fig. 5-7. Standard inclined letters. (Adapted from ASME Y14.2M.)

American National Standard Drafting Manual ASME Y14.2M. The manual is published by the American National Standards Institute, 11 West 42nd Street, New York, NY 10036.

According to the ANSI Standard: "Lettering on drawings must be legible and suitable for easy and rapid execution." In some industrial drafting rooms, vertical letters are required. In others inclined letters are used. The student should learn both before he or she completes training, starting with the vertical.

5.5 UNIFORMITY IN LETTERING. One of the main requirements for good lettering is uniformity, Fig. 5-8(a). Never mix uppercase and lowercase letters as at (b). Use guide lines to prevent irregularities as at (c) to (f). Avoid thick-and-thin strokes as at (g) and (h). The background *areas* between letters should appear equal as at (a) and not as at (j). The spaces between words should not be too small or too large as at (k), but should be equal. Spacing of letters and words is further discussed in Sec. 5.10.

5.6 PENCIL TECHNIQUE. Anyone can learn to letter well if he or she follows certain steps. He or she must (1) learn the *shapes,* (2) learn the *strokes,* (3) learn the rules of *spacing,* and (4) *practice* with a real determination to

improve. "Practice" without real *effort* to improve is useless. Basically, lettering is a form of freehand sketching.

Most drawings today are reproduced directly from pencil tracings. Thus, it is most important to learn to do good pencil lettering, Fig. 5-9. Above all, the letters must be uniform and legible. Use a fairly soft pencil, such as HB, F, or H, sharpened to a sharp conical point, Fig. 5-10(a). For a detailed explanation

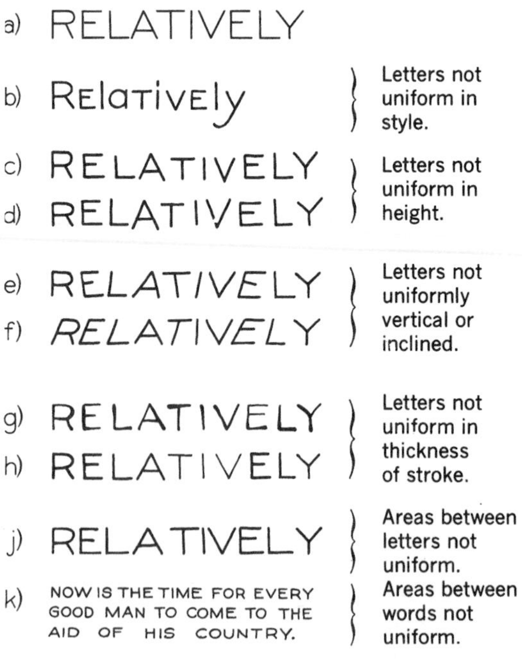

a) RELATIVELY

b) Relatively — Letters not uniform in style.

c) RELATIVELY
d) RELATIVELY — Letters not uniform in height.

e) RELATIVELY
f) *RELATIVELY* — Letters not uniformly vertical or inclined.

g) RELATIVELY
h) RELATIVELY — Letters not uniform in thickness of stroke.

j) RELATIVELY — Areas between letters not uniform.

k) NOW IS THE TIME FOR EVERY GOOD MAN TO COME TO THE AID OF HIS COUNTRY. — Areas between words not uniform.

Fig. 5-8. Uniformity in lettering.

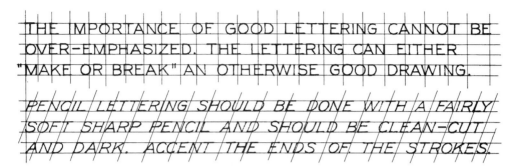

Fig. 5-9. Example of pencil lettering (full size).

of pencils and how to sharpen them, see Secs. 4.7 to 4.9. Hold the pencil naturally, as in writing, Fig. 5-10(b), with your forearm on the drawing board. *Never letter with your forearm off the board.* Make the letters clean-cut and dark. If your fingers begin to cramp, stop and rest for a few seconds.

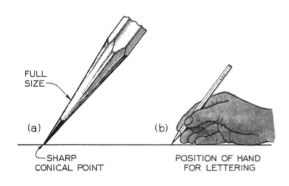

Fig. 5-10. Lettering pencil.

Make the beginnings and ends of all strokes *definite by accenting* or "bearing down" on the pencil. Shift the position of the pencil frequently. This will prevent the lead from wearing down in one place and producing dull lettering.

For information on ink lettering, see Sec. 8.21.

5.7 GUIDE LINES. For vertical letters, draw both horizontal and vertical guide lines, Fig. 5-11(a) and (b). The horizontal guide lines are needed to keep the letters exactly the same height. The vertical guide lines are spaced at random. They are not used to space the letters but only to keep them vertical. Use a "needle-sharp" hard pencil (4H to 6H). Draw the lines so lightly that they can barely be seen when viewed at arm's length.

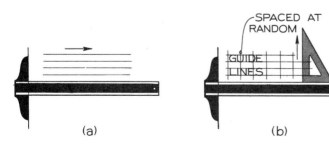
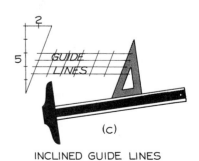

HORIZONTAL GUIDE LINES   VERTICAL GUIDE LINES   INCLINED GUIDE LINES

Fig. 5-11. Guide lines.

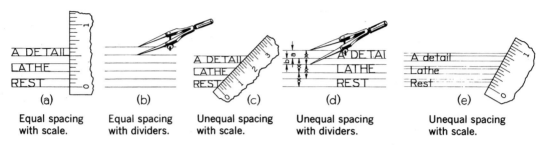

(a) Equal spacing with scale. **(b)** Equal spacing with dividers. **(c)** Unequal spacing with scale. **(d)** Unequal spacing with dividers. **(e)** Unequal spacing with scale.

Fig. 5-12. Spacing of guide lines.

For inclined letters, draw the inclined guide lines at a slant of about 68° with horizontal, or parallel to the hypotenuse of a right triangle whose sides are 2 and 5 units, respectively, Fig. 5-11(c).

The most common spacing for horizontal guide lines is to make the letters $\frac{1}{8}''$ (.12″) high and the space between lines of lettering $\frac{1}{8}''$ (.12″). Place the scale in a vertical position. Mark light dashes at right angles to the scale, Fig. 5-12(a). Another method is to space the guide lines with the bow dividers, (b).

Slightly better appearance results if the space between lines of lettering is less than the height of the letters (but never less than half the height of the letters). Place the scale

diagonally, (a). Set off, for example, 4 units for the height of the letters and 3 units for the height of the spaces. Another method is to use the bow dividers as at (b), making spaces x equal to a + b. Guide lines for lowercase letters may be spaced as shown at (c).

The *Ames Lettering Guide,* Fig. 5-13(a), is highly recommended. Heights are varied as the central disc is turned to one of the numbers at the bottom of the disc. These numbers indicate the heights of letters in 32nds of an inch. Thus, to draw guide lines for letters $\frac{3}{16}''$ (.19″) high, set the disc at number 6, since $\frac{3}{16}'' = \frac{6}{32}''$. The *Ames Lettering Guide* is also available with metric graduations for desired metric spacing.

5.8 VERTICAL CAPITAL LETTERS. The standard vertical capital alphabet is shown in Fig. 5-14. Each letter is shown in a grid 6 units high. You can compare the widths of the letters to the heights by counting the number of squares across each letter. With the exception of the I, which has no width, and the W, which is the widest letter of the alphabet, all letters are either 5 or 6 units wide. You can easily remember which letters are 6 units wide simply by recalling the name TOM Q. VAXY, Fig. 5-15. Each of the letters in his name is 6 units wide. All others in the alphabet, except the I and W, are 5 units wide.

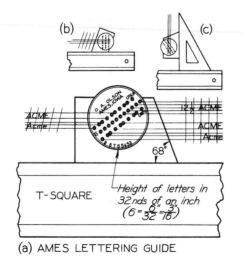

Fig. 5-13. A guide-line device.

Fig. 5-14. Vertical capital letters and numerals.

In addition to the widths of the letters, you must also learn the order of strokes, as indicated in Fig. 5-14. Note that horizontal strokes are made from left to right. Vertical strokes are drawn downward. The best way to learn the proportions and strokes of the letters is to practice sketching them large on cross-section paper. Make each letter 6 squares high.

5.9 LEFT-HANDERS. Are you left-handed? If so, you can learn to letter as well as anyone else. Many of the very best drafters are left-handed. The most important step is to learn

Fig. 5-15. Tom Q. Vaxy.

91

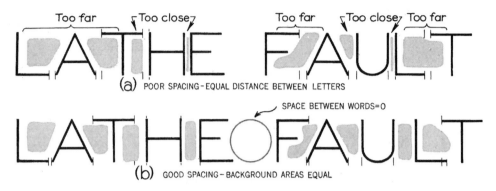

Fig. 5-16. Spacing letters and words.

the correct shapes and proportions. You can do this as well as any right-hander. The habits of left-handers vary so much that no standard system of strokes can be used by all. Therefore, instead of using the strokes indicated in Fig. 5-14, work out a system of strokes of your own. However, do not adopt any strokes in which the pencil tends to dig into the paper. When you do ink lettering, such strokes cannot be used.

5.10 SPACING OF LETTERS AND WORDS. Do not space letters *equal distances* apart, Fig. 5-16(a). Instead, space them by eye so that the background areas between letters *appear* approximately equal, (b). Certain letters that have straight sides, such as the H and E, must be spaced apart to avoid small areas between them, as shown at (a). Other letters, such as the L and T, are of such shape that they may actually be overlapped to produce good spacing. The lower stroke of the

L may be slightly shortened when an A or J follows.

Most beginners space letters too far apart and words too close together, Fig. 5-17. Letters within a word should be evenly but compactly spaced. Words should be separated from each other by a space at least equal to the letter O. Make each word stand out as a separate unit, like the printed words in this book. An example of good spacing is shown in Fig. 5-9.

For a discussion of titles on drawings, see Sec. 16.3.

5.11 VERTICAL NUMERALS. All numerical values on drawings are expressed by various combinations of the ten basic numerals in Fig. 5-14. Therefore, it is worth your time to study carefully the proportions and strokes of these characters. All numerals, except the 1, are five units wide. The 3 and the S are both based on the shape of the 8. This shape is

Fig. 5-17. Spacing letters and words.

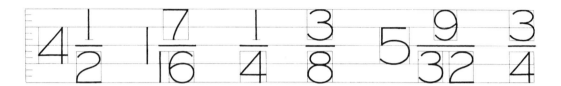

Fig. 5-18. Vertical numerals and fractions.

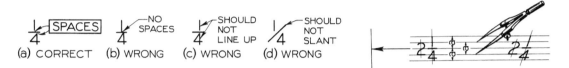

Fig. 5-19. Fractions (full size).   **Fig. 5-20.** Spacing with bow dividers (full size).

Fig. 5-21. Inclined capital letters and numerals.

made up of a small ellipse centered over a larger ellipse. The 6 and 9 are alike, except reversed. Both fit into the elliptical 0 (zero).

Whole numbers and fractions are shown in Fig. 5-18. It is common practice in dimensioning to make whole numbers $\frac{1}{8}$″ (.12″) high, and fractions double this, or $\frac{1}{4}$″ (.25″) high. The numerator and denominator are each slightly less than $\frac{1}{8}$″ (.12″) high. A clear space must be left between them and the fraction bar, as shown in Fig. 5-19.

Horizontal guide lines for numerals and fractions can be spaced with dividers, Fig. 5-20, with spaces *a* made equal. They can also be drawn with guide-line devices, Fig. 5-13. Vertical guide lines should also be used, as shown.

5.12 INCLINED CAPITAL LETTERS. The standard inclined capital alphabet is shown in Fig. 5-21. The grids and letters are the same as for the vertical capitals. However, they are leaned

to the right. The proportions and strokes correspond to those used for making vertical letters. In addition to horizontal guide lines, draw inclined guide lines at random with the triangle as shown in Fig. 5-11(c), or with guide-line devices as in Fig. 5-13. The correct and incorrect ways of making certain letters that usually give difficulty are shown in Fig. 5-22. The letters O, C, Q, G, and D are elliptical in shape. The ellipses should definitely *lean to the right,* as shown for the O and G. The letters V, A, W, X, and Y are balanced equally about an imaginary inclined center line, as shown for the A in the figure.

5.13 INCLINED NUMERALS. Inclined letters and numerals are shown in Fig. 5-21. The numerals are drawn like the vertical numerals in Fig. 5-14, except that they are leaned to the right. Whole numbers and fractions are shown in Fig. 5-23. It is common practice to make whole numerals $\frac{1}{8}$″ (.12″) high and fractions

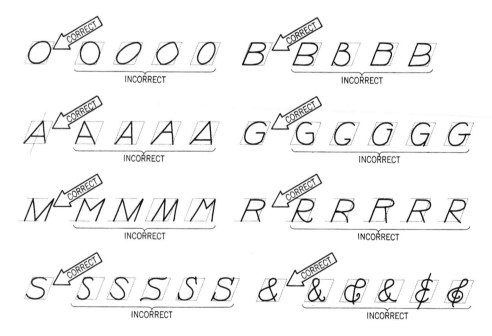

Fig. 5-22. Correct and incorrect inclined capitals.

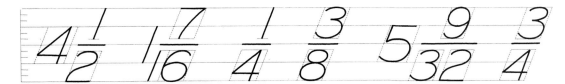

Fig. 5-23. Inclined numerals and fractions.

twice this, or $\frac{1}{4}''$ (.25″) high. Horizontal guide lines are drawn in a similar manner to those for vertical numerals, as shown in Fig. 5-20. Inclined guide lines should be used, as shown.

5.14 VERTICAL LOWERCASE LETTERS. These letters are used mostly on maps and very seldom on machine drawings, Fig. 5-24. They are formed of straight lines and circles with a few variations. The main body of the letter is

two-thirds the height of the capital letter. Three horizontal guide lines should be drawn, Figs. 5-12(e) and 5-13. In addition, vertical guide lines should be drawn at random. To learn the proportions and strokes, practice these letters on cross-section paper.

5.15 INCLINED LOWERCASE LETTERS. The inclined lowercase letters are shown in Fig. 5-25. The grids and letters are the same as for

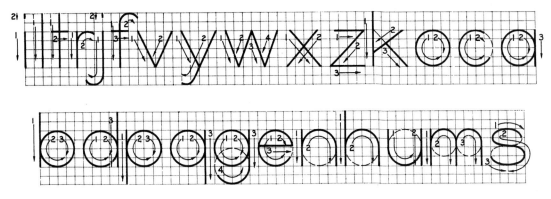

Fig. 5-24. Vertical lowercase letters.

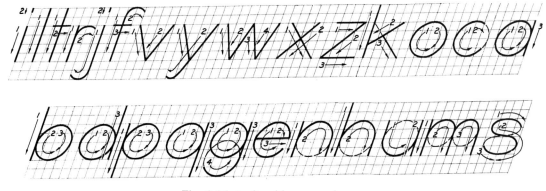

Fig. 5-25. Inclined lowercase letters.

Keuffel & Esser Co.

Fig. 5-26. Leroy lettering instrument.

Kroy, Inc.

Fig. 5-27. Kroy lettering machine.

the vertical lowercase letters, except that they are leaned to the right. The circles become ellipses. These should lean definitely to the right. Be careful to make the letters v, y, w, and x equally balanced about an imaginary inclined center line. Guide lines are drawn in a similar manner to those for vertical lowercase letters, as described in Figs. 5-12(e) and 5-13. However, inclined instead of vertical guide lines are used.

5.16 LETTERING DEVICES. A number of instruments are available for lettering. The *Leroy Lettering Instrument* and other similar instruments, Fig. 5-26, are used with a special guide or template to form the letters. These instruments, especially the Leroy, are used in industry, particularly for lettering titles and drawing numbers.

Other lettering devices, like the Kroy Lettering Machine, create letters on a tape. This tape can be applied to drawings and artwork. The Kroy, Fig. 5-27, is operated by dialing a type disc to the desired letter or character and pressing the print button. Letters are automatically spaced. Various sizes and styles of lettering can be produced by changing type discs.

Dry-transfer, press-on lettering, and special lettering typewriters may also be used to produce uniform, mechanically produced lettering.

5.17 FILLED-IN LETTERS. You can make large letters for titles, posters, and so forth by drawing the outlines, using the T-square, triangles, and compass, and some freehand work for curves. Then fill in the letters, Fig. 5-28. "Block letters," which can be easily drawn with the T-square and triangles, are shown in Fig. 5-29. Gothic and Roman alphabets are shown in Figs. 5-30 and 5-31. First, draw the grids lightly in pencil, or use the bow dividers to set off the unit distances, as shown. Then draw the letters in pencil. Finally, ink them in. Then erase the grid lines. The thicknesses of the stems may be varied from one-tenth to one-fifth the height of the letters, as desired. One-seventh has been adopted in Figs. 5-29 to 5-31.

5.18 SPECIAL PENS. Technical fountain pens are used for drafting, ruling, mechanical and freehand lettering, writing, and commercial art work. Among the more prominent pens are the Koh-I-Noor *Rapidograph, Unitech,* and *Mars* pens.

Use compass   Add spurs if desired   Leave in outline if desired

Fig. 5-28. Filled-in letters.

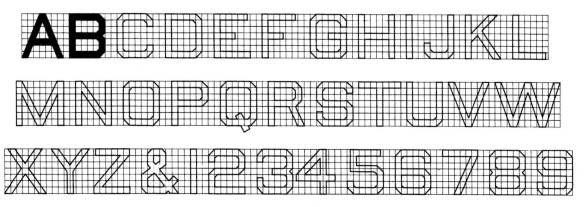

Fig. 5-29. Block letters.

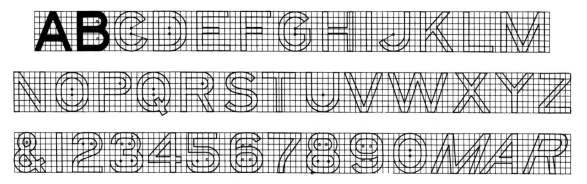

Fig. 5-30. Gothic capital letters.

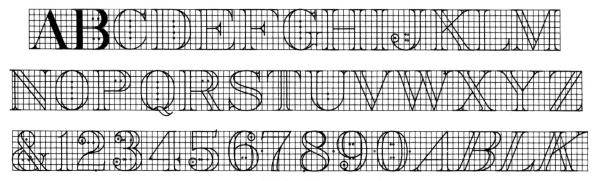

Fig. 5-31. Modern Roman capitals.

(a)

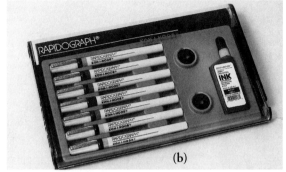
(b)

Koh-I-Noor Rapidograph, Inc.

Fig. 5-32. Technical drawing pen set.

A typical technical drawing pen set, Fig. 5-32(a), contains pens with several line widths, varying from extra fine to wide, Fig. 5-32(b). The pens are fairly easy to use and provide a uniform line width.

5.19 LETTERING WITH CAD. Lettering with a CAD system is easy. The size and style of the text may be pre-defined for the whole drawing or for individually selected lines. When choosing text font styles, a variety of options are available. Some of the more common options include the font style, height and width of the text, the rotation angle, and the oblique angle. Additional options allow text to be placed horizontally, vertically, backwards, and upside down, Fig. 5-33.

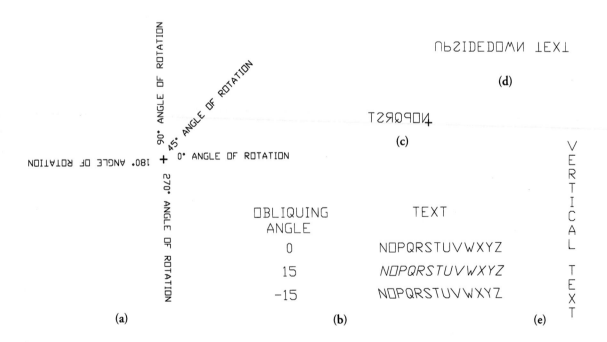

Fig. 5-33. CAD lettering options.

⊥EFT JUSTIFIED TEXT

Fig. 5-34. The start point option being used to generate left-justified text.

Text is usually placed on an individual layer and assigned a color. The layer option allows text to be turned off or on for viewing on screen or printing. One of the advantages of creating text with a CAD system is the lack of guide lines. A pick point is used to start the text. The computer then automatically places the text, Fig. 5-34.

Many CAD software packages now include spell checkers and text editing as part of their editing functions. However, editing text on a CAD system is not always as easy as creating the text. In many cases, you may have to retype the text string to correct a misspelled word or an improper phrase. Checking your work carefully before plotting the drawing eliminates errors and saves time.

While it is easy to set and use a type style and size (font), the rules of drafting still apply. It is important to follow the conventions of lettering when placing text on a drawing or design.

LETTERING PRACTICE

To learn the proportions and the strokes of the letters, sketch them six squares high on cross-section paper, preferably paper with $\frac{1}{8}$″ (.12″) divisions, Fig. 5-35. A variety of lettering pads or lettering practice booklets are available. In these, the layouts are already printed. It is only necessary to fill in the lettering as indicated. These pads and booklets are excellent for the beginning student. They do not require a complete layout before lettering.

For those classes where prepared lettering sheets are not available, the layouts in Figs. 5-36 to 5-43 are included here. For all of these, use Layout A in the Appendix. Draw light horizontal guide lines as shown. Then *add light vertical or inclined guide lines* the entire height of the sheet, using a sharp 4H pencil. Fill in all blank spaces with letters or words. Use an HB pencil for all $\frac{3}{8}$″ (.38″) letters and an F pencil for the smaller letters. The last line on each sheet may be lettered first lightly in pencil and then in ink over the pencil if the instructor so assigns. Omit all spacing dimensions.

If assigned by the instructor, construct drawings using decimal-inch or metric scales by converting given fractional dimensions to decimal-inch or metric equivalents. Refer to Table 20 in the Appendix.

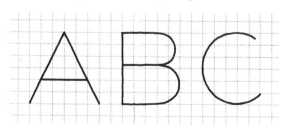

Fig. 5-35. Practice on cross-section paper.

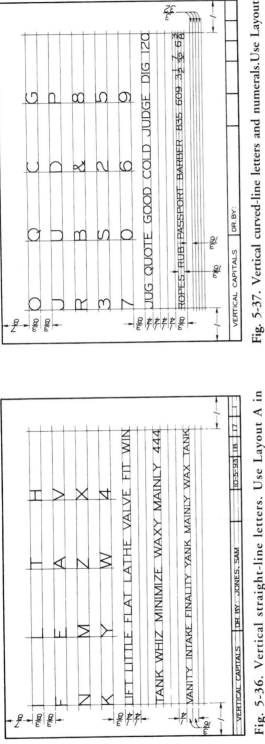

Fig. 5-36. Vertical straight-line letters. Use Layout A in Appendix.

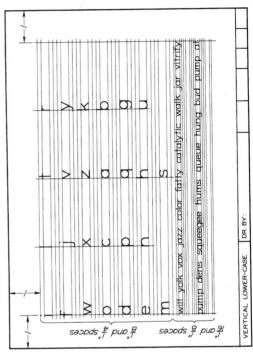

Fig. 5-37. Vertical curved-line letters and numerals. Use Layout A in Appendix.

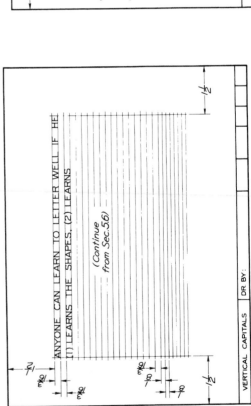

Fig. 5-38. Vertical capital lettering. Use Layout A in Appendix.

Fig. 5-39. Vertical lowercase lettering. Use Layout A in Appendix.

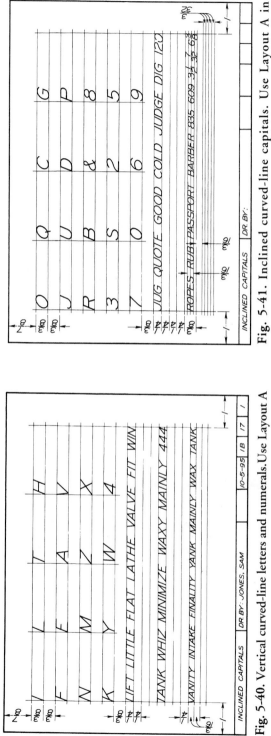

Fig. 5-40. Vertical curved-line letters and numerals. Use Layout A in Appendix.

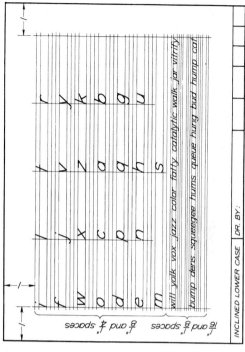

Fig. 5-41. Inclined curved-line capitals. Use Layout A in Appendix.

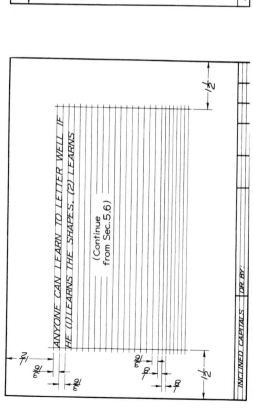

Fig. 5-42. Inclined capital lettering. Use Layout A in Appendix.

Fig. 5-43. Inclined lowercase lettering. Use Layout A in Appendix.

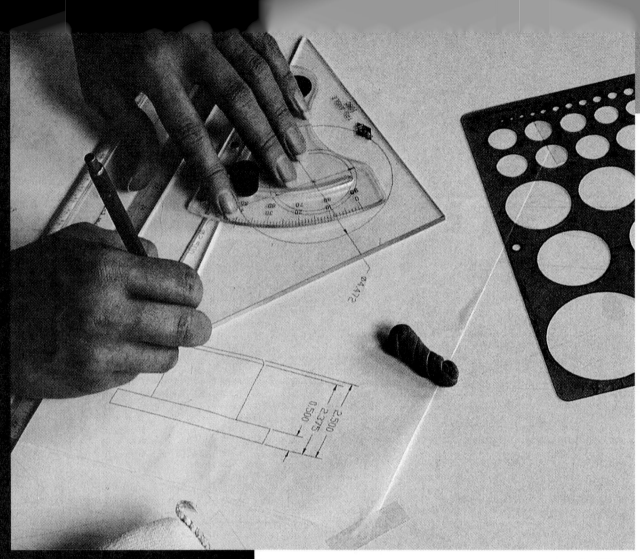

Arnold & Brown

CHAPTER 6

OBJECTIVES

After studying this chapter, you should be able to:

- Enlarge or reduce a drawing sheet by use of a diagonal.

- Bisect horizontal and vertical lines.

- Divide a line into equal parts.

- Bisect an angle.

- Draw regular polygons.

- Draw arcs in various relationships to straight lines.

- Draw ellipses.

CHAPTER 6

Geometry of Technical Drawing

This chapter discusses geometry as it applies to drafting. It refers especially to those geometric constructions that are useful in making technical drawings. It discusses drawing these shapes with T-squares, triangles, and other manual drawing instruments, as well as with a CAD system.

Among the many fields requiring knowledge of geometric constructions is aeronautical engineering. The aeronautical engineer works on aircraft designs of all types, including small private planes, passenger aircraft, planes capable of carrying huge cargo payloads, and those intended for military applications.

Structural analysis, aerodynamic studies, engine design, materials selection, and computer modeling and simulation are all part of the work done by the aeronautical engineer. Accurate drawings play a crucial role in communicating aircraft design ideas and concepts.

CAREER LINK

To learn more about careers in aeronautical engineering, contact the American Institute of Aeronautics and Astronautics, Inc., Suite 500, 1801 Alexander Bell Drive, Reston, VA 20191-4344.

6.1 GEOMETRY IN DRAFTING. To make drawings, and in some cases to solve problems by means of lines, certain geometrical constructions are frequently used. You should learn some of these now. Others are included here for reference when you need them later. This is not mathematician's geometry, but drafter's geometry. All of these constructions are easy to draw with the T-square, triangles, and drawing instruments. They are based mostly on *plane geometry*. Plane geometry is a subject that you may study elsewhere from a mathematical point of view.

In drawing these constructions, accuracy is important. Use a sharp medium-hard lead (2H or 3H) in your pencil and compass. Draw construction lines lightly. Make all given and required lines thin but dark.

6.2 GEOMETRIC SHAPES. The most common geometric shapes in technical drawing are shown in Fig. 6-1, mainly for reference. Study each figure carefully. Learn the terms that are used.

A common symbol for *angle* is ∠ (singular) or ∠⁄ₛ (plural). There are 360 degrees (360°) in a full circle. A degree is divided into 60 minutes (60′). Each minute is divided into 60 seconds (60″).

A *triangle* (symbols: △ singular, △⁄ₛ plural) is a plane (flat) figure, bounded by three sides. The sum of the interior angles always equals 180°. The triangle is the basis for the study of *trigonometry,* a branch of mathematics of much value to the engineer and drafter.

Quadrilaterals have four sides. *Regular polygons* have equal sides. *Regular solids* have faces that are regular polygons. *Prisms* have plane faces parallel to an imaginary axis. A prism having bases that are parallelograms is called a *parallelepiped. Pyramids* have plane triangular faces that intersect at a common point called the *vertex. Cylinders* and *cones* are said to be single-curved surfaces. They have straight-line elements. The *sphere* is like a round ball. The *torus* is like a "doughnut."

6.3 GEOMETRIC CONSTRUCTIONS. The constructions shown in Figs. 6-2 to 6-23 are the ones most often used by the drafter. Two symbols that are commonly used in drafting are ₡ (which means "center line") and ⊥ (which means "perpendicular"). See Figs. 6-13 and 6-14.

6.4 CAD AND GEOMETRY. CAD systems excel in creating geometric entities. Most functions that can be generated manually are produced automatically by CAD. The CAD software systems are capable of developing geometric shapes in two dimensions, as three-dimensional wire frames, or solid models. The variety of design techniques allows the CAD operator to create virtually any geometric shape. Angles are developed using the polar coordinate system with radius and angle as the input information. Most systems allow up to eight decimal places of accuracy.

Lines, circles, ellipses, or any geometric entity can be divided into equal parts by selecting the line and choosing the number of parts required. Polygons are created automatically by choosing the number of sides and indicating whether the polygon should inscribe or circumscribe a circle of specified radius.

CAD systems allow the drawing of lines, arcs, and circles that are attached to each other in various ways. For example, the CAD system can be directed to draw an entity so that it is tangent to another, or it might be perpendicular, parallel, along, at the midpoint, at the end of a line, in a quadrant of a circle, or at an intersection.

(continued on page 113)

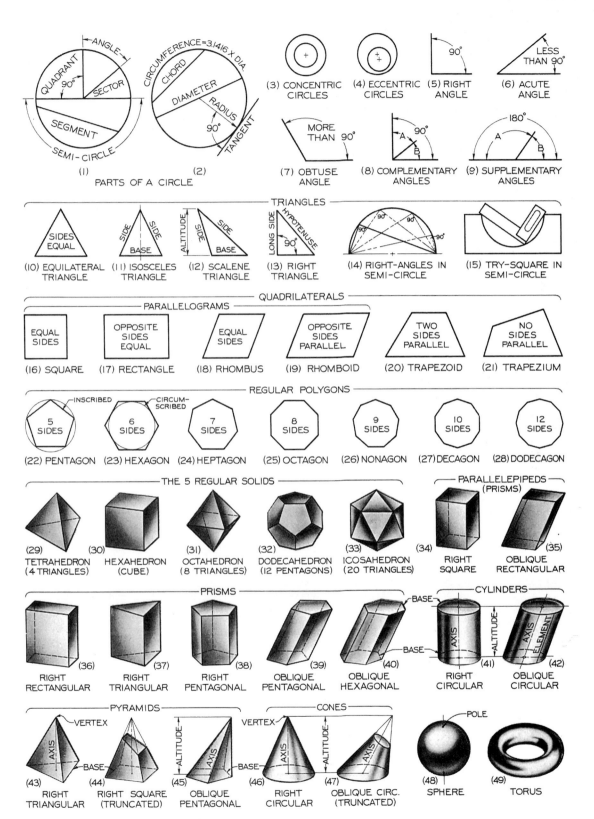

Fig. 6-1. Circles, angles, plane figures, and solids.

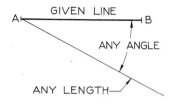
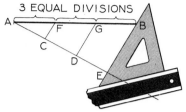
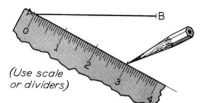

1. Draw light construction line at any convenient angle from either end of line.

2. With scale or dividers, set off the required number of equal divisions.

3. Join E to B, and draw DG and CF parallel to EB.

Fig. 6-2. To divide a line into equal parts.

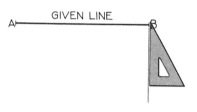
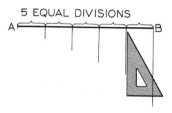
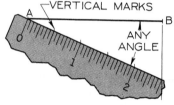

1. Draw vertical line at A or B of indefinite length.

2. Adjust scale with zero at A and 5th division on vertical line. Mark each division.

3. Draw vertical lines to AB to divide AB into 5 equal parts.

Fig. 6-3. To divide a line into equal parts.

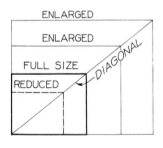
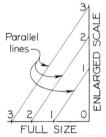
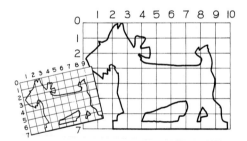

1. Enlarging or reducing a drawing sheet or any square or rectangle by use of diagonal.

2. Enlarging or reducing one or more measurements.

3. Draw grid of equal squares on drawing to be copied. Draw similar but larger grid on new sheet. Sketch lines on new grid through same grid intersections as on small drawing.

Fig. 6-4. Enlarging or reducing.

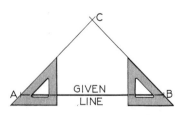
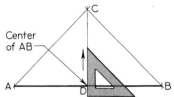

1. Draw 45° lines from ends A and B to locate C.

2. Draw vertical line through C to locate center D.

3. Draw 45° lines from A and B.

4. Draw horizontal line through C with T-square.

Fig. 6-5. To bisect horizontal and vertical lines.

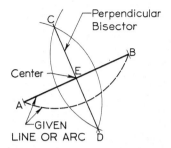

1. Draw equal arcs from end points A and B to intersect at points C and D, using radius more than half AB.
2. Draw perpendicular bisector through C and D.

Fig. 6-6. To bisect a line or an arc.

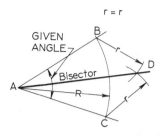

1. Strike arc R of any convenient radius, locating points B and C.
2. Strike equal arcs r to locate D.
3. Draw bisector through A and D.

Fig. 6-7. To bisect an angle.

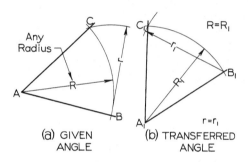

1. Draw AB in new location A_1B_1.
2. Strike arc R and make $R_1 = R$.
3. Strike arc r and make $r_1 = r$.
4. Draw AC in new location A_1C_1.

Fig. 6-8. To copy an angle in a new location.

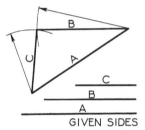

1. Draw one side, as A, in new location.
2. Strike arcs from ends of A, equal to B and C.

Fig. 6-9. Constructing a triangle from given sides.

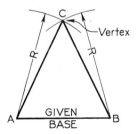

1. From ends A and B, construct any equal arcs R, to locate C.
2. Draw CA and CB.

Fig. 6-10. Constructing isosceles triangles.

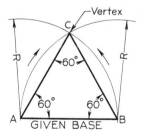

1. From ends A and B, construct arcs with radii equal to AB, to locate C.
2. Draw CA and CB.

Fig. 6-11. Constructing equilateral triangles.

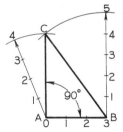

1. From ends of 3-unit base AB, strike 4-unit and 5-unit arcs to locate C.
2. Draw CA and CB.

Fig. 6-12. Constructing a right triangle.

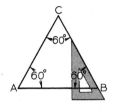

(a) EQUILATERAL TRIANGLE
Draw equal 60° angles, as shown.

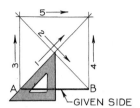

(b) SQUARE
Draw lines in order shown.

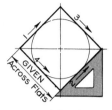

(c) SQUARE
1. Draw ₵'s and circle.
2. Draw 45° tangents.

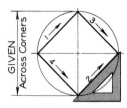

(d) SQUARE
1. Draw ₵'s and circle.
2. Draw 45° lines.

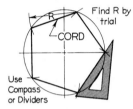

(e) PENTAGON
1. Divide circle into five equal parts.
2. Draw chords.

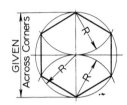

(f) HEXAGON
1. Draw circle and arcs R = radius.
2. Draw sides.

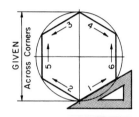

(g) HEXAGON
1. Draw ₵'s and circle.
2. Draw sides as shown.

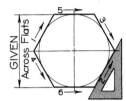

(h) HEXAGON
1. Draw ₵'s and circle.
2. Draw tangent sides.

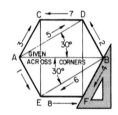

(j) HEXAGON
Draw lines in order shown.

(k) OCTAGON
1. Draw ₵'s and circle.
2. Draw 45° diagonals.
3. Connect intersections.

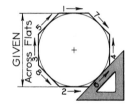

(m) OCTAGON
1. Draw circle.
2. Draw tangents.

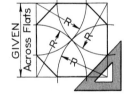

(n) OCTAGON
1. Draw square.
2. Draw arcs and sides.

Fig. 6-13. To draw regular polygons.

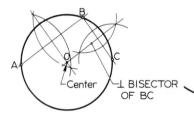

1. Connect A, B, and C.
2. Draw ⊥ bisectors.
3. Draw circle with center O, through A, B, and C.

Fig. 6-14. To draw a circle through three given points.

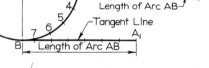
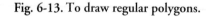

1. From A on arc, set off equal divisions till near B.
2. Set off same divisions on line. (Use bow dividers.)

Fig. 6-15. To lay off an arc on a straight line.

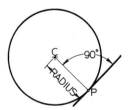

1. Erect ⊥ to line at P equal to radius desired.
2. Draw tangent circle, through P, using center C.

Fig. 6-16. To draw a circle tangent to a line.

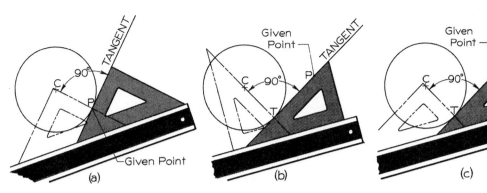

1. Move triangle until one side passes through C and tangent point P.
2. Slide triangle on T-square until other side passes through P.
3. Draw tangent line.

1. Move triangle until hypotenuse passes through P and just touches circle.
2. Revolve triangle about 90° corner, and mark T.
3. Revolve triangle back, and draw tangent line.

1. Move triangle until side passes through P and just touches circle.
2. Slide triangle until other side passes through C. Mark tangent point T.
3. Slide triangle back, and draw tangent line.

Fig. 6-17. To draw a line tangent to a circle.

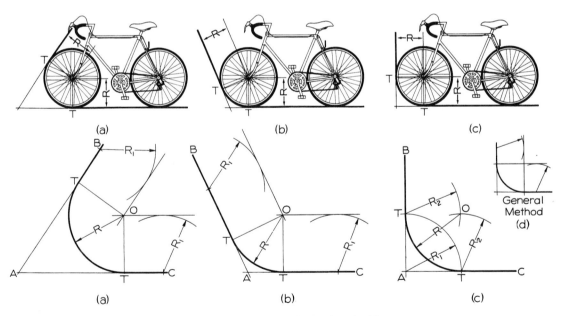

Instructions for (a) and (b):
1. Strike arcs R_1 equal to given radius R.
2. Draw construction lines parallel to given lines AB and AC, to intersect at O, center of arc.
3. Construct perpendiculars to lines AB and AC through point O to locate points of tangency T.
4. Draw tangent arc R from T to T, with center O.

Instructions for (c):
1. Strike arc R_1 = radius R.
2. Strike arcs R_2 = radius R, to locate center O of arc.
3. Draw tangent arc R from T to T, with center O.

Fig. 6-18. To draw an arc tangent to two straight lines.

109

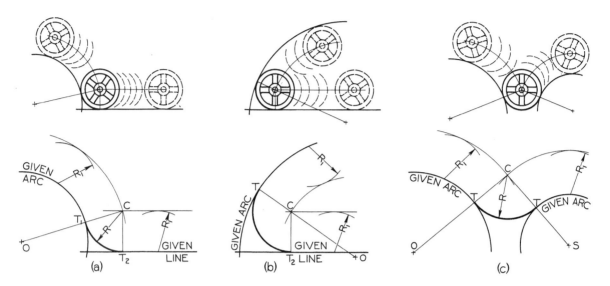

Instructions for (a) and (b):
1. Strike arcs R_1 = radius R (given radius).
2. Draw construction arc parallel to given arc, with center O.
3. Draw construction line parallel to given line.
4. From intersection C, draw CO to get tangent point T_1, and drop perpendicular to given line to get point of tangency T_2.
5. Draw tangent arc R from T_1 to T_2 with center C.

Instructions for (c):
1. Strike arcs R_1 = radius R.
2. Draw construction arcs parallel to given arcs, using centers O and S.
3. Join C to O and C to S to get tangent points T.
4. Draw tangent arc R from T to T, with center C.

Fig. 6-19. To draw an arc tangent to straight lines and arcs.

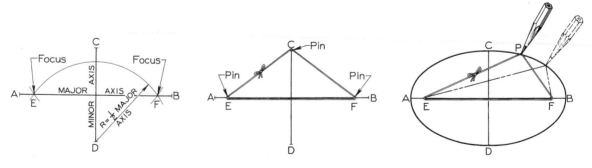

1. From C and D, strike arcs as shown, locating foci E and F.

2. Place pins at E, F, and C, and tie string around them without slack.

3. Remove pin C. Move pencil, keeping string without slack.

Fig. 6-20. To draw a "pin and string" ellipse.

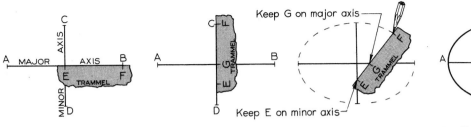

1. On trammel (scrap of paper) mark off EF = half major axis AB.

2. On trammel, set off GF = half minor axis CD.

3. Move E along minor axis and G along major axis. Mark points at F.

4. Sketch smooth curve through points. Then use French curve for final ellipse.

Fig. 6-21. To draw a trammel ellipse.

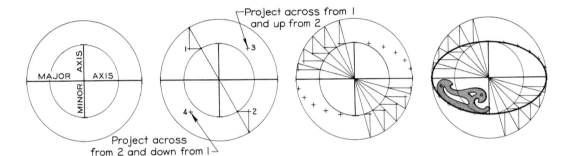

1. Draw light construction circles as shown, using the axes as diameters.

2. Draw any diagonal. At intersections with circles draw lines parallel to axes, as shown, to get 1, 2, 3, and 4.

3. Draw as many additional diagonals as needed, each diagonal producing 4 points.

4. Sketch smooth curve through points. Then use French curve for final ellipse.

Fig. 6-22. To draw a concentric-circle ellipse.

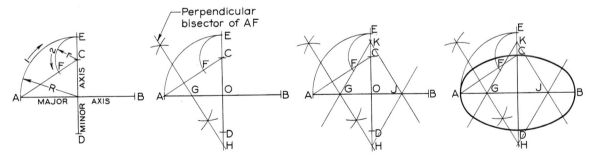

1. Draw diagonal AC. Draw arcs 1 and 2 to locate F.

2. Draw perpendicular bisector of AF to locate centers G and H.

3. With compass, make OJ = OG and OK = OH, and draw lines as shown.

4. With compass, draw small arcs from centers G and J and large arcs from centers K and H.

Fig. 6-23. To draw an approximate ellipse with compass.

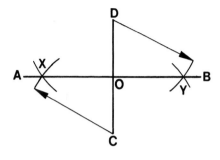

1. Set compass at distance OB. Place compass on D. Scribe arc Y and arc X. Repeat at point C.

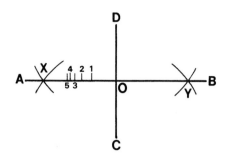

2. Set off distances along the major axis. Distances become closer as you move out on the major axis. The greater the divisions, the more precise the elipse. In our case, we are going to use 5 divisions.

USE COMPASS. TAKE DISTANCE B TO 1.

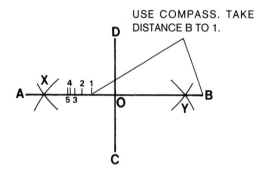

3. Set compass at distance B1.

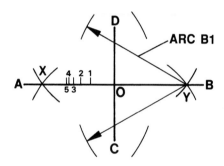

4. Place compass on Y and scribe arc B1. Repeat this by placing your compass on X and scribing an arc in all quadrants.

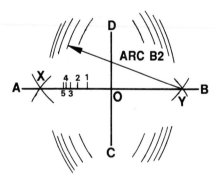

5. Repeat for ARCS B2, B3, B4, B5, and scribe from point Y. Repeat process from point X.

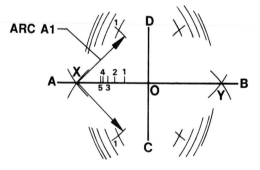

6. We can now repeat this process, but this time we will work from the other end. Use your compass. Set it at distance A1. Place compass at X and scribe ARCS.

Fig. 6-24. To draw a foci ellipse.

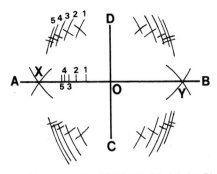

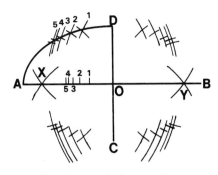

7. Repeat this process for ARCS A2, A3, A4, A5. Place compass at X and scribe ARCS 1, 2, 3, 4, 5. Repeat at point Y.

8. Using your French curve, draft a smooth curve through the points. Repeat in each quadrant.

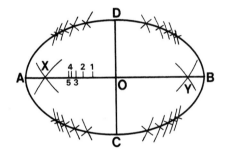

9. Complete your ellipse and clean up.

Fig. 6-24. To draw a foci ellipse (continued).

Solid objects such as a cylinder and sphere are created using 3D wire frame and solid modeling. After the sizes and variables have been chosen, the computer draws the object to specification, Fig. 6-25. To produce a cylinder with a radius of 1.25″ and a length of 3.5″, the following sequence might be used:

Draw – Cylinder
Radius 1.25
Length 3.5

The cylinder is automatically drawn to the specifications, Fig. 6-25.

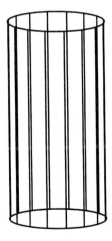

Fig. 6-25. A wire frame cylinder.

PROBLEMS

The problems that follow are included to provide practice in using geometric constructions. Those problems in Figs. 6-26 to 6-28 cover the basic constructions. Those problems in Figs. 6-29 to 6-44 involve tangencies. Accuracy and clean drawing are the objectives. Dimensions may or may not be included, as assigned by the instructor. If they are assigned, the student should study Sec. 5.11 or 5.13 on lettering and Secs. 10.1 to 10.12 on dimensioning conventions.

Use sheet Layout A in the Appendix. Divide into four equal parts, as shown in Fig. 6-26. Other problems that may be assigned are given in Figs. 6-27 and 6-28. All are printed one-third size. Apply your dividers directly to these problems. Step off all spacing measurements triple size. Lettering is $\frac{1}{8}''$ (0.125") high. The steps in drawing the Conveyor Link, Fig. 6-29, are shown in Fig. 6-30. First lay out the sheet, using Layout C in the Appendix. Draw the main center lines, as shown in step I. Then draw the arcs, circles, and straight lines as shown in steps II to V. All lines should be light construction lines through step V. Finally, heavy-in the final lines as shown in step VI.

If assigned by the instructor, construct drawings using decimal-inch or metric scales by converting given fractional dimensions to decimal-inch or metric equivalents. Refer to Table 20 in the Appendix.

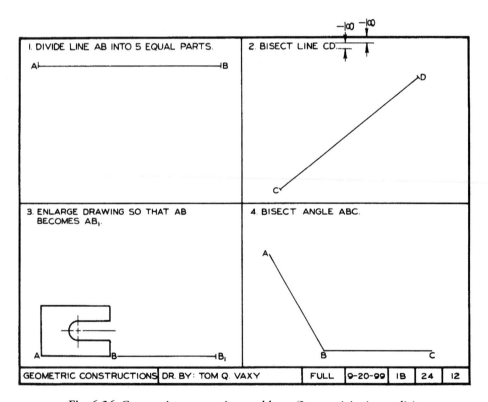

Fig. 6-26. Geometric construction problems (Layout A in Appendix).

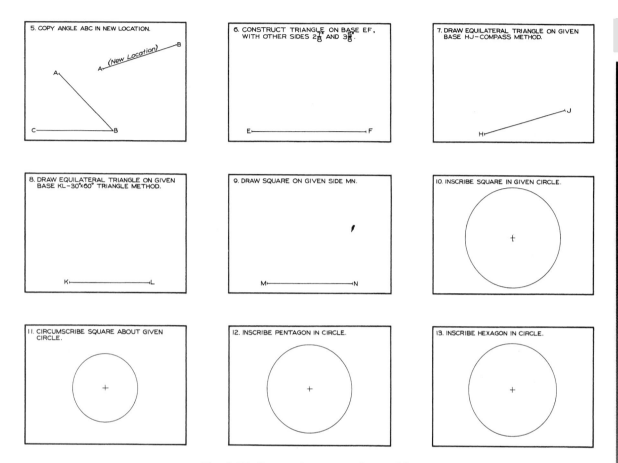

Fig. 6-27. Geometric construction problems.

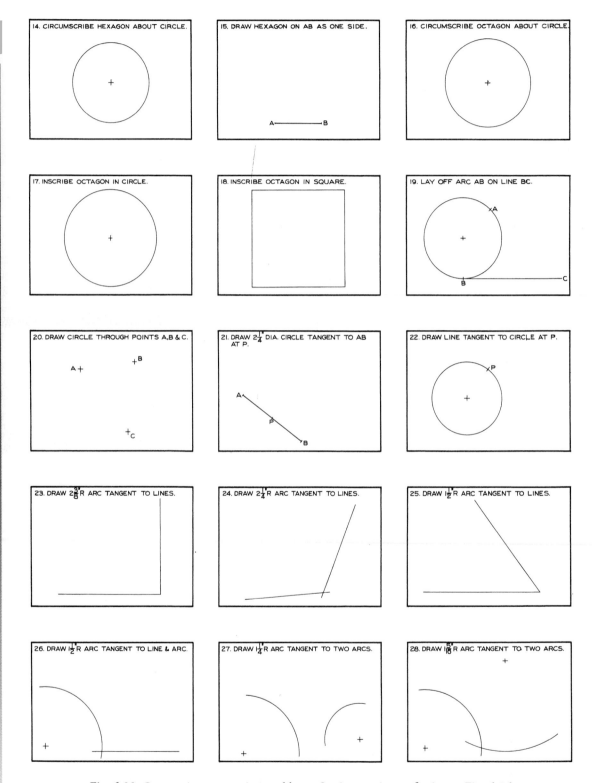

14. CIRCUMSCRIBE HEXAGON ABOUT CIRCLE.

15. DRAW HEXAGON ON AB AS ONE SIDE.

A———B

16. CIRCUMSCRIBE OCTAGON ABOUT CIRCLE.

17. INSCRIBE OCTAGON IN CIRCLE.

18. INSCRIBE OCTAGON IN SQUARE.

19. LAY OFF ARC AB ON LINE BC.

20. DRAW CIRCLE THROUGH POINTS A,B & C.

21. DRAW $2\frac{1}{4}''$ DIA. CIRCLE TANGENT TO AB AT P.

22. DRAW LINE TANGENT TO CIRCLE AT P.

23. DRAW $2\frac{3}{8}''$R ARC TANGENT TO LINES.

24. DRAW $2\frac{1}{4}''$R ARC TANGENT TO LINES.

25. DRAW $1\frac{1}{2}''$R ARC TANGENT TO LINES.

26. DRAW $1\frac{1}{2}''$R ARC TANGENT TO LINE & ARC.

27. DRAW $1\frac{1}{4}''$R ARC TANGENT TO TWO ARCS.

28. DRAW $1\frac{5}{8}''$R ARC TANGENT TO TWO ARCS.

Fig. 6-28. Geometric construction problems. See instructions referring to Fig. 6-26.

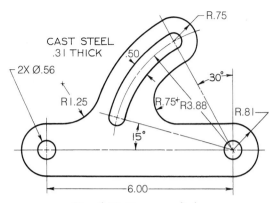

Fig. 6-29. Conveyor link.

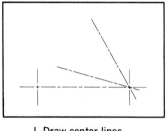

I. Draw center lines.

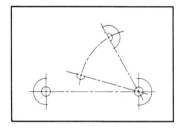

II. Draw circular arcs.

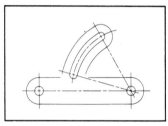

III. Draw connecting straight lines and arcs.

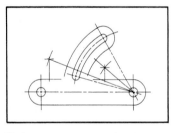

IV. Locate centers and points of tangency of tangent arcs.

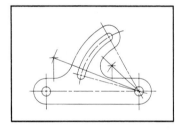

V. Draw tangent arcs.

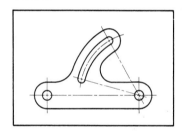

VI. Heavy-in all visible object lines.

Fig. 6-30. Steps in drawing conveyor link.

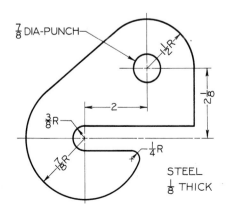

Fig. 6-31. Cover plate (Layout C in Appendix).

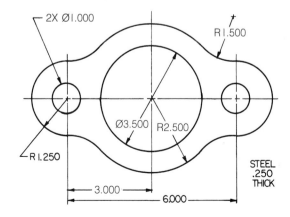

Fig. 6-32. Gasket (Layout C in Appendix).

117

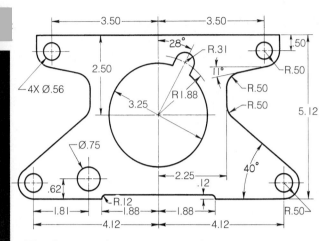

Fig. 6-33. Buick transmission gasket. Use Layout C in Appendix.

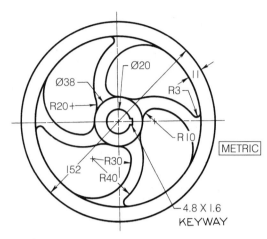

Fig. 6-34. Handwheel. Use Layout C in Appendix.

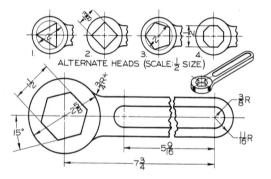

Fig. 6-35. Closed-end wrench.(Draw with type head assigned by instructor.) Use Layout C in Appendix.

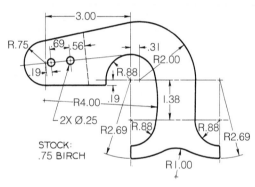

Fig. 6-36. Keyhole saw handle. Use Layout C in Appendix.

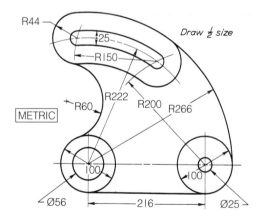

Fig. 6-37. Quadrant for lathe. Use Layout C in Appendix.

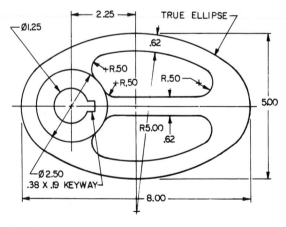

Fig. 6-38. Elliptical cam. Use Layout C in Appendix.

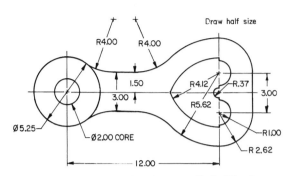

Fig. 6-39. Locomotive truck swing link. Use Layout C in Appendix.

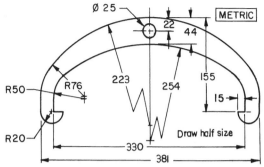

Fig. 6-40. Clamp for laundry machine. Use Layout C in Appendix.

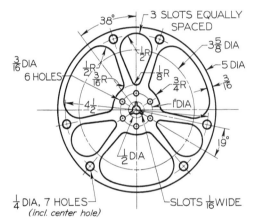

Fig. 6-41. Movie film reel. Use Layout C in Appendix.

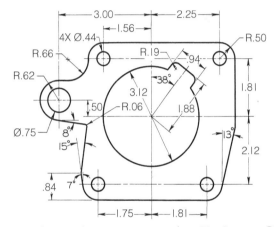

Fig. 6-42. Buick transmission gasket. Use Layout C in Appendix.

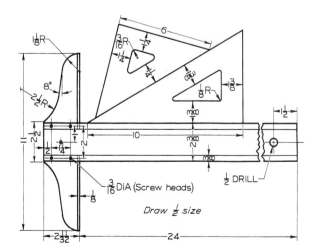

Fig. 6-43. T-square and triangles. Use Layout C in Appendix.

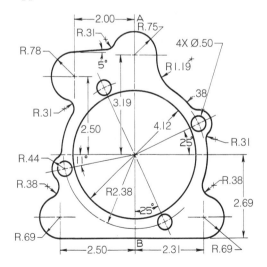

Fig. 6-44. Buick rear transmission gasket. Use Layout C in Appendix.

119

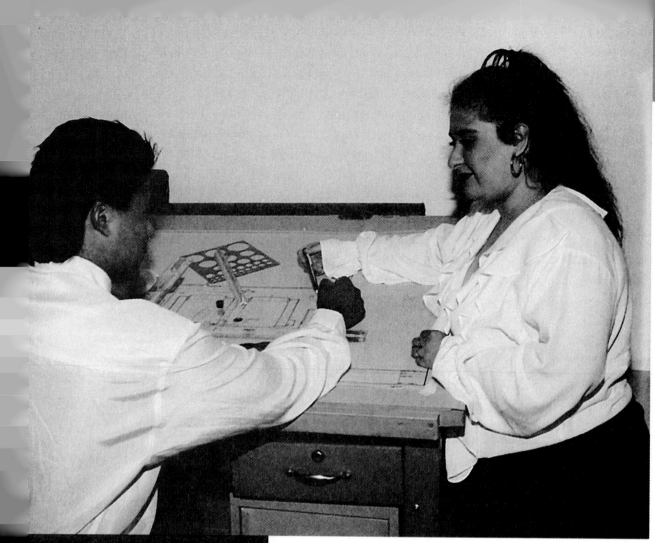

CHAPTER 7

OBJECTIVES

After studying this chapter, you should be able to:

- Explain the different views of an object.
- Explain how an object can be revolved to illustrate different views.
- Use the idea of a "glass box" to explain views.
- Identify the "necessary views" of a given object that is to be drawn.
- Identify objects that usually require only two views.
- Sketch three or more views of an object.
- Place views correctly on the page.

CHAPTER 7
Views of Objects

This chapter presents information on preparing different views of an object. The different views allow the person reading the drawing to visualize the object in three dimensions. The ability to draw views correctly is one of the keys to success in technical drawing. For example, product designers must be able to create view drawings as well as pictorial illustrations.

Product designers work with materials as diverse as cloth, leather, plastic, glass, metal, and ceramics. They develop functional, durable, and attractive designs for products used in our homes and on the job—everything from kitchen appliances, to airplane seats, to computer keyboards.

Product designers must know how to choose the view, or views, that will highlight particular features of an object. To communicate the appearance of the finished product, the designer may use a solids modeling feature of a 3D CAD program to show how the product would look in different colors, surface textures, or patterns.

CAREER LINK

To learn more about careers in design, contact the National Association of Schools of Art and Design, 11250 Roger Bacon Drive, Suite 21, Reston, VA 20190.

7.1 PICTURES AND VIEWS. An architect's sketch is made to show clients how their new home will look when built. Such a sketch is shown in Fig. 7-1. This drawing shows how the house will *appear* from one position. However, it provides no practical information as to the size of the house. It provides no information on the shapes, arrangement, and sizes of the rooms. It does not give details of windows, doors, fireplace, kitchen cabinets, and so forth. The contractor will need all this information, and much more, to build the house. Further, the clients themselves will not agree to build until they can see exact drawings that show every detail unmistakably. In fact, the complete drawings (and specifications) are needed to figure costs. No clients would build a house until they and the contractor could agree on the price.

To describe the exterior of the house completely, a series of *views* must be drawn. Each view shows the house from a different point of view. You can walk around the house and view it from all four sides. If you fly over the house in a helicopter or airplane, you can look down and get a top view.

First, refer to Fig. 7-2(a). Suppose you stand at a distance in front of the house and look directly toward the front. (Theoretically you would be standing at an infinite distance.) Imagine a glass plane between you and the house. This plane is parallel to the house. The view on the plane is what you would see. It is called a *front view* or *front elevation*. This view shows the true *width* and *height* of the house, but not the *depth*. More exactly, the view is obtained by extending perpendicular *projectors* from all points on the house to the plane. Collectively, the piercing points of all these perpendiculars form the view.

Next, take a position looking toward the right side of the house, (b). The view seen from here is the *right-side view* or *right-side elevation*. It shows the true *height* and *depth* of the house, but not the *width*.

Finally, take a position (or imagine it) above the house and looking down, (c). The resulting view is projected up to the plane, which is parallel to the ground. This view is called the *top view* or *plan*. This view shows the true *width* and *depth* of the house, but not the *height*.

Fig. 7-1. Architect's sketch of a house.

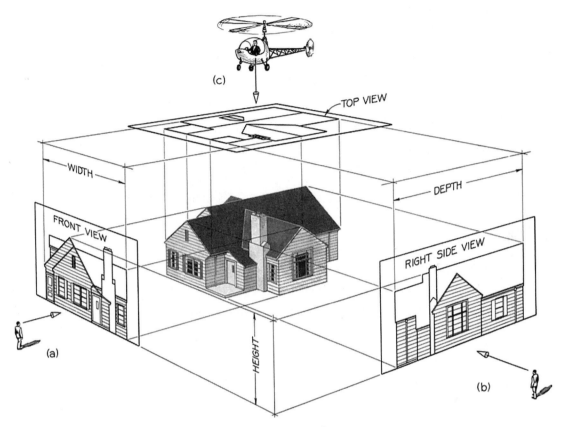

Fig. 7-2. Three views of a house.

When these three views are accurately drawn to scale (usually $\frac{1}{4}'' = 1'\text{-}0''$), they can be dimensioned fully. Thus they will provide detailed information to the clients, the contractor, and the workers. Drawings of buildings are large. Because of this, architects usually draw each view on a separate sheet, as shown in Fig. 7-3.

Actually, for an object as complicated as a house, additional views and sections would be needed to provide complete information. The three views here are sufficient to show how views are obtained.

7.2 REVOLVING THE OBJECT. In Fig. 7-2 the object to be drawn is very large. You can obtain views by *shifting your position* while the house remains stationary. If the object is small, such as a machine part, you can remain in your chair and *shift the object* to the desired position.

Fig. 7-3. Front view.

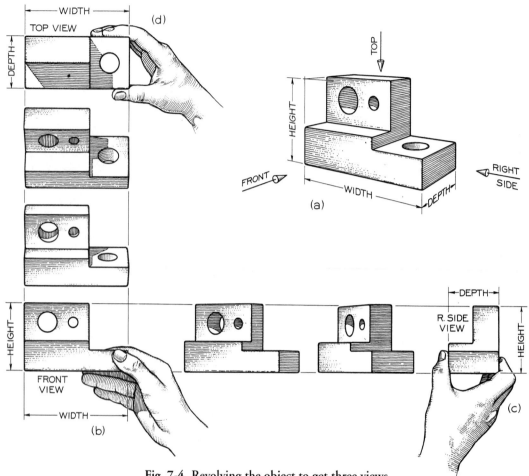

Fig. 7-4. Revolving the object to get three views.

Suppose you wish to draw three views of the Holder for an Offset Press, Fig. 7-4(a). The arrows indicate the directions in which you will view the object. Note the three principal dimensions: *width, height,* and *depth.*

Hold the object in your hand so that you look perpendicularly toward the front of the object, (b). *The front view shows the width and the height, but not the depth.*

To get the right-side view, revolve the object and bring the right side toward you, (c). *The right-side view shows the height and the depth, but not the width.*

To get the top view, revolve the object,

bringing the top toward you, (d). *The top view shows the width and depth, but not the height.*

Note that each view shows two dimensions, but not the third dimension. Thus, a three-dimensional object requires at least two views to describe it (except as explained in Secs. 7.6 and 7.7). Often three or more are needed, as will be shown later.

7.3 THE GLASS BOX. A more scientific explanation of how views are obtained is presented in Fig. 7-5. As shown at (a), the plane is placed parallel to the object. Perpendiculars are extended to the plane from all points on

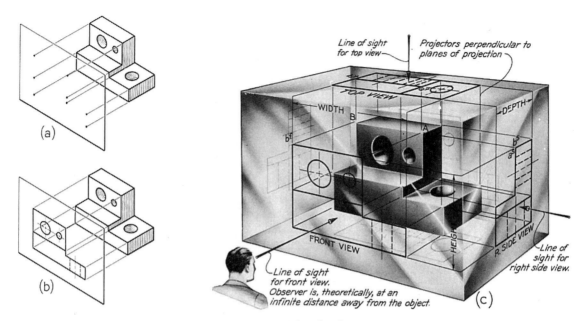

Fig. 7-5. The glass box.

the object. As shown at (b), the piercing points of these *projectors* are then connected to form the view, or *projection.*

Since any rectangular object has six sides, we can obtain six views by placing six planes parallel to the six sides. Together, these planes form the "glass box," as shown at (c). The arrows indicate the directions in which the object is viewed to get the three regular views.

7.4 HIDDEN LINES. One of the big advantages that a view has over a photograph or an artist's drawing is that hidden parts of the object can be shown by means of dashed lines, called *hidden lines,* Fig. 4-11. In Fig. 7-6, the edge AB on the object would not be visible as you look in the direction of the arrow. Therefore, its view 9-12 is represented by a hidden line. Likewise, the contours of hole C are invisible. They are projected as hidden lines 10-13 and 11-14. Hole E is projected as hidden lines 3-4 and 5-6. Note how other hidden lines are projected in this figure.

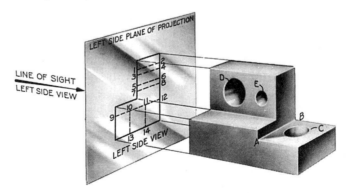

Fig. 7-6. Hidden edges and contours.

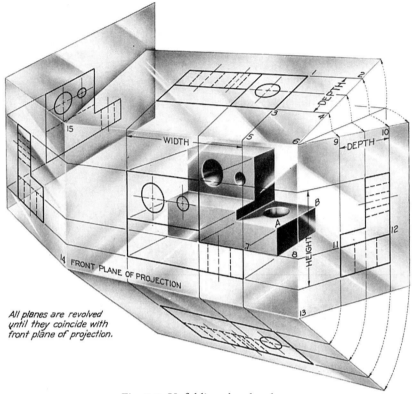

All planes are revolved until they coincide with front plane of projection.

Fig. 7-7. Unfolding the glass box.

7.5 UNFOLDING THE GLASS BOX. You may wonder, "How can I get the views from the planes onto my flat sheet of drawing paper?" You do this, as shown in Fig. 7-7, by unfolding the glass box until all planes lie in the same plane as the front plane. The top, bottom, and both side planes are hinged to the front plane. The rear plane is hinged to the left-side plane.

Note, at the upper right of Fig. 7-7, that distances 4-2 and 9-10 are equal. This explains why the *depth* in the top view must always be the same as the *depth* in the side view. Observe also why the *width* must always be the same in the top, front, and bottom views. The *height* must always be the same in the rear, left-side, front, and right-side views.

The six planes, after being revolved into one plane, are shown in Fig. 7-8. The top view is directly over the front view. The bottom view is directly under the front view. The right-side view is directly to the right of the front view. The left-side view is directly to the left of the front view. The rear view is directly to the left of the left-side view. This orientation is known as *third-angle projection.*

7.6 ELIMINATION OF VIEWS. Clearly, all of the possible six views may not be needed to describe completely the shape of the object, as shown in Fig. 7-9. In practice, *only those views that are necessary to describe the shape of the object should be drawn.*

To choose the necessary views, examine the object to see what shapes must be described. For example, in Fig. 7-9(a), the object has two right-angled notches and three holes whose shapes must be described. These are all shown clearly in the six views, but with considerable duplication.

The front view shows the shapes of a right-angled notch and two holes. These are also shown in the rear view, but the rear view has a hidden line. The front view is preferred. The rear view is crossed out.

Both the top and bottom views show the shape of the hole at the right end. However, the top view is preferred because it has fewer hidden lines. The bottom view is, therefore, crossed out.

Both the left-side and right-side views show the right-angled notch. However, the right-side view is preferred because it has fewer hidden lines. The left-side view is, therefore, crossed out.

The remaining three views, the front, top, and right-side, are the necessary views of this object, as indicated by the border line around them in Fig. 7-9(b). These are often referred to as the "three standard views." They are the views most commonly required.

7.7 CHOICE OF VIEWS. Keep in mind that your job is to select views that are necessary to describe each contour or shape of the object. For example, in Fig. 7-10(a), if the sheet-metal part is viewed in the direction of the arrow, all the essential shapes are seen at once. Only the thickness is not seen. A one-view drawing is sufficient if the thickness is given in a note. A second view showing the thickness may be added, producing a two-view drawing.

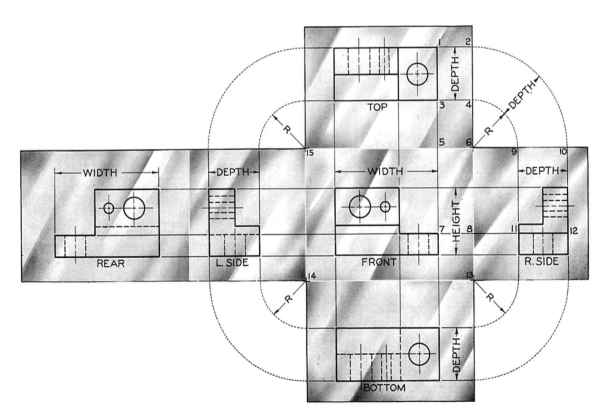

Fig. 7-8. The glass box unfolded.

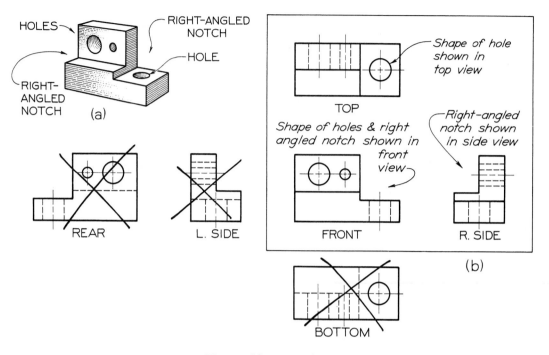

Fig. 7-9. Necessary views.

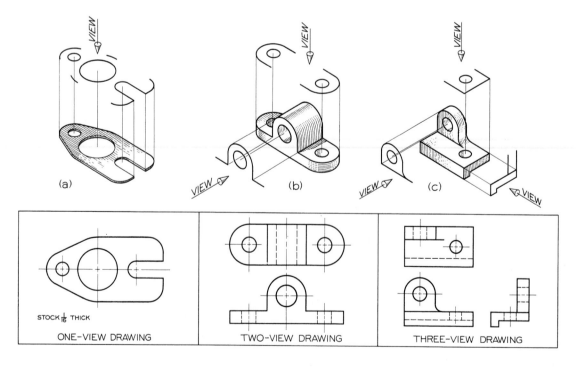

Fig. 7-10. Necessary views.

If the object shown at (b) is viewed in two different directions, as shown by the arrows, all essential shapes are seen. A two-view drawing is required.

If the object shown at (c) is viewed in three different directions, as shown by the arrows, all essential shapes are seen. A three-view drawing is required.

7.8 TWO-VIEW DRAWINGS. In Fig. 7-11(a) a machine part is shown in the front-view and right-side-view positions. These views are enough to show all essential contours and shapes. The corresponding two-view drawing is shown at (b). Note the use of center lines. Also observe that no shading is used on the drawing. Note that hidden lines show interior shapes that are not seen by looking at the object itself at (a).

Some typical examples of objects that require only two views are shown in Fig. 7-12. At (a) the top view is omitted because it is

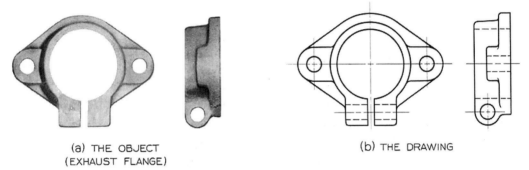

(a) THE OBJECT
(EXHAUST FLANGE)

(b) THE DRAWING

Fig. 7-11. Two views of an exhaust flange.

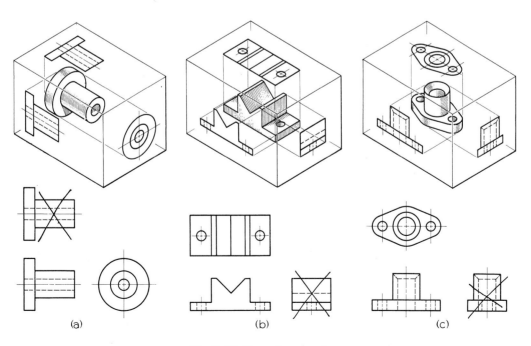

(a) (b) (c)

Fig. 7-12. Two-view drawings.

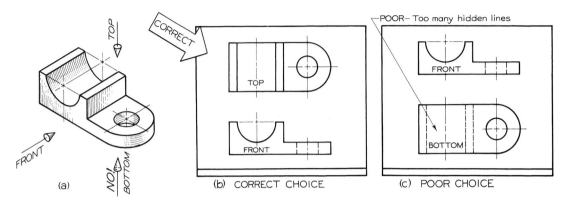

Fig. 7-13. Views with fewest hidden lines.

a duplication of the front view. At (b) and (c) the side views are not needed because they show no shapes not already given in the front and top views.

In choosing between two views that give the same information, such as between the top and bottom views of the object in Fig. 7-13(a), choose the view that contains the least number of hidden lines, as shown at (b) and not as shown at (c). Note that the views at (c) could be called the front and top views. Still, the lower view would not be desirable.

Sometimes, when there is little or nothing to choose between the top and side views, the views selected should be those that will space best on the sheet. For example, in Fig. 7-14(a), either a top or a side view could be used. If the side view is used, the result is a well-spaced drawing, as shown at (b). If the top view is used (c), the drawing will be poorly spaced.

7.9 SKETCHING TWO VIEWS. The best way to learn the principles of shape description is to make freehand sketches of many objects. Start by sketching simple shapes. Then gradually work up to more complicated shapes. Freehand sketching technique has been discussed in Chapter 2, to which you are referred for review.

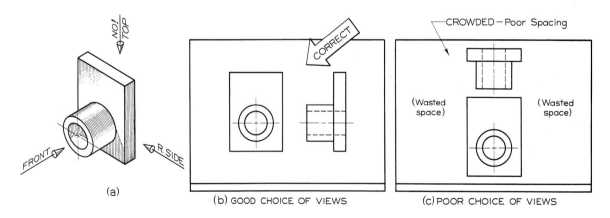

Fig. 7-14. Effect of spacing on choice of views.

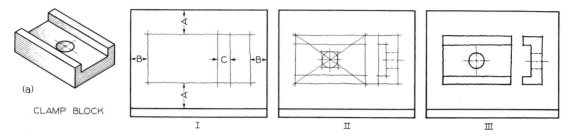

Fig. 7-15. Sketching two views.

To sketch two views of a simple object, such as the Clamp Block in Fig. 7-15(a), proceed as follows. Use either plain paper or cross-section paper.

I. Sketch the enclosing rectangles for the two views, making spaces A about equal. Make spaces B about equal, and make space C about equal or slightly less than spaces B.

II. Sketch diagonals in front view to locate center. Construct circle and other details.

III. Dim all construction lines with the eraser. Then heavy-in all final lines, making them clean-cut and dark.

7.10 THREE-VIEW DRAWINGS. Figure 7-16(a) shows a Bracket in the front-view, top-view, and right-side-view positions required to show all the essential shapes of the object. The corresponding three-view drawing is shown at (b). Note the use of center lines and hidden lines and the absence of any shading.

As shown in Figs. 7-7 and 7-8, the relative positions of the views depend upon their locations when the glass box is unfolded. However, errors in this are so common among beginners that this principle needs to be re-emphasized here.

One of the worst mistakes you can make in technical drawing is to draw a view out of place!

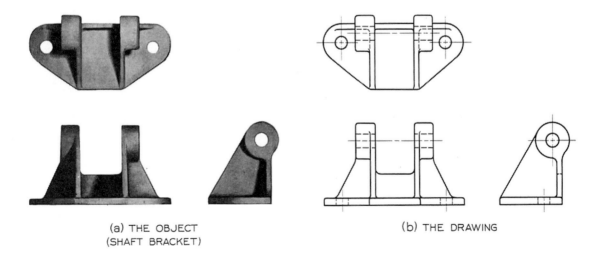

(a) THE OBJECT
(SHAFT BRACKET)

(b) THE DRAWING

Fig. 7-16. Three views of a bracket.

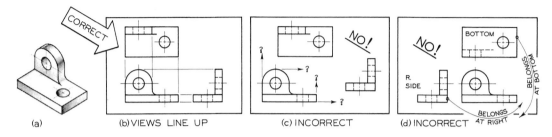

Fig. 7-17. Position of views.

For example, the object shown in Fig. 7-17(a) requires front, top, and right-side views. As shown at (b), the right-side and top views must "line up" with the front view. Never draw the views out of alignment, as shown at (c). Also, never draw the views in reversed positions, as shown at (d), even though they do line up with the front view.

7.11 SKETCHING THREE VIEWS. To sketch the three views of the Stop Clamp, Fig. 7-18(a), proceed as follows:

I. Sketch the enclosing rectangles for the three views, making spaces A about equal. Make space B about equal or slightly less than either of spaces A. Make spaces C about equal. Make space D about equal or slightly less than either of spaces C. A scrap of paper may be used, as shown, to transfer the depth from the top to the side view, and to transfer other equal distances. It is very

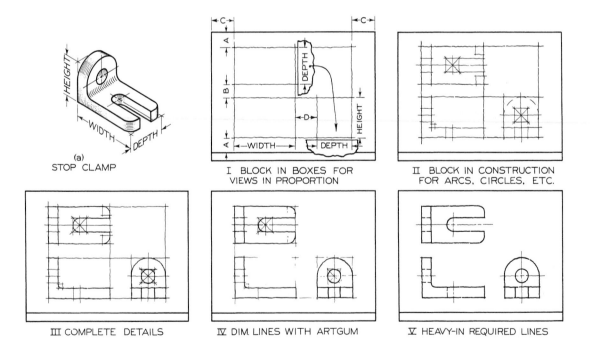

Fig. 7-18. Sketching three views.

important to draw these rectangles in correct proportion, Sec. 2.7. The entire sketch depends upon your having the large main shapes of the object in correct relative proportion.

II. and III. Block in main shapes lightly. Then add smaller details.

IV. Dim all lines with the eraser.

V. Heavy-in all final lines, clean-cut and dark.

7.12 UPRIGHTNESS OF VIEWS. Objects that we are accustomed to seeing in a certain position, such as a house, Fig. 7-2, a telephone, or a chair, should be drawn in their normal upright position. However, many objects, particularly machine parts, may be drawn in any convenient position. See Figs. 7-11(b) and 7-16(b). It is customary to draw screws, bolts, shafts, and similar parts in a horizontal position. See Figs. 16-6 and 16-11.

7.13 SIDE VIEW BESIDE TOP VIEW. If three views of a wide flat object, such as a card tray, are drawn with the side plane of the glass box hinged to the front plane, like the glass box in Fig. 7-7, the side view may extend too far, as shown in Fig. 7-19(a). Even if the three views can be drawn within the border, a large wast-

ed space will be left in the upper-right corner of the sheet. In such cases, the side plane may be hinged to the top plane of the glass box. It can then be revolved upward, as shown at (b). This places the side view beside the top view, as shown at (c). The spacing is thereby improved.

7.14 CENTER LINES. A *center line* (symbol ₵) is used to indicate an axis of a symmetrical part, Figs. 7-11 and 7-16. As such, the center line becomes a skeleton about which a view or a symmetrical feature is drawn. Center lines are necessary in dimensioning. Many important dimensions are given to them.

Center lines, Fig. 7-11, should be thin enough to contrast well with the visible lines and hidden lines. The dashes and spaces should be carefully spaced by eye. The long dashes should be $\frac{3}{4}''$ to $1\frac{1}{2}''$ (.750″ - 1.500″) long. The short dashes should be about $\frac{1}{8}''$ (.125″) long, with spaces of about $\frac{1}{16}''$ (.062″) between. Short dashes should occur in open places on the drawing. In circular views, crossed center lines are drawn, with the small dashes crossing at the center. Center lines should extend uniformly about $\frac{1}{4}''$ (.250″) outside the view or circular part, as shown in Figs. 7-11 and 7-16.

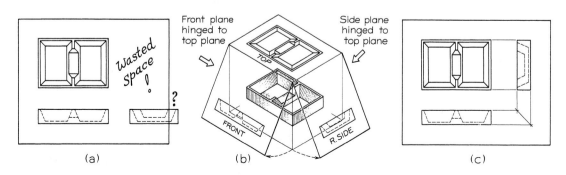

Fig. 7-19. Side view beside top view.

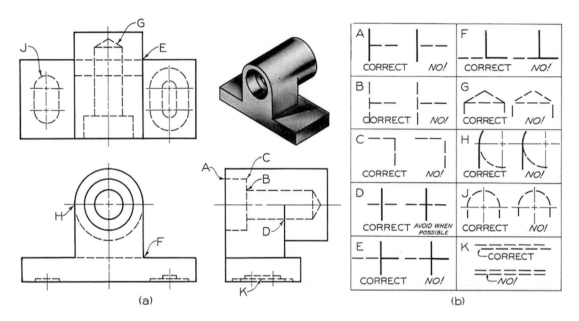

Fig. 7-20. Hidden lines.

7.15 HIDDEN LINES. Poorly made hidden lines can easily make an otherwise good drawing very sloppy in appearance and hard to "read." Hidden lines are used to show invisible parts that otherwise could not be shown at all. They are just as important as visible lines. Make the dashes thin, and dark, about $\frac{1}{8}$" (.125") long, and with spaces about $\frac{1}{32}$" (.031") — all carefully estimated by eye. See Fig. 7-11.

The correct methods of drawing hidden-line dashes are illustrated in Fig. 7-20(a). Special attention should be given to the places marked A, B, C, etc. The correct and incorrect methods for each of these, full size, are shown at (b). Study each example carefully. In drawings of complicated objects, hidden lines that are not necessary for clearness should be omitted. However, beginners should draw all hidden lines until experience

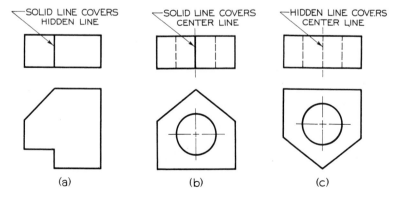

Fig. 7-21. Lines that coincide.

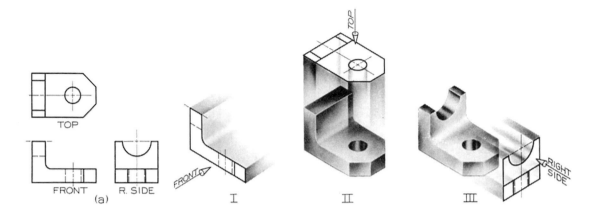

Fig. 7-22. Visualizing an object from given views.

teaches them which lines are not necessary and may be omitted.

7.16 LINES THAT COINCIDE. When a visible line coincides with (lies in front of or behind) either a hidden line or a center line, the visible line is shown, Fig. 7-21(a) and (b). When a hidden line coincides with a center line, (c), the hidden line is shown.

7.17 VISUALIZING THE VIEWS. Suppose we are given the three-view drawing in Fig. 7-22(a) to "read" or visualize. The front view tells us that the object is L-shaped as shown at I. Certain hidden lines and solid lines in the front view are still not clear. However, these will be explained by the other views. The top view tells us that two corners of the base are chamfered and that a hole is drilled through the base, as shown at II. Certain solid lines and hidden lines still are not clear. The right-side view tells us that the left end of the object has a semicircular cut.

7.18 PROGRESSIVE CUTS. An excellent way to learn to read views and to understand the relationship between a picture of an object and the views of the object is to make a series of drawings. These drawings would show the progressive cuts that a machinist might make on a piece of raw stock to complete some object, Fig. 7-23.

The first drawing, I, would show the three views of the piece of raw stock. In each succeeding drawing, the various cuts are shown on the views, as shown in II to VI. Instead of making six different drawings, all of these steps could be taken in a single drawing to produce the result shown at VI.

7.19 COMPUTER GRAPHICS. Views of an object may be developed in two ways when using a CAD system. One way is to draw the object in the traditional method by developing each view separately (2D drawing). The other way is to draw the object as a three-dimensional model (3D drawing). The 3D

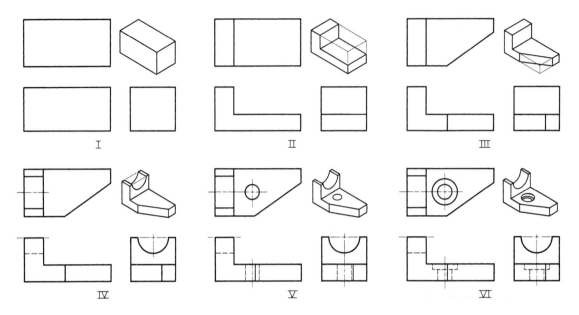

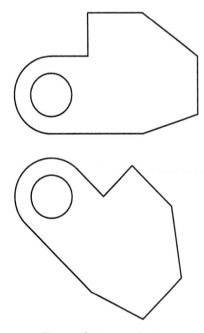

I

II

III

IV

V

VI

Fig. 7-23. Progressive cuts.

Fig. 7-24. A rotated view.

drawing can be rotated in space so that any view can be displayed.

The traditional 2D method allows the CAD drafter to draw a view quickly and copy the required parts to other views. Changes to the separate views are made to conform to drafting conventions. Hidden lines, center lines, cutting plane lines, and any special line applications are automatically drawn by specifying the layer containing the line or by changing the line type.

Developing a three-dimensional (3D) model of an object allows the flexibility of choosing the required views and placing them in the appropriate position. The 3D modeling method gives the drafter the option of displaying any desired view. More importantly, it allows advanced CAD operators to check the fit between mating parts. To choose a right, top, or end view, the drafter rotates the 3D model to the appropriate angle and saves the results. Each view is then placed in the correct position and saved as a complete drawing, Fig. 7-24.

The drafter then edits any unwanted lines and places hidden lines, center lines, and any other lines to conform to drafting conventions. To fully describe the object, dimensions and annotations are then added, Fig. 7-25. The 3D models also allow for a pictorial presentation if it is required.

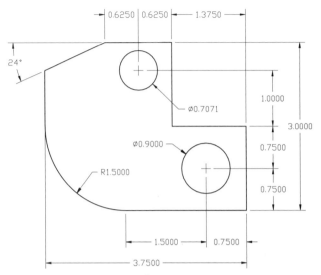

Fig. 7-25. A dimensioned view.

PROBLEMS

A variety of objects to be sketched are shown in Figs. 7-26 to 7-30. The items are to be sketched either on cross-section paper or on plain paper. Sketch Layout B in the Appendix, freehand. Sketch two problems per sheet. Pictorial sketches, either in isometric or in oblique, may be drawn along with the sketches of views. These may be sketched on separate paper or in the upper right corner of each space on the sheet. Before making any pictorial sketches, study Secs. 17.3 to 17.5, 17.18, and 17.19.

Additional pictorial sketching problems are available at the end of Chapter 17.

Use $8\frac{1}{2}'' \times 11''$ cross-section paper or plain drawing paper. Sketch border and title strip of Layout B in the Appendix. Sketch the necessary views of two assigned problems from this. Each grid space on the isometrics is $\frac{1}{4}''$ (.250"). Sketch the views full size. An example sketch is shown in Fig. 7-26.

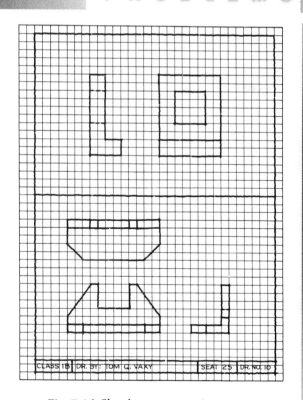

Fig. 7-26. Sketch on cross-section paper.

If plain paper is used, count the isometric grid spaces to determine the proportions. However, make the sketches by eye and not with the aid of the scale.

Make all required lines clean-cut and dark, using a soft pencil. Make center lines thin and hidden lines medium in thickness, but dark in both cases. For a discussion of freehand sketching techniques, see Chapter 2.

Additional sketching problems may be assigned from Figs. 8-26 to 8-28.

If assigned by the instructor, construct drawings using decimal-inch or metric scales by converting given fractional dimensions to decimal-inch or metric equivalents. Refer to Table 20 in the Appendix.

Fig. 7-27. Sketching problems.

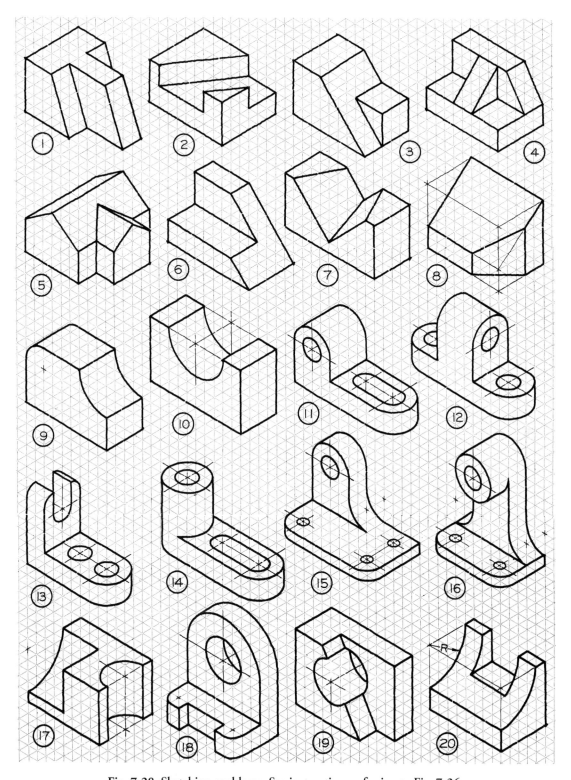

Fig. 7-28. Sketching problems. See instructions referring to Fig. 7-26.

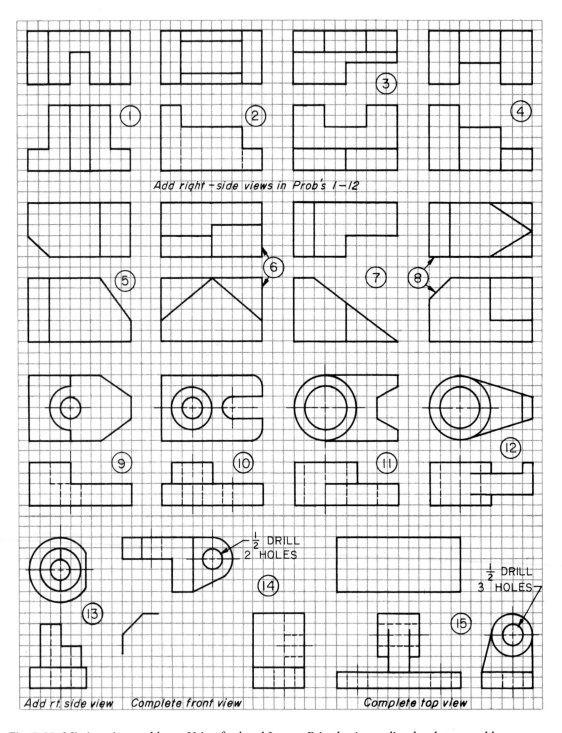

Fig. 7-29. Missing-view problems. Using freehand Layout B in the Appendix, sketch two problems per sheet as in Fig. 7-26. Sketch the given views, and then add the third view in each problem. Use cross-section paper or plain paper, as assigned. In most cases, spacing between views can be improved by spacing views farther apart than shown here. Each grid space = $\frac{1}{4}''$ (.250").

140

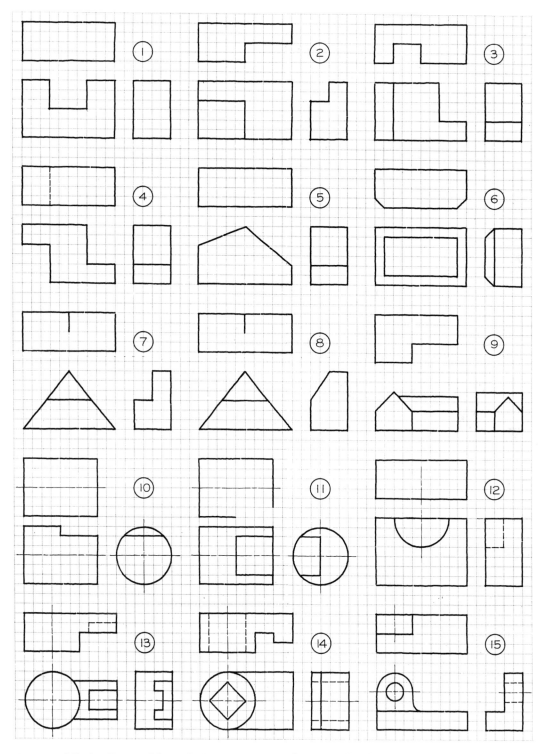

Fig. 7-30. Missing-line problems. Using Layout B in the Appendix, divided as in Fig. 7-26, sketch the views of two assigned problems, adding all missing lines. Each grid space = $\frac{1}{4}''$ (.250″). In most cases, spacing between views can be improved by spacing views farther apart.

IMPORTANT TERMS

blueprinting   microfilming

diazo-dry
process

CHAPTER 8

OBJECTIVES

After studying this chapter, you should be able to:

- Demonstrate good techniques in the drawing of lines.
- Draw two views mechanically.
- Draw three views mechanically.
- Plot points on a curve.

CHAPTER 8
Techniques and Applications

This chapter discusses various drawing techniques. It also discusses methods for reproducing drawings for distribution or storage. Knowledge of these topics is important in all fields of drawing, whether the product being designed is a skyscraper, a heart valve, or the latest fashion in jeans.

You may not have thought of a fashion designer as a "technical" field, yet the designer must have detailed knowledge of human design factors, fabrics, color, and design principles such as balance and proportion.

The fashion designer will work not only with clothing styles but also with accessories such as shoes, purses, hats, belts, and ties. Drawings in this profession range from pattern drafting to pictorial sketches and full-color display renderings. There are specialized CAD programs available for the fashion designer.

CAREER LINK

To learn more about careers in design, contact the National Association of Schools of Art and Design, 11250 Roger Bacon Drive, Suite 21, Reston, VA 20190.

8.1 PENCIL DRAWING. Blueprinting was introduced in the United States at the Philadelphia Centennial Exposition in 1876. Before this, drawings were generally made on imported handmade white paper, usually in ink and often in colors. Only the original drawing was available. This was often made by the owner of the shop, who handed it to a mechanic as a guide to production. Much was left to the imagination of the mechanic. However, in any case, the person who made the drawing was usually at hand to supervise the construction.

As industry developed, more people, often in different places, needed the instructions on the drawing. It was at about this time that the blueprint process was introduced. This process allowed duplicates to be made at low cost. It became the practice to make pencil drawings on detail paper and then to trace the drawings in ink on tracing cloth. As reproduction methods and transparent tracing papers were improved, much time was saved by making the drawings directly in pencil with black lines on the tracing paper and making blueprints from these. This did away with the need for preliminary pencil drawings.

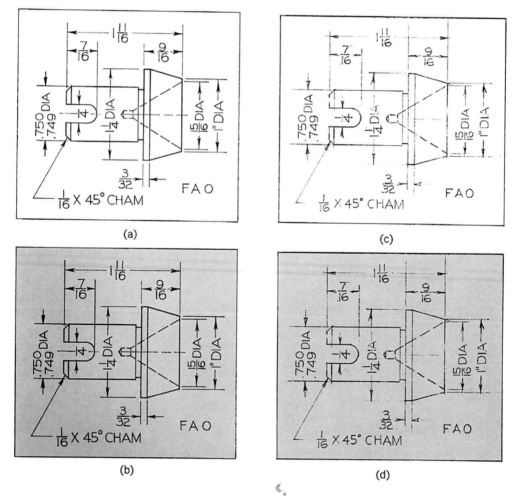

(a)

(b)

(c)

(d)

Fig. 8-1. Pencil technique.

Today, most drawings are made directly in pencil or on a CAD system. Skill in drafting therefore depends largely on the ability to make good pencil or CAD drawings.

8.2 PENCIL TECHNIQUE. Drafters are said to have "good technique" if they know how to use their tools correctly — especially the pencil. Lines and lettering are clean-cut and dense black. The drawing is said to have "snap" or "punch." Good pencil technique consists of the following, Fig. 8-1:

1. *Light construction lines.* These should be so light that they can hardly be seen at arm's length. If so made, they will not show on the blueprint.
2. *Dense black lines.* A fairly soft pencil should be used. Enough pressure should be applied to produce black lines. See Secs. 4.7 to 4.10.
3. *Correct line thicknesses and dash lengths.* The different types of lines should contrast well. Visible lines should be thick enough to make the views stand out clearly. See the Alphabet of Lines, Fig. 4-11.
4. *Accented ends of dashes.* Bear down on the pencil at the beginning and end of each dash (hidden lines and center lines).
5. *Sharp, clean corners.* Be careful not to overrun corners or leave gaps. Make each corner square and sharp.
6. *Smooth tangencies.* Locate centers carefully, Figs. 6-17 to 6-19. Make sure your compass lead is sharpened properly, Fig. 4-35.

A pencil drawing with good technique is shown in Fig. 8-1(a). A reproduction of it is shown at (b). At (c) is shown a drawing with poor technique. Below it at (d) is shown the corresponding blueprint. In general, good

Fig. 8-2. Holding tracing up to the light.

technique consists in line work and lettering that will produce the clearest, most legible reproductions.

When a drawing is made directly on tracing paper or vellum, both of which are quite thin and transparent, a sheet of drawing paper should be placed underneath. This serves as a "backing sheet" to provide a smooth, firm surface. A white backing sheet is recommended so you can clearly see the density of the lines and lettering.

An excellent way to test the density or blackness of your lines and lettering is to hold the tracing up to the light, Fig. 8-2.

Steps in pencil drawing are illustrated in Figs. 8-9 and 8-12.

8.3 REPRODUCTION PROCESSES. An essential part of the designer's or drafter's education is a thorough knowledge of reproduction techniques and processes. Specifically, he or she should be familiar with the various processes available for the reproduction of drawings:

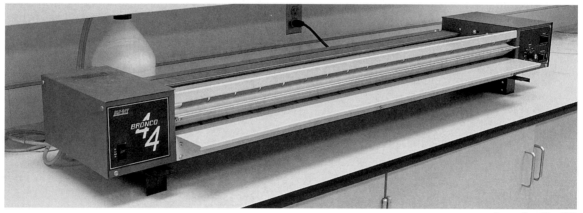

Fig. 8-3. Whiteprinters use ammonia to make prints of drawings.

blueprint, diazo, microfilm, etc. Also helpful is a knowledge of industrial printing and duplicating methods: *lithography, xerography,* and so on.

Each of these processes has very definite advantages and disadvantages. In some, such as blueprinting, the original must be transparent. In others, such as xerography, opaque or translucent originals may be used. In addition to copies, some reproducing equipment can make enlargements or reductions of the original, which often are extremely useful. A general familiarity with all of these reproduction processes is necessary. The reproduction method selected for a specific project will be dependent on the type of original used, number of copies required, size and appearance of copy desired, and the cost.

8.4 BLUEPRINTING. The oldest of the several processes used today for reproducing drawings is *blueprinting.* It is still occasionally used, largely because of its low cost.

Blueprint paper is ordinary paper coated on one side with a chemical preparation sensitive to light. The coated surface is a pale green color. The tracing is placed flat against the coated surface of the blueprint paper. Light is allowed to pass through the tracing onto the sensitized paper. The light passes through the transparent background areas of the tracing, but not through the black pencil or ink lines of the tracing. Thus, the blueprint paper is *exposed* everywhere except where the lines are. After a few seconds or minutes of exposure to light (depending upon the "speed" of the paper and the intensity of the light), the blueprint paper is washed in clear water. This washes away the chemicals not affected by light—that is, those covered by the lines of the drawing. The background is a deep blue. The lines are the white paper itself.

Modern blueprint machines are available in which cut sheets are fed through for exposure only and then washed in a separate washer. Where large quantities of blueprints are made, a *continuous blueprint machine* is used. This machine combines exposure on a continuous role of blueprint paper with washing and drying in a single operation.

8.5 DIAZO-DRY PROCESS. A widely used method of duplicating technical drawings is the *diazo-dry process.* It also uses paper coated with special chemicals. The paper is exposed to light through a tracing in the same manner as in blueprinting. It is then developed by contact with ammonia vapors. These chemicals produce a white background with lines in black, blue, maroon, etc., depending upon the paper used.

A wide range of machine models is available for both exposure and development, Fig. 8-3. You can make excellent prints quickly by simply feeding cut sheets through and waiting a few seconds for the finished dry prints to come out. This process is very popular in industry and has largely replaced blueprinting. A number of different companies make ammonia-process machines and paper supplies.

Safety Note. Modern diazo and blueprint equipment is designed to emit very little odor. Many machines have self-contained filtering systems that neutralize the ammonia fumes associated with print processing, so outside ventilation may not be necessary. However, older machines *do* require outside ventilation. Regardless of whether a self-circulating or outside-vented system is being used, it is important to make sure the system is working properly. When refilling the printers with developer, always wear chemical-resistant plastic gloves and safety goggles. Be careful to avoid spills. Turn on ventilating fans or blowers and, if possible, open windows to eliminate buildup of fumes. Dispose of empty containers or old, leftover chemicals in accordance with the OSHA-approved safety regulations followed by your school or business.

8.6 MICROFILMING AND DIGITAL TECHNIQUES.

The process known as *microfilming* reduces drawings and records to $\frac{1}{15}-\frac{1}{16}''$ of their original size. It is still widely used in many phases of engineering and business, Fig. 8-4. Microfilming stores copies of drawings and records

Arnold & Brown

Fig. 8-4. Microfilm.

on 16-mm, 35-mm, 70-mm, and 105-mm film, either in rolls or on individual frames. The 16-mm and 35-mm frames are usually mounted on aperture cards. The 70-mm and 105-mm frames are generally stored in envelopes. For quick reference, the film can be viewed on special reader-printers, Fig. 8-5. These also furnish instant full-size copies of the drawing or record being projected.

The obvious advantage of microfilming is the great reduction in the space required for storing engineering and business documents. When less space is needed, there's also less cost involved.

The microfilm process, however, is much more versatile than mere physical reduction of documents. Information recorded on film produces reasonably permanent records that resist deterioration. Also, microfilm can be used with various types of specially designed automatic equipment (cameras, reader-printers, etc.) to provide an automatic information filing and retrieval system.

Digital techniques are rapidly replacing microfilming as a compact and convenient

3M Engineering Document System Division

Fig. 8-5. Microfilm reader-printer.

Bonnie Kamin/PhotoEdit

Fig. 8-6. Wide-format printers can print large digitally stored drawings.

means of drawing storage and retrieval. Drawings produced on a CAD system can be stored on the computer's hard drive or on a floppy disk or recordable CD-ROM. Drawings produced by hand can also be scanned and placed into the computer's memory. When a copy of any digitally stored drawing is desired, it is easy to call the drawing up on the computer and print it using any of a vari-

ety of printers, such as a laser or ink-jet printer, Fig. 8-6. Printers are available that will handle paper widths as great as 62 inches.

Plotters can also be used. (See Fig. 3-9.) Today, both printers and plotters are capable of producing high quality color output.

8.7 MECHANICAL DRAWING OF TWO VIEWS.

The first problem in making any mechanical

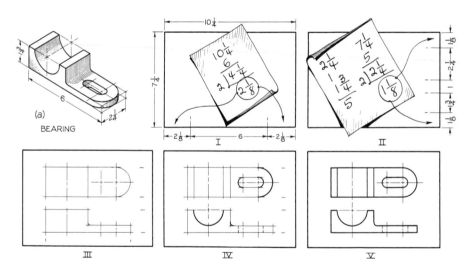

Fig. 8-7. Two-view mechanical drawing.

drawing is spacing the views on the sheet. No lines should be drawn until this has been done properly. To draw two views mechanically, Fig. 8-7, proceed as follows:

I. *Figure horizontal spacing.* We have a working space $10\frac{1}{4}''$ wide, and the width of the object is $6''$. Subtract $6''$ from $10\frac{1}{4}''$, leaving $4\frac{1}{4}''$. Divide $4\frac{1}{4}''$ by 2 to get the space on each side. Set this off with the scale, making sure that the $6''$ is accurate.

II. *Figure vertical spacing.* We have a working space $7\frac{1}{4}''$ high. The top view requires a vertical space of $2\frac{1}{4}''$, and the front view a height of $1\frac{3}{4}''$. Assume a space between views that will look well, say $1''$. Then add $2\frac{1}{4}''$, $1''$, and $1\frac{3}{4}''$, to get $5''$. Subtract $5''$ from $7\frac{1}{4}''$ to get $2\frac{1}{4}''$. Divide $2\frac{1}{4}''$ by 2, to get $1\frac{1}{8}''$, the space at the top and bottom, as shown.

III. *Block in the views* with light construction lines. Let construction lines cross at corners. *Do not draw construction lines between the views.* Use a hard pencil.

IV. *Draw arcs.* These arcs can be put in heavy at once, or if desired you can draw them lightly first. Project the edge of the large arc up to the top view. Project the edges of the small arcs down to the front view. Draw the hidden lines heavy.

V. *Heavy-in all final lines.* Use a medium-soft pencil. Make all lines dense black. Be sure to have three distinct thicknesses of lines: *heavy* for visible lines, *medium* for hidden lines, and *thin* for center lines. Heavy-in arcs and circles first.

NOTE: If dimensions are to be added, Secs. 10.1 to 10.21, allowances in spacing should be made in steps I and II.

8.8 TRANSFERRING DEPTH DIMENSION. Several methods are used in mechanical drawing to transfer the depth dimension from the top to the side view or the reverse, Fig. 8-8. At (a) is shown the use of the 45° *mitre line.* This is recommended for elementary work to emphasize the idea that the depths of the top and side views are the same.

There can be some cumulative error in projecting by means of the mitre line. Thus the professional drafter is more likely to use the dividers, (b), or the scale, (c). This is especially true in the case of complicated drawings where there are a great many transfers to be made between the top and side views.

8.9 SPACING THREE VIEWS. If three views of the Control Block in Fig. 8-9(a) are to be drawn, the first problem is to determine the spacing of

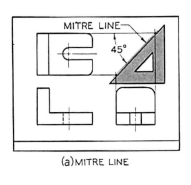
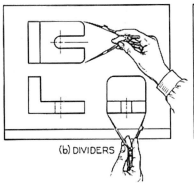
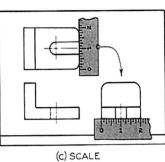

(a) MITRE LINE   (b) DIVIDERS   (c) SCALE

Fig. 8-8. Transferring depth dimension.

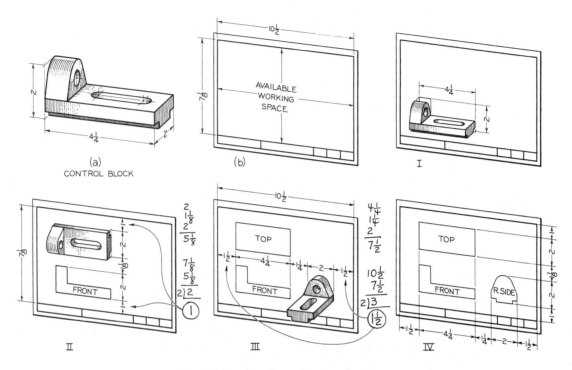

Fig. 8-9. Spacing three views on the sheet.

the three required views. If you use Layout C in the Appendix, you will have a *working space* $10\frac{1}{2}''$ wide $\times 7\frac{1}{8}''$ high, Fig. 8-9(b).

As shown at I, the front view will occupy a space $2''$ high. As shown at II, the top view will occupy a space $2''$ high. Assume a space between the top and front views that will look well, say $1\frac{1}{8}''$. Then add $2''$, $1\frac{1}{8}''$, and $2''$, to get $5\frac{1}{8}''$. Subtract $5\frac{1}{8}''$ from $7\frac{1}{8}''$ to get $2''$. Divide $2''$ by 2 to get $1''$, the space at the top and at the bottom.

As shown at III, the front and top views occupy spaces $4\frac{1}{4}''$ wide. The right-side view requires a space $2''$ wide. Assume a space between the front and right-side views that will be appropriate, say $1\frac{1}{4}''$. Then add $4\frac{1}{4}''$, $1\frac{1}{4}''$, and $2''$ to get $7\frac{1}{2}''$. Subtract $7\frac{1}{2}''$ from $10\frac{1}{2}''$ to get $3''$. Divide $3''$ by 2 to get $1\frac{1}{2}''$, the space on each side. Note that the distances between the top and front views and between the front and side views do not have to be equal.

8.10 MECHANICAL DRAWING OF THREE VIEWS.
To draw mechanically three views of the Control Block, Fig. 8-10(a):

I. *Figure vertical spacing.* Place scale in vertical position with full-size scale to the left. Mark short dashes perpendicular to scale.

II. *Figure horizontal spacing.* Place scale horizontally with full-size scale up. Mark short dashes perpendicular to scale.

III. *Block in the views lightly,* using a hard pencil. Let construction lines cross at corners. Do not draw construction lines between the views. Construct all three views together, not one at a time. Draw construction for all points of tangency. Draw hidden lines in final weight.

IV. *Heavy-in arcs and circles.* Use medium soft lead in compass.

V. *Heavy-in straight lines.* Use a medium soft pencil. Make all lines dense black.

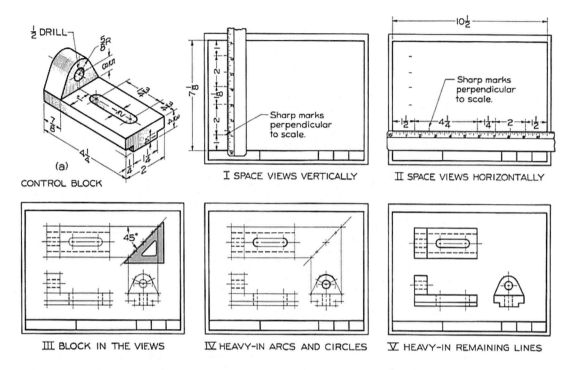

I SPACE VIEWS VERTICALLY II SPACE VIEWS HORIZONTALLY

III BLOCK IN THE VIEWS IV HEAVY-IN ARCS AND CIRCLES V HEAVY-IN REMAINING LINES

Fig. 8-10. Mechanical drawing of three views.

Be sure to have three distinct thicknesses of lines: *heavy* for visible lines (it may be necessary to run the pencil back and forth over the lines to make them heavy enough), *medium* for hidden lines, and *thin* for center lines.

NOTE: If dimensions are to be added, Secs. 10.1 to 10.21, make allowances in spacing in steps I and II.

8.11 POINTS. Any given point will show in all views. As shown in Fig. 8-11, each point has a front view, a top view, and a side view. If several points are lined up in a row perpendicular to the plane for a view, they will appear as one point, as shown for points A, B, and C in the front view.

If two views of an object are given and a third view is required, it will be helpful to

number the corners and to project the points one by one and connect them to get the required view. Suppose a right-side view of the object in Fig. 8-12(a) is required, with the front and top views given, as shown at I. Number the four corners of surfaces A and B as shown at (a) and at I. Notice that if a point is

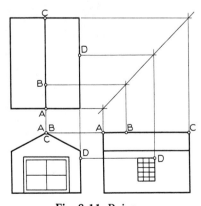

Fig. 8-11. Points.

151

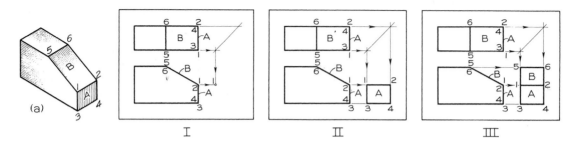

Fig. 8-12. Projecting points.

visible in a given view, the number is placed *outside* the corner. If the point is invisible, the number is placed *inside*. The four corners 1, 2, 3, 4 of surface A are projected as shown at II. The four corners 1, 2, 5, 6 of surface B are projected to complete the view, as shown at III.

8.12 LINES. Let your drawing pencil represent a line. If you hold it so you can look at it at right angles, you will see it in its *true length*. If you look at it at an angle, it will appear *foreshortened* (shorter than true length). As shown in Fig. 8-13(a), if the pencil is perpendicular to a plane of the glass box, it will be parallel to the other two planes. The line will appear true length on the planes to which it is parallel (top and front views) and as a point on the plane to which it is perpendicular, (b).

If the line is parallel to one plane but inclined to the other two, (c), it will show true length on the plane to which it is parallel (top view) and foreshortened on the planes to which it is inclined, (d).

If the line is oblique to all three planes, (e), it will be foreshortened in all three views, (f).

8.13 PLANES. You can use this book for the study of plane surfaces in another way than reading about it here. Lay the book flat on the table, parallel to the table edges. Look down perpendicularly toward the cover. The cover will appear true size, as shown in the top view of Fig. 8-14(a). Now stoop down and look parallel to the cover to see the front view in which the surface appears as a line. The cover will also appear as a line in the side view, as shown.

Now lift the cover so that it makes an

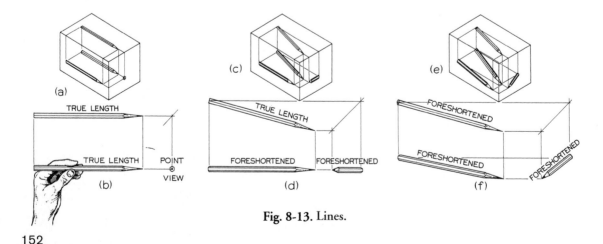

Fig. 8-13. Lines.

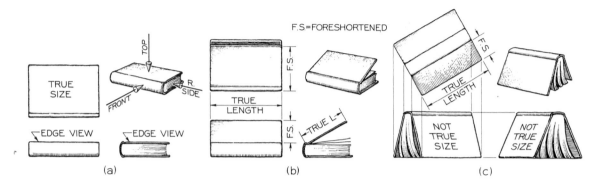

Fig. 8-14. Planes.

angle, as shown at (b). The edge view (right side) of the cover shows one dimension of the cover in true length. The front view shows one dimension true length and the other foreshortened. The same is true in the top view.

Finally, set the book up on edge, as shown at (c), in an oblique position. In the top view, one dimension of the cover appears true length (because it is parallel to the top plane of projection), and one is foreshortened. Both dimensions are foreshortened in the front and side views.

8.14 ANGLES. You can use a triangle to demonstrate how angles appear when viewed from different directions. Refer to, for exam-

ple, Fig. 8-15(a). If you look perpendicularly toward your 45° triangle, you will see it (front view) in its true size and shape, all three angles being true size. The other two views show the angles as zero.

If you move the lower right corner away from you, (b), the front view remains true height, but the width is foreshortened. The 90° angle remains true size, but the other two angles appear larger or smaller. This process is carried further in (c) and (d), as shown. Notice what happens to the angles in each case.

8.15 CURVED SURFACES. The various standard geometric shapes are illustrated in Fig. 6-1. The shapes used in engineering depend upon

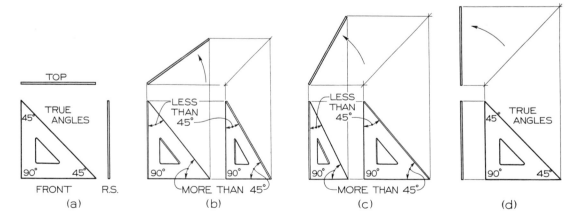

Fig. 8-15. Angles.

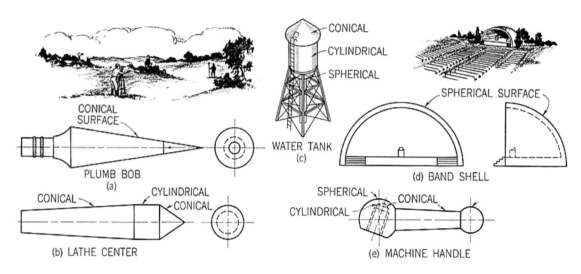

Fig. 8-16. Curved surfaces.

the requirements and upon shop processes. Rounded surfaces are quite common. They are easily produced on rotary-cutting machines, such as the *lathe* and the *drill press,* Secs. 11.11 and 11.12. The most common rounded shapes are the cylinder, cone, and sphere. Some applications are shown in Fig. 8-16.

8.16 PLOTTING POINTS ON CURVE. When a plane intersects a cylindrical surface, a curved edge is produced. For example, consider three views of a piece of quarter-round molding, as shown in Fig. 8-17(a). If the molding is cut at an angle by a plane, as shown at (b), a curved edge is formed that will appear as a plotted curve in the top view. As shown at (b), mark

points on the curve in the side view, using enough points to define the curve clearly. The points need not be spaced equally. Then, (c), project each point to the front and top views. Finally, (d), draw a smooth curve through the points in the top view, using the irregular curve, Sec. 4.30.

Another example of plotting a curve is shown in Fig. 8-18. The procedure is similar to that for Fig. 8-17.

8.17 INTERSECTIONS AND TANGENCIES. When a curved surface is *tangent* to a plane surface, no line should be shown where they join, Fig. 8-19(a). When a curved surface *intersects* a plane surface, a definite edge is formed, (b).

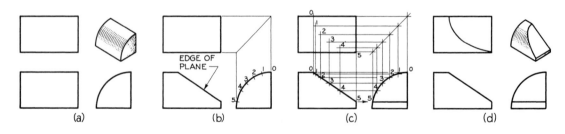

Fig. 8-17. Plotting points on curve.

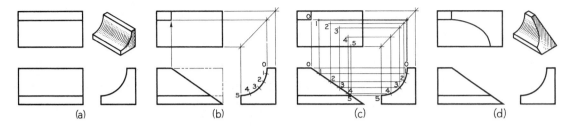

Fig. 8-18. Plotting points on curve.

In the case shown at (c) no lines would appear in the top view. At (d) a vertical surface in the front view produces a line in the top view. Various applications of these principles are shown at (e) to (h). To locate the point of tangency A in (g), refer to Fig. 6-17(c).

8.18 INTERSECTIONS OF CYLINDERS. In Fig. 8-20(a) is shown the intersection of a small cylinder with a large cylinder. The intersection is so small that it is shown merely by a straight line. At (b) the intersection is larger. It is approximated by drawing an arc whose radius r is the same as the radius R of the large cylinder.

At (c) the intersection is still larger. The true curve is constructed. Points are selected

at random on the circle in the side view. These are then projected to the top and front views to locate points on the curve in the front view. These are then connected with a smooth curve with the aid of the irregular curve, Sec. 4.30.

At (d) both cylinders are the same size. The resulting true intersection is drawn with straight lines, as shown.

8.19 PARTIAL VIEWS. Sometimes a partial view (incomplete view) will serve the purpose as well as a full view, Fig. 8-21. At (a) the *half view* shown is understood to be symmetrical with the missing half. If a full view were drawn, the views would have to be drawn to a

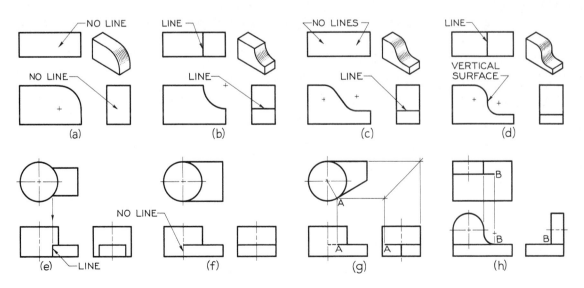

Fig. 8-19. Intersections and tangencies.

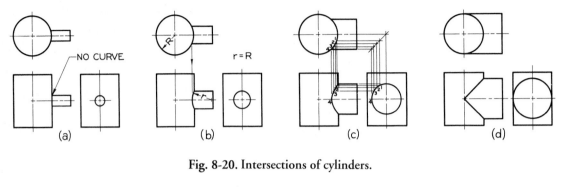

Fig. 8-20. Intersections of cylinders.

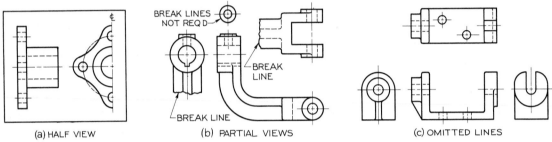

(a) HALF VIEW (b) PARTIAL VIEWS (c) OMITTED LINES

Fig. 8-21. Partial views.

smaller scale to fit on the sheet. In this case, it makes sense to draw partial views.

At (b) are shown one complete view and several partial views. Break lines, Fig. 4-11, are used in two views. They are not needed in the circular view of the boss at the top.

At (c) are shown one complete view (front) and three partial views. Study the top and side views. Note the lines that are omitted. Including

these lines would not add to the clarity of the drawing. Note that each side view includes only the features of one end of the object, leaving the features of the other end to be shown by the other side view. See also Sec. 13.8.

8.20 LEFT-HAND AND RIGHT-HAND DRAWINGS.
It is common in industrial drafting to design individual parts to function in opposite pairs.

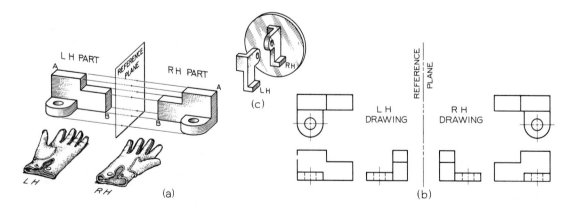

Fig. 8-22. Right-hand and left-hand parts.

These opposite parts, such as the pedals on a bicycle, may be exactly alike but simply placed opposite one another. When this is possible it is always done. It is more economical to make two parts exactly alike than to make two different parts. But opposite parts often cannot be exactly alike, as, for example, the gloves in Fig. 8-22(a), or a pair of shoes. Similarly, the right fender of an automobile cannot have the same shape as the left fender. Do not think that a left-hand part is simply a right-hand part turned around. Such parts will be opposite and different.

A left-hand part is referred to as an LH part. A right-hand part is referred to as an RH part. Two opposite machine parts are shown in Fig. 8-22(a) on opposite sides of an imaginary reference plane. Every point in one part, as A or B, is exactly opposite the corresponding point in the other part and the same distance from the reference plane. At (b) are shown LH and RH drawings of the same part. It will be seen that the drawings are also symmetrical with respect to a reference plane line between them.

The drafter can take any drawing or any object and see what the opposite part would look like by holding it to a mirror, (c). If you look at yourself in a mirror, you will see your hair parted on the opposite side from reality. A drawing can be held faced against a window pane. The reverse image can be traced on the back. This will be the drawing of the opposite part. A tracing can be run through a blueprint machine upside down. Although the lettering will be reversed and readable only in a mirror, the print will be that of the opposite part.

8.21 INKING. With the advent of computer-aided drafting, inking is seldom used today. A

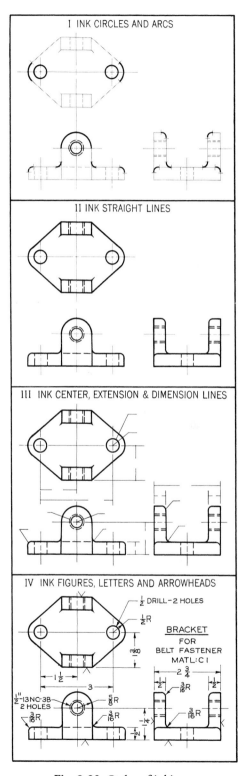

Fig. 8-23. Order of inking.

CAD drawing may be plotted in ink in very little time. Thus, it is not necessary to spend the time to produce hand-drawn ink drawings. However, you should be aware of the techniques of inking.

Black waterproof *drawing ink* is used when it is necessary to give a fine appearance to a drawing or to make it more permanent. *Ink tracings* produce better prints than pencil tracings, but they take more time to make.

Display drawings and others that are not to be reproduced by blueprinting or similar methods may be inked directly on the pencil lines. The original pencil drawing should be drawn lightly. In such cases, the paper should be white and of a quality to take ink well.

After all ink lines are dry, any uncovered pencil lines are erased with the soft eraser.

The steps in inking a drawing are shown in Fig. 8-23.

I. Ink all arcs and circles, centering the ink lines over the pencil lines.

II. Ink all straight lines, first doing the horizontal, then the vertical, and finally the inclined lines. It is much easier to join straight lines to arcs than the reverse.

III. Ink all center lines, dimension lines, and extension lines.

IV. Ink in all arrowheads and lettering. It is best to draw guide lines for lettering directly on the tracing paper or film.

PROBLEMS

The following problems are primarily for mechanical drawing, but any of them would be equally suitable for freehand sketching. It is hoped that some of the earlier problems assigned will be drawn in pencil on tracing paper and that prints will be made to show students whether their lines are dark enough. If desired, any of these problems, after being drawn mechanically, may be traced in ink on vellum or tracing film.

Any of the problems may be dimensioned, if desired by the instructor. In that case the student should study Sec. 5.11 or 5.13, on lettering, and Secs. 10.1 to 10.21, on dimensioning.

Figures 8-26, 8-27, and 8-28 consist of problems in which two views are given and a third view is to be added. You are to make sketches or mechanical drawings of problems assigned by your instructor. Use Layout C in either case, as shown in Figs. 8-24 and 8-25. See Layout C in the Appendix. When sketches are required, isometric or oblique sketches may be included, if assigned, Fig. 8-24. For pictorial sketching, see Secs. 17.3, 17.4, 17.5, 17.18, and 17.19. In sketching, estimate all dimensions. Use scale only for the title strip and border.

If mechanical drawings are assigned, center the drawings in the working space, Secs. 8.9 and 8.10. Make lines clean and dark.

If assigned by the instructor, construct drawings using decimal-inch or metric scales by converting given fractional dimensions to decimal-inch or metric equivalents. Refer to Table 20 in the Appendix.

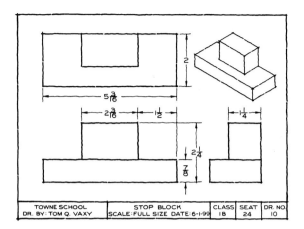

Fig. 8-24. Sketch.

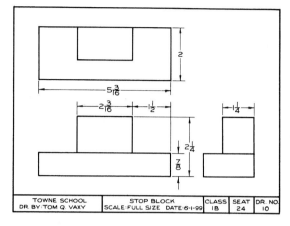

Fig. 8-25. Mechanical drawing.

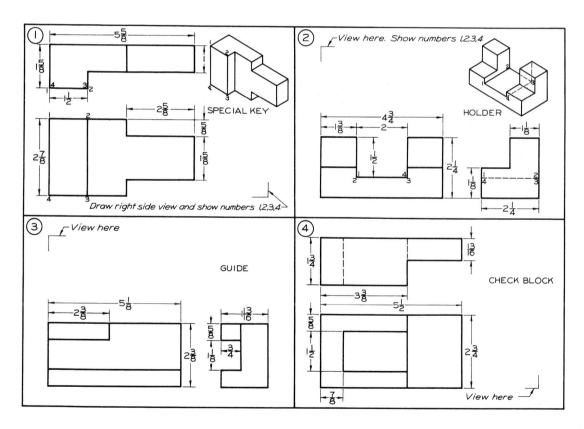

Fig. 8-26. Third-view problems. Omit pictorial drawings and instructional notes. Move titles to title strip. Use Layout C in the Appendix for each problem.

159

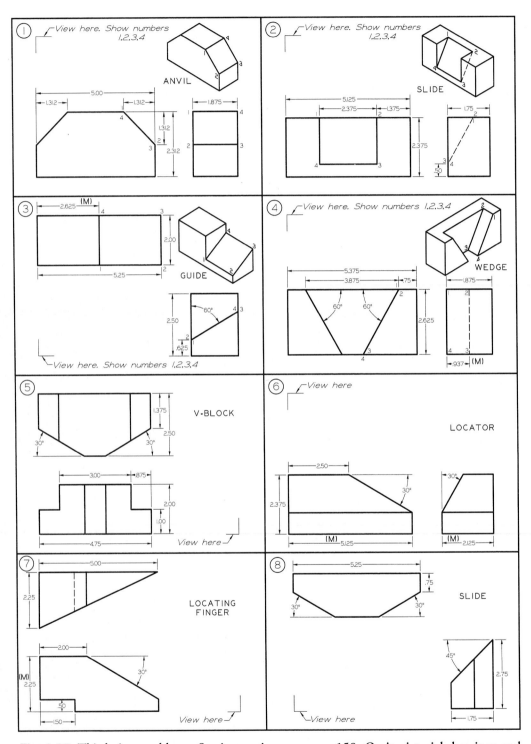

Fig. 8-27. Third-view problems. See instructions on page 158. Omit pictorial drawings and instructional notes. Move dimensions marked (M) to the new views, and move titles to title strip. Use Layout C in the Appendix for each problem.

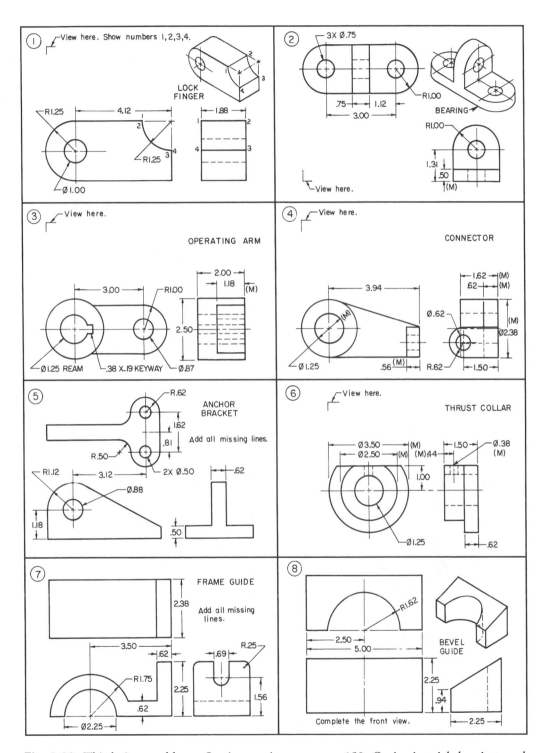

Fig. 8-28. Third-view problems. See instructions on page 158. Omit pictorial drawings and instructional notes. Move dimensions marked (M) to the new views, and move titles to title strip. Use Layout C in the Appendix for each problem.

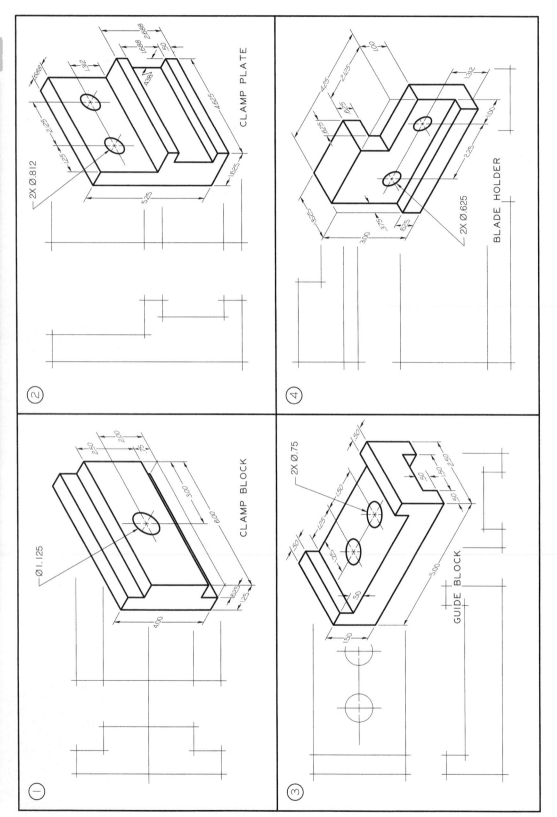

Fig. 8-29. Working drawing problems. Using Layout C in the Appendix, make mechanical drawings. Omit dimensions unless assigned. Omit pictorial drawings. Move titles to title strip.

162

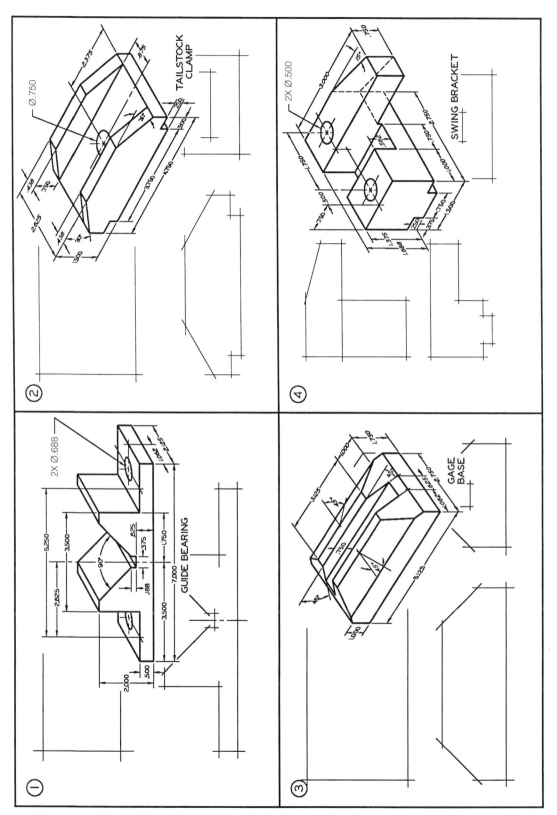

Fig. 8-30. Working drawing problems. Using Layout C in the Appendix, make mechanical drawings. Omit dimensions unless assigned. Omit pictorial drawings. Move titles to title strip.

163

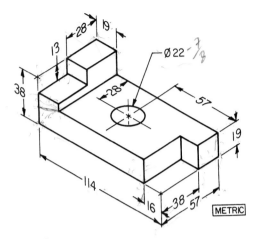

Fig. 8-31. Bearing plate. Using Layout C in the Appendix, draw necessary views with instruments. Omit dimensions unless assigned.

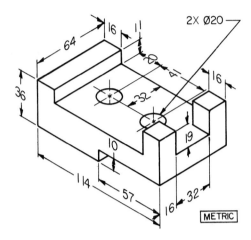

Fig. 8-32. Cross base. Using Layout C in the Appendix, draw necessary views with instruments. Omit dimensions unless assigned.

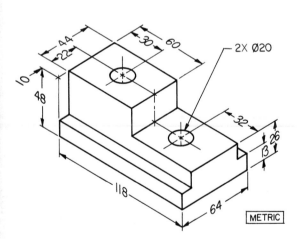

Fig. 8-33. Cutter holder shoe. Using Layout C in the Appendix, draw necessary views with instruments. Omit dimensions unless assigned.

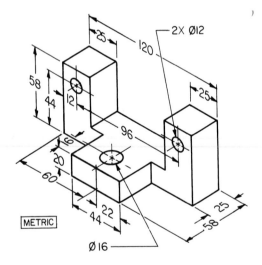

Fig. 8-34. Starting catch. Using Layout C in the Appendix, draw necessary views with instruments. Omit dimensions unless assigned.

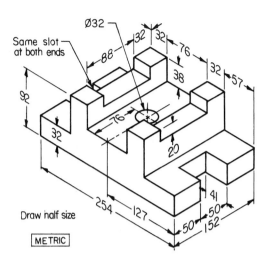

Same slot at both ends
Ø32
Draw half size
METRIC

Fig. 8-35. Fixture base. Using Layout C in the Appendix, draw necessary views with instruments. Omit dimensions unless assigned.

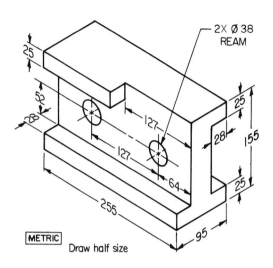

2X Ø 38 REAM
METRIC
Draw half size

Fig. 8-36. Bed plate. Using Layout C in the Appendix, draw necessary views with instruments. Omit dimensions unless assigned.

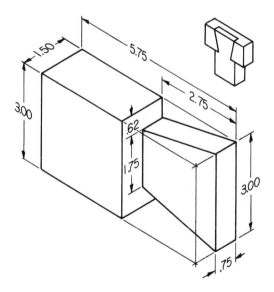

Fig. 8-37. Lap dovetail. Using Layout C in the Appendix, draw necessary views with instruments. Omit dimensions unless assigned.

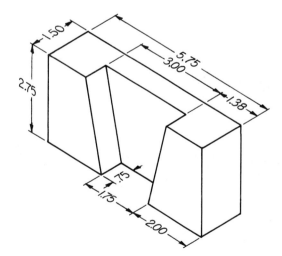

Fig. 8-38. Lap dovetail. Using Layout C in the Appendix, draw necessary views with instruments. Omit dimensions unless assigned.

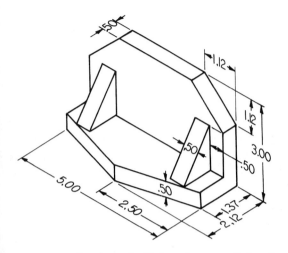

Fig. 8-39. Book end. Using Layout C in the Appendix, draw necessary views with instruments. Omit dimensions unless assigned.

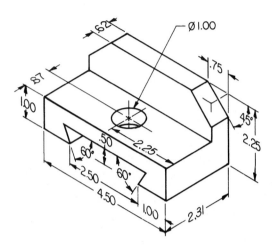

Fig. 8-40. Dovetail slide. Using Layout C in the Appendix, draw necessary views with instruments. Omit dimensions unless assigned.

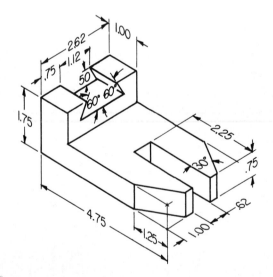

Fig. 8-41. Dovetail finger. Using Layout C in the Appendix, draw necessary views with instruments. Omit dimensions unless assigned.

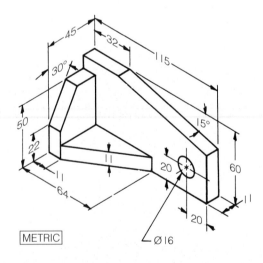

METRIC

Fig. 8-42. Switch bracket. Using Layout C in the Appendix, draw necessary views with instruments. Omit dimensions unless assigned.

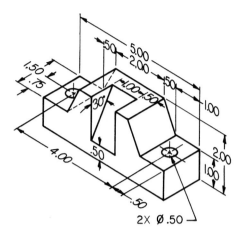

Fig. 8-43. Adjustor block. Using Layout C in the Appendix, draw necessary views with instruments. Omit dimensions unless assigned.

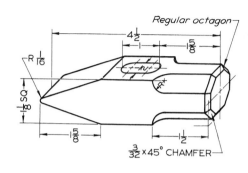

Fig. 8-44. Riveting hammer head. Using Layout C in the Appendix, draw necessary views with instruments. Omit dimensions unless assigned.

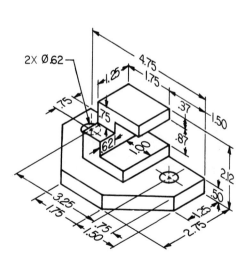

Fig. 8-45. Flipper dog. Using Layout C in the Appendix, draw necessary views with instruments. Omit dimensions unless assigned.

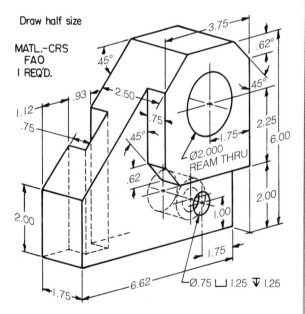

Fig. 8-46. Guide base. Using Layout C in the Appendix, draw necessary views with instruments. Omit dimensions unless assigned.

167

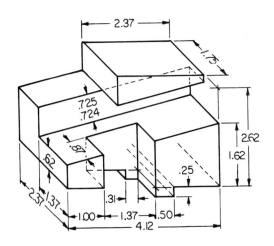

Fig. 8-47. Holder base. Using Layout C in the Appendix, draw necessary views with instruments. Omit dimensions unless assigned.

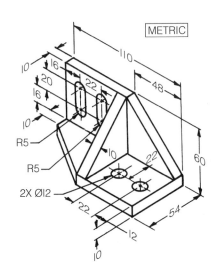

Fig. 8-48. Support bracket. Using Layout C in the Appendix, draw necessary views with instruments. Omit dimensions unless assigned.

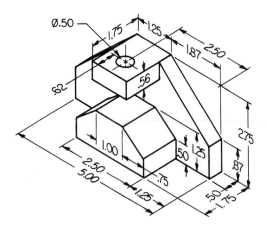

Fig. 8-49. Jig block. Using Layout C in the Appendix, draw necessary views with instruments. Omit dimensions unless assigned.

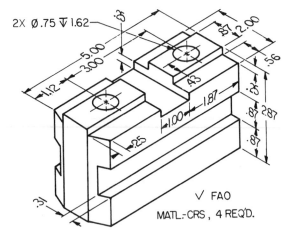

Fig. 8-50. Chuck jaw blank. Using Layout C in the Appendix, draw necessary views with instruments. Omit dimensions unless assigned.

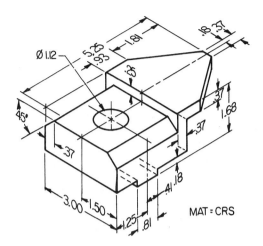

Fig. 8-51. Switch dog. Using Layout C in the Appendix, draw necessary views with instruments. Omit dimensions unless assigned.

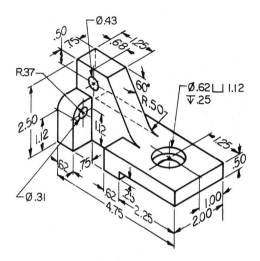

Fig. 8-52. Roller rest bracket. Using Layout C in the Appendix, draw necessary views with instruments. Omit dimensions unless assigned.

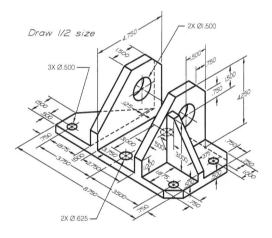

Fig. 8-53. Control bracket. Using Layout C in the Appendix, draw necessary views with instruments. Omit dimensions unless assigned.

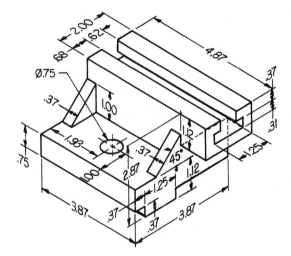

Fig. 8-54. Tool post block. Use Layout D in the Appendix.

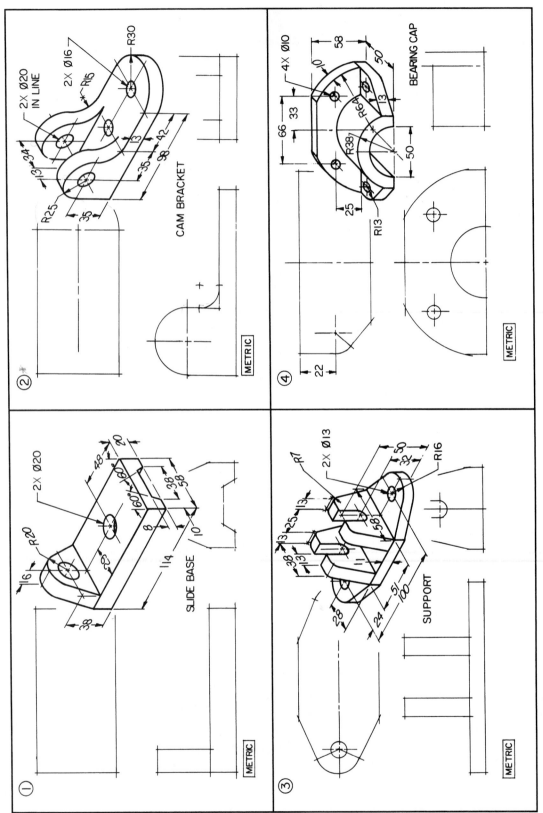

Fig. 8-55. Working drawing problems. Using Layout C in the Appendix, make mechanical drawings. Omit pictorial drawings. Move titles to title strip. Omit dimensions unless assigned.

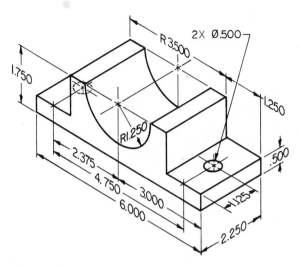

Fig. 8-56. Trunion block. Using Layout C in the Appendix, draw necessary views with instruments. Omit dimensions unless assigned.

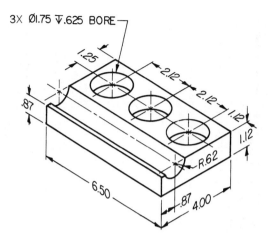

Fig. 8-57. Tool stand. Using Layout C in the Appendix, draw necessary views with instruments. Omit dimensions unless assigned.

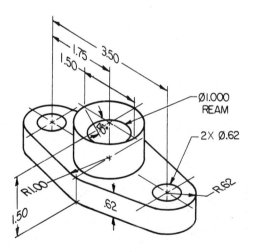

Fig. 8-58. Packing gland. Using Layout C in the Appendix, draw necessary views with instruments. Omit dimensions unless assigned.

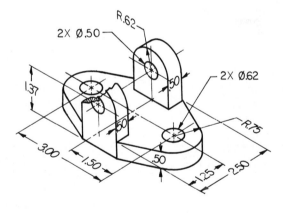

Fig. 8-59. Fastener bracket. Using Layout C in the Appendix, draw necessary views with instruments. Omit dimensions unless assigned.

171

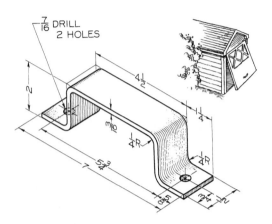

Fig. 8-60. Garage door handle. Using Layout C in the Appendix, draw necessary views with instruments. Omit dimensions unless assigned.

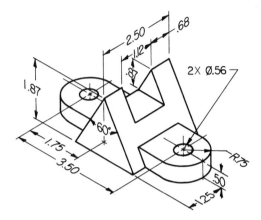

Fig. 8-61. Wedge base. Using Layout C in the Appendix, draw necessary views with instruments. Omit dimensions unless assigned.

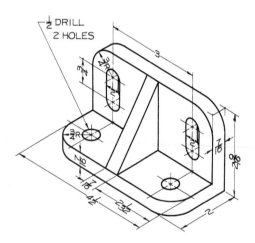

Fig. 8-62. Base bracket. Using Layout C in the Appendix, draw necessary views with instruments. Omit dimensions unless assigned.

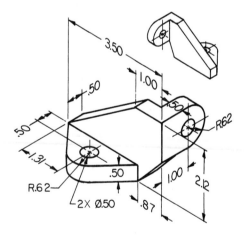

Fig. 8-63. Angle bracket. Using Layout C in the Appendix, draw necessary views with instruments. Omit dimensions unless assigned.

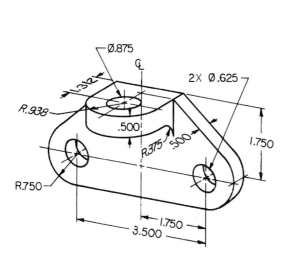

Fig. 8-64. Guide. Using Layout C in the Appendix, draw necessary views with instruments. Omit dimensions unless assigned.

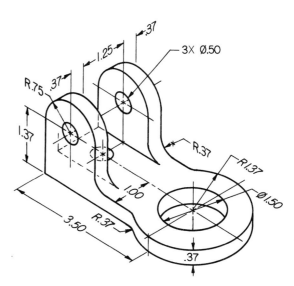

Fig. 8-65. Actuator base. Using Layout C in the Appendix, draw necessary views with instruments. Omit dimensions unless assigned.

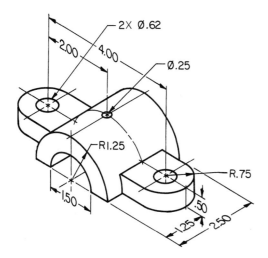

Fig. 8-66. Bearing cap. Using Layout C in the Appendix, draw necessary views with instruments. Omit dimensions unless assigned.

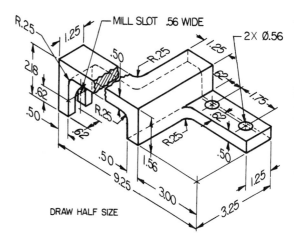

Fig. 8-67. LH hook (for Buick). Using Layout C in the Appendix, draw necessary views with instruments. Omit dimensions unless assigned.

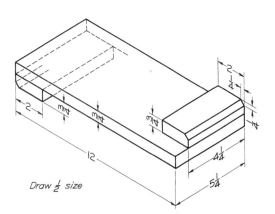

Fig. 8-68. Bench hook. Using Layout C in the Appendix, draw necessary views with instruments. Omit dimensions unless assigned.

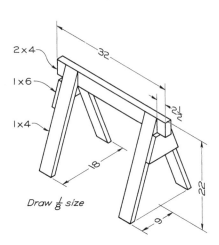

Fig. 8-69. Sawhorse. Using Layout C in the Appendix, draw necessary views with instruments. Omit dimensions unless assigned.

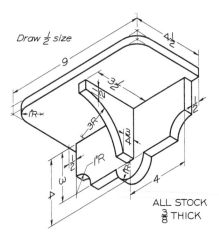

Fig. 8-70. Shelf. Using Layout C in the Appendix, draw necessary views with instruments. Omit dimensions unless assigned.

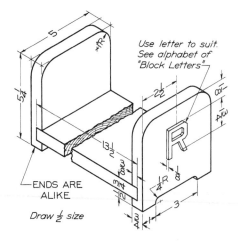

Fig. 8-71. Book rack. Using Layout C in the Appendix, draw necessary views with instruments. Omit dimensions unless assigned.

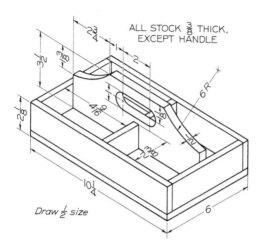

ALL STOCK ¾ THICK, EXCEPT HANDLE

Draw ½ size

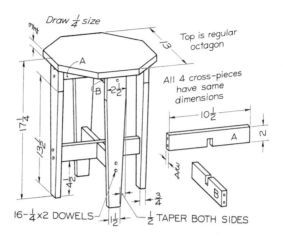

Draw ¼ size

Top is regular octagon

All 4 cross-pieces have same dimensions

16-¼×2 DOWELS   ½ TAPER BOTH SIDES

Fig. 8-72. Nail box. Using Layout C in the Appendix, draw necessary views with instruments. Omit dimensions unless assigned.

Fig. 8-73. Taboret. Using Layout C in the Appendix, draw necessary views with instruments. Omit dimensions unless assigned.

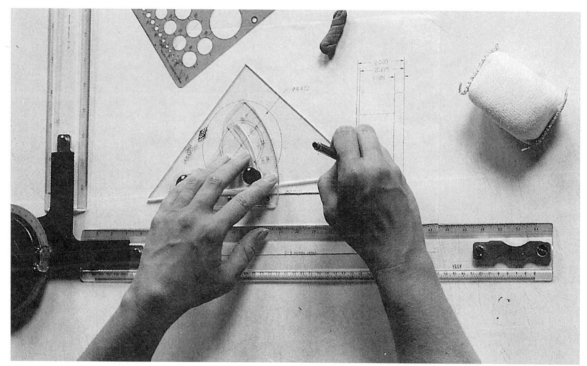

Arnold & Brown

Fig. 8-74. Skill in technical drawing results from the careful development of good technique.

Gontier/The Image Works

CHAPTER 9

IMPORTANT TERMS

icons

drawing window

menu

command line

cursor location
coordinates

grid

snap

object snap

layer

ortho

view

objects

trim

blocks

patterns

OBJECTIVES

After studying this chapter, you should be able to:

- Perform basic drawing operations on a CAD system.
- Save a CAD drawing properly.
- Exit from a CAD program correctly.

CHAPTER 9

Using a CAD System

Engineering design companies hire people who add value to their organization. By learning the skills that high-tech companies want, you can increase your value in the work force. The more skills you have, and the more specialized they are, the greater your value as an employee.

One common employer request is knowledge of CAD software. Every CAD software program has different commands and features. However, the basic capabilities and even many of the commands are similar for many CAD programs. For example, most CAD programs have commands such as CREATE, EDIT, SAVE, DELETE, COPY, and PRINT.

This chapter describes basic CAD operations. By understanding how to use CAD programs, you can become a valuable member of the design team.

CAREER LINK

Many of the companies that are hiring drafters are looking for someone with CAD experience. You can read about job openings on the Internet. Use a search engine and key in "CAD and career."

Screen image © Autodesk, Inc.

Fig. 9-1. Double-clicking on a Windows 95, Windows 98, or Windows NT desktop icon will start the selected program.

9.1 INSTALLING CAD SOFTWARE. CAD software is usually distributed on a CD-ROM (compact disk, read-only memory). The software files on the CD-ROM must be transferred onto your computer's hard disk so that you can run the CAD program. This process is called software installation. The installation process includes setup options. With these options you can specify where on the hard disk you want the program files to be stored and which extra files you want installed, such as sample files, demos, and tutorials.

In most classrooms, students do not need to install software. However, if your teacher asks you to install software, be sure to read and follow the instructions that come with it. Installation procedures vary. Even if you have installed earlier versions of the software, check carefully. The new version may have different procedures.

9.2 STARTING THE CAD PROGRAM. Once the CAD software has been installed on your computer, you begin by starting the program. Most CAD programs in schools run on computers that use the Microsoft Windows operating system. There are two ways to start a program in Windows. If an *icon* (graphic symbol) for the CAD software you want to run is on your desktop, you can double-click it, Fig. 9-1. If not, you click on the Start button and then choose the CAD program from the menus that appear.

9.3 THE DRAWING WINDOW. Once the CAD software program has been started, the main CAD graphics screen will appear on the monitor, Fig. 9-2. The large blank area in the middle of the screen is the drawing window. The *drawing window* is the central portion of the screen where you can create, view, and edit a drawing. You can make your drawing very small so that you can see all of it in the window. You can also magnify the drawing so that you can see one part of the drawing in detail. You can open more than one window at a time. You can also place multiple windows next to each other so that you can see two parts of a drawing at once.

Understanding how to manipulate your drawing in the drawing window is an important part of learning CAD. Every program has slightly different viewing commands. Some of the more common ones are ZOOM, SCROLL, WINDOW, AUTOSIZE, PAN, and DISPLAY, Fig. 9-3 (see page 180).

9.4 MENUS, ICONS, AND THE COMMAND LINE. In the border areas around the drawing window you will see various tools that you can use to create a drawing. The three basic types of software tools are menus, icons, and the command line.

The *menu* items are usually listed at the top of the window. When you click on a menu item such as Draw, a "pull-down" menu appears below it. On this menu is a list of draw commands, Fig. 9-4. When you select a command from the list, the computer will carry out the command.

The icons represent frequently used commands and functions. When you click on an icon, the computer will carry out the command that the icon represents. If you don't know what command the icon represents, you can hold the mouse cursor over the icon for a moment, and the name of the icon will appear.

The *command line* is usually at the bottom of the window. It allows you to type the name of a command. Before Windows, there were only two ways to input commands. You could type them in the command line or select

them from icons on a digitizing tablet. For experienced users who know the command names, the command line can sometimes be an easier way to issue commands. Most new users prefer to use the menus and icons.

The command line is also the place where the CAD program talks back to you. If the software detects an error, or if the software needs additional information, it will prompt you with information or questions on the command line.

Many commands require you to provide additional information. For example, to draw a circle the system might prompt you for the method of drawing the circle, the diameter of the circle, and the location of the circle's center.

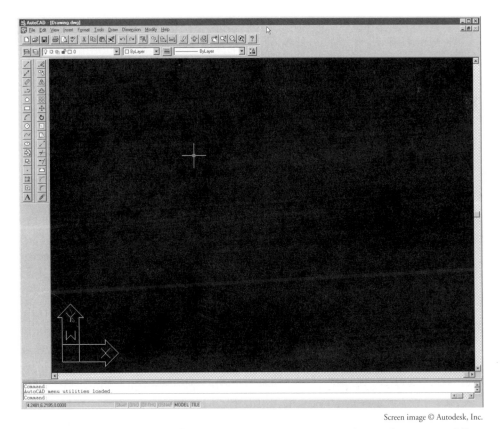

Screen image © Autodesk, Inc.

Fig. 9-2. The main CAD graphics screen contains menu items, icons, the command line, and the drawing window.

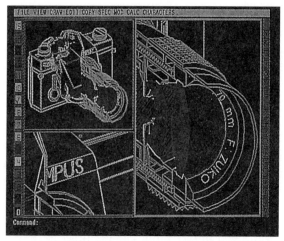

Fig. 9-3. The ZOOM command lets you make objects appear larger or smaller.

When responding to prompts, you can type the answer from the keyboard or select an answer or location with the mouse or puck. To select an option with the mouse, move the mouse until the desired menu item is highlighted. Then click the left mouse button to select the item.

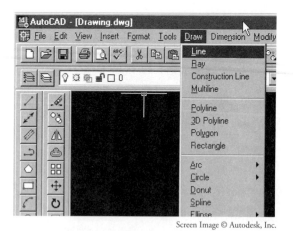

Fig. 9-4. When you select the Draw menu item, a list of all the drawing commands is displayed. Right arrows indicate additional menu choices. Note also the icons down the left side of the screen.

All drawing is performed by issuing CAD commands. The commands tell the computer what you want to draw, where you want to draw it, and how you want to draw it.

9.5 STATUS DISPLAYS. The CAD screen has areas that show current position, drawing parameters, and other information about status.

The cursor on a CAD system usually has horizontal and vertical crosshair lines. When you move the mouse on the desk or the puck on its tablet, the crosshairs move in the drawing window.

The *cursor location coordinates* indicate the location of the crosshairs. The coordinates follow the standard Cartesian coordinate system. The lower left-hand corner of the drawing window is usually the origin, coordinate (0,0). When the tracking option is turned on, the cursor location coordinates are visible.

In the AutoCAD software, these coordinates are shown in the status bar at the bottom of the screen. They change as the mouse or puck is moved, Fig. 9-5.

The status bar also lets you turn various drawing options on and off. Some of the commonly used options in the status bar are GRID, SNAP, and OSNAP.

The *grid* is a horizontal and vertical pattern of dots on the screen. It allows you to estimate distance. For example, if grid spacing is set to 0.25″, you can draw a 1″ line by counting off four dots.

Snap puts "gravity" on the grid dots or on any specified distances. When *snap* is turned on, the crosshairs will jump from snap point to snap point. This makes it easy to draw extremely accurate lines without typing in their length on the keyboard.

Object snap (OSNAP) puts gravity on specified parts of an object, such as the midpoint

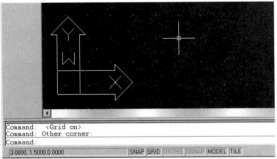

Fig. 9-5. The CAD cursor is usually shown as crosshairs in the drawing window. The coordinates of the cursor are shown below the command line. The status bar shows which options are turned on and off.

or endpoint. This makes it easier to connect lines to each other with precision.

A *layer* is a sheet of electronic drawing paper on which you can draw. On an architectural drawing, one layer might contain the floor plan, another layer the electrical details, and a third layer the dimensions. You can choose to see one or more drawing layers at one time. However, you can only draw or edit on the active layer, and only one layer can be active at a time.

Ortho restricts crosshair movement to either vertical or horizontal directions. This makes it easier to draw lines at exactly 0° and 90°.

View is the orthographic view you are looking at in the drawing window. The top, front, right side, and isometric views are the most common views. However, any projection view can be created using CAD, Fig. 9-6.

By manipulating these options, you can quickly create an extremely accurate drawing. You can control hundreds of option settings. However, not all CAD programs have the same option names. Where is the status display on the CAD program you are using?

9.6 DRAWING LINES. Drawing with CAD consists of creating lines, circles, and arcs. These basic geometric shapes are called *objects*, or entities. By creating objects of the correct size and shape and placing them in the right location, you can create any drawing you want.

One nice feature of CAD is that if you make a mistake, you can electronically erase the object or undo the command. You do not need to use an eraser and erasing shield as with traditional drawing methods.

To create a line with the Windows version of AutoCAD, you select Draw from the main menu. Then select Line from the pull-down menu. Most CAD programs will ask you what kind of line, where to start it, and how long you want it to be. Lines can be created at specific angles, tangent to circles, parallel or perpendicular to another line, or in the middle of a given line. The line can be any length and width you choose. In AutoCAD you can click the left mouse button in the drawing window to start the line, move the cursor, and then click again to end the line.

When connecting two lines using CAD, you do not have to be too careful about making the ends of the lines touch. The object snap feature will connect the lines perfectly. Another feature, called *trim*, can be used to cut the lines to length if one line overlaps another. In fact, it is often faster to draw overlapping lines and trim them to size later. With CAD you can use your technical drawing skills to make quick work of a complicated drawing.

9.7 DRAWING CIRCLES AND ARCS. Drawing circles is similar to drawing other objects. Select the command options from the menus,

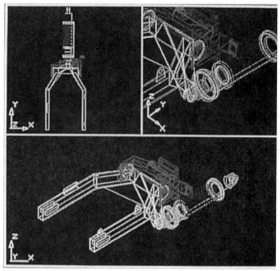

CADKEY, Inc.

Fig. 9-6. Multiple views in CAD.

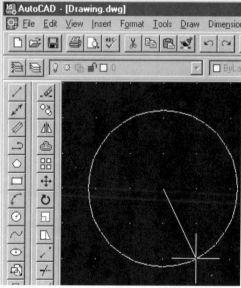

Screen Image © Autodesk, Inc.

Fig. 9-7. Dragging (holding the left mouse button while moving the mouse) changes the radius of the circle.

icons, or command line. Locate the position in the drawing window with the mouse or keyboard.

Here are the commands for creating a circle in the Windows version of AutoCAD.
1. Select **Draw** from the main menu.
2. Select **Circle** from the pull-down menu list.
3. Select **Center**, **Radius** to define the circle by specifying its center and radius. *Note that there are several ways of defining a circle.*
4. Move the mouse into the drawing window to locate the center point. Click the left mouse button.
5. Move the mouse again to define the radius. Fig. 9-7. *Notice how the circle changes size on the screen as you move the mouse.*

To create a circle of a specific radius, such as 1.25 inches, move the cursor into the command line and type the dimension.

Every object is drawn using the same basic procedure. Select the drawing command(s), establish the location and size, and proceed to the next command. When you have become more familiar with CAD, you can use shortcut command sequences to redo a previous command or repeat a command sequence.

9.8 GROUPING OBJECTS. Objects can also be assembled in groups and given names so that you can recall them later as one group. These groups are called *blocks* or *patterns*. After you have created a library of commonly used blocks, you can draw much faster. Bolts, washers, plumbing fixtures, steel beams, tables, and chairs are just a few examples of blocks you can create and recall later.

9.9 CONNECTING OBJECTS. There are several methods of connecting two objects. The easiest method is the mouse. With grid and object snap turned on, mouse position can be very accurate. However, without these aids, the mouse cannot be positioned accurately enough to connect two objects.

For accurate connection of two objects, you should use a more precise method such as the Endpoint, Midpoint, Center, or Nearest object snap or other selection options. For example, if you select Endpoint instead of picking a point using the mouse, you can place the object precisely at the endpoint of another object. By selecting Midpoint, you can place the object exactly on the midpoint of the other object.

Here are the steps for connecting a line to the midpoint of another line when using AutoCAD for Windows.

1. Select **Draw** from the main menu.
2. Select **Line** and draw a line several inches long in the drawing window. Fig. 9-8 (top).
3. Select **Draw**, **Line** again, and begin the new line to the right of the first line.
4. On the command line, type **Mid** at the "To Point:" prompt. Press Enter.
5. Point the mouse anywhere along the first line and click the left mouse button. Fig. 9-8 (bottom).

When connecting objects, be sure to use one of these precision placement methods instead of using the mouse to place objects. Keep in mind that lines may appear to connect on the screen, but in fact may be slightly off. This is because the screen may have a resolution of about .0139″, while the CAD software program may maintain accuracy to .000001″. This means that the CAD software is 13,900 times more accurate than what you see on the screen.

9.10 EDITING OBJECTS. Most CAD programs have an EDIT or MODIFY command. These

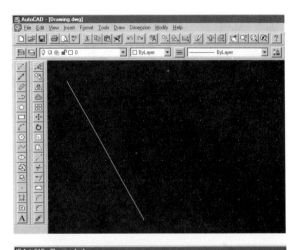

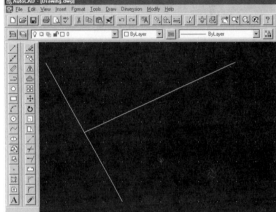

Screen Images © Autodesk, Inc.

Fig. 9-8. Using connecting options such as the Midpoint object snap along with create commands such as Draw->Line results in very precise drawings.

commands allow you to change objects that you have already drawn. The EDIT command allows you to change the size, location, and properties of the object, such as color and line type. It also allows you to copy and stretch an object.

In Windows CAD programs, editing is often performed directly with the mouse. When you select a line, little square boxes called handles appear at the ends and midpoint of a line. For example, in the Windows version of AutoCAD, when you select the

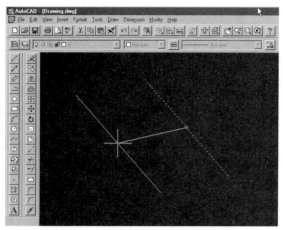

Screen Image © Autodesk, Inc.

Fig. 9-9. Windows CAD programs allow the user to select objects and drag them across the drawing window.

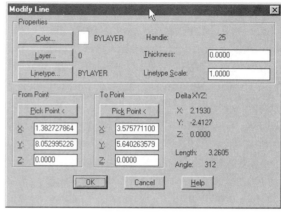

Screen Image © Autodesk, Inc.

Fig. 9-10. Properties dialog boxes allow the user to specify color, layer, line type, line thickness, and position with ease and precision.

midpoint handle you can drag the line to a new location. Fig. 9-9. When you select one of the endpoint handles, you can drag the endpoint to a new location.

The CAD software maintains a list of properties or settings that describe each object in a drawing. These properties include Color, Layer, Line type, Line width, and various other options. These properties determine the appearance and visibility of the object in the drawing. The AutoCAD Modify Line dialog box shows the current settings for the selected line. Fig. 9-10. When you change the settings in the dialog box, the properties of the line will automatically change when you click the OK button.

9.11 DIMENSIONING OBJECTS. Every CAD program has a dimensioning feature. It automatically measures various features in a drawing and displays the measure as a dimension. Many different options are available for selecting a dimension.

For basic CAD, you need to learn how to dimension horizontal, vertical, and inclined lines. You also need to learn how to dimension circles, arcs, and angles.

For horizontal and vertical lines, use the linear dimensioning feature. For inclined lines, we use aligned dimensioning. This feature draws the dimension lines parallel to the inclined line. When dimensioning a CAD drawing, plan the location of dimensions ahead of time. Then place all horizontal dimensions at one time, followed by the vertical and aligned dimensions.

Circles and arcs are dimensioned in one of two ways. You can specify either diameter dimensioning or radius dimensioning.

Angular dimensions are dimensions that describe the angle between two lines or surfaces. CAD programs can measure angles with extreme precision.

As with manual drafting, CAD drawings use leaders to annotate various parts. They are used, for example, to describe the size of a hole, Fig. 9-11.

184

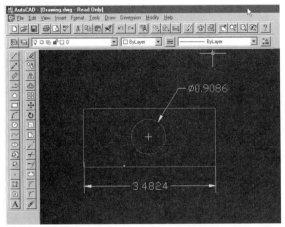

Screen Image © Autodesk, Inc.

Fig. 9-11. CAD programs can dimension objects quickly and with precision.

9.12 DELETING OBJECTS. Lines and other objects will sometimes need to be erased from a drawing. Sometimes this is because of mistakes. Sometimes it is because the drawing needs to be modified. In either case, the DELETE or ERASE command will remove unwanted objects.

After you have selected the DELETE command, the software will prompt you for a method of selecting objects for deletion. In the Windows version of AutoCAD, you delete an object as follows:

1. Select the entity you want to delete.
2. Press the **Delete** key.

You must be careful how you tell the computer what to delete. There are many options for deleting objects. You can delete multiple objects by selecting them in sequence. You can delete everything within a defined space called a window. You can also delete everything on an entire layer.

9.13 SAVING A DRAWING. Saving a drawing copies the images you have created onto the hard disk or other storage medium. You

should get into the habit of saving every 15 minutes. The image in the drawing window is temporarily stored in RAM. If the power goes out, the work you did since the last save will be lost. If you save your drawing every 15 minutes, the most you can lose is 15 minutes of work.

To save a file in AutoCAD for the first time:

1. Select **File** from the main menu.
2. Select **SAVE**.
   A dialog box will appear where you can type a new file name.
3. Type in a new file name, and click **OK**.

If you have already saved the file once, you can select SAVE instead of SAVE AS.

When working on different computers that are not networked, you may wish to save your drawings on a removable medium such as a Zip™ disk or 3.5″ floppy diskette. This way, you can work on any computer you wish, taking your files from one computer to another. Saving your files on a Zip disk or diskette also gives you a backup copy.

9.14 THREE-DIMENSIONAL (3D) DRAWING. Some CAD software programs can be set for either two-dimensional (2D) or three-dimensional (3D) drawing. In two-dimensional drawing, all lines are created on the flat plane of the screen. There is no depth. Objects are simply located with (X,Y) coordinates.

With three-dimensional drawing, every point on an object has three coordinates, (X, Y, Z). If you are drawing in three dimensions, make sure you know what view you are drawing in. You also should know how to find the coordinates for every object.

9.15 PRINTING AND PLOTTING A DRAWING. After you have completed a drawing, the other members of the design team (or your instructor) need to review and approve your work. The easiest way of doing this is to print or plot the drawing on paper.

Every CAD software program comes with printer and plotter drivers. These drivers convert the CAD drawing language into printer or plotter commands. There is a different driver for each printer or plotter.

There are two ways of printing or plotting a drawing: printing to scale or printing to paper size. If you want someone to be able to measure the printed drawing, it must be printed to scale. To print at full size, the scale factor would be 1.0000. To print at half size, the scale factor would be 0.5000. If you wanted to print a drawing at the scale of 1/8″ = 1′-0, the scale factor would be 1/96 or 0.0104.

These scale factors must be entered manually in the print dialog boxes. The printer has no way of knowing whether you want the drawing to scale or not, and if you do, what scale it should print at.

9.16 EXITING THE CAD SOFTWARE PROGRAM.
You should never turn off the computer with a software program running. Many CAD software programs open multiple files during a drawing session. To make sure all files are closed properly, issue the proper command sequence to quit and exit a drawing session. In AutoCAD select File and then Exit, or type quit on the command line.

9.17 A CAD DRAWING EXERCISE.
The best way to learn CAD is to use it. The following exercise provides a step-by-step method for creating a V-block and rod. This example uses AutoCAD software. You can use any software program you want. Just change the com-

mands for the software that you are using. The completed drawing is shown in Fig. 9-12.

9.18 STARTING THE PROGRAM.
Begin by starting the AutoCAD software program. If the software is already running, skip to the next section.

1. Turn the computer on.
   The Windows desktop will appear on the screen.
2. Double-click the AutoCAD icon on the desktop, or select
   Start->Programs->AutoCAD.
   The AutoCAD graphics screen will appear. The drawing area is the large blank area in the middle of the screen. By default, the main menu command choices are in the upper left corner. The command line and coordinate display are at the lower left corner.
3. If the Startup dialog box appears, select the **Start from Scratch** button and click **OK**.

9.19 SETTING PARAMETERS.
Before drawing, you need to set some drawing parameters. Examples are grid, snap, and object snap (OSNAP).

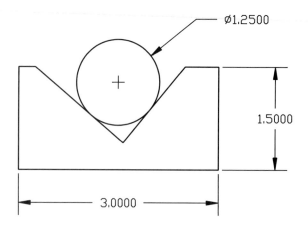

Fig. 9-12. The V-block and rod.

186

1. Click on the command line at the bottom of the screen.
2. Type **grid** and press **Enter.**
   The prompt at the bottom of the screen asks for new grid spacing.
3. Type **.25** and press **Enter.**
4. Type **snap** and press **Enter.**
   The prompt at the bottom of the screen asks for a new snap spacing.
5. Type **.25** and press **Enter.**
   Move the mouse around the drawing area. Notice how the cursor is snapping on every quarter-inch grid mark on the screen.
6. Many CAD commands can be issued in more than one way. From the main menu, select **Tools->Drawing Aids.** The grid and snap settings can also be changed in the Drawing Aids dialog box. Click **OK** to close the dialog box.
7. From the status display at the bottom of the screen, double-click on **OSNAP.** If the Osnap Settings dialog box appears, select endpoint and click **OK.**
   The status boxes for OSNAP, GRID, SNAP, and so on change from gray to black when these features are turned on.
8. If the Osnap Settings dialog box does not appear, select **Tools->Object Snap Settings** from the main menu. Click on **Endpoint** (if it is not already checked) and then click **OK.**

9.20 CREATING A RECTANGLE. Start the V-block by creating a rectangle. Then you can add the two inclined lines to create the V-shape in the block.

1. Pick the **Rectangle** icon or type **rectang** on the command line.
2. Click the left mouse button at location (2.00, 2.00, 0.00).
   The cursor location coordinates are dis-

played on the status bar in the lower left-hand corner of the screen.

3. Move the mouse to location (5.00, 3.50, 0.00) and click the left mouse button.
   A rectangle should appear on the screen.

9.21 CREATING THE TWO INCLINED LINES.
1. Turn off OSNAP by double-clicking on the **OSNAP** box at the bottom of the screen.
2. Pick the **Line** icon or type **line** at the command line.
3. Begin the first angled line by clicking the left mouse button at location (2.25, 3.50).
   This location is one-quarter inch to the right of the upper left-hand corner of the rectangle.
4. Type **@2<320** and press **Enter** and then **Escape.**
   The @ symbol tells AutoCAD the polar coordinates are relative to the selected first point. The 2 indicates the length of the line in inches. The 320 represents the angle in degrees. Note that 0 degrees is at 3 o'clock, not 12 o'clock, and is measured counterclockwise.
5. Draw a second line at a 220° angle by selecting the Line icon again. Click the left mouse button at location (4.50, 3.50).
6. Type **@1.75<230** and press **Enter** and then **Escape.**
   You should see two angled lines on the rectangle.

9.22 TRIMMING LINES. The lines you have just drawn overlap and must be trimmed. The V-notch also has a horizontal line at the top that must be broken (erased) between the two inclined lines. CAD can automatically trim

187

lines so that every corner is perfect. To make selecting the angled lines easier, we will first turn the snap off.

1. Double-click on the SNAP box at the bottom of the screen to turn snap off.
2. Pick the **Trim** icon or type **trim** at the command line.
3. The prompt at the bottom of the screen asks you to select objects to use as cutting lines. You want to trim several different lines, so type **all** to select all the objects on the screen. You could also use the mouse to select objects individually.
4. The prompt asks you to select objects again. This time, you are selecting parts of lines to trim away. Using the mouse, pick the points indicated by the small squares in Fig. 9-13. The parts of the lines you selected should disappear.

9.23 CREATING A CIRCLE. There are several methods of drawing circles and arcs with CAD. Circles and arcs are often drawn tangent to one or more lines. For this assignment, you will draw a 1.250″ diameter circle tangent to both angle lines of the V-block.

1. Pick the **Circle** icon or type **circle** on the command line.
2. Type **t** to select the TTR (tangent, tangent, radius) option.
3. Click on the left angle line (first tangent).

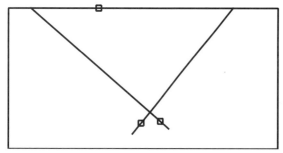

Fig. 9-13.

4. Move the mouse and click on the second angle line (second tangent).
5. Type the radius (**0.625**) and press **Enter**. A circle of diameter 1.250 will appear tangent to both angle lines.

9.24 DIMENSIONING THE DRAWING. Now add basic dimensions. The software automatically sizes dimensions so you don't have to measure each line and type its length.

1. If the CAD software you are using has dimensioning icons, display the icons. (They may be on a dimensioning "toolbar" in some software. In some versions of AutoCAD, for example, you select **View->Toolbars** on the main menu. Then select the **Dimension** toolbar from the list.)
2. Pick the **Linear Dimension** icon or type **dimlin** and then **Enter** on the command line.
3. Press **Enter** and then pick the bottom line of the V-block.
   The dimension appears at the cursor.
4. Move the cursor away from the V-block. Pick a point to position the dimension below the V-block as shown in Fig. 9-12.
5. Pick the **Linear Dimension** icon again (or type **dimlin** and **Enter**).
6. Press **Enter** and pick the right side of the V-block.
7. Move the cursor to the right and pick a point to position the vertical dimension as shown in Fig. 9-12.
8. Now dimension the circle. Pick the **Diameter Dimension** icon or type **dimdiameter** on the command line.
9. Pick a point anywhere on the circle (not inside it).
   The dimension appears at the cursor.

10. Move the cursor and note how the appearance of the dimension changes. Pick a point to position the dimension as shown in Fig. 9-12. The drawing is now complete.

9.25 SAVING THE DRAWING FILE. You should save your work every 15 minutes while you are drawing. You also should save your drawing before you print it or exit the software program. To save an AutoCAD drawing, proceed as follows:

1. Select **File->Save** from the main menu. The first time you do this, AutoCAD will prompt you for a name. Type a new name for the file and click **Save**.
2. After that, just select **File->Save** from the main menu each time you want to save that drawing.

Remember: Using the SAVE command will save the drawing without changing its name. Therefore the previous version of the drawing will be overwritten. If you want to keep the older version of the drawing, you must give the new version its own name.

9.26 EXITING AUTOCAD. Always exit a computer program by using the proper procedure for that program. Never leave a program running unattended on a computer. Never turn off the computer while a program is running.

To exit AutoCAD, proceed as follows:

1. From the main menu, select **File->Exit** or type **QUIT** in the command line.

PROBLEMS

PROBLEM 9-A. Start the CAD program that is available in your CAD lab. Start a new drawing and name it PROBLEM 9-A. Note the positions of the command lines, menus, and status display in this software. Then create the drawing shown in Fig. 9-14. You will need the LINE and CIRCLE commands to complete the drawing. For example, in AutoCAD, follow these steps.

1. Set the grid and snap to .5.
2. Enter the **LINE** command. Draw the lines according to the dimensions shown in the figure. Start the drawing at point A and proceed in a clockwise direction.
3. Enter the **CIRCLE** command. When the prompt asks for the center point of the circle, position the mouse or puck so that the center of the circle is positioned 1.5 units above and 2 units to the left of the bottom right corner of the object. Indicate that you wish to use diameter. Then enter the diameter for the circle when prompted. Dimension the object.
4. Create a layer named DIM for the dimensions. Make DIM the current drawing layer.
5. Use the linear dimensioning command to create the horizontal dimensions.
6. Use the diameter dimensioning command to dimension the circle.

Do not worry if the dimensions on your drawing do not look exactly like those shown in Fig. 9-14. The numbers should correspond. However, the physical appearance is controlled by system variables in the individual CAD programs. For example, the dimensions in your drawing may have four decimal places, whereas those in Fig. 9-14 have only two. Save the final version of the drawing.

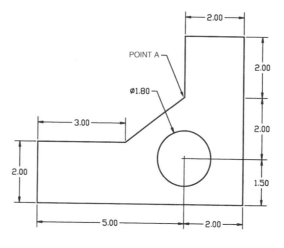

Fig. 9-14.

PROBLEM 9-B. Using the CAD system, draw and dimension the object shown in Fig. 9-15. For this problem, you will need to use your software's LINE, CIRCLE, and ARC commands. Create the lines first. Next, snap the endpoints of the arc to the ends of the two appropriate lines. (Hint: Snap to the right endpoint first to form the arc correctly.) Then create the circles. Dimension the drawing, using a procedure similar to that described for PROBLEM 9-A. Save the drawing as PROBLEM 9-B.

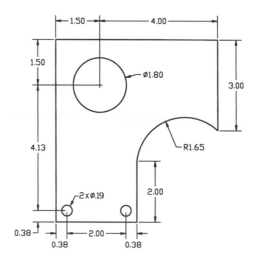

Fig. 9-15.

PROBLEM 9-C. Using the CAD system, draw and dimension the object shown in Fig. 9-16. Use the procedures you used in Problems 9-A and 9-B. (Hint: Start by creating a circle with a radius of 1.50. Trim the circle to create the arc on the right side of the object.) Save the drawing as Problem 9-C.

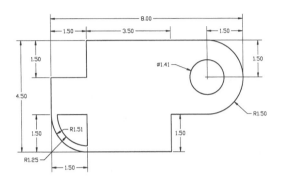

Fig. 9-16.

PROBLEM 9-D. Plot or print the drawings you created for 9-A, 9-B, and 9-C. Inspect the plotted drawings. Do they look like you expected? If part of a drawing did not plot, find out why. Then correct the problem. Plot the drawing again.

PROBLEM 9-E. Open the Problem 9-B drawing. Plot or print the drawing *without* plotting the dimensions. Do not erase the dimensions, however. Instead, freeze the layer on which you placed the dimensions.

Chris Salvo/FPG International

CHAPTER 10

OBJECTIVES

After studying this chapter, you should be able to:

- Identify the four types of lines used in dimensioning.
- Demonstrate the correct methods of lettering decimal dimension figures.
- Demonstrate the correct dimensioning of arcs and angles.
- Demonstrate the correct placement of dimensions.

CHAPTER 10
Dimensioning

This chapter discusses the standard practices of proper dimensioning. Dimensions on drawings give information about sizes and locations. Since production workers follow these drawings to make the product, accurate dimensioning is very important.

In some fields, accurate dimensioning can even be a life-and-death matter. One example is aerospace engineering. Aerospace engineers are responsible for the design and manufacture of traditional aircraft as well as spacecraft capable of flying beyond the earth's atmosphere. Problems concerning aerodynamics, structural design, and materials used in aircraft all require the aerospace engineer to be able to make, use, and interpret complicated drawings. Such drawings must be made to exacting standards and checked carefully to insure correctness. The use of computer-aided drawing and design programs makes this task much easier, allowing for rapid examination of possible design changes.

CAREER LINK

To learn more about careers in aerospace engineering, contact the American Institute of Aeronautics and Astronautics, Inc., Suite 500, 1801 Alexander Bell Drive, Reston, VA 20191-4344.

10.1 COMPLETE DESCRIPTION OF OBJECTS. A drawing must be complete so that the object represented can be made from it exactly as intended by the drafter or designer. Thus, the drawing must tell two complete stories. It tells these by means of (1) *views* and (2) *dimensions and notes.* Views describe the shape of the object. Dimensions and notes give sizes and other manufacturing information.

The drawing shows the object in its completed state. Whether the views are drawn full size or to scale, Sec. 4.19, the dimensions must be the actual dimensions of the completed object. The job of the shop is to produce the object exactly as shown on the drawing. If the drawing is wrong, the object will be made incorrectly. Such errors may be expensive. (Suppose 10,397 pieces have been made before the error is caught.) A serious error may cost a drafter his or her job.

Professional drafters must understand the fundamental manufacturing processes (Chapter 11) so they can issue correct instructions to the workers through dimensions and notes. However, most of the problems of elementary dimensioning are covered by a few simple rules. You will have little difficulty if you carefully follow these rules and apply them with common sense. You should also keep in mind that someone must actually make the object from your drawing.

Remember, the dimensions are at least as important as the views of the object. Correctness is absolutely necessary. Do not make the mistake of simply giving the dimensions you use to make the drawing. Give the dimensions you want the workers to use in making the part. Examples of good dimensioning are shown in Figs. 17-47 to 17-52.

10.2 LEARNING TO DIMENSION. First, you must learn the correct *conventions,* or standard practices of dimensioning. These include the types of lines, the spacing of dimensions, correct arrowheads, and so forth,

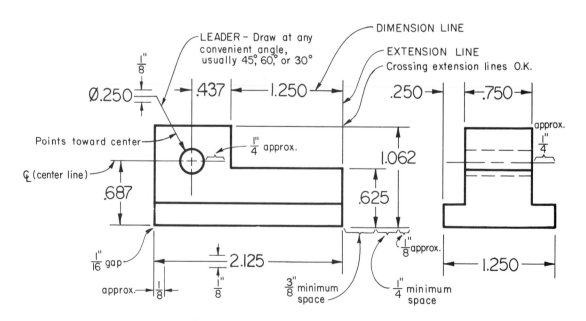

Fig. 10-1. Dimensioning conventions.

Fig. 10-1. Note the strong *contrast* between the thick visible lines of the drawing and the thin lines used in dimensioning. The views should stand out clearly from the dimensions.

The four types of lines used (see the Alphabet of Lines, Fig. 4-11) are the *extension line, dimension line, center line,* and *leader.* These are all drawn thin but dark, with a sharp medium hard pencil, such as H or 2H.

The extension line, Fig. 10-1, "extends" from the object, with a gap of about $\frac{1}{16}''$ (.062") next to the object. It continues to about $\frac{1}{8}''$ (.125") beyond the outermost arrowhead. Never leave a gap where extension lines cross any other line.

A dimension line, Fig. 10-1, has an arrowhead at each end indicating the extent of the dimension. A gap is left (except in architectural and structural drawing) near the middle for the dimension figure. On small drawings, dimension lines are spaced at least $\frac{3}{8}''$ (.375") from the object and at least $\frac{1}{4}''$ (.250") apart. The spacing must be uniform throughout the drawing.

Center lines, Fig. 10-1, are used to indicate axes of symmetry. They are also used in place of extension lines for locating holes and other features. "Wild" center lines spoil a drawing. Make center lines end about $\frac{1}{4}''$ (.250") outside the hole or feature. See also Sec. 7.14.

A leader, Figs. 10-1 and 10-2, is a thin solid line that "leads" from a note or dimension and is terminated by an arrowhead touching the part to which attention is directed. Leaders are straight *inclined* lines (never vertical or horizontal), that are usually drawn at 45°, 60°, or 30° with horizontal. However, they may be drawn at any convenient angle, Fig. 10-2. A short horizontal "shoulder" should extend from the mid-height of the lettering. The inclined line should be drawn so that if

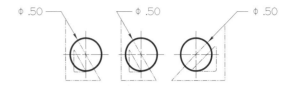

Fig. 10-2. Leaders.

extended it will pass through the center of the circle. Leaders may extend from either the beginning or end of a note, Fig. 10-2.

10.3 ARROWHEADS. *Arrowheads,* Fig. 10-3, are drawn with two sharp strokes toward or away from the point. The length should be about $\frac{1}{8}''$ (.125") and the width about one-third the length. For better appearance, they may be filled in, as shown at (b). Avoid sloppy, careless arrowheads such as those marked NO at the bottom of the figure.

10.4 DIMENSION FIGURES. The correct lettering of dimension figures is explained in Secs. 5.11 and 5.13. These sections should be studied thoroughly. Incorrect or unclear dimension figures can lead to costly mistakes in manufacturing. Thus, great care should be exercised to letter them properly. The standard height for whole numbers and decimals is $\frac{1}{8}''$ (.125"). The standard height for fractions is double

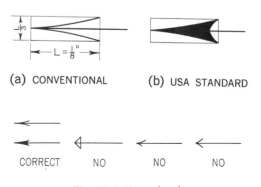

(a) CONVENTIONAL   (b) USA STANDARD

CORRECT   NO   NO   NO

Fig. 10-3. Arrowheads.

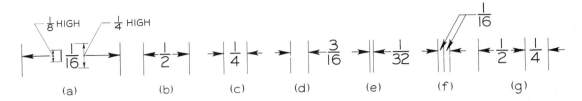

Fig. 10-4. Common fraction dimension figures (full size).

this, or $\frac{1}{4}''$ (.250"), as shown in Fig. 10-4(a). Clear spaces should be left on each side of the fraction bar, Fig. 5-19(a). A typical dimension is shown in Fig. 10-4(a). Methods of giving dimensions to avoid crowding in small places are illustrated at (b) to (g). *Never letter a dimension figure over any line of the drawing.*

The correct methods of lettering decimal dimension figures are shown in Fig. 10-5. The numerals in all cases are made $\frac{1}{8}''$ (.125") high, regardless of whether they appear on one line or two lines. The space between lines of numerals is $\frac{1}{16}''$ (.062"), or $\frac{1}{32}''$ (.031") above and below the dimension line, as shown at (b) and (d).

When the maximum and minimum limits are specified, one directly above the other, the maximum limit is always placed above the minimum limit. That portion of the dimension line between the lines of numerals is omitted, as shown at (b) to (f). When both limits are specified on one horizontal line, as with a leader or note, the minimum limit is always placed first. A short dash is placed between the limits.

Guide lines should be used for all dimension figures. All numerals and decimal points should be carefully lettered and properly spaced.

10.5 INCH MARKS. As explained in Sec. 4.19, dimensions of machine parts are expressed in inches *understood.* It is common practice to omit all inch marks on a drawing when all dimensions are expressed in inches, except in

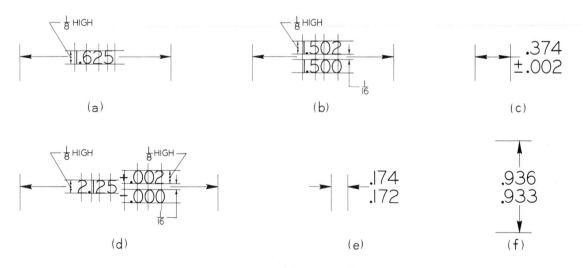

Fig. 10-5. Decimal dimension figures.

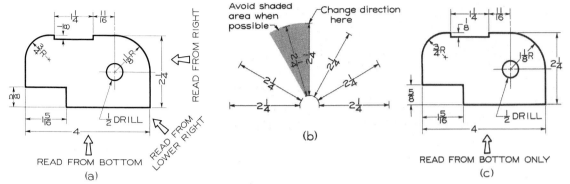

Fig. 10-6. Direction of dimension figures.

situations where this may lead to misinterpretation of dimensions or notes. A one-inch dimension would be shown as 1″. A one-inch drill note would be shown as Ø1″ DRILL, where Ø is the symbol for diameter.

10.6 DIRECTION OF DIMENSION FIGURES. The two systems of reading direction for dimen-

sions on drawings are the *aligned* and *unidirectional* systems.

In the aligned system, Fig. 10-6(a), dimensions are lettered so as to be read from the bottom of the sheet, or from the right side, or between these positions. Sometimes a dimension must be lettered to read slightly from the left, such as the ¾R at the upper left at (a).

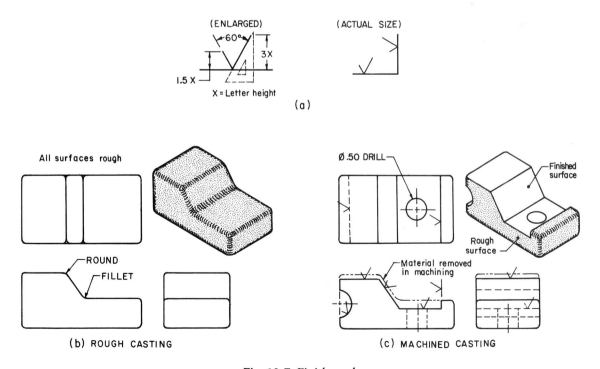

Fig. 10-7. Finish marks.

However, this condition should be avoided where possible. The shaded zone at (b) shows the directions of dimensions that should be avoided if possible. Notes, such as $\varnothing\frac{1}{2}$, should always be lettered horizontally.

In the unidirectional system, Fig. 10-6(c), all dimensions are lettered to read from the bottom. This system is the preferred ANSI standard widely used in industry.

10.7 SURFACE TEXTURE SYMBOLS. A *finish mark* is a symbol to indicate that a surface is to be *finished,* or machined. A finish mark on a drawing tells the patternmaker that he or she must allow extra material on the pattern. This will provide extra metal on the casting to be removed in machining. The same finish mark tells the machinist to machine the surface.

The ✓-type is the ANSI standard shown in Fig. 10-7(a).

A finish mark is shown on the edge view of a surface to be finished. It is repeated in every view where the surface appears as a line, including hidden lines and curved lines. The point of the ✓ should point inward toward the solid metal like a cutting tool.

In Fig. 10-7(b) are shown three views and a pictorial of a rough casting before it is machined. At (c) the casting has been machined. The ✓-type finish marks are

shown. As explained in Sec. 10.1, the drawing shows the part in its completed state.

Finish marks are not needed for drilled holes or for any other holes where the machining operations are clearly shown in a note, Sec. 10.24. If a part is to be finished all over, omit all finish marks. Letter a general note on the drawing, such as FINISH ALL OVER or FAO. No finish marks or notes are needed if a part is machined from cold finished stock.

10.8 DIMENSIONING ANGLES. Angles are drawn with the triangles or with the aid of the protractor, Fig. 4-19. They are indicated by degrees, Fig. 10-8(a) to (f), or by two dimensions as shown at (g). When degrees are given, the circular dimension lines are drawn with the compass center at the vertex of the angle, (f). The lettering is done on horizontal guide lines.

10.9 DIMENSIONING ARCS. Arcs are dimensioned in the views in which their true shapes appear by giving the radius, as shown in Fig. 10-9. The letter R, for radius, is always lettered before the figures, as shown.

The dimension figure and the arrowhead should be inside the arc, as shown at (a) and (b), where there is sufficient space. If the

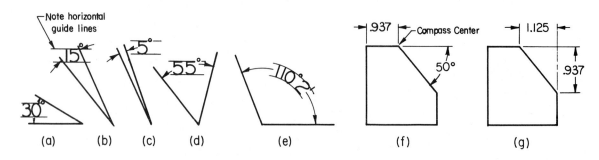

Fig. 10-8. Dimensioning angles.

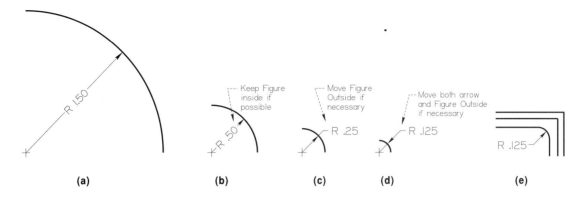

Fig. 10-9. Dimensioning arcs (full size).

space is too crowded, the figure is moved outside the arc, (c). If necessary the arrowhead is also moved outside, (d). Where lines of the drawing interfere, the method at (e) is used.

10.10 FILLETS AND ROUNDS. Fillets and rounds, Sec. 11.3, are dimensioned as shown in Fig. 10-9. It is not necessary to give the radius of every fillet or round, but only a few typical radii, as shown in Fig. 11-2. If all fillets and rounds are uniform in size, dimen-

sions may be omitted. A note can be added to the drawing as follows:

ALL FILLETS R.125 AND ROUNDS R.125
or   ALL FILLETS & ROUNDS R.125

10.11 PLACEMENT OF DIMENSIONS. Correct and incorrect methods of placing dimensions are shown in Figs. 10-10 and 10-11. As shown in Fig. 10-10(a), the smaller dimensions should be placed nearest to the object

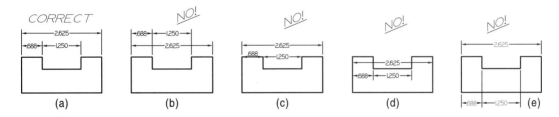

Fig. 10-10. Placement of dimensions.

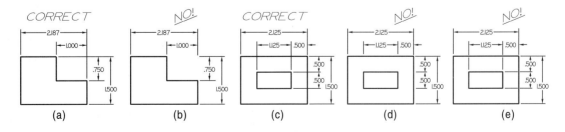

Fig. 10-11. Crossing lines.

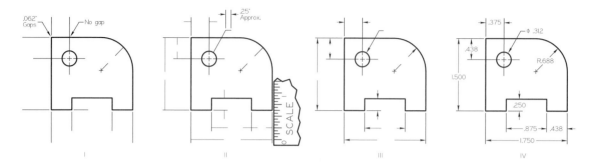

Fig. 10-12. Steps in applying dimensions.

and lined up in chain fashion. Note that a complete chain of these is unnecessary. The overall dimensions are always farthest from the view. Avoid placing the dimensions as at (b), where dimension lines must cross extension lines. Never place a dimension to coincide with a line of the drawing or join end to end with a line of the drawing, (c). Avoid placing dimensions on a view, (d), unless some advantage is gained thereby. In certain cases, such as the radii in Fig. 10-6(a), and on complicated drawings where it is difficult to find places for dimensions, some dimensions must be placed within the view, Fig. 16-14. As a general rule, however, *never place a dimension on a view unless something in clearness or directness of application is gained thereby.* Avoid placing dimensions so that extension lines cross lines of the drawing unnecessarily, (e).

Extension lines may cross each other freely, as in Fig. 10-11(a). They should never be shortened as at (b). In many cases, extension lines must cross the lines of the drawing as at (c). The extension lines should never be shortened as at (d) or have gaps at crossing points as at (e).

10.12 STEPS IN APPLYING DIMENSIONS. The steps in placing dimensions on a view are

shown in Fig. 10-12.

I. Draw extension lines dark and sharp. Use a medium-hard pencil (as 2H) with a sharp conical point. Extend the center lines of the hole, to be used in the same manner as extension lines.

II. Use the scale to space the dimensions at least $\frac{3}{8}''$ (.375″) from the object and $\frac{1}{4}''$ (.250″) apart; $\frac{3}{8}''$ (.375″) if space is available). Draw the dimension lines dark and sharp, leaving gaps for the dimension figures.

III. Draw all arrowheads about $\frac{1}{8}''$ (.125″) long and very narrow.

IV. Add all dimension figures and lettering. Draw guide lines as shown in Sec. 5.11.

Note the contrast at IV between the heavy visible lines of the view and the dimensions. The view should stand out clearly from the dimensions.

10.13 GEOMETRIC BREAKDOWN. Any machine or structure, when broken down into its basic shapes, is found to be made up largely of simple geometric shapes. These shapes are the prism, cylinder, cone, and sphere. For example, the Bearing Bracket in Fig. 10-13(a) is made up of simple basic shapes, as shown at (b). Even the holes are regarded as cylinders.

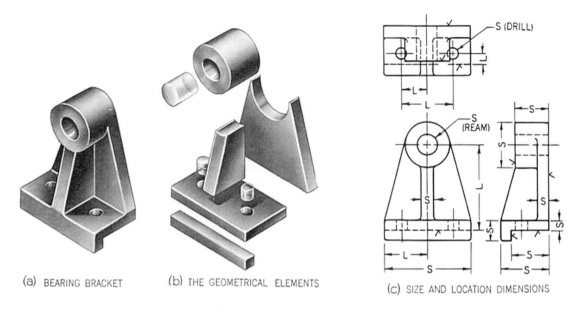

Fig. 10-13. Size (S) and location (L) dimensions.

Three views of the Bearing Bracket are shown at (c). This illustrates the two main types of dimensions indicated by the letters S and L:

S = Size Dimensions, which give the sizes of the geometric shapes.

L = Location Dimensions, which locate the geometric shapes with respect to each other.

In actual practice, of course, the dimension figures are given instead of the letters S and L.

10.14 SIZE DIMENSIONS OF PRISMS. The simplest geometric shape is the prism, Fig. 10-14(a), or modifications of it. Three dimensions are required, the *width, height,* and *depth.* Only two views are required to give these dimensions, as shown at (b) and (c). In either case, the width (4.50″) and the height (3.00″) are given on the front view. The depth (1.75″) is given on the top view or side view.

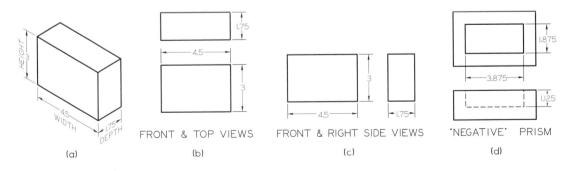

Fig. 10-14. Size dimensions of prisms.

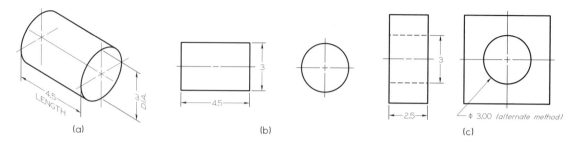

Fig. 10-15. Size dimensions of cylinders.

In the example at (b), the 4.50″ dimension should be between the views, as shown. It should not be above the top view or below the front view, or across either view, Sec. 10.11. The 3.00″ and 1.75″ dimensions should be lined up together, as shown. Similarly, at (c) the 3.00″ dimension should be between views. The 4.50″ and 1.75″ dimensions should be lined up together. However, these may be placed above the views, if desired.

10.15 SIZE DIMENSIONS OF CYLINDERS. The cylinder, Fig. 10-15(a), is the next most common shape. It is usually seen as a shaft or a hole. Dimension a cylinder by giving both its diameter and length in the *rectangular* view, as shown at (b) and (c). Never give the radius of a cylinder, Sec. 11.10, or give the diameter in the circular view, either diagonally across the circle or between extension lines from the circle. An exception is made when dimensioning a hole where the manufacturing operation is given in a note, as in Fig. 10-16(c)

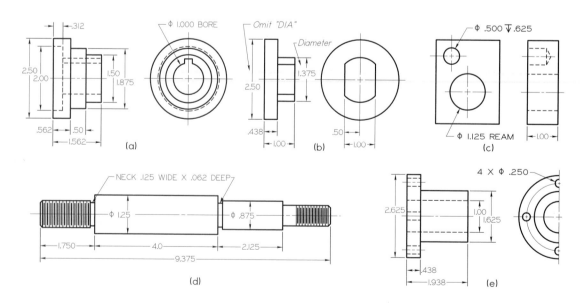

Fig. 10-16. Decimals, diameter symbol.

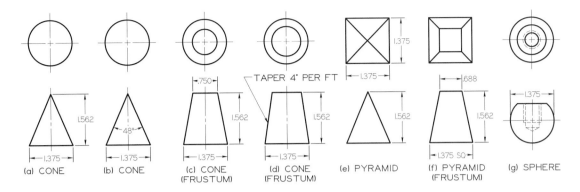

Fig. 10-17. Size dimensions of cones, pyramids, and spheres.

and (e), or in dimensioning a large hole or a bolt circle, where the diagonal diameter is permitted, Fig. 10-19. See also Sec. 10.24.

Several applications of size dimensions of cylinders are shown in Fig. 10-16. At (a) the pulley is made up of a number of concentric cylinders. All diameters except one are given in the rectangular view. Some are given at the right and some at the left so as to place the diameters as close to the shapes as possible. Note that the dimension figures are "staggered" to provide ample space for each figure.

At (b) is shown a case where Ø is given to show clearly that the small piece is cylindrical in spite of the two flat surfaces. At (c) is shown the use of notes where it is desired to specify the operations. In all such cases the *diameter,* and not the radius, is given. At (d) is shown a cylindrical shaft in which the symbol Ø on

each diameter is given to make an end view unnecessary. At (e) is shown a bearing in which a half-view is used to save space. This can be done only when both halves are the same.

10.16 SIZE DIMENSIONS OF CONES, PYRAMIDS, AND SPHERES. Dimension a cone by giving both the diameter of the base and the altitude in the triangular view, Fig. 10-17(a). Sometimes it is desirable to give the diameter and the angle, (b), or give two diameters and the altitude, (c), or one diameter and the *taper* by note, (d). *Taper per foot* means the difference in diameter in one foot of length.

Dimension a pyramid by giving the dimensions of the base in the square or rectangular view, and the altitude in the other view, (e). If the base is square, only one dimension for the base is necessary if marked SQ, as shown at (f).

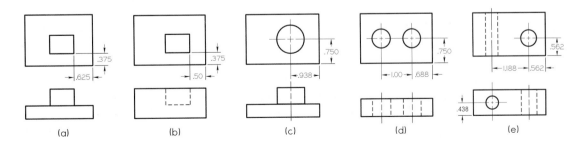

Fig. 10-18. Location dimensions.

203

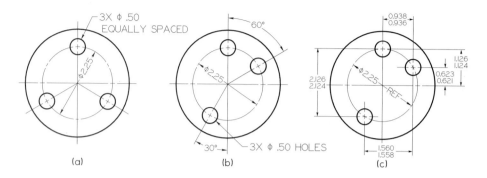

Fig. 10-19. Locating holes about a common center.

Dimension a sphere by giving its diameter, as shown at (g).

10.17 LOCATION DIMENSIONS. Location dimensions, Sec. 10.13, are used to locate geometric shapes with respect to each other after the size dimensions for these shapes have been given. Rectangular shapes are located from surface to surface, Fig. 10-18(a) and (b). Cylinders are located from their center lines, (c) and (d). Location dimensions for holes should be given in the circular view of the holes, if possible, as at (d). Otherwise, they should be given as at (e).

If holes are equally spaced about a common center, Fig. 10-19(a), the diagonal diameter of the *circle of centers,* or *bolt circle,* is

given. The equal spacing is indicated in the note. If the holes are unequally spaced, it is necessary to give angles to one of the two main center lines, (b). Where great accuracy is required, the holes should be located as shown at (c), where the bolt circle diameter is given only for reference.

Symmetry of an object about one or two center lines is very common, in which case dimensions are given from those main center lines, Fig. 10-20. At (a) one small hole is located from the main center line A. The other small hole is located from it. The two small holes must be tied together. No dimension should be given at B.

At (b), in addition to the location dimensions from the two main center lines, the

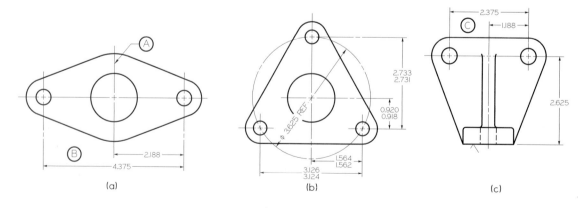

Fig. 10-20. Locating holes on center.

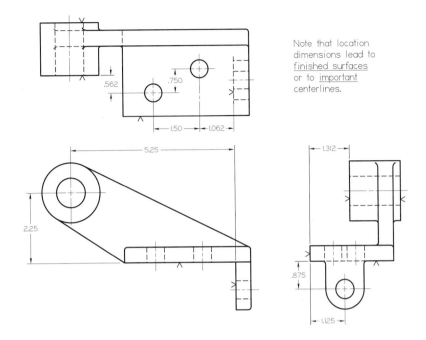

Fig. 10-21. Location dimensions to finished surfaces.

Note that location dimensions lead to <u>finished surfaces</u> or to <u>important</u> centerlines.

diagonal diameter is given for reference in laying out the holes in the shop. At (c) the vertical center line is the main axis of symmetry. Note the omission of the unnecessary dimension at C.

A typical example of location dimensions on a three-view drawing of a bracket is shown in Fig. 10-21. Location dimensions should lead to finished surfaces or to important center lines, as shown.

10.18 CONTOUR DIMENSIONING. *Dimensions should be given in the views where the shapes are shown,* Fig. 10-22(a). When manufactur-

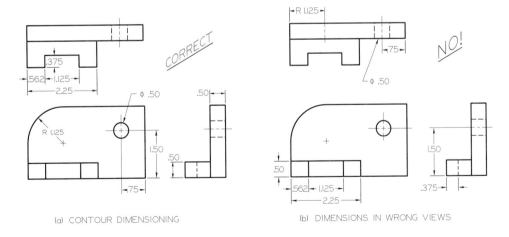

(a) CONTOUR DIMENSIONING

(b) DIMENSIONS IN WRONG VIEWS

Fig. 10-22. Contour dimensioning.

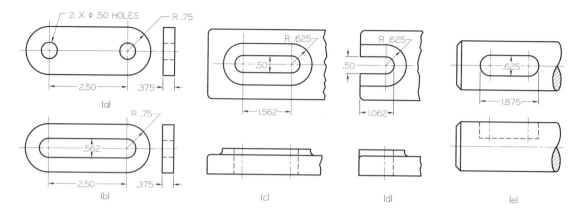

Fig. 10-23. Dimensioning rounded-end shapes.

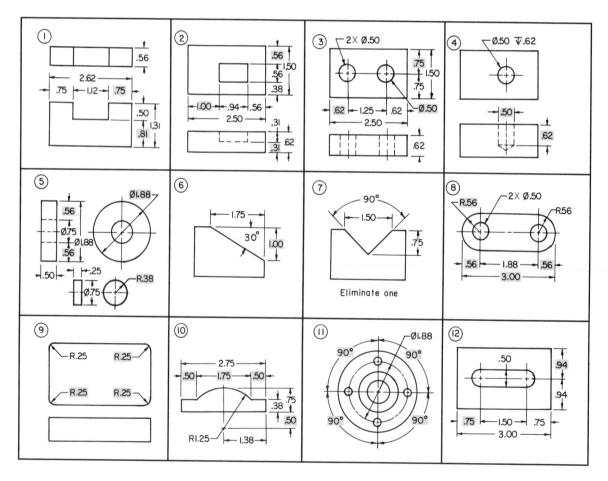

Fig. 10-24. Superfluous dimensions.

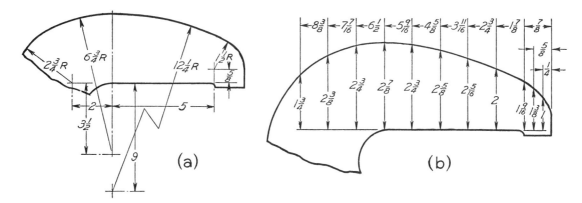

Fig. 10-25. Dimensioning curves.

ing workers read the drawing, they will naturally look for the size description close to the shape description. The meaning of a dimension is always clearest in the view where the corresponding shape is given. Violations are shown at (b). In general, avoid dimensioning to hidden lines where possible.

10.19 ROUNDED ENDS. Rounded-end shapes are dimensioned according to the manufacturing methods used, Fig. 10-23. At (a) and (b) the shapes, to be cast or to be cut from sheet or plate, are dimensioned as they would be laid out in the shop. This is done by giving the center-to-center distance and the radii of the ends. At (c) the pad on a casting with a milled slot is dimensioned from center to center conforming to the total travel of the milling cutter. At (d) the full length of the milled slot represents the total travel of the milling cutter. At (e) the method of dimensioning a Pratt and Whitney keyseat is shown. Here the width and the total length are given in conformity with the standard key sizes. In general, where great accuracy is required, the overall length of rounded-end shapes should be given. As a general rule, give diameters of slots rather than radii of the ends, (b) to (e).

10.20 SUPERFLUOUS DIMENSIONS. Before giving any dimension, ask yourself the question, "What is this dimension for?" Is the dimension really needed? Is it the best way to tell exactly how the part is to be made? In Fig. 10-24 are shown several cases of superfluous dimensioning. In practically every case additional and unnecessary dimensions are given because the drafter was not thinking of manufacturing requirements. Give dimensions in the most direct and simple way. Never give a dimension more than once in the same view or in different views. Avoid duplication. *If you cannot give a definite reason for a dimension, omit it.*

10.21 DIMENSIONING CURVES. If a continuous curve is made up of a series of circular arcs, the various radii are given, Fig. 10-25(a). Note that for extremely large radii, such as the radius $12\frac{1}{4}$, the center may be drawn closer to the arc as shown. In such cases, the main part of the dimension line is drawn toward the actual center.

If a curve is not made up of circular arcs or if the use of radii is not desired, the curve is dimensioned by a series of dimensions at regular intervals, (b). Note accumulation of values in the upper line of dimensions.

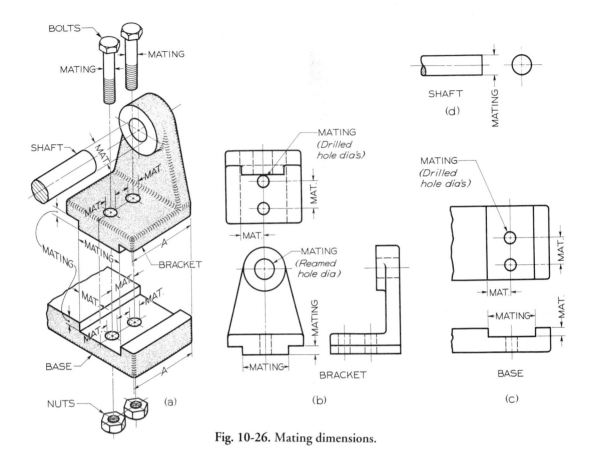

Fig. 10-26. Mating dimensions.

10.22 MATING DIMENSIONS. When two or more parts fit together, they are called *mating parts*. On two mating parts certain dimensions must correspond to make the parts fit together. These are called *mating dimensions*. For example, Fig. 10-26(a), a bracket fits into a base. Two bolts are used to fasten the parts together. The dimensions of the projection under the bracket must correspond to those on the slot in the base. Also, the diameters of the bolts must correspond to the sizes of the drilled holes in the bracket and in the base. The spacing of the drilled holes must correspond in the two mating parts. The diameter of the shaft must correspond to the reamed hole in the bracket.

The drawings of the principal parts are shown at (b) to (d). Mating dimensions are those that correspond in the separate drawings of the mating parts. They are needed for the accurate fitting of the parts. Dimensions that agree on two mating parts, such as dimension A in Fig. 10-26(a), but which are not essential for accurate fitting, are not mating dimensions.

Also, keep in mind that the actual *values* of two *corresponding* mating dimensions, while close, may not be exactly the same. For example, the width of the slot in the base may be $\frac{1}{32}''$ (.031'') or $\frac{1}{64}''$ (.016'') or several thousandths larger than the width of the projection of the base that fits in the slot. However, these are mating dimensions figured from a single basic width.

10.23 PATTERN AND MACHINE DIMENSIONS. The bracket in Fig. 10-26 was machined from a rough casting. Bear in mind that the patternmaker and the machinist both follow the same working drawing, which shows the completed object. Some of the dimensions are used by the patternmaker, some only by the machinist, and some by both, Fig. 10-27(a).

The patternmaker uses only those dimensions needed to make the wood pattern from which the rough workpiece is molded in sand, Sec. 11.2. Castings are not exactly uniform or accurate. Thus pattern dimensions are rough, to about the nearest $\frac{1}{16}''$ (.062"). They are given in whole numbers and fractions.

The machinist is interested only in the rough workpiece from the foundry and in the dimensions needed to machine the various holes and surfaces, Sec. 10.25. The completely dimensioned working drawing as used by both the patternmaker and the machinist is shown in Fig. 10-27(b).

10.24 NOTES. To supplement the ordinary dimensions, *notes* are often needed to supply additional information, usually in connection with required manufacturing processes. Notes should be brief. They should be carefully worded so as to permit only one interpretation. The wording and form of such notes are fairly well standardized. These standards should be carefully followed. *Notes should always be lettered horizontally on the sheet.* There are two kinds of notes, as follows:

General Notes — These give general information about the drawing as a whole, as:

FINISH ALL OVER
FILLETS & ROUNDS R.125
BREAK SHARP EDGES

General notes like these should be lettered where space is available in the lower portion of the sheet. In machine drawings the title strip will carry many general notes, such as

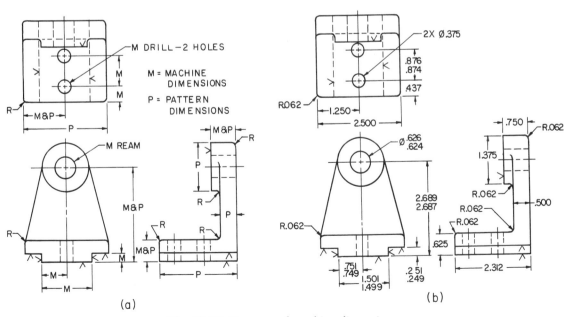

Fig. 10-27. Pattern and machine dimensions.

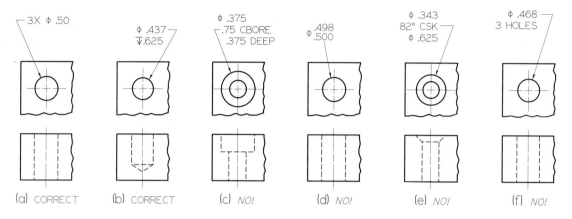

Fig. 10-28. Leaders to notes.

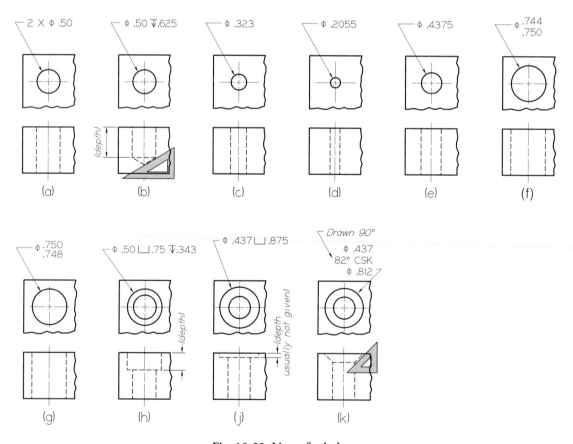

Fig. 10-29. Notes for holes.

material, number of pieces required, and general tolerances.

Local Notes — These apply to specific operations to be performed. They are connected by a leader, Fig. 10-2, to the appropriate point on the drawing. Typical examples are:

2 x Ø.25
.125 x 45° CHAMFER
MEDIUM KNURL

Always attach leaders at the front of the first word of a note, or after the last word as shown in Fig. 10-28(a) and (b). Never attach them to any other part of a note as shown from (c) to (f).

Methods of forming the most common types of holes are discussed in Secs. 11.11 and 11.12. Of these, the most common is the *drilled hole.* A hole cut entirely through a piece is called a *through hole,* Fig. 10-29(a). One that is not through a piece is called a *blind hole,* (b). The bottom of a drilled hole is conical. It is drawn with the 30° × 60° triangle, as shown. Drill depth, usually given in the drill note, *does not include the conical point.*

"Fractional-size" drills are available in drill sizes of $\frac{1}{16}''$ diameter to $3\frac{1}{2}''$ diameter. Diameter increments are $\frac{1}{64}''$ for drills of $\frac{1}{64}''$ to $1\frac{3}{4}''$ diameter, $\frac{1}{32}''$ for drills of $1\frac{3}{4}''$ to $2\frac{1}{4}''$ diameter, and $\frac{1}{16}''$ for drills $2\frac{1}{4}''$ to $3\frac{1}{2}''$ diameter. In addition, there are many "numbered" and "lettered" drills in the small-diameter range. See Table 3 in the Appendix.

It is common practice to give the decimal sizes for all diameters. This is recommended by the American National Standards Institute.

Typical notes for reamed, bored, counterbored, spotfaced, and countersunk holes are

TABLE 10-A. COMMONLY USED SYMBOLS

Symbol	Meaning
Ø	Diameter
⎿⏌	Counterbore
∨	Countersink
⊤↓	Depth
×	Number of Times or Places

shown from (f) to (k). Note that countersunk holes, (k), are drawn conventionally at 90° to approximate the angle of 82° (actual).

Today the preferred method for specifying written notes is by use of symbols. Some of the commonly used symbols are shown in Table 10-A.

The ANSI Standards on dimensioning practices recommend that, when practicable, a finished part should be defined without indicating any specific manufacturing methods. For example, only the diameter of a hole would be given, without indicating whether it should be drilled, reamed, etc. However, where such information is essential to the description of engineering or manufacturing requirements, it should be clearly indicated on the drawing or other accompanying documents.

Methods of producing holes in the shop are discussed in Secs. 11.11 and 11.12. Examples of many notes commonly used on drawings are shown in Fig. 10-30. See the Glossary for definitions of various manufacturing operations.

211

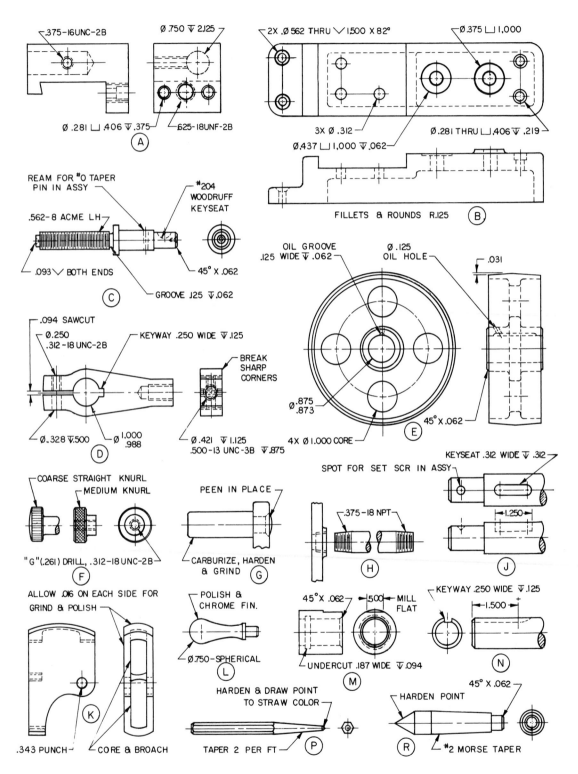

Fig. 10-30. Notes. (See also Fig. 10-29 for notes for holes and Fig. 15-23 for thread notes.)

10.25 DECIMAL DIMENSIONS. Some parts, even though machined, do not need to be more accurate than the nearest $\frac{1}{64}''$ (.016″). These parts include certain railroad car parts and agricultural machinery parts. These drawings are dimensioned entirely with whole numbers and common fractions. Other parts, such as certain parts of a milling machine or a sewing machine, must be so accurately made that every dimension must be given in decimals, Fig. 10-27(b). Most parts, however, require only certain key dimensions to be highly accurate while the others can be "rough." The average drawing, then, contains some rough dimensions in common fractions and some accurate dimensions in decimals. In architectural drawing or drawings of woodwork, all dimensions are given in whole numbers and fractions.

The complete decimal system is increasingly used, especially in the automotive and aircraft industries. In this system, all dimensions are stated in decimals, Fig. 10-31. Complete decimal dimensioning is also preferred by the American National Standards Institute. Thus, measurements formerly expressed in common fractions are expressed in decimals to two or more places, as 2.33 or 5.625, etc. See Table 20 in the Appendix. A two-place dimension is preferred for decimal dimensioning, as 2.12 or 7.62, but always in increments of .02″. When divided by two, the result is still a two-

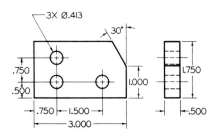

Fig. 10-31. Complete decimal dimensioning.

place decimal. For still more accurate dimensions, in which the micrometer and other accurate measuring tools are used, two-place decimal dimensions in other than .02″ increments and decimals to three, four, or more places are used, Sec. 11.10.

10.26 LIMIT DIMENSIONS. When a part must be replaced on your car, you can easily obtain and install a new part that will fit and run properly. This is possible by means of *interchangeable manufacturing*. This means making similar parts nearly enough alike so that any pair of mating parts will fit satisfactorily.

This sounds simple. Even though the parts are made in different, widely separated factories, it might be thought that all the workers have to do is make each measurement to exact size as shown by the dimension on the drawing. It is not that easy. It is impossible to make anything to *exact size*. Parts can be made to very close dimensions, even to a mil-

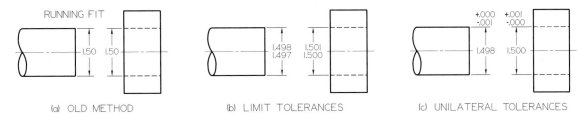

Fig. 10-32. Tolerance dimensions.

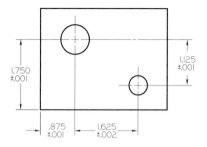

Fig. 10-33. Bilateral tolerances.

lionth of an inch, but the expense becomes increasingly greater as we require greater accuracy. It would be foolish to require airplane-engine accuracy for a part of a hand-operated meat grinder. The grinder would then cost so much that no one would buy it.

The old method of giving dimensions of two mating parts was to give the same dimensions on both parts in simple whole numbers and common fractions. A note indicated the kind of fit desired, as shown in Fig. 10-32(a). The machinist was then depended upon to produce the parts so that they would fit together properly. In the example, the machinist would make the hole close to $1\frac{1}{2}''$ in diameter. He or she would then make the shaft, say .003″ (three thousandths inch) less in diameter. It would not matter if the diame-

ter of the hole were several thousandths more or less than $1\frac{1}{2}''$. The machinist could always make the diameter of the shaft about .003″ less than that of the hole and thus obtain the desired "running fit." In quantity production, however, this would not work. The sizes would vary considerably. The parts would not be interchangeable; that is to say, not every shaft would fit in every hole.

To make sure that every part will fit properly in assembly, it is necessary to indicate a permissible amount of "oversize" or "undersize" for each dimension. Let us suppose that the hole in Fig. 10-32(b) may be 1.500″ or 1.501″ or anywhere in between. These figures are called *limits.* The difference between them is called *tolerance,* in this case .001″. Similarly, the shaft is assigned limits of 1.498″ and 1.497″. The largest shaft (1.498″) will fit in the smallest hole (1.500″) with a minimum air space of .002″. This minimum space is called *allowance.* Notice at (b) that for both the shaft and hole the larger figure is on top.

Another way to indicate the same tolerances is to give a decimal figure followed by plus and minus values, as in Fig. 10-32(c). The hole is given the basic size of 1.500″, with the provision that it may be as much as

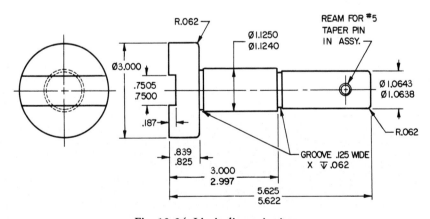

Fig. 10-34. Limit dimensioning.

.001″ larger. The shaft is given a basic size of 1.498″ (which leaves an allowance of .002″ between the parts) with the provision that it may be as much as .001″ smaller. Thus, the loosest fit would be 1.501″ minus 1.497″, or .004″, which is thought to be close enough.

Tolerances are often given in both directions (plus or minus) from a basic size when a variation in one direction is no more critical than in another, such as in locating the holes in the object shown in Fig. 10-33.

A typical drawing containing limit dimensions is given in Fig. 10-34. Note that the dimensions at the bottom are all given from one common surface. This is called *baseline dimensioning*. It is often used to avoid difficulties resulting from the accumulation of tolerances in a chain of dimensions. Never give a complete chain of tolerance dimensions and also an overall tolerance dimension. One dimension in the chain, or the overall dimension, should be omitted or changed to common fractions.

10.27 DEFINITIONS OF TERMS. It is vitally important that any tolerance dimensioning clearly and specifically indicate the sizes and locations of all features of an object or assembly. It is equally important that the terms used in tolerance dimensioning be clearly understood before any detailed study of this subject is attempted.

The following definitions have been adopted from the ANSI Standards (ANSI B4.1 and ANSI Y14.5):

Nominal Size — The designation that is used for the purpose of general identification. In Fig. 10-32(b) the nominal size of both hole and shaft is 1½″.

Basic Size — That size from which the limits of size are derived by the application of allowances and tolerances. In Fig. 10-32(b) the basic size is the decimal equivalent of the nominal size 1½″, or 1.500″.

Actual Size — An actual size is a measured size.

Design Size — That size from which the limits of size are derived by the application of tolerances. When there is no allowance, the design is the same as the basic size. In Fig. 10-32(b) the design size of the hole is 1.500″ (diameter of smallest hole), and that of the shaft is 1.498″ (diameter of largest shaft).

Tolerance — The total permissible variation of a size, or the difference between the limits of size. In Fig. 10-32(b) the tolerance on either the hole or shaft is the difference between the limits, or .001″.

Limits of Size — The maximum and minimum sizes. In Fig. 10-32(b) the limits for the hole are 1.500″ and 1.501″. For the shaft they are 1.498″ and 1.497″.

Allowance — An intentional difference between the maximum material limits of mating parts. It is a minimum clearance (positive allowance) or maximum interference (negative allowance) between mating parts. In Fig. 10-32(b) the allowance is the difference between the smallest hole, 1.500″, and the largest shaft, 1.498″, or + .002″, and thus a *clearance fit*. If the allowance is negative, the result will be an *interference fit*. See Sec. 10.28.

Unilateral Tolerance — A tolerance in which variation is permitted in only one direction from the design size. In Fig. 10-32(c) the design size of the hole is 1.500″. The tolerance .001″ is all in one direction — toward a larger size. The design size of the shaft is 1.498″. The tolerance .001″ is again all in one direction, but in this case toward a smaller size. In a unilateral tolerance either the plus or the minus value must be zero.

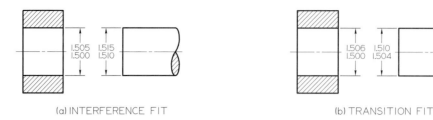

Fig. 10-35. Fits between cylindrical parts.

Bilateral Tolerance — A tolerance in which variation is permitted in both directions from the design size. In Fig. 10-33 the horizontal distance between the holes is 1.625 ± .002. The design size is 1.625″. The tolerance is .004″, or a variation of .002″ in each direction from the design size.

10.28 GENERAL TYPES OF FITS. A fit is the general term used to signify the range of tightness that may result from the application of a specific combination of allowances and tolerances in the design of mating parts. Fits are of three general types: *clearance, transition,* and *interference.*

A clearance fit is one having limits of size so given that a clearance always results when mating parts are assembled. In Fig. 10-32(b) the difference between the largest shaft, 1.498″, and the smallest hole, 1.500″, is .002″. This difference is the allowance. In this case it is positive, so the resulting fit is a clearance fit.

An interference fit is one having limits of size so given that an interference always results when mating parts are assembled. In Fig. 10-35(a) the smallest shaft is 1.510″. The largest hole is 1.505″, resulting in an interference of metal (negative allowance) of a minimum of .005″.

A transition fit is one having limits of size so given that either a clearance or an interference may result when mating parts are assembled. In Fig. 10-35(b) the smallest shaft, 1.504″, will fit in the largest hole, 1.506″, with a clearance of .002″. However, the largest shaft is 1.510″. The smallest hole is 1.500″, resulting in an interference of metal (negative allowance) of .010″.

10.29 BASIC HOLE AND BASIC SHAFT SYSTEMS. To specify the dimensions and tolerances of an internal and an external cylindrical surface so that they will fit together as desired, it is necessary to begin calculations by assuming either the minimum hole size or the maximum shaft size.

A *basic hole system* is a system of fits in which the design size of the hole is regarded as the basic size, and the allowance is applied to the shaft.

In Fig. 10-36(a) the minimum hole size, 1.500″, is considered as the basic size. An

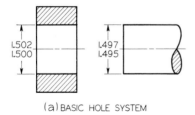

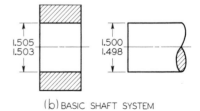

Fig. 10-36. Basic hole and basic shaft systems.

allowance of .003″ is assumed. It is *subtracted* from the basic hole size to obtain the diameter of the maximum shaft, 1.497″. A tolerance of .002″ is decided upon. This is applied to both the hole and the shaft to obtain the maximum hole of 1.502″ and the minimum shaft of 1.495″. The minimum clearance is the difference between the smallest hole, 1.500″, and the largest shaft, 1.497″, or .003″. The maximum clearance is the difference between the largest hole, 1.502″, and the smallest shaft, 1.495″, or .007″. To obtain the maximum shaft size of an interference fit, the desired allowance, or maximum interference, would be *added* to the basic hole size.

A *basic shaft system* is a system of fits in which the design size of the shaft is the basic size, and the allowance is applied to the hole.

In Fig. 10-36(b) the maximum shaft size, 1.500″, is considered as the basic size. An allowance of .003″ is decided upon and *added* to the basic shaft size to obtain the diameter of the minimum hole, 1.503″. A tolerance of .002″ is decided upon and applied to the hole and shaft to obtain the maximum hole, 1.505″, and the minimum shaft, 1.498″. The minimum clearance is the difference between the smallest hole, 1.503″, and the largest shaft, 1.500″, or .003″. The maximum clearance is the difference between the largest hole, 1.505″, and the smallest shaft, 1.498″, or .007″. To obtain the minimum hole size of an interference fit, the desired allowance would be *subtracted* from the basic shaft size.

To convert a basic hole size to a basic shaft size, subtract the allowance for a clearance fit, or add it for an interference fit. To convert a basic shaft size to a basic hole size, add the allowance for a clearance fit, or subtract it for an interference fit.

10.30 METRIC DIMENSIONING. In the United States, technical drawings have usually shown dimensions in feet (′) and inches (″). Often such drawings show dimensions only in the total number of inches (″).

In the early days of manufacturing, parts of an inch were shown as fractions, such as $\frac{1}{4}″$, $\frac{1}{8}″$, and $\frac{1}{32}″$. Dimensions on modern machine drawings, however, usually show parts of an inch in decimals, as in .250″, .125″, .062″, and .031″. Dimensions on architectural and structural drawings still use feet and inches, and fractions of an inch, because closer accuracy is relatively unimportant.

We usually think of the distance from one point to another in feet and inches. In contrast, the rest of the world uses the International System (SI), which originated in France in the 1790s.

In SI the meter (m) is the standard unit of length. A meter is divided into one thousand equal parts, called millimeters (mm). The metric unit of measurement most often used on technical drawings for the mechanical industries is the millimeter. To show long distances, as on maps, SI uses the kilometer (km), which is one thousand times as long as the meter. These units are related as follows:

1 mm = 1 millimeter ($\frac{1}{1000}$ of a meter)
1 m = 1 meter = 1 000 mm
1 km = 1 kilometer = 1 000 m = 1 000 000 mm

The International System (SI) uses spaces instead of commas with numbers having more than four numerals, as follows: 15 847, *not* 15,847; 648 034, *not* 648,034; and 4 368 952, *not* 4,368,952.

SI also uses a zero before a decimal quantity smaller than 1 as : 0.046, *not* .046; 0.333, *not* .333; and 0.789, *not* .789.

217

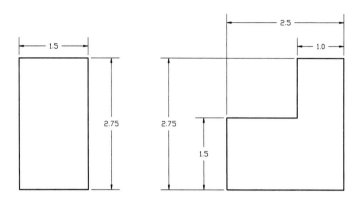

Fig. 10-37. Examples of CAD dimensioning.

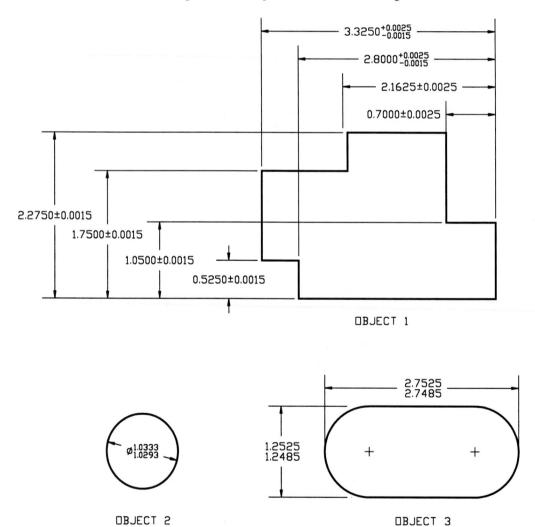

Fig. 10-38. Examples of tolerances in CAD dimensioning.

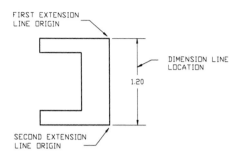

FIRST EXTENSION
LINE ORIGIN

DIMENSION LINE
LOCATION

1.20

SECOND EXTENSION
LINE ORIGIN

Fig. 10-39. Picking dimensions.

The United States is the last major country to put the International System to use. It may be some time before we use only metric dimensioning on technical drawings. In most cases the techniques of dimensioning practice in Chapter 10 apply as well to International System (SI) dimensioning as they do to fractional inch or decimal inch dimensioning.

Some companies use the *dual dimensioning* system, which shows both metric and decimal inch dimensions on the same drawing. The millimeter value is usually placed above the inch dimension, and the two are separated by a horizontal line.

When technical drawings are dimensioned in the units of only one system, a conversion table may be added as a reference to change the units from one system to the other; that is, millimeters to inches, or vice versa. For convenience, a conversion table is usually made for only the dimensions found on a particular drawing.

The metric equivalents for length are:

U.S. to SI Metric
1 inch = 25.4 millimeters = 2.540 centimeters
1 foot = 0.304 meter
1 yard = 0.914 meter
1 mile = 1.609 kilometers

SI Metric to U.S.
1 millimeter = .039 inch
1 centimeter = .394 inch

1 meter = 3.281 feet or 1.094 yards
1 kilometer = .621 mile

Table 20 in the Appendix shows millimeter equivalents for fractional and decimal parts of an inch.

Although it has often been necessary to convert from one system to another, the United States will use metric units more and more in the future. Learn to think meters and millimeters when technical drawings call for metric dimensioning. That will be easier than converting back and forth from feet and inches.

10.31 DIMENSIONING IN CAD. The use of CAD has made it easier to place dimensions on a drawing quickly and accurately. Any drawing is easily dimensioned by selecting two lines or an arc or circle. CAD software allows the user to choose the mode of dimensioning—metric, architectural, scientific, decimal, or dual—and the exact placement of the dimension, Fig. 10-37.

To specify the mode and style of dimensioning, the CAD user sets the dimensioning variables. For example, the variables could be set so that the upper and lower limits for each dimension are shown. Tolerances, geometric tolerancing, and architectural style with ticks are also allowed in most CAD programs. The number of decimal places desired, as well as a vast array of specific conventions, can be defined by the CAD operator, Fig. 10-38.

For linear dimensions, the user selects the origins of two extension lines and then picks the location of the dimension. The software automatically generates the extension lines, dimension lines, and arrowheads, Fig. 10-39.

Dimensions for angles, arcs, and circles are also quickly produced. In the case of angular dimensions, the user picks the two lines repre-

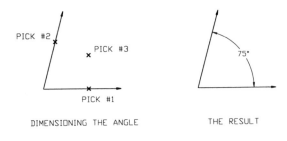

Fig. 10-40. Angular dimensions.

senting the angle to be dimensioned. The user then picks the location of the dimension. The software places the arcs, arrowheads, and degree symbols in the desired location, Fig. 10-40.

Arcs and circles can also be picked, and the location of the dimension is then chosen by the user. As with manual drafting, arcs are dimensioned with a radius, and circles are dimensioned with a diameter symbol.

Leaders can be used to point to the place where a dimension or note applies.

Figure 10-37 shows a drawing with completed dimensions. Changes can still be made at this stage. For example, the user could specify a different text style or switch to metric dimensions. If the user stretches or shrinks the length of a line, the CAD system makes the corresponding change in the dimension figure.

PROBLEMS

Problems given on the following pages are presented to afford practice in elementary dimensioning. They are to be drawn. Then the full-size dimensions are to be added, as shown in Figs. 10-41 and 10-42. Various lines have been omitted. The student is to add these to make a complete drawing.

For more practice in dimensioning, use the problems assigned from any previous chapter, especially Chapter 8.

Figures 10-43, 10-44, and 10-45, which follow, consist of drawings in which certain lines are missing and in which dimensions are omitted. They are to be drawn mechanically, as assigned by your instructor. Add all missing lines and dimension fully, as shown in Fig. 10-42, using Layout C in the Appendix. These problems are printed one-fourth size. Make your drawings full size by stepping off each measurement four times with the dividers. Add dimensions and notes to agree with your full-size drawings. Use heavier visible lines in the finished drawing, Fig. 10-42. Move all titles to title strips.

If assigned by the instructor, construct drawings using decimal-inch or metric scales by converting given fractional dimensions to decimal-inch or metric equivalents. Refer to Table 20 in the Appendix.

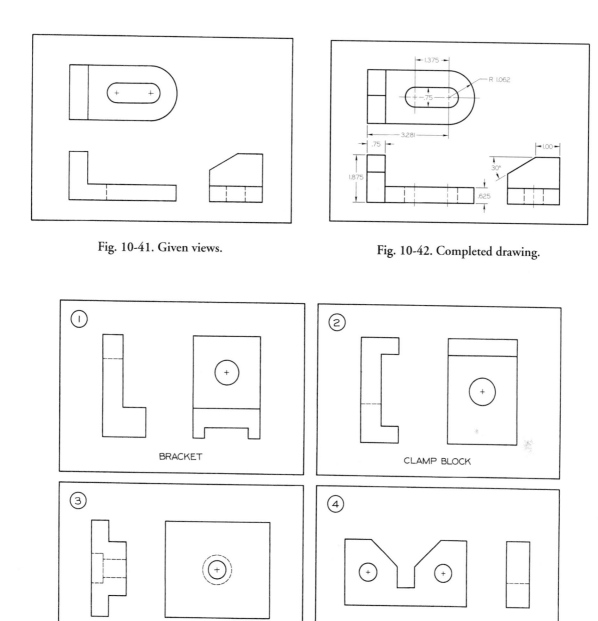

Fig. 10-41. Given views.

Fig. 10-42. Completed drawing.

① BRACKET

② CLAMP BLOCK

③ SLIDE

④ HOLDER

Fig. 10-43. Dimensioning problems. Redraw with instruments. Add missing lines and dimension fully.

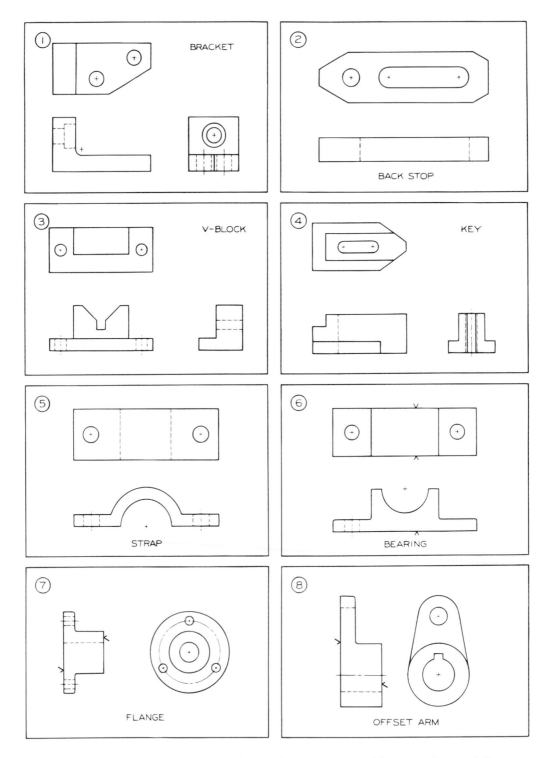

Fig. 10-44. Dimensioning problems. Redraw with instruments. Add missing lines and dimension fully. See instructions on page 220.

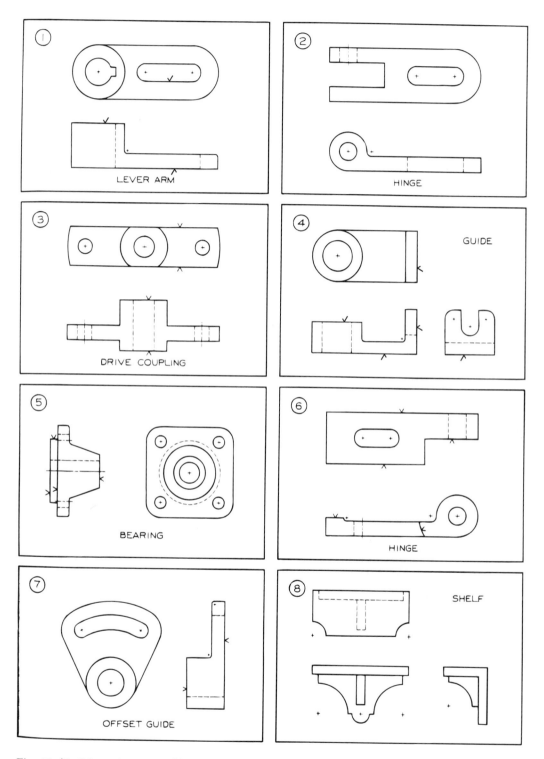

Fig. 10-45. Dimensioning problems. Redraw with instruments. Add missing lines and dimension fully. See instructions on page 220.

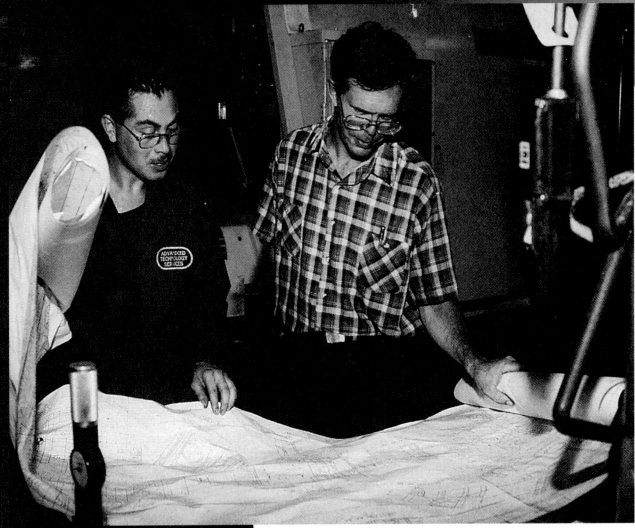

Arnold & Brown

CHAPTER 11

IMPORTANT TERMS

casting	counterboring
forging	spotfacing
welding	countersinking
pattern drawing	shaper
pattern	milling machine
shrink rule	grinding machine
lathe	surface broach
turning	heat treating
knurling	rapid prototyping
drilling	
reaming	

OBJECTIVES

After studying this chapter, you should be able to:

- Identify the three stages in the manufacture of a machine part.
- Identify the processes by which parts are manufactured.
- Identify the stages in the casting process.
- List the measuring instruments used in a manufacturing facility.

CHAPTER 11

Manufacturing Processes

This chapter discusses the use of technical drawings in manufacturing. Technical drawings provide manufacturers with the information they need to make a part. The manufacturing processes discussed include casting, forging, welding, shaping, milling, grinding, turning on the lathe, and broaching.

There are many careers in manufacturing. For example, the industrial engineer is involved with industrial manufacturing processes, assembly lines, material inventory control, facility design, and management of manufacturing sites. Industrial engineers decide how workers can best interact with machines. They also determine how to make a manufacturing plant most efficient by careful choice of factors such as placement of assembly lines, loading docks, storage and flow of materials, lighting and other employee comfort issues. Flow charts and diagrams, floor layouts, and other more detailed plans of manufacturing facilities are among the types of drawings the industrial engineer will make and use.

CAREER LINK

To learn more about careers in industrial engineering, contact JETS-Guidance, 1420 King Street, Suite 405, Alexandria, VA 22314 or visit their web site at *www.asee.org/jets*

Fig. 11-1. The working drawing is a plan that shows how the product is to be made.

11.1 MANUFACTURING PROCESSES. The information an individual needs to produce a part in a manufacturing department is most often in the form of a print of a working drawing, Fig. 11-1.

Most modern manufacturing facilities receive this information in the form of numerical data generated by CAD. This information may be transmitted on computer diskettes, magnetic tapes, or via direct tie-in to the computer through modems or hard wiring. The production drawing provides all of the necessary dimensional details concerning the finished part. It also gives complete instructions to the manufacturer as to how the object is to be made. It therefore follows that the designer and drafter of the part must have at least a working knowledge of the various processes that could be used to produce

the part. The best training is, of course, extensive experience with manufacturing processes. The purpose of this chapter is to provide the basic knowledge of the various manufacturing processes needed by the drafter and designer.

In general, the production of a machine part consists of three main stages: *rough forming, finishing,* and *assembling.* There are three important types of rough forming. These are *casting* (in a *foundry*), *forging* (in a *forge shop*), and *welding* (in a *welding shop*).

Finishing (such as the final sizing of holes and other machined features) is largely done in the *machine shop.* In the *assembly shop,* the various parts are put together to form the completed machine. In many cases, this involves further machining operations that are best done in assembly.

In general, the part drawing shows the part in its completed state. Each of the different manufacturing shops "picks off" from the drawing the information it needs. In the case of complicated castings, however, a special *pattern drawing* may be made. This drawing gives only the information needed in the pattern shop. Due to the high initial cost of forging dies, it is common to make special drawings for the forge shop, Fig. 11-10.

11.2 SAND CASTING. A typical drawing as received in the shops is shown in Fig. 11-2. To produce the part shown, a rough *sand casting* must be made, as shown in Fig.11-3(a). The part is then *machined,* as shown at (b). The casting is made by pouring molten metal (iron, steel, aluminum, brass, or some other metal) into a cavity in damp sand. The metal is allowed to cool until it hardens. Then it is removed from the sand. The cavity is made by placing a model of the object, called a *pattern,* in the sand and then withdrawing it. This leaves an imprint of the model in the sand.

Patterns are usually constructed from a durable and dimensionally stable wood. (Basswood and mahogany are often used.) In some cases, duplicate patterns are made of metal or from casting resins with very low shrink characteristics. A wood pattern of the bearing in Fig. 11-2 is shown in Fig. 11-3(c).

Since shrinkage occurs when metals cool, patterns are made slightly oversize. The pat-

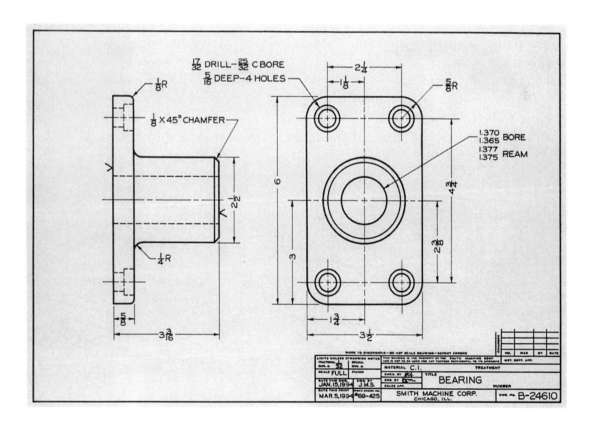

Fig. 11-2. A detail working drawing.

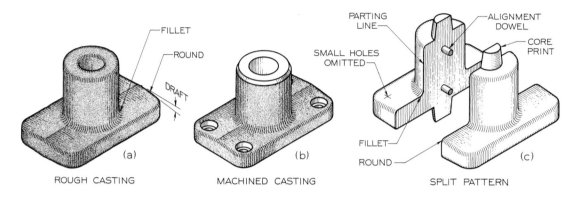

Fig. 11-3. A casting and a pattern.

ternmaker uses a *shrink rule.* The units on a shrink rule are slightly oversize. Since various metals shrink different amounts, a different rule is used for each metal. For example, cast iron shrinks $\frac{1}{8}''$ (.125″) per foot. The detail drawing, Fig. 11-2, shows the object in its final state and does not allow for shrinkage. This is taken care of entirely in the pattern shop by making the pattern oversize.

Draft, or relief angle, is the taper given to a pattern to permit it to be easily withdrawn from the sand without damaging the shape of the mold. Draft is taken care of by the patternmaker. It is not shown on the drawing unless it is also a feature of the design. Only a slight draft on each side of the flat base is needed for this object, Fig. 11-3(a).

The patternmaker must also make the pattern oversize in certain places to provide additional metal for each surface that is to be machined. For example, the bottom surface of the base of the casting of Fig. 11-2 is to be *finished* or machined. The patternmaker must provide extra thickness to the bottom of the

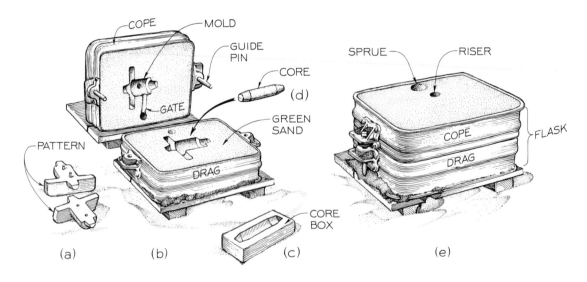

Fig. 11-4. Sand molding.

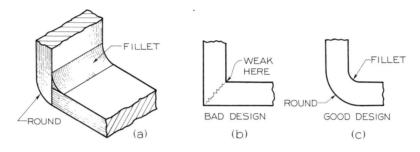

Fig. 11-5. Fillets and rounds.

base so that there will be from 1/16″ (.062″) to 1/8″ (.125″) of metal to be removed. For details on proper use of finish marks, see Sec. 10.7.

The sand is contained in a two-part box called a *flask,* Fig. 11-4(e). The upper part is the *cope.* The lower part is the *drag.* The mold is full when molten metal can be seen in the *riser.* The riser also allows hot gases to escape. As cooling occurs and the metal shrinks, the extra metal in the riser flows back into the mold. For more complicated work, one or more intermediate boxes (*cheeks*) are used between the cope and drag. Patterns are often split, Fig. 11-3(c), so that one half can be placed in the cope and the other in the drag. The molds in the cope and drag have been formed by ramming sand around each half of the pattern, Fig. 11-4(b). A *sprue stick,* or round peg, is placed in position during the ramming. It is then removed to leave a hole through which molten metal is poured. A trench (the *gate*) is formed to lead from the hole to the casting. The sand packed around the pattern is called green sand because it is not dried, baked, or cured before the mold is used.

Green sand is often not strong enough to form some shapes. In such cases *dry sand cores* are used. Dry sand cores are made by ramming a prepared mixture of sand and a binding substance into a *core box,* Fig. 11-4(c).

The core is then removed and baked in a core oven to make it hard. The most common use of a core is to extend it through a casting to form a *cored hole.* The *core prints* of the pattern, Fig. 11-3(c), form openings in the sand to support the ends of the core in Fig. 11-4(b). When the metal is poured into the mold, it flows around the core, leaving a hole in the casting. When the casting has cooled, it is removed from the sand. The dry sand core is broken out. The casting is then cleaned. Any projecting fins are ground off.

The rough casting is sent to the machine shop. There the cored hole will be enlarged by *boring* and *reaming.* The top and bottom surfaces will be machined. The small holes will be drilled and counterbored, Fig. 11-3(b).

11.3 FILLETS AND ROUNDS. As shown in Figs. 11-3(a) and 11-5, a rounded inside corner on a casting is called a *fillet.* A rounded outside corner is called a *round.* Rounds are formed on a pattern by simply rounding the edges with a plane or sandpaper. Fillets are formed by gluing on preformed leather or wooden strips. They can also be made by forming wax into the patterns sharp corners using a round-nosed hot tool similar to a soldering iron. Sharp corners should be avoided on a casting. Inside corners in particular should have fillets as large as possible. It is difficult to obtain such sharp corners in the casting process. Also, such sharp corners are weak. They are likely to yield cracks and other failures in the cast part.

All fillets and rounds are shown on the drawing, Fig. 11-2. They are drawn to scale with the use of a compass for large fillets or with the aid of a circle template for smaller features. For methods of dimensioning, see Sec. 10.10.

11.4 RUNOUTS. Various ways of showing intersecting and tangent surfaces were reviewed in Sec. 8.17. If fillets and rounds are present, an exact representation is not only difficult but usually unnecessary. Conventional methods should be used in such cases, as shown in Fig. 11-6. In the cases shown from (a) to (c), the points of tangency must be found in the top view. They must then be projected to the other views where *runouts* are carefully drawn freehand, or with the aid of a circle template. At (d) the inter-

secting member is elliptical, and the runouts are turned inward. At (e) and (f) the runouts depend upon the shape of the top of the triangular rib. If runouts are large, as at (d), an irregular curve should be used. Otherwise they should be drawn freehand.

11.5 CONVENTIONAL EDGES. It is often necessary to draw lines representing rounded edges when, strictly speaking, no edges exist. For example, the upper top views in Fig. 11-7 would in each case be completely blank as shown, if lines were not used to represent rounded edges. The drawings would then be misleading. Therefore, if an edge has only a small radius, a line should be shown. If the radius is large, as X in (b), no line should be shown. It is best to follow this rule: *Draw lines for rounded edges whenever such lines make the drawing clearer.*

11.6 FORGING. A *forging* is produced by hammering heated bars or billets of metal

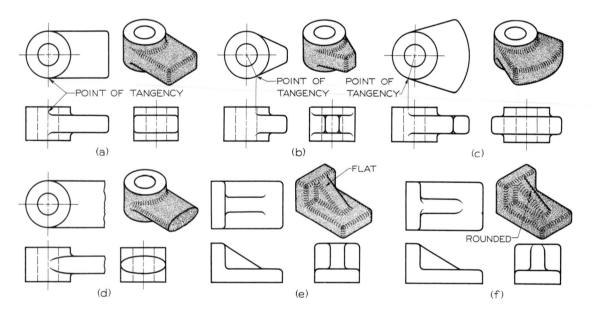

Fig. 11-6. Runouts.

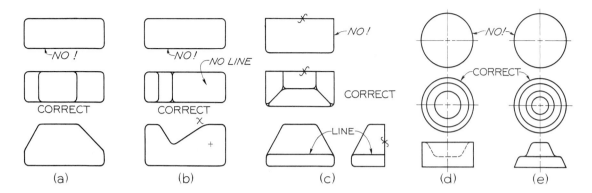

Fig. 11-7. Conventional edges.

between *dies*. Forging presses, or hammers, are used to exert the tremendous pressure that forces the plastic metal into the dies. If the pressure used during the forming process was exerted by the impact of heavy blows, the process is referred to as *drop forging*. If the pressure is from a slow squeezing action, then it is referred to as *press forging*. The forging dies have machined cavities and projections of the shapes desired.

The dies for forging an automobile connecting rod are shown in Fig. 11-8. The several stages of forging are shown in Fig. 11-9.

Interstate Drop Forge Co.

Fig. 11-8. Forging dies.

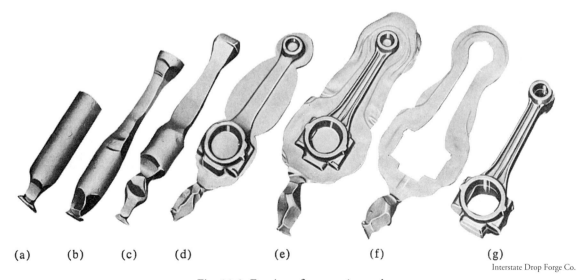

(a)   (b)   (c)   (d)   (e)   (f)   (g)

Interstate Drop Forge Co.

Fig. 11-9. Forging of connecting rod.

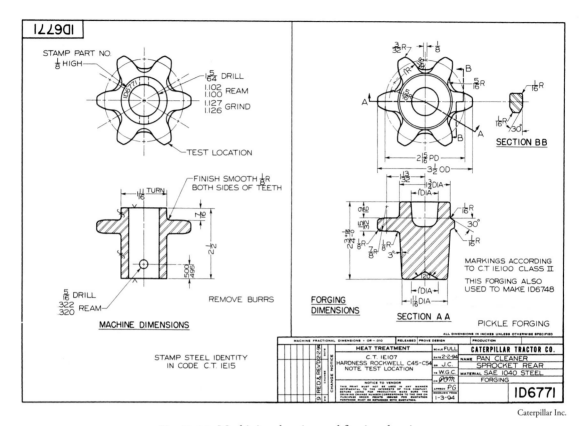

Fig. 11-10. Machining drawing and forging drawing.

The rough stock, shown at (a), is gradually formed through the shapes shown to (e), where large fins (called *flash*) are formed. These must be trimmed off as shown at (f). The final forging, shown at (g), is ready for machining.

The advantages of forgings over sand castings are that forgings are much stronger, tougher, and less brittle.

The working drawing as it comes to the forge shop may show the finished machine part. In this case, the diemaker provides the necessary forging and machining allowances. Separate forging drawings are often made. These show the rough forging with only the dimensions needed in the forge shop. As in sand casting, a draft (relief angle) must be provided for forgings so they will withdraw easily from the dies. In most cases, this draft is from 5° to 7°. It may be less under special conditions. Separate machining and forging drawings of a pan cleaner sprocket are shown in Fig. 11-10.

11.7 DIE CASTING. *Die casting* is the fastest of the casting processes. It is often used where rapid production and economy are essential. Thus, it is used in making automobile carburetors, door handles, and toys. Die castings are formed by forcing molten metal, usually a zinc alloy, into cavities between metal dies. Die castings are much more accurate than sand castings. They usually require little or no machining.

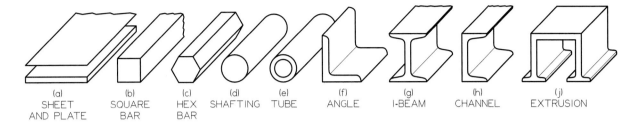

Fig. 11-11. Common stock forms.

11.8 STOCK FORMS. Many standardized structural shapes are available in stock sizes for use in the fabrication of parts or structures. Among these are bars of various shapes, flat stock, rolled structural shapes, and extrusions, Fig. 11-11. Modern graphite composite materials can be vacuum-formed into these shapes. The tube shown at (e) is often round, square, or rectangular. It can be formed by seamless extrusion or by rolling flat stock and welding the seam.

Many of the smaller machine parts, especially screw-machine work (which requires the toughness and strength of rolled metal), are machined altogether from these stock forms. Some examples are shown in Fig. 16-11. On drawings of such parts, finish marks (or "finish all over" notes) are usually omitted. The parts are understood to be finished.

11.9 WELDING. *Welding* is the fusion or joining of two pieces of metal by means of heat, with or without the application of pressure. *Welded* structures are usually built up from stock forms, such as plates, tubing, and angles. These parts are cut to shape, assembled, and welded together. Often, heat-treating and machining operations are performed after the parts are joined. Table 14 in the Appendix shows various welding symbols and processes. Welding drawings are discussed in Sec. 16.16.

11.10 MEASUREMENTS. *The machinists steel rule,* or *scale,* is a commonly used measuring tool in the shop, Fig 11-12(a). The smallest division on one scale of this rule is $\frac{1}{64}''$ (.016"). Such a scale is used for common fractional dimensions. Many machinists rules also have a decimal scale with the smallest division of .016". This is used for dimensions given on the drawing using the decimal system. The *outside spring calipers* are used to check the nominal size of outside diameters, (b). The measurement is then read off on the steel scale, (c). The *inside spring calipers* are likewise used to measure nominal inside diameters, (d) and (e). The outside calipers can also be used to measure the nominal distance between holes (center to center distance), as shown at (f). One use for the combination square is to measure the nominal height of an object, as shown at (g).

For dimensions that require more precise measurements, the *vernier caliper,* (h) and (j), or the *micrometer caliper,* (k), may be used. By means of the *vernier* graduations, measurements may be made to four decimal places, or $\frac{1}{10,000}''$.

Computerized measuring devices have broadened the range of accuracy previously attainable. Figure 11-13 illustrates an ultra-precision electronic digital readout micrometer and caliper. This contains integral microprocessors. The hand-held printer/recorder provides a hard-copy output of measurements. The printer also calculates and lists statistical values of the measurements. These assist machinists in maintaining a high level of quality.

Note that diameters are easily measured, as shown in 11-12(b), (d), (h), and (k). Diameters of cylinders, and not radii, should

therefore be given on drawings, as shown in Fig. 11-10. See Sec. 10.15 for correct techniques of dimensioning cylinders.

11.11 THE LATHE. Cylinders, cones, and other rounded shapes are often machined on a *lathe* (pronounced layth). A long piece of stock is usually held between centers. The left end of the stock is fastened securely to a *face plate* by

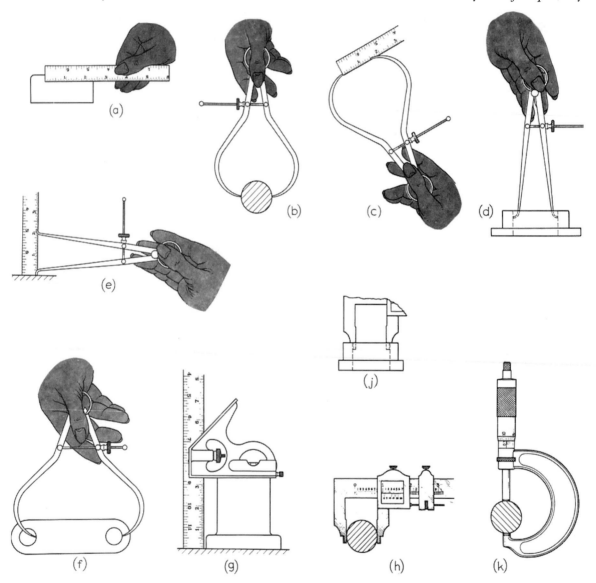

Fig. 11-12. Taking measurements.

Fowler Computerized Measuring System (Digitrix II + Printer/Calculator + Max-Cal).

Fred V. Fowler Co., Inc.

Fig. 11-13. Computerized measurement system.

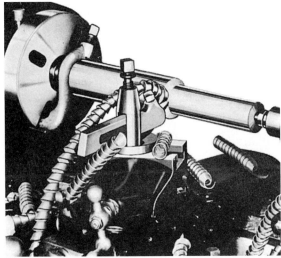

South Bend Lathe Works

Fig. 11-14. Turning.

means of a "dog." The right end turns freely, Fig. 11-14. A short piece of stock is usually

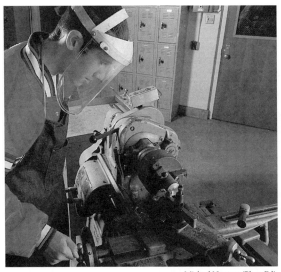

Michael Newman/PhotoEdit

Fig. 11-15. Facing.

South Bend Lathe Works

Fig. 11-16. Drilling on the lathe.

held in a *chuck* (essentially a revolving vise), Fig. 11-15. The cutting of an external cylindrical surface is called *turning*, Fig. 11-14. The cutting of a flat surface is called *facing*, Fig. 11-15. Other common operations performed on the lathe include *drilling*, Fig. 11-16, *boring*, Fig. 11-17, and *reaming*, Fig. 11-18. In addition, threads are often cut on the lathe. In all of these, the stock rotates and the cutting tool is fed into or along the stock as required.

235

Fig. 11-17. Boring on the lathe.

Fig. 11-18. Reaming on the lathe.

Knurling, Fig. 11-19, is also done on the lathe by using a knurling tool in the tool holder and forcing it against the revolving work. The result is a roughened surface com-

Fig. 11-19. Knurling on the lathe.

posed of crossing diagonal grooves or of parallel grooves lengthwise of the piece. These are common on thumbscrews and handles of various kinds to provide a better grip. Knurls may be fine, medium, or coarse, depending upon the roughness required.

Some of the most common lathe operations performed by a tool bit held in a tool holder or boring bar are shown in Fig. 11-20.

11.12 How Finished Holes Are Made. The *drilling* of a common drilled hole on a *drilling machine,* or *drill press,* is shown in Fig. 11-21. The revolving drill, Fig. 11-22(a), is fed into the work, which remains stationary. The drill is also used in the lathe, Fig. 11-16, the milling machine, and other machines. The drill does not produce a very accurate hole

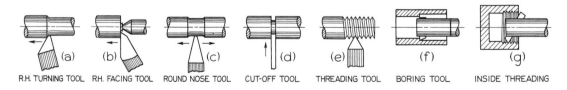

(a) R.H. TURNING TOOL (b) R.H. FACING TOOL (c) ROUND NOSE TOOL (d) CUT-OFF TOOL (e) THREADING TOOL (f) BORING TOOL (g) INSIDE THREADING

Fig. 11-20. Lathe operations.

either in roundness or in straightness. For greater accuracy, the drill is followed by *boring,* Figs. 11-17 and 11-22(b), or by *reaming,* Figs. 11-18 and 11-22(c), (d), and (e). Boring enlarges the drilled hole slightly and makes it rounder and straighter. Reaming enlarges the drilled or bored hole slightly and improves its surface quality. Good practice is to drill, bore, and then ream to produce an accurate hole.

Counterboring is the cutting of an enlarged cylindrical portion at the top of a hole, Fig. 11-22(f). This is usually done to receive the head of a fillister-head or a socket-head screw, Figs. 15-31 to 15-33.

Spotfacing is simply the cutting of a shallow counterbore, usually about $\frac{1}{16}''$ (.062″) deep or

Michael Newman/PhotoEdit

Fig. 11-21. Drilling.

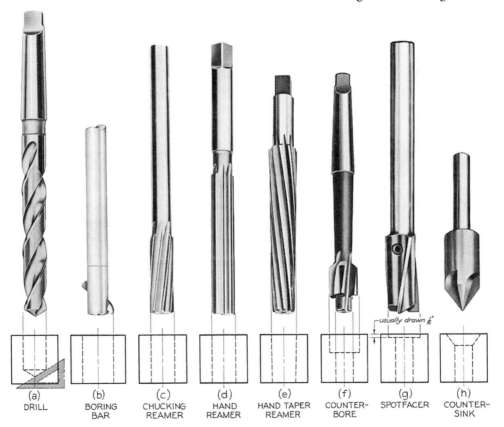

Fig. 11-22. Types of holes.

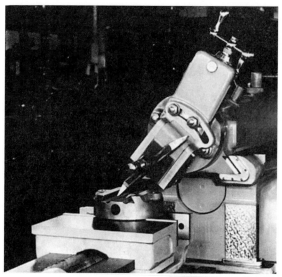

Cincinnati Shaper Co.

Fig. 11-23. The shaper.

Cincinnati Milling Machine Co.

Fig. 11-24. The milling machine.

deep enough to get under the "scale" on the rough surface, Fig. 11-22(g), or to finish off the top of a boss to form a bearing surface. The depth of a spotface is usually not indicated in the note on the drawing, but is left to the manufacturer. It is commonly drawn $\frac{1}{16}''$ (.062″) deep. A spotface provides an accurate bearing surface for the underside of a bolt or screw head.

Countersinking is the cutting of an enlarged conical portion at one end of a hole, Fig. 11-22(h). It is usually done to receive the head of a flat-head screw, Figs. 15-31 to 15-33. To simplify drafting, the countersink is usually drawn with a 90° included angle, even though the note calls for, say, 80° or 82°.

Tapping is the threading of a small hole with one or more *taps,* Fig. 15-18.

11.13 THE SHAPER. In the *shaper,* Fig. 11-23, the stock is held in a vise. A single-pointed, non-rotating cutting tool similar to

that used in the lathe cuts as it is forced forward in a straight line past the stationary work. In the figure, the tool is moving forward on a cutting stroke. The work remains stationary as the tool cuts. The tool then returns to the starting position to take another cut. Then the work is "fed," or moved slightly to the side. This places a fresh portion of the work in the path of the tool for the next cut.

11.14 THE MILLING MACHINE. In the *milling machine* the work is fastened to the table by means of clamps or a vise. It is then fed into a rotating milling cutter, Fig. 11-24. By using cutters of many different shapes, a wide variety of cuts is possible on the milling machine. Some of the most common milling operations are shown in Fig. 11-25.

The accuracy of mills and many other types of machine tools is dependent on the degree of accuracy with which the machine is set up and maintained. A laser-based mea-

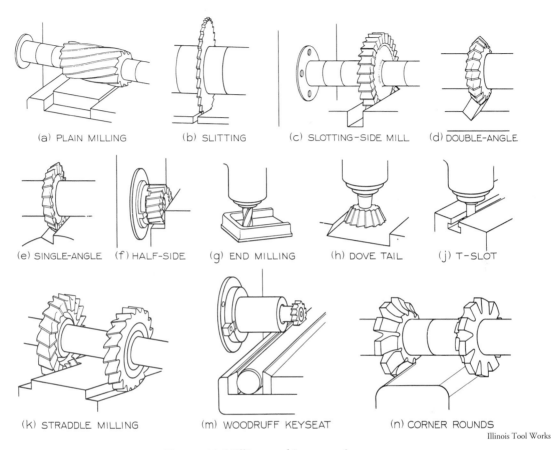

(a) PLAIN MILLING   (b) SLITTING   (c) SLOTTING–SIDE MILL   (d) DOUBLE-ANGLE

(e) SINGLE-ANGLE   (f) HALF-SIDE   (g) END MILLING   (h) DOVE TAIL   (j) T–SLOT

(k) STRADDLE MILLING   (m) WOODRUFF KEYSEAT   (n) CORNER ROUNDS

Illinois Tool Works

Fig. 11-25. Milling machine operations.

surement system, Fig. 11-26, is used in the calibration of the positioning accuracy of numerically controlled machine tools such as milling machines.

11.15 THE GRINDING MACHINE. The *grinding machine* removes a small amount of metal to bring the work to a very fine and accurate finish. In *surface grinding,* Fig. 11-27, the work is moved past the revolving grinding wheel. The operator can dial into the electronic memory the precise amount of metal to be removed during each of the grinding passes.

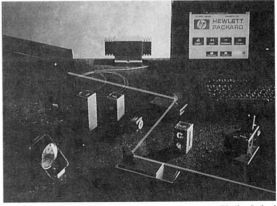

Hewlett-Packard

Fig. 11-26. Laser machine tool calibration system.

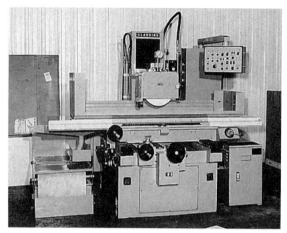

Fig. 11-27. Electronic sequence-controlled surface grinder.

The machine automatically compensates for the loss of abrasive on the grinding wheel.

In *cylindrical grinding*, the work revolves slowly on centers and is moved past the grinding wheel. On a drawing, a surface to be ground is indicated with a letter G on the edge view of the surface or by a note GRIND, Fig. 16-10.

11.16 BROACHING. The *broaching machine* uses a long cutting tool called a *broach*. The broach has a series of teeth that gradually increase in size. The broach is forced through a hole or over a surface to produce a desired shape. Thus, a drilled hole may be changed to a triangular, square, hexagonal, or some other shape, Fig. 11-28. A surface may be machined with a flat *surface broach*. As the broach is forced through or across the work, each succeeding tooth bites deeper and deeper until the final teeth form the required hole or surface.

11.17 HEAT-TREATING. The process of changing the properties of metals by heating and cooling is called *heat-treating*. *Annealing* and *normalizing* are generally used to soften metal. They involve heating followed by slow cooling. *Hardening* requires heating and then rapid cooling (quenching) in oil, water, or other substances. There are many other kinds of heat treatment, such as *tempering*, *case-hardening*, and *carburizing*.

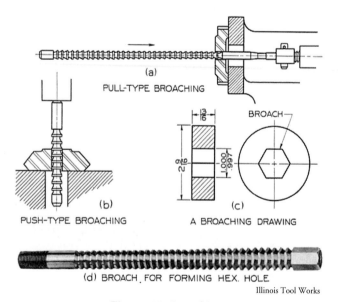

Fig. 11-28. Broaching.

11.18 MANUFACTURING NOTES. The various operations described in this chapter are specified on the drawing in the form of notes. For a complete discussion of notes, see Sec. 10.24. A large variety of notes are illustrated in Figs. 10-29 and 10-30.

11-19 AUTOMATION. *Automation* is the term applied to systems of automatic machines and processes. These machines and processes are essentially the same as previously discussed. However, they also have mechanisms to control the sequence of operations, movement of tools, flow, and so forth. Operator interaction is seldom required once the equipment has been "set up."

Most of these machines contain built-in computers to control all their functions, Fig. 11-27. Industrial robots, Fig. 11-29, are often used to take the place of human operators for the tedious load and unload cycles of the machine.

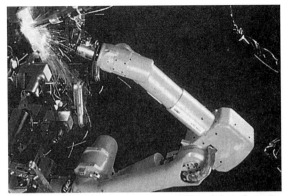

GMFanuc Robotics Corporation

Fig. 11-29. An industrial robot.

Single machining centers similar to the one shown in Fig. 11-30 eliminate the need for multiple shaping, planing, and boring machines. These centers also do away with the need for moving machined parts from machine to machine. The automatic tool changer on this machine stores 48 different cutters weighing up to 120 pounds each.

Ingersoll Milling Machine Co.

Fig. 11-30. Vertical and horizontal turning center.

11.20 PLASTICS PROCESSING. The plastics industry represents one of the major manufacturing segments. The two main families of plastics are known as thermosetting and thermoplastic. Thermosetting plastics, as their name implies, will take a set when molded. They will not soften when reheated. Thermoplastics, however, will soften whenever heat is reapplied.

Typical plastic processing operations include:

- *Extrusion.* Used in the manufacture of pipe, profiles, and film for bags.
- *Blow Molding.* Used in the production of bottles, automotive ductwork, hollow toy components, and door panels.
- *Injection Molding.* Used to manufacture products such as housings for electronic implements (toys, games, household appliances, tools), automotive components, food storage containers, components for medical applications (instrumentation and replacement parts), and a variety of other items, Fig. 11-31.
- *Thermoforming.* Used primarily in the manufacture of thin-walled packages for the food industry (egg cartons, bakery and cookie packages), and cosmetic packaging and displays. Heavier gauge material is also thermoformed into industrial components such as liners and door facings for refrigerators.

11.21 CAD IN MANUFACTURING. During the design phase of product development, data from a 3D CAD drawing can be used to directly generate a *prototype,* or actual physical model, of the object. The process is called *rapid prototyping.* Stereolithography is one form of rapid prototyping. The object is built up from a series of thin cross sections made of a light-sensitive plastic resin. The resin is hardened by exposure to a laser beam. Another method uses a laser beam to fuse together powdered metal to form a very strong sample part.

Van Dorn Demag Corporation

Fig. 11-31. Injection molding machine.

The X, Y, and Z coordinates of a CAD drawing can be used to direct the movements of a variety of machines. These include milling machines, drill presses, punches, lathes, and grinders. The process of developing designs on a CAD system and producing them on computerized machines is called CAD/CAM.

Converting the vector-based CAD drawing to a specific tool movement is called *computer numerical control,* or CNC. The CAD drawings must be extremely accurate. All lines must meet at precise points. Such accuracy is necessary because the design geometry is used to develop the tool path for the machine. If breaks are present in the design, the tool will not continue on the proper path.

Another form of manufacturing is *computer-integrated manufacturing,* or CIM. In CIM, each part is designed on a CAD system. It is then sent to a particular machine, which makes the part. The part is run through an automated assembly line controlled by a computer. The automotive industry uses this approach to manufacturing.

PROBLEMS

The following problems are given to provide practice in drawing objects that have rough surfaces and machined surfaces, and to give experience with dimensioning and shop notes. Add finish marks on all machined surfaces.

Most problems can be drawn full size on Layout C, shown on page 558 in the Appendix. If a reduced scale or a larger sheet, or both, are needed, it is so noted on each problem.

If assigned by the instructor, construct drawings using decimal-inch or metric scales by converting given fractional dimensions to decimal-inch or metric equivalents. Refer to Table 20 in the Appendix.

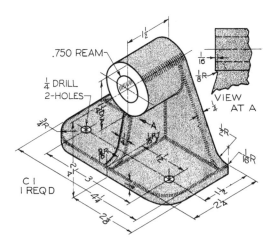

Fig. 11-32. Shaft bracket. Use Layout C in the Appendix.

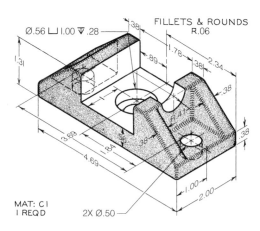

Fig. 11-33. Clamp block. Use Layout C in the Appendix.

243

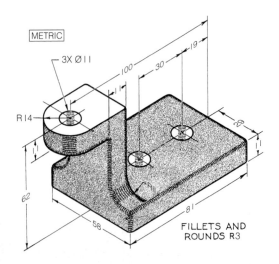

Fig. 11-34. Swivel. Use Layout C in the Appendix.

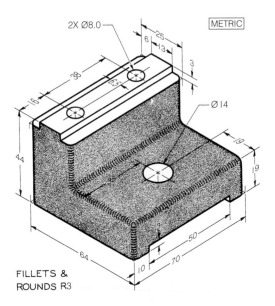

Fig. 11-35. Control base. Use Layout C in the Appendix.

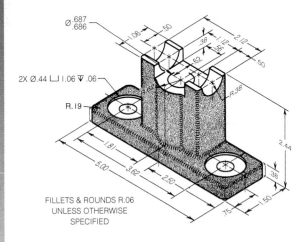

Fig. 11-36. Plunger bracket. Use Layout C in the Appendix.

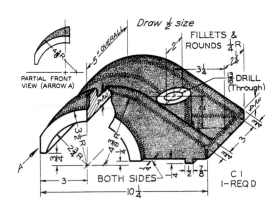

Fig. 11-37. Tailstock cap. Use Layout C in the Appendix.

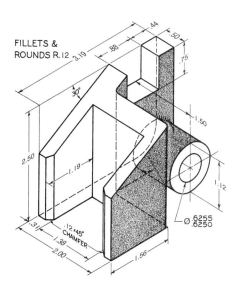

FILLETS &
ROUNDS R.12

Fig. 11-38. Double shifter yoke. Use Layout C in the Appendix.

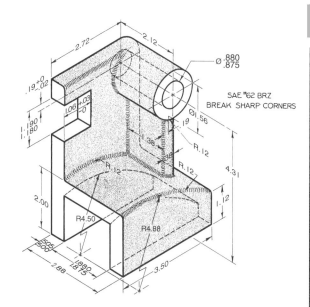

SAE.#62 BRZ
BREAK SHARP CORNERS

Fig. 11-39. Double shifter. Use Layout C in the Appendix.

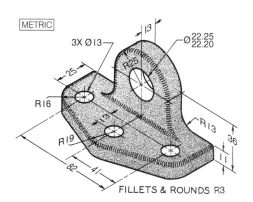

METRIC

3X Ø13

FILLETS & ROUNDS R3

Fig. 11-40. RH pipe bracket. Use Layout C in the Appendix.

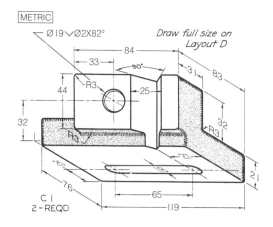

METRIC

Ø19∨Ø2X82°

Draw full size on
Layout D

C I
2 - REQD

Fig. 11-41. Control bracket.

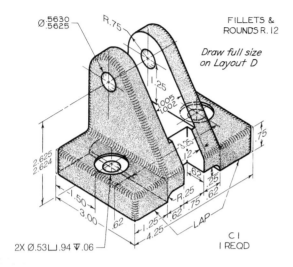

Fig. 11-42. Shifter link bracket.

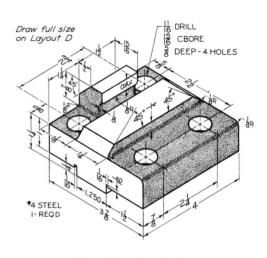

Fig. 11-43. Rest block.

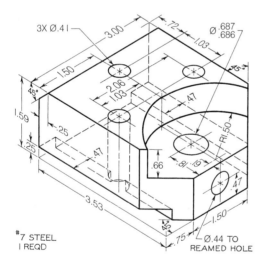

Fig. 11-44. Lock bolt block. Use Layout C in the Appendix.

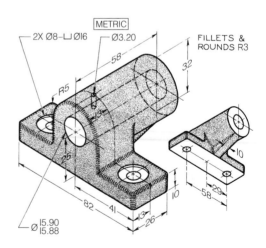

Fig. 11-45. Gear bracket, RH. Use Layout C in the Appendix.

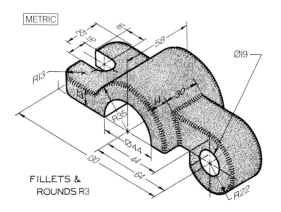

Fig. 11-46. Gum roll cap. Use Layout C in the Appendix.

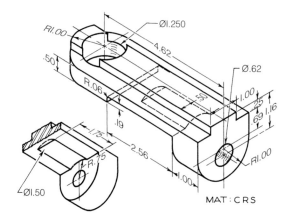

Fig. 11-47. Tie bracket. Use Layout C in the Appendix.

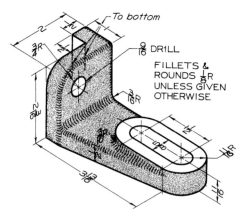

Fig. 11-48. Adjustable arm. Use Layout C in the Appendix.

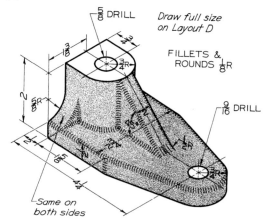

Fig. 11-49. Support.

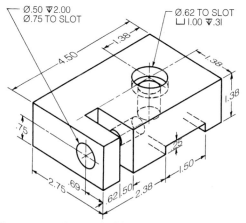

Fig. 11-50 Drive holder. Use Layout C in the Appendix.

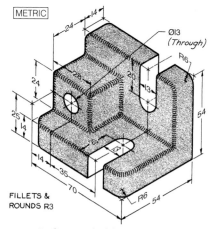

Fig. 11-51. Indicator holder. Use Layout C in the Appendix.

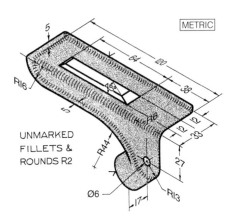

Fig. 11-52. Grinder guide. Use Layout C in the Appendix.

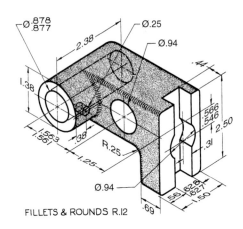

FILLETS & ROUNDS R.12

Fig. 11-53. Shifter block. Use Layout C in the Appendix.

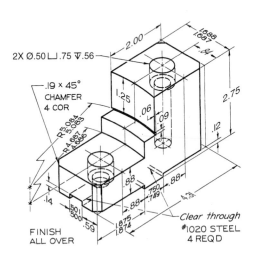

FINISH ALL OVER

Fig. 11-54. Chuck jaw. Use Layout C in the Appendix.

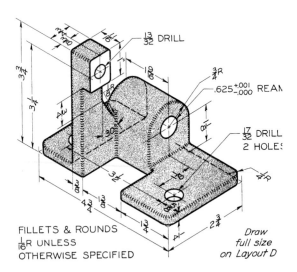

FILLETS & ROUNDS $\frac{1}{16}$R UNLESS OTHERWISE SPECIFIED

Draw full size on Layout D

Fig. 11-55. Secondary base.

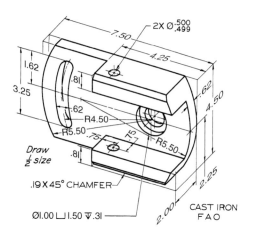

Fig. 11-56. Clapper box. Use Layout C in the Appendix.

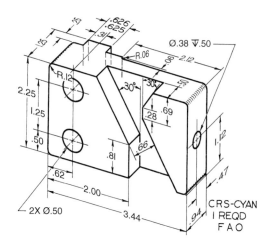

Fig. 11-57. Control dog. Use Layout C in the Appendix.

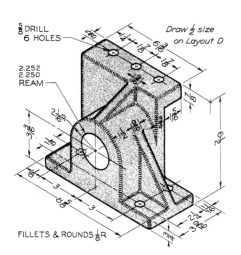

Fig. 11-58. Bracket.

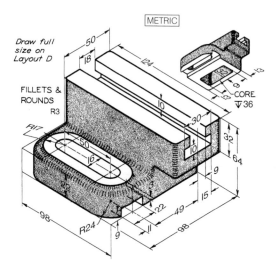

Fig. 11-59. Tool post block.

249

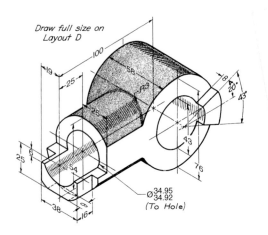

Fig. 11-60. Adjustable head.

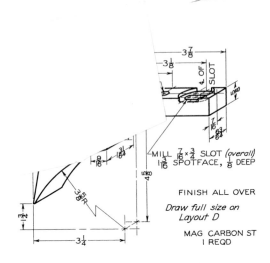

Fig. 11-61. Chip breaker shoe.

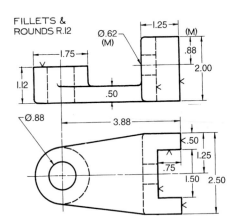

Fig. 11-62. Upper guide. Use Layout C in the Appendix. Move dimensions (M) to better locations.

Given: Front and bottom views.
Req'd: Front, top, and right-side views.

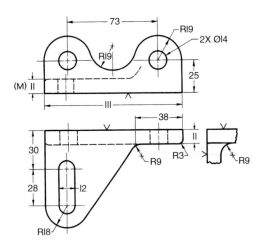

Fig. 11-63. Bracket. Use Layout C in the Appendix. Move dimensions (M) to better locations.

Given: Front and top views.
Req'd: Front to be top view; new front and right-side views.

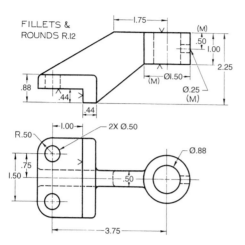

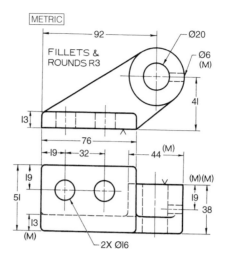

Fig. 11-64. Guide bearing. Use Layout C in the Appendix. Move dimensions (M) to better locations.

Given: Front and bottom views.
Req'd: Front, top, and right-side views.

Fig. 11-65. LH bearing. Use Layout C in the Appendix. Move dimensions (M) to better locations.

Given: Front and bottom views.
Req'd: Front, top, and right-side views.

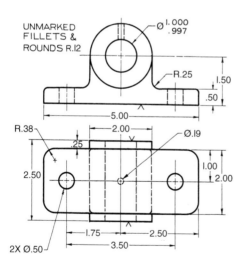

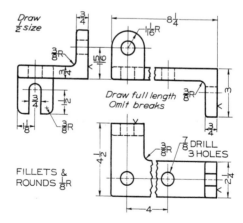

Fig. 11-66. Bearing. Use Layout C in the Appendix.

Given: Front and bottom views.
Req'd: Front, top, and right-side views.

Fig. 11-67. Holder strap. Use Layout C in the Appendix.

Given: Front, bottom, and left-side views.
Req'd: Front, top, and right-side views.

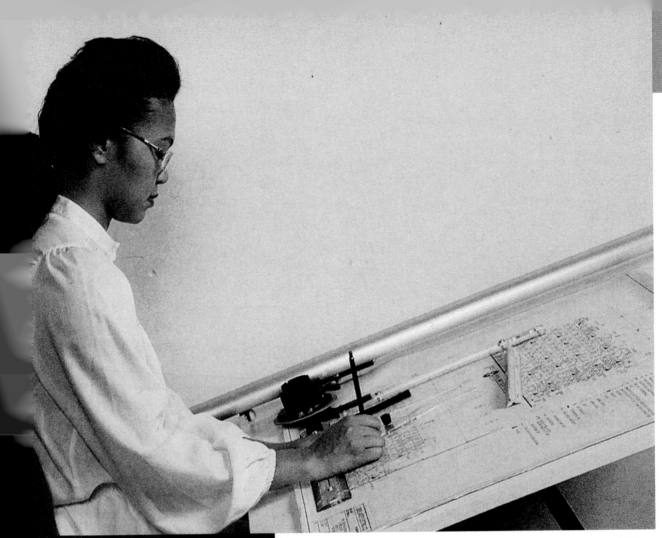

Arnold & Brown

CHAPTER 12

OBJECTIVES

After studying this chapter, you should be able to:

- Explain the reasons for showing objects in sectional views.
- Correctly draw section lines in sectional views.
- Correctly draw the various sectional views discussed in the chapter.

CHAPTER **12**

Sectional Views

Chapter 12 explains the drawing of sectional views. The sectional views discussed include full sections, half sections, broken-out sections, revolved sections, removed sections, and offset sections.

Sectional views are often needed to show the inside of a mechanical part, such as a flange or bearing. They are also used in architectural drawings to show how components fit together. For example, a theater designer may draw a sectional view of a stage set.

Theater designers may be involved with many different aspects related to the production of a live stage play or concert, the setting for a television production or movie, or the movie theater or concert hall itself. Stage sets, backdrops, platforms, props, costumes, sound and lighting systems—all are part of the theater designer's work. He or she must be creative and imaginative, with an artistic flair. The ability to make sketches and drawings expressing these thoughts and ideas on paper is very important to the theater designer.

CAREER LINK

To learn more about careers in theater design, contact the National Association of Schools of Art and Design, 11250 Roger Bacon Drive, Suite 21, Reston, VA 20190

12.1 FULL SECTIONS. The basic method of representing an object by its views was described in Chapter 7. Hidden parts of the object were shown with hidden lines, Sections 7.4 and 7.15. However, if an object has internal shapes that would require hidden lines, particularly if it has a complicated interior, as an automobile engine block does, we can obtain a clearer picture of the internal shapes by drawing the object cut apart.

For example, imagine an object to be cut in half, as shown in Fig. 12-1(a). The half nearest the observer is then pulled away, as shown at (b), exposing the back half. Here we see an imaginary *cutting plane* that might have been passed through the object instead of the hacksaw blade. At (c) the back half of the object is shown in front-view and top-view positions. The corresponding two-view drawing with the front view in section is shown at (d). The top view shows the entire object with the cutting plane shown edgewise. The front view shows the back half of the object only, with the front half removed.

The sectional view is called a *full section,* because the cutting plane has passed fully through the object. The edge view of the cutting plane is represented in the top view by a *cutting-plane line,* Fig. 4-11. The arrowheads at the ends show the direction of sight for the section. The parallel section lines indicate the surfaces cut by the cutting plane.

The cutting-plane line is shown in Fig. 12-1(d) for illustration. In practice, it is omitted in cases such as this where the location of the cutting plane is obvious.

12.2 SECTION LINING. Section lines should be drawn at an angle of 45°, as shown in Fig. 12-2(b), unless there is an advantage in using a different angle. Use a 0.3-mm mechanical pencil or a medium-grade drafting pencil, such as a 2H, with a *sharp* conical point, Fig. 4-8(c) and (d). Make section lines dark and *very thin,* to contrast well with the thick visible lines. *Space the section lines by eye.* Keep the spacing uniform by continually looking back and comparing with the spacing of the

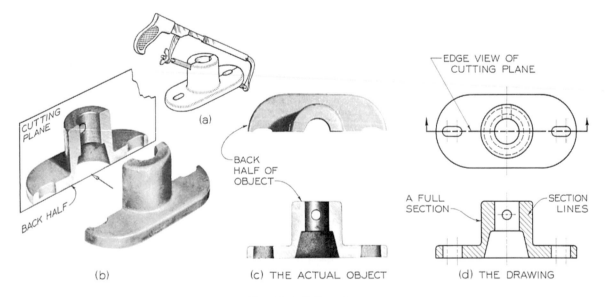

Fig. 12-1. Full section.

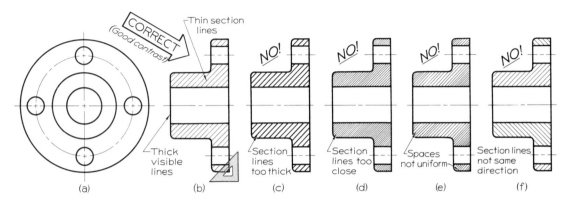

Fig. 12-2. Section-lining technique.

first lines drawn. The spaces between lines may vary from as little as $\frac{1}{16}''$ (.062") for very small sections to $\frac{1}{8}''$ (.125") or more for large sections. For average drawings, a spacing of $\frac{3}{32}''$ (.094") is about right.

Avoid thick section lines (or thin visible lines) that would not permit section lines to contrast well with the visible lines, Fig. 12-2(c). In general, space the lines well apart as at (b), and not crowded closely as at (d). Keep the spacing uniform. Especially avoid the tendency to crowd the lines closer together as the section lining approaches a corner or a small area, (e). All sectioned areas in a view of a single piece must be sectioned in the same direction, as at (b), and not as at (f). Section lining in different directions is understood to indicate entirely different parts, as in assembly drawing, Sec. 16.13.

The direction of visible outlines around a sectioned area sometimes makes it necessary to change the direction of section lines, Fig. 12-3. At (a) the section lines have been drawn at approximately 45° with the main outlines. If the section lines are drawn parallel or perpendicular to prominent visible lines, as at (b) and (c), the results are highly unsatisfactory.

The ANSI recommends that on a detail drawing the general-purpose (cast-iron) section lining, Fig. 12-2, be used for all materials. The exact material specifications will generally be more detailed than by the section-lining symbol alone. Thus they will be given in a note or title strip. Symbolic section lining, Fig. 16-18, may be used on assembly drawings in situations where it may be desirable to show a distinction between different materials. Otherwise, the cast-iron symbol

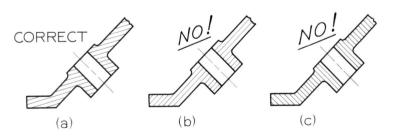

Fig. 12-3. Direction of section lining.

255

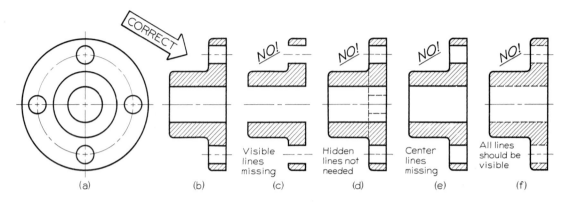

Fig. 12-4. Visible, hidden, and center lines.

would be used for all materials. For assembly sections, see Sec. 16.13.

12.3 VISIBLE, HIDDEN, AND CENTER LINES. As a rule, visible edges and contours behind the cutting plane should be shown, Fig. 12-4(b). If all visible lines are not shown, a disconnected and confusing drawing results, as shown at (c). Since sections are drawn to replace hidden-line representation, hidden lines are unnecessary and tend to confuse the drawing, (d). Hidden lines are always omitted except in special cases that require them for clearness. Center lines should not be omitted as at (e). No sectioned area can ever be bounded by

hidden lines as at (f), as the edges of the cut surfaces are always visible.

12.4 HALF SECTIONS. If the cutting plane is passed only halfway through an object, Fig. 12-5(a), and then the quarter of the object in front of the cutting plane is removed, (b), the resulting section is a *half section,* (c). Thus, a half section has the advantage of showing in a single view both inside and outside shapes. Hence, its usefulness is limited largely to symmetrical objects. Hidden lines are usually omitted from the unsectioned half, as shown at (c), unless they are needed for clearness or for dimensioning of the inside shapes. The

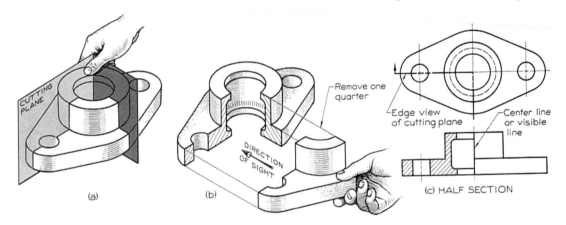

Fig. 12-5. Half section.

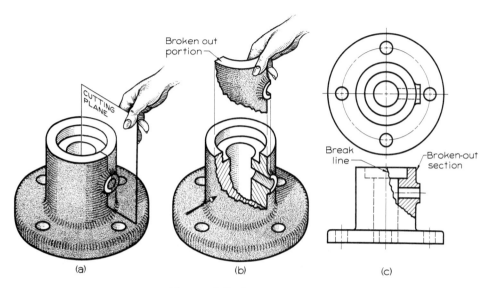

Fig. 12-6. Broken-out section.

greatest usefulness of the half section is in assembly drawing, Fig. 16-17(b) and (c). There it is often desirable to show both inside and outside construction of a symmetrical assembly in a single view.

Notice, in Fig. 12-5(c), how the cutting plane line is shown, with one arrow to indicate the direction of sight. Usually, the location of the cutting plane is obvious. It is not necessary to show the cutting-plane line. It is

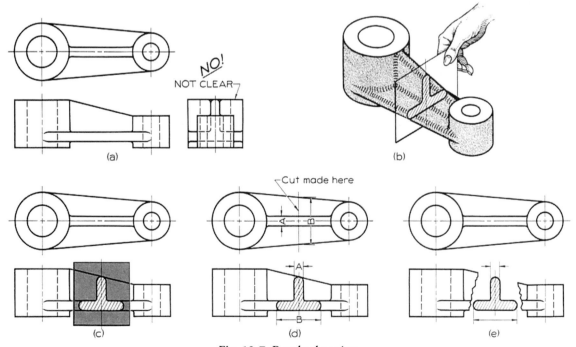

Fig. 12-7. Revolved section.

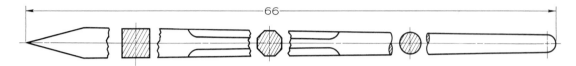

Fig. 12-8. Revolved sections.

shown here only for illustration. The line separating the sectioned half from the unsectioned half may be either a center line or a visible line.

12.5 BROKEN-OUT SECTIONS. Often only a small part of a view needs to be sectioned to show some detail of inside construction. For example, Fig. 12-6(a), an imaginary cutting plane is passed through the part to be sectioned. At (b) the plane and the broken-out portion of the object in front of it are removed. At (c) is shown the *broken-out section,* limited by a freehand break line, Fig. 4-11. The cutting-plane line need not be shown in the top view, as the location of the cut is obvious.

12.6 REVOLVED SECTIONS. Figure 12-7(a) shows three views of a Lever Arm in which the right-side view shows the T-shape of the central portion. However, the view is very confusing because of the number of lines. If a cutting plane is passed at right angles through the T-shaped central portion, as shown at (b), and then revolved into position as shown at

(c), the result is a *revolved section,* (d). Note that distances A and B on the section are taken from A and B in the top view. The revolved section shows the T-shape clearly.

If desired, the view may be broken away on each side of a revolved section, as shown at (e). This often makes the section stand out better and provides clearer dimensioning, as shown.

Revolved sections applied to a Lining Bar are shown in Fig. 12-8. In this case, breaks were used not only to make the sections stand out. They were used also to shorten the object so that the drawing would fit on the paper. Note that the actual full length is indicated by the 66″ dimension. For a further discussion of breaks, see Sec. 12.12.

In Fig. 12-9, a correct revolved section for an I-shaped arm is shown at (a). Some common errors are shown at (b), (c), and (d).

12.7 REMOVED SECTIONS. A *removed section* is a section that is moved from its normal position to some more convenient position on the sheet, Fig. 12-10. In such cases, the removed section may be drawn to a larger

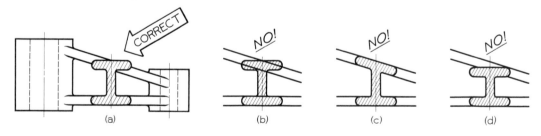

Fig. 12-9. Correct and incorrect revolved sections.

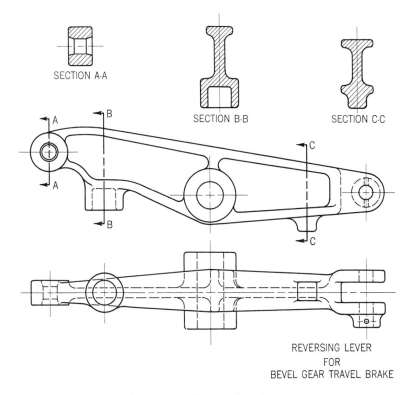

Fig. 12-10. Removed sections.

scale if desired. In this way some small detail can be magnified and dimensioned more clearly than at the scale of the main drawing. We need not show all visible lines behind each cutting plane, but only those near the section and needed for clearness.

For example, in Section B-B the back rim of the hole is shown. However, the rest of the object to the left of the cutting plane is not shown. Similarly, in Section C-C all visible lines beyond the cutting plane are omitted.

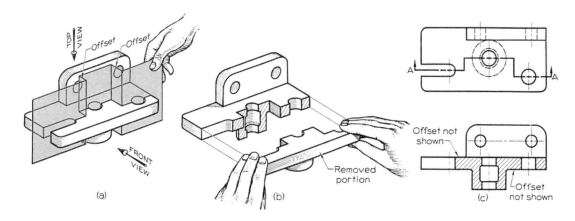

Fig. 12-11. Offset section.

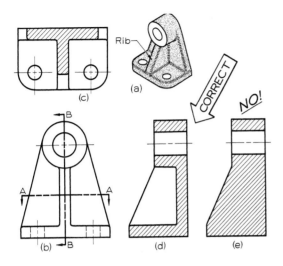

Fig. 12-12. Ribs in section.

12.8 OFFSET SECTIONS. In order to include in a single section several features of an object that are not in a straight line, the cutting plane may be bent or "offset" so as to pass through these features. Such a section is called an *offset section*. For example, in Fig. 12-11(a), it is desired to include in the section the slot at the left end, the central hole, and the hole at the right end. The cutting plane is shown with offsets to include these. The front portion of the object is then removed, as shown at (b). The offset section is shown at (c). The edge view of the bent cutting plane is shown by the cutting-plane line A-A in the top view. *Note that the bends, or offsets, are not shown in the sectional view.*

A removed section may be drawn in any convenient place on the sheet. Thus, it is necessary to label each cutting plane using capital letters at the ends, as A-A, B-B, etc. You also need to place a note under each section, as SECTION A-A, SECTION B-B, and so on.

A removed section should not be rotated on the paper; that is, it should always be drawn with its lines parallel to what they would be in the normal position.

12.9 RIBS AND SPOKES IN SECTION. A *rib* or *web* is a thin flat part of an object used for bracing or adding strength, as shown in Fig. 12-12(a). If a cutting plane, A-A, is passed crosswise through a rib, (b), the rib will be section-lined as shown in the top view at (c). If a cutting plane, B-B, is passed "flatwise" through the rib, (b), the rib should not be section-lined, but should be left entirely blank as at (d). The reason for this is that if

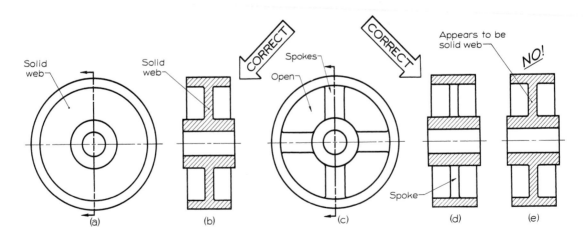

Fig. 12-13. Solid web and spokes in section.

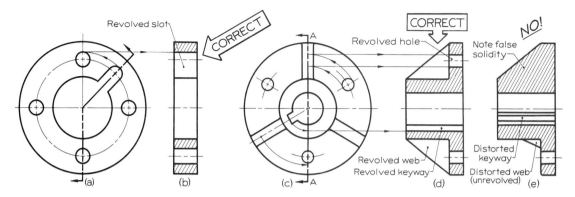

Fig. 12-14. Revolved features.

the rib is section-lined, as at (e), there is a false impression of thickness or solidity.

Figures 12-13(a) and (b) show a front view and a section of a wheel that has a solid web between the hub and rim. The web is sectioned because the cutting plane cuts crosswise through it. The section gives a correct appearance of solidity. A similar wheel is shown at (c), but it has spokes instead of a solid web. In the correct section, (d), *the spokes are not section-lined.* The reason for this is that if the spokes are section-lined, as at (e), the section gives a false appearance of solidity and not the open effect of spokes. Note that (e) and (b) are exactly alike.

12.10 REVOLVED FEATURES. Clearness can often be improved, or a section may include more information, if certain features are sectioned in a revolved position. For example, in Fig. 12-14(a), the cutting plane can be bent so as to pass through the slot. The slot is then revolved to the upright position and projected across to the section, as shown at (b).

Figure 12-14(c) shows a flange with three ribs, three holes, and a keyway. The straight cutting plane A-A includes only one rib and one hole and not the keyway. However, we can revolve a rib, a hole, and the keyway into the cutting plane as shown, and then project across to the section, as at (d). The incorrect

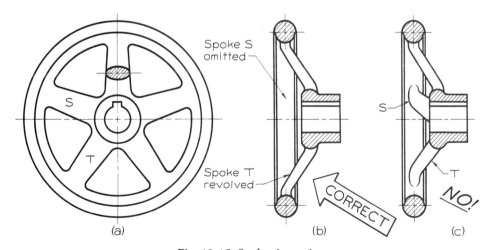

Fig. 12-15. Spokes in section.

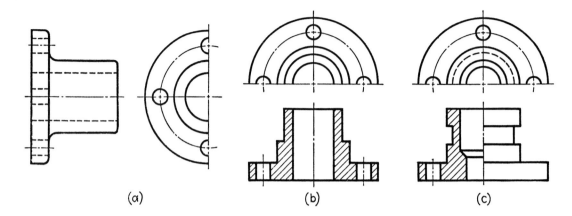

Fig. 12-16. Half views.

section is shown at (e), where the upper rib is section-lined, and the various features are not revolved. The view is misleading and does not give enough information. The lower rib and the keyway are shown in unsatisfactory and distorted positions. Note that features may be revolved into the straight cutting plane, or the plane may be bent to pass through these features and then revolved.

When a wheel has an odd number of spokes, such as three or five, Fig. 12-15(a), a straight cutting plane will produce a confus-

ing section, as shown at (c), in which spokes S and T are distorted. The correct section, (b), is made by revolving spoke T into the imaginary cutting plane and eliminating spoke S from the section.

12.11 HALF VIEWS. If space is limited, a half view may be drawn of a symmetrical object, Fig. 12-16. Note that the "near half" of the circular view is drawn at (a). At (b) and (c) the "far half" is shown in each case. See also Sec. 8.19.

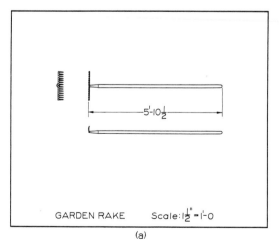

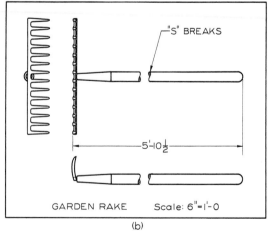

Fig. 12-17. Use of conventional breaks.

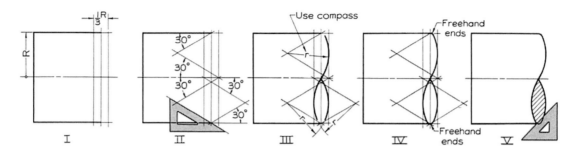

Fig. 12-18. Steps in drawing "S" break for solid shaft.

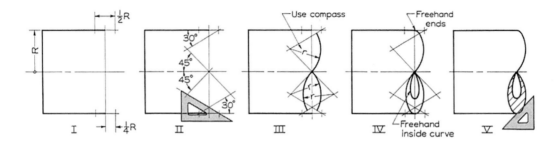

Fig. 12-19. Steps in drawing "S" break for tubing.

12.12 CONVENTIONAL BREAKS. Long objects, such as a garden rake, Fig. 12-17(a), may appear too small on the drawing or require a very large sheet of paper. A larger scale can easily be used if the long portion is "broken" and a considerable length removed, (b). The broken ends are drawn in a standard way and are called *conventional breaks*. Parts thus broken must have the same section throughout or be tapered uniformly. Note that the full-length dimension at (b) shows the actual length. The steps in drawing the "S" break

used in Fig. 12-17 are shown in Fig. 12-18.

The steps in drawing the "S" break for tubing are shown in Fig. 12-19. The methods shown in Figs. 12-18 and 12-19 are to be used for diameters of approximately 1″ or over where the compass can easily be used. For smaller diameters the "S" breaks should be carefully drawn freehand. You can also use one of the special "S" break templates manufactured for this purpose.

Ordinary breaks for metal and wood are shown in Fig. 12-20.

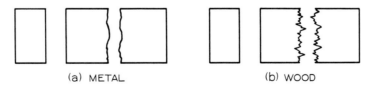

(a) METAL   (b) WOOD

Fig. 12-20. Rectangular metal and wood breaks.

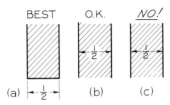

Fig. 12-21. Dimensions and section lines.

12.13 DIMENSIONS ACROSS A SECTION.
Dimensions for a section should be placed outside the section if possible, Fig. 12-21(a). Where it is necessary to have a dimension on a sectioned area, an opening in the section lining should be left for the dimension figure, as at (b). Dimensions should never be lettered over section lining, as shown at (c).

12.14 COMPUTER GRAPHICS.
Sectional drawings can be drawn by either 2D or 3D sectioning. The 2D method is similar to manual drafting in that the sections are drawn as a two-dimensional solution. All sections are shown with hidden lines removed and cutting lines in place. The cutting plane line is drawn with a preset thickness and line style. Arrowheads are placed at the appropriate ends of the cutting plane line. This method is used to develop all section views. A hatch pattern is chosen from a library of many types that are accepted ANSI standards, Fig. 12-22. Conventional break lines can be called up and placed where required.

The desired pattern is chosen. The angle and spacing of lines is designated. A window or fence is then placed around the area to be hatched. When the enter key is used, the hatch is automatically placed. Fig. 12-23.

The 3D method of sectioning on a CAD system uses geometry developed on a solids model. The cutting plane line is placed at the

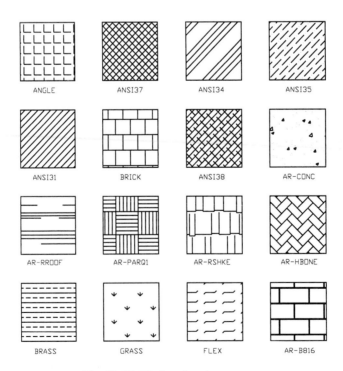

Fig. 12-22. Various hatch patterns.

point where the section is desired. Any type of section may be chosen. The model is divided along the cutting-plane line. The resulting piece is then hatched using the method previously described, Fig. 12-24. Sections produced by this method can be rotated, removed, revolved, or placed in any of the accepted drafting standards.

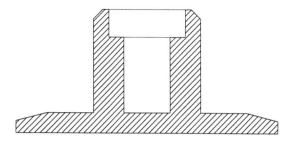

Fig. 12-23. A hatched area in a CAD drawing.

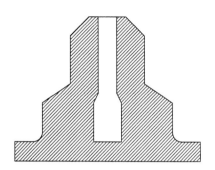
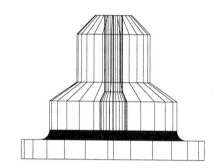

Fig. 12-24. Sectioning on a CAD system using geometry developed as a solids model.

PROBLEMS

The following problems are intended to be drawn with instruments or freehand, as required by the instructor. All are to be drawn on Layout C in the Appendix, except Figs. 12-38, 12-41, and 12-42, which require larger paper. All problems are to be dimensioned unless otherwise assigned by the instructor.

If assigned by the instructor, construct drawings using decimal-inch or metric scales by converting given fractional dimensions to decimal-inch or metric equivalents. Refer to Table 20 in the Appendix.

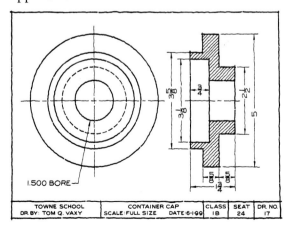

Fig. 12-25. Sketch.

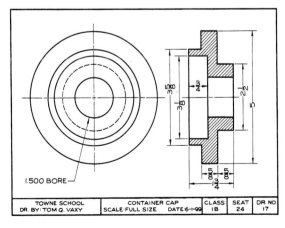

Fig. 12-26. Mechanical drawing.

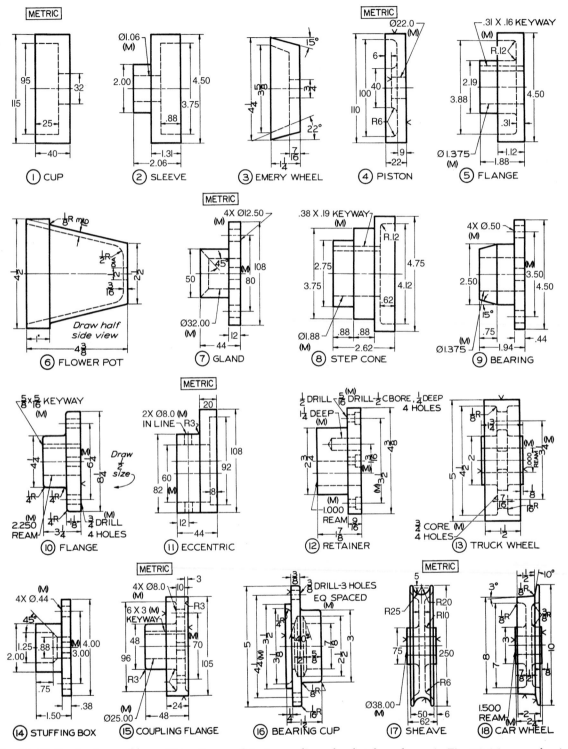

Fig. 12-27. Sectioning problems. Using Layout C in Appendix make sketch as shown in Fig. 12-25 or mechanical drawing as shown in Fig. 12-26 of problem assigned from above. In each case draw a circular view and a full or half section as assigned. Vertical dimensions are diameters. Dimensions marked (M) should be moved to circular view. Holes are equally spaced. Give part names in title strips.

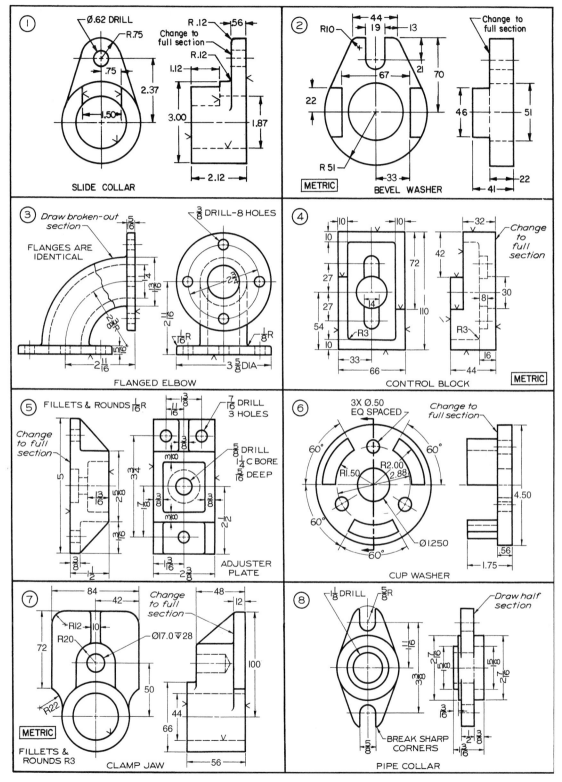

Fig. 12-28. Sectioning problems. Using Layout C in Appendix make sketch, Fig. 12-25, or mechanical drawing, Fig. 12-26. Section as indicated. Omit instructional notes. Give part names in title strips.

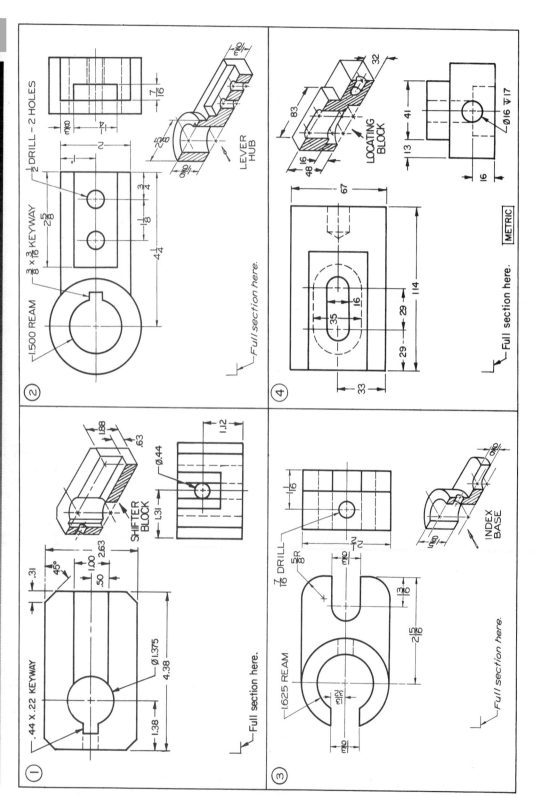

Fig. 12-29. Sectioning problems. Using Layout C in Appendix make sketch, Fig. 12-25, or mechanical drawing, Fig. 12-26, of assigned problem, showing the given views plus a section as indicated. Omit pictorial drawings and instructional notes. Move dimensions from pictorial drawing to the sectional views. Give part names in title strips. Note: Problem 2 uses an alternate position for the side view. See Fig. 7-19.

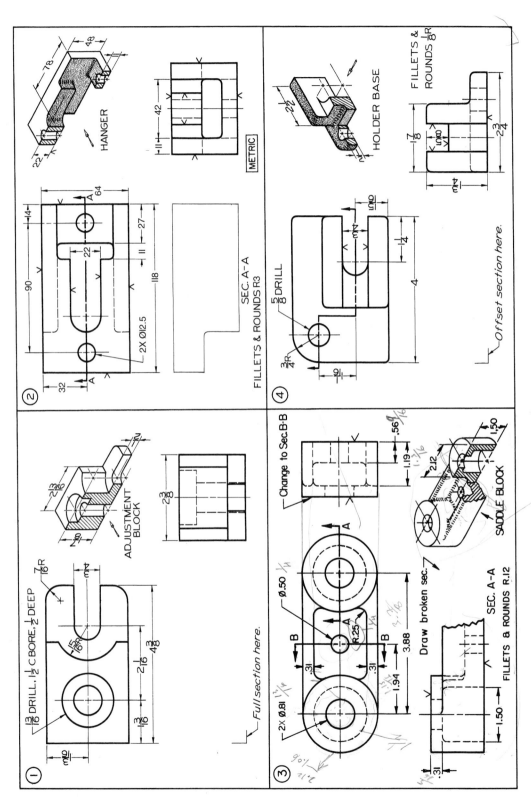

Fig. 12-30. Sectioning problems. Using Layout C in Appendix make sketch, Fig. 12-25, or mechanical drawing, Fig. 12-26, of assigned problem, showing the given views plus a section as indicated. Omit pictorial drawings and instructional notes. Move dimensions from pictorial drawings to the sectional views. Give part names in title strips.

269

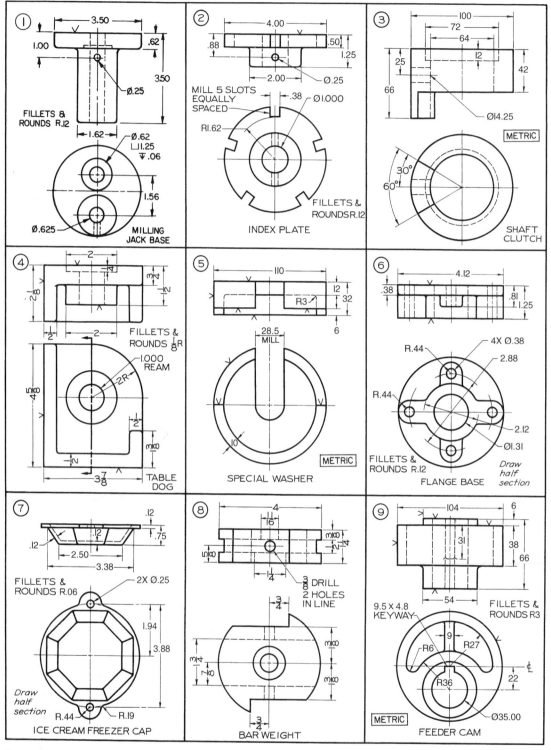

Fig. 12-31. Sectioning problems. Using Layout C in Appendix make sketch, Fig. 12-25, or mechanical drawing, Fig. 12-26. Omit given top view, draw front view as shown, and draw side view in full section, except in problems 6 and 7, which require half sections. Omit instructional notes. Give part names in title strips. Move dimensions to new locations as necessary.

SECTIONAL VIEWS

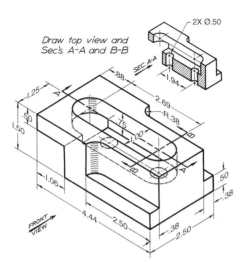

Draw top view and
Sec's. A-A and B-B

Fig. 12-32. Guard block (Layout C).

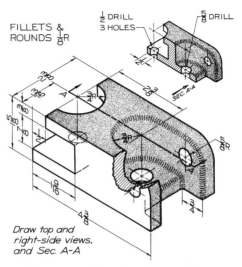

FILLETS &
ROUNDS ⅛R

Draw top and
right-side views,
and Sec. A-A

Fig. 12-33. Pivot base (Layout C).

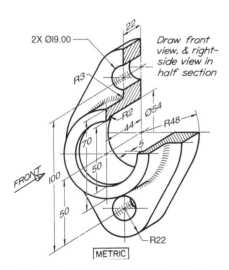

Draw front
view, & right-
side view in
half section

2X Ø19.00

METRIC

Fig. 12-34. Packing gland (Layout C).

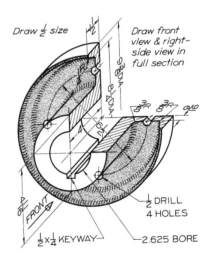

Draw ½ size

Draw front
view & right-
side view in
full section

½ DRILL
4 HOLES

2.625 BORE

½ X ¼ KEYWAY

Fig. 12-35. Flange (Layout C).

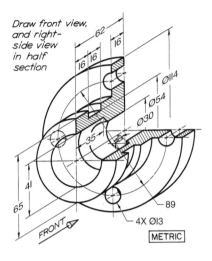

Draw front view, and right-side view in half section

62
16
16 16
Ø114
Ø54
Ø30
35
41
65
89
4X Ø13

FRONT

METRIC

Fig. 12-36. Stuffing box (Layout C).

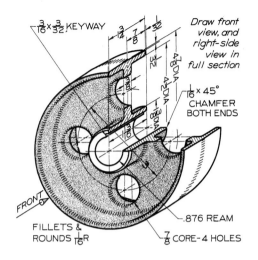

$\frac{3}{16} \times \frac{3}{32}$ KEYWAY

Draw front view, and right-side view in full section

$\frac{3}{4}$
$\frac{7}{8}$
32
$4\frac{7}{8}$ DIA
$4\frac{1}{2}$ DIA
$1\frac{3}{8}$ DIA

$\frac{1}{16} \times 45°$ CHAMFER BOTH ENDS

.876 REAM

$\frac{7}{8}$ CORE-4 HOLES

FRONT

FILLETS & ROUNDS $\frac{1}{16}$ R

Fig. 12-37. Pulley (Layout C).

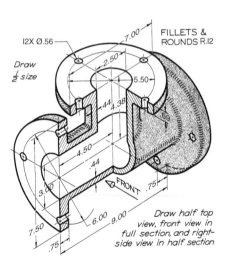

12X Ø.56

FILLETS & ROUNDS R.12

7.00
2.50
5.50

Draw $\frac{1}{2}$ size

.44 4.38
.50
4.50
.44
3.00
.75
7.50
.75
6.00
9.00

FRONT

Draw half top view, front view in full section, and right-side view in half section

Fig. 12-38. Flanged tee (Layout D).

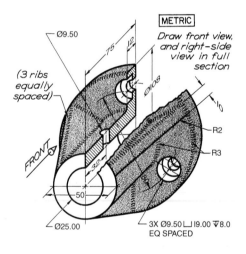

METRIC

Ø9.50

Draw front view, and right-side view in full section

.75
12
Ø108

(3 ribs equally spaced)

10
R2
R3
32
.50
Ø25.00

3X Ø9.50 ⌴ 19.00 ▽8.0 EQ SPACED

FRONT

Fig. 12-39. Bearing (Layout C).

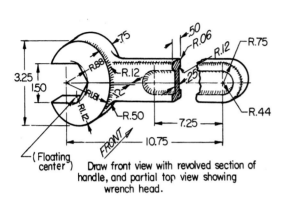

Fig. 12-40. Wrench (Layout C).

Draw front view with revolved section of handle, and partial top view showing wrench head.

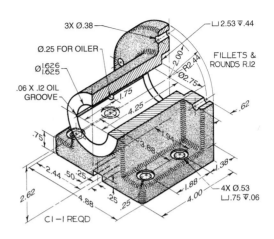

Fig. 12-41. Head-end bearing (Layout E).

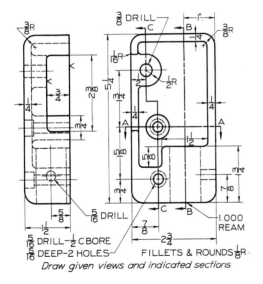

Draw given views and indicated sections

Fig. 12-42. Clamp guard (Layout D).

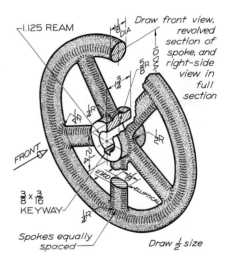

Draw front view, revolved section of spoke, and right-side view in full section

Fig. 12-43. Valve handwheel (Layout C).

Mark Richards/PhotoEdit

CHAPTER 13

OBJECTIVES

After studying this chapter, you should be able to:

- Draw each of the three ordinary auxiliary views.
- Revolve an auxiliary view drawing.
- Demonstrate the method for finding the true length of an oblique line.

274

CHAPTER 13
Auxiliary Views

Auxiliary views are important in many product designs. One field that makes use of auxiliary views is automotive design.

An automotive designer/stylist focuses on the appearance of motor vehicles, including automobiles, trucks, vans, and sport-utility vehicles. Interior and exterior styling are both important. The final design must look appealing so that consumers will want to buy the product. It must also be functional, meet safety standards, and be cost-effective to build and maintain. Human factors, materials, and production technology all must be considered.

Before the final design of a vehicle is approved, thousands of sketches and drawings will be made of every feature and detail—seats, instrument panel, body, wheels, and other styling features. Mechanical components, including engine, drivetrain, exhaust, steering, braking, and electrical systems, must also be considered. These all require drawings, including, in many cases, auxiliary views.

CAREER LINK

To learn more about preparing for a career in automotive design, contact the National Association of Schools of Art and Design, 11250 Roger Bacon Drive, Suite 21, Reston, VA 20190.

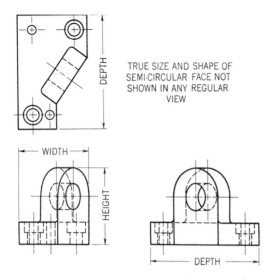

TRUE SIZE AND SHAPE OF
SEMI-CIRCULAR FACE NOT
SHOWN IN ANY REGULAR
VIEW

Fig. 13-1. Regular views.

Fig. 13-1, the true size and shape of the semi-circular face of the angle bracket are not shown in any of the views. Since the inclined surface is not parallel to any plane of the glass box, the circles project as ellipses. The result is not only confusing but difficult to draw.

13.2 THE AUXILIARY VIEW. To obtain a true-size view of the inclined face, we must view the object at right angles to that face, Fig. 13-2(a), through a special inclined *auxiliary plane* parallel to it. A view obtained in this manner is an *auxiliary view.* In this case, the auxiliary plane is perpendicular to the top plane of the glass box and hinged to it. When this auxiliary plane is unfolded, Fig. 13-2(b), the auxiliary view is shown in its correct position on the drawing. Note that the true size and shape of the inclined face are shown in the auxiliary view. Actually, only these two views are needed to describe the shape of this object. The complete auxiliary-view drawing is shown in Fig. 13-3. We know it is complete because all dimensions can be shown on these views. The object can be built in the shop.

13.1 INCLINED SURFACES. Up to this point we have considered the various *regular views* of objects. These views, projected upon the planes of the "glass box," Fig. 7-5, were enough to describe clearly the shape of the object.

However, if the object has an inclined surface that needs to be shown true size, the regular views are not enough. For example, in

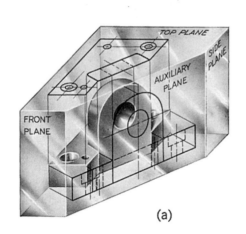

(a)

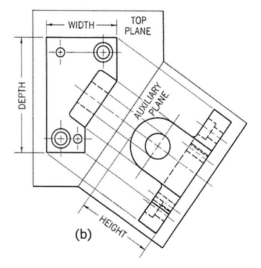

(b)

Fig. 13-2. The auxiliary plane.

13.3 THE THREE AUXILIARY VIEWS. The three types of ordinary auxiliary views are the *depth auxiliary, height auxiliary,* and *width auxiliary,* Fig. 13-4. They are named according to the principal dimensions of the object shown in the auxiliary view. They are also frequently referred to as *primary auxiliary views.*

Thus, the auxiliary view in Fig. 13-3 is a height auxiliary view because the auxiliary view shows the principal dimension, *height.* The auxiliary plane is always hinged to the regular plane to which it is perpendicular. Note that in each auxiliary view, one main dimension is projected directly from a regular view, and one is transferred with dividers (or scale) from another regular view.

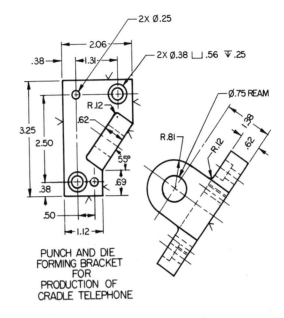

PUNCH AND DIE
FORMING BRACKET
FOR
PRODUCTION OF
CRADLE TELEPHONE

Fig. 13-3. Minimum required views.

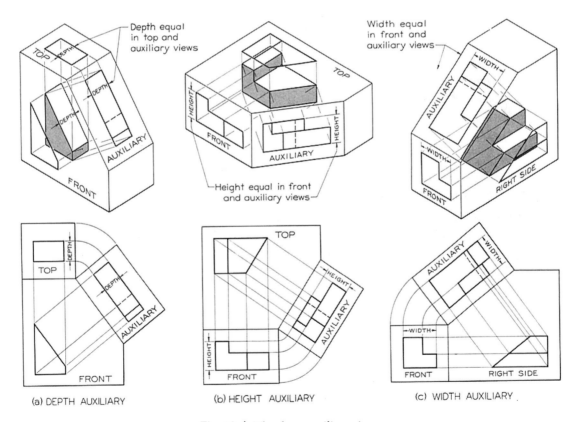

(a) DEPTH AUXILIARY (b) HEIGHT AUXILIARY (c) WIDTH AUXILIARY

Fig. 13-4. The three auxiliary views.

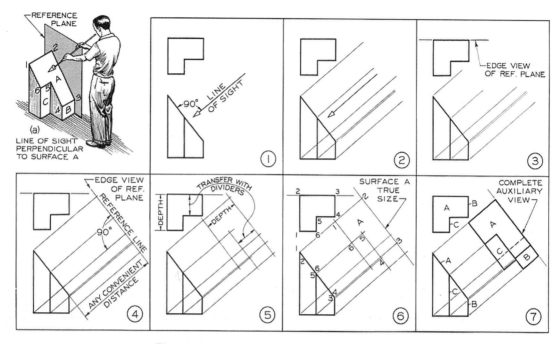

Fig. 13-5. Steps in drawing a depth auxiliary view.

13.4 To Draw a Depth Auxiliary View. The steps in drawing a simple front auxiliary view are shown in Fig. 13-5. As shown at (a), the line of sight is assumed to be perpendicular to surface A, in order to show the true size and shape of that surface. The steps are:

1. Draw arrow perpendicular to inclined surface A in the front view.

2. Draw light projection lines from all points of the object parallel to the arrow. Slide the triangle on the T-square, as shown in Fig. 13-6(a).

3. Assume a *reference plane* to contain the back face of the object. Draw *reference line* in top view. This is the edge view of the reference plane.

4. Draw reference line (edge view of refer-

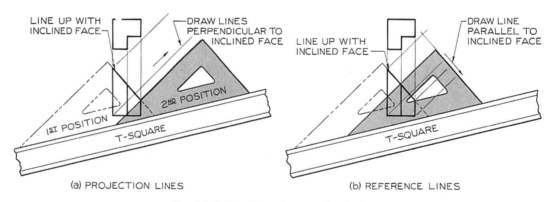

Fig. 13-6. Parallel and perpendicular lines.

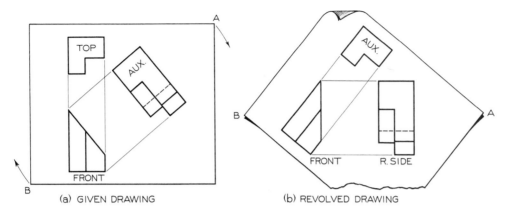

Fig. 13-7. Revolving a drawing.

ence plane) perpendicular to projection lines. Slide triangle on T-square as shown in Fig. 13-6(b).

5. Transfer, with dividers or scale, the depth dimensions from the top view to the auxiliary view. Measurements are made in both cases perpendicular to the reference line.

6. Project points 1, 2, 3, 4, 5, and 6. Join them to produce the auxiliary view of the inclined face A.

7. Complete the auxiliary view by projecting the remaining points to obtain surfaces B and C.

13.5 REVOLVING THE DRAWING. Figure 13-7(a) is an auxiliary-view drawing. At (b) the same drawing is revolved until the auxiliary view is directly to the right of the front view, becoming a right-side view. The top view then becomes an auxiliary view. The views at (b) are exactly the same as at (a). Only the names

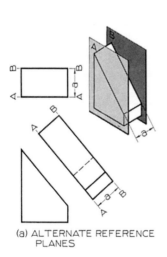

(a) ALTERNATE REFERENCE PLANES

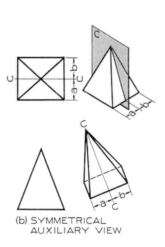

(b) SYMMETRICAL AUXILIARY VIEW

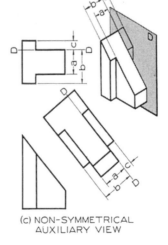

(c) NON-SYMMETRICAL AUXILIARY VIEW

Fig. 13-8. Reference planes.

279

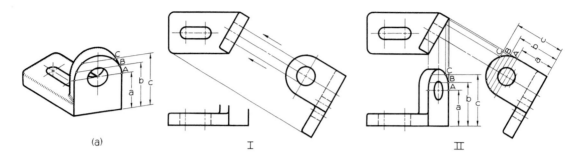

Fig. 13-9. Plotted curves.

of the views are changed. To understand an auxiliary view, or to draw one, it may be helpful to revolve the drawing as shown. In this way the auxiliary view becomes a regular view.

13.6 REFERENCE PLANES. Locate the reference plane in the position where it can be most conveniently used. For the object shown in Fig. 13-8(a), the reference plane may coincide with either the front or the back surface, as shown. Usually, where the object is symmetrical, the reference plane is assumed to run through the center of the object, (b). In other cases it may be convenient to assume the reference plane to be coinciding with some inside surface, (c). Remember that the reference line represents the edge view of the reference plane.

In the auxiliary view it must always be perpendicular to the projection lines.

13.7 PLOTTED CURVES. An auxiliary view is often needed to draw the regular views. See, for example, the object in Fig. 13-9(a). The top and auxiliary views and part of the front view can be readily drawn, as shown at I. Note that in the top view the hidden lines for the hole are projected from the hole in the auxiliary view.

To complete the front view, II, select any points, as A, B, and C, on the curve in the auxiliary view. Project them to the top view and then down to the front view, as shown. Transfer the heights a, b, and c from the auxiliary view to the front view with dividers.

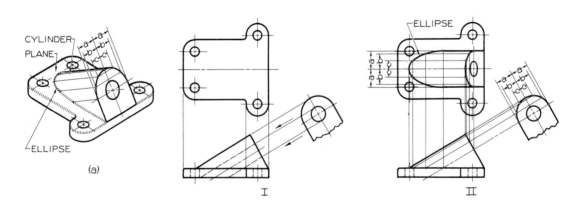

Fig. 13-10. Plotted curves.

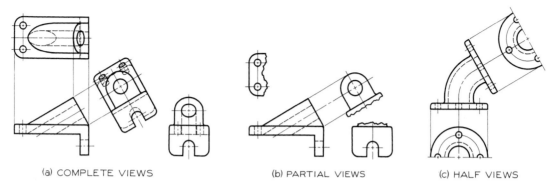

(a) COMPLETE VIEWS   (b) PARTIAL VIEWS   (c) HALF VIEWS

Fig. 13-11. Partial views.

Note that each height dimension can be used to transfer two points on opposite sides of the semicircle in the auxiliary view. In addition to points A, B, and C, as many additional points should be selected as necessary to define the curve accurately. Sketch smooth curves through the points in the front view and apply the irregular curve, Sec. 4.30.

Figure 13-10(a) shows a case where an auxiliary view is needed to plot the curve of intersection between a cylinder and a plane. At I, the front view, the incomplete top view, and a partial auxiliary view (see Sec. 13.8) are shown. Note that in the front view the hidden lines for the large inclined hole are projected from the auxiliary view.

Draw light construction lines across the auxiliary view to locate points on the curve, as shown at II. Project these points to the front view and then up to the top view, as shown. Transfer pairs of equal distances *a, b,* and *c* from the auxiliary view to the top view with dividers, as shown. Then sketch smooth curves through the points in the top view and apply the irregular curve, Sec. 4.30.

13.8 PARTIAL VIEWS. Figure 13-11(a) gives complete front, top, right-side, and auxiliary views of an Angle Bearing. You can give a complete shape description, while also omitting the drawing of difficult curves, by drawing *partial views,* as shown at (b). Note the

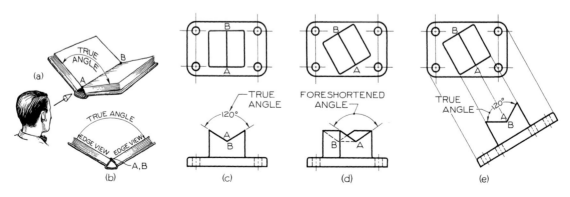

Fig. 13-12. Angles.

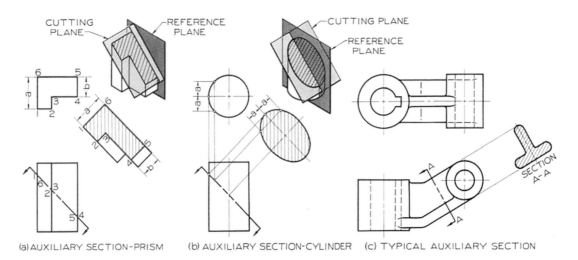

(a) AUXILIARY SECTION-PRISM   (b) AUXILIARY SECTION-CYLINDER   (c) TYPICAL AUXILIARY SECTION

Fig. 13-13. Auxiliary sections.

use of break lines, Fig. 4-11, to limit the partial views.

For simple symmetrical objects, *half views* may be drawn, as shown at (c). Note the use of center lines instead of break lines to limit the half views. See also Secs. 8.19 and 12.11.

13.9 ANGLES. As has been shown in Sec. 8.14, you can view an angle in its true size only when your line of sight is perpendicular to the plane of the angle. To see the true angle between two planes, you must look parallel to the line of intersection of the planes. For example, Fig. 13-12(a), the true angle between the leaves of the book is seen if you look parallel to line AB. As shown at (b), line AB will appear as a point. The two planes will appear as lines.

At (c) a V-block drawing shows how the two inclined surfaces intersect at line AB. Since this line shows as a point in the front view, the true 120° angle is shown.

At (d) the V-block is turned so that line AB does not appear as a point in the front view. Therefore, because the angle does not show

true size, this drawing is unsatisfactory.

At (e) an auxiliary view is drawn so that the projection lines are parallel to line AB. The line shows as a point, and the true angle is shown. In industrial drafting, auxiliary views are used more for showing true angles than for showing true sizes of surfaces.

13.10 AUXILIARY SECTIONS. An auxiliary section is a section on an auxiliary plane — that is, a section viewed at an angle. The entire object may be shown in the auxiliary-section view, Fig. 13-13(a), or the cross-hatched surface alone may be shown, (b). A typical example of an auxiliary section in machine drawing is shown at (c). Note that there is not enough space to show a revolved section of the arm (see Sec. 12.6). Note also that visible lines behind the section are omitted.

13.11 TRUE LENGTH OF LINE. The true length of any oblique line can easily be found by means of an auxiliary view. For example, refer to Fig. 13-14. Find the true length of hip rafter 1-2. It is only necessary to draw an auxiliary view with the direction of sight perpendicular to the line in any view, as shown. The true length may also be found by revolution, Fig. 14-7(c).

13.12 SECONDARY AUXILIARY VIEWS. A *secondary auxiliary view* is any auxiliary view that is projected from a primary auxiliary view. A specific example of a secondary auxiliary view is shown in Fig. 13-15. Here it is required to draw an auxiliary view that will show the true size and shape of an oblique surface, such as surface (plane) 1-2-3. The true size (TS) of a plane will appear only in a view that has a line of sight normal (perpendicular) to that plane. Thus, it is first necessary to draw a primary auxiliary view that will show surface 1-2-3 as a line (edge view). To do this, a line of sight, arrow P, is selected that is parallel to a true length (TL) line of surface 1-2-3 in the front view. Thus, arrow P is drawn parallel to line 1-3 of the front view, since the resulting primary auxiliary view will produce a point view of line 1-3, as explained in Sec. 13.9, and therefore an edge view (EV) of surface 1-2-3. The reference plane X, used to draw the primary auxiliary view, is assumed to coincide with the back surface of the object. Reference line X-X is drawn in the top view and primary auxiliary view. To complete the primary auxiliary view, all depth measurements, such as *a,* are transferred with dividers from the top view to the primary auxiliary view with respect to reference line X-X.

To draw the secondary auxiliary view, a line of sight, arrow S, is selected that is perpendicular to the edge view of surface 1-2-3 in the primary auxiliary view. Reference plane Y is assumed cutting through the object to

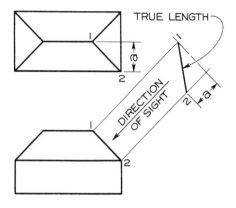

Fig. 13-14. True length of line.

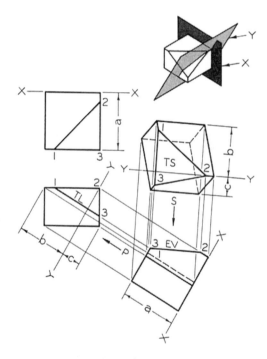

Fig. 13-15. Secondary auxiliary views.

simplify the transfer of measurements. Reference line Y-Y is drawn in the front view and the secondary auxiliary view. Note carefully the transfer of measurements to either side of reference line Y-Y in both views and

283

their relative locations with respect to the primary auxiliary view. For example, dimension *b* is transferred to the side of Y-Y *away* from the primary auxiliary view in both views.

13.13 COMPUTER GRAPHICS. Auxiliary views can be produced on a CAD system in much the same manner as in manual drafting. In two-dimensional CAD, the auxiliary plane is placed in the desired position. Lines are projected perpendicular to the plane. CAD allows distances to be set and lines to be copied to other positions, speeding the

process. The true size and shape of any line or shape can be quickly determined. Commands and menu picks give information about length of line, circumference, area, and perimeter.

Three-dimensional CAD models allow the drafter to view the object from any angle, Fig. 13-16. An auxiliary view is obtained by viewing the object perpendicular to the surface or line desired. The view is then saved as a separate view and placed in the proper position. Partial auxiliaries can be developed by eliminating lines and shapes that are not required.

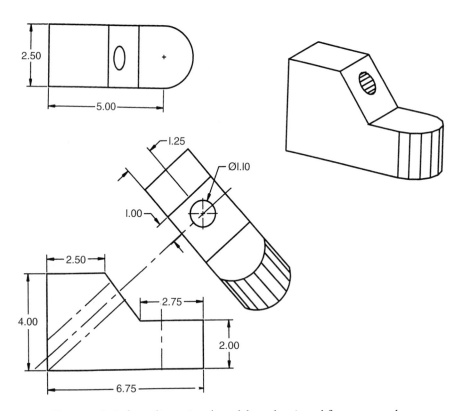

Fig. 13-16. A three-dimensional model can be viewed from any angle.

PROBLEMS

Elementary auxiliary-view problems are given in Figs. 13-17 to 13-19. They are to be drawn either freehand or mechanically, with or without dimensions, as assigned. The problems that follow are practical drafting-room problems such as would be encountered by the industrial drafter. All problems are to be drawn on Layout C in the Appendix unless otherwise indicated.

If assigned by the instructor, construct drawings using decimal-inch or metric scales by converting given fractional dimensions to decimal-inch or metric equivalents. Refer to Table 20 in the Appendix.

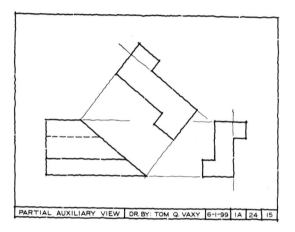

Fig. 13-17. Sketch.

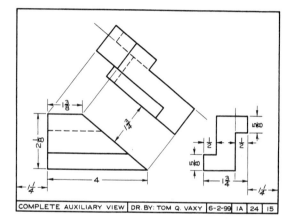

Fig. 13-18. Mechanical drawing.

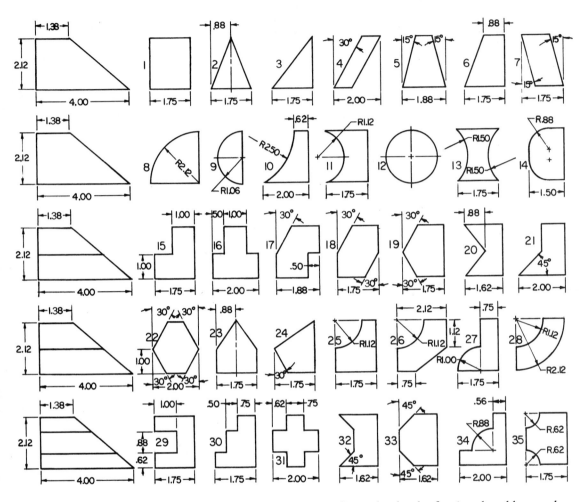

Fig. 13-19. Auxiliary-view problems. Using Layout A in Appendix, make sketch of assigned problem as shown in Fig. 13-17 or mechanical drawing as in Fig. 13-18. In either case, draw complete auxiliary view, or partial auxiliary view of inclined face only, as assigned.

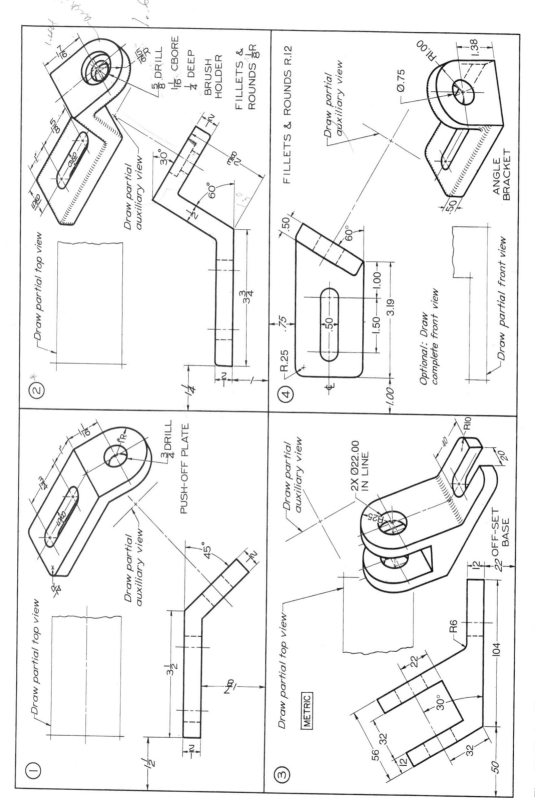

Fig. 13-20. Auxiliary-view problems. Using Layout C in Appendix, make complete mechanical drawing of problem assigned. Omit pictorial drawings, spacing dimensions, and instructional notes. Include complete dimensions, moving dimensions from pictorials to the views.

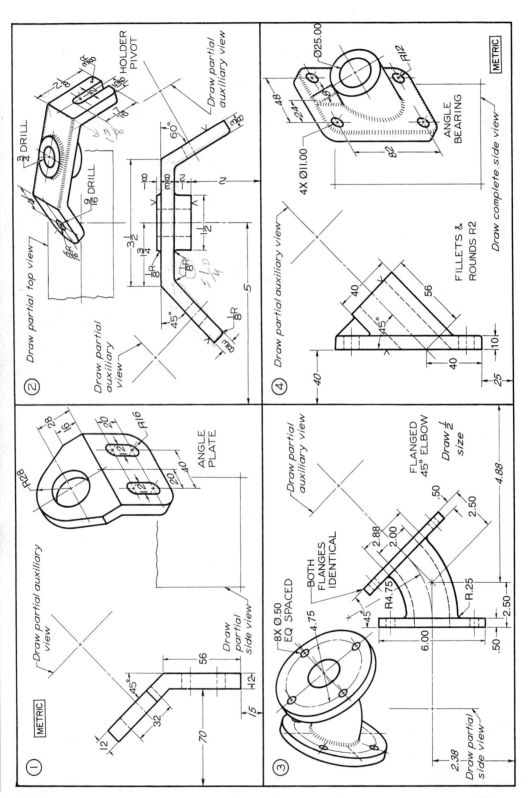

Fig. 13-21. Auxiliary-view problems. Using Layout C in Appendix, make complete mechanical drawing of problem assigned. Omit pictorial drawings, spacing dimensions, and instructional notes. Include complete dimensions, moving dimensions from pictorials to the views.

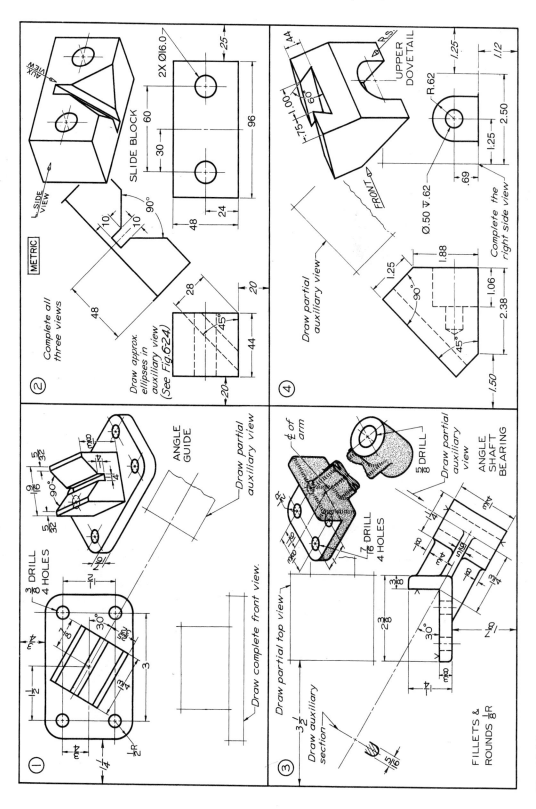

Fig. 13-22. Auxiliary-view problems. Using Layout C in the Appendix, make complete mechanical drawing of problem assigned. Omit pictorial drawings, spacing dimensions, and instructional notes. Include complete dimensions, moving dimensions from pictorials to the views.

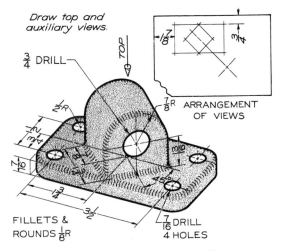

Draw top and auxiliary views.

¾ DRILL

FILLETS & ROUNDS ⅛R

DRILL 4 HOLES

Fig. 13-23. Bearing (Layout C).

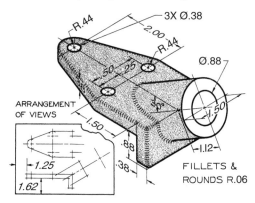

Draw front, top, and partial auxiliary views.

3X Ø.38

R.44

Ø.88

FILLETS & ROUNDS R.06

Fig. 13-24. Holder bracket (Layout C).

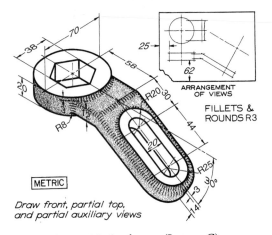

FILLETS & ROUNDS R3

METRIC

Draw front, partial top, and partial auxiliary views

Fig. 13-25. Angle arm (Layout C).

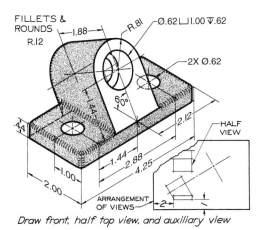

FILLETS & ROUNDS R.12

Ø.62 ⌴1.00 ▽.62

2X Ø.62

HALF VIEW

Draw front, half top view, and auxiliary view

Fig. 13-26. End bearing (Layout C).

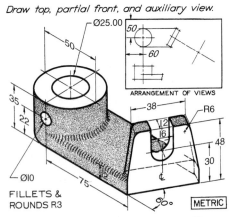

Draw top, partial front, and auxiliary view.

Ø25.00

R6

FILLETS & ROUNDS R3

METRIC

Fig. 13-27. Contact arm (Layout C).

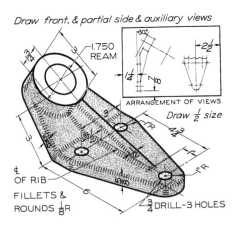

Draw front, & partial side & auxiliary views

1.750 REAM

Draw ½ size

¢ OF RIB

FILLETS & ROUNDS ⅛R

¾ DRILL-3 HOLES

Fig. 13-28. Shaft bracket (Layout C).

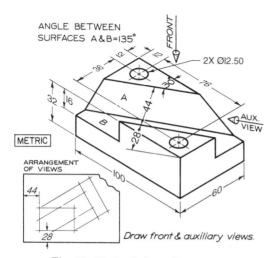

Fig. 13-29. Angle base (Layout C).

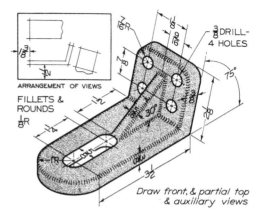

Fig. 13-30. Spar bracket (Layout C).

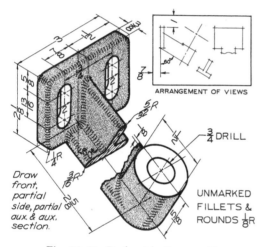

Fig. 13-31. Rod guide (Layout C).

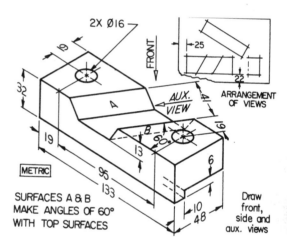

Fig. 13-32. Slotted support (Layout C).

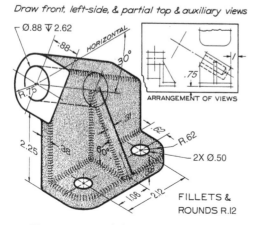

Fig. 13-33. Angle bearing (Layout C).

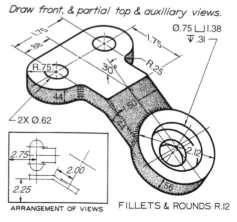

Fig. 13-34. Spacing lever (Layout C).

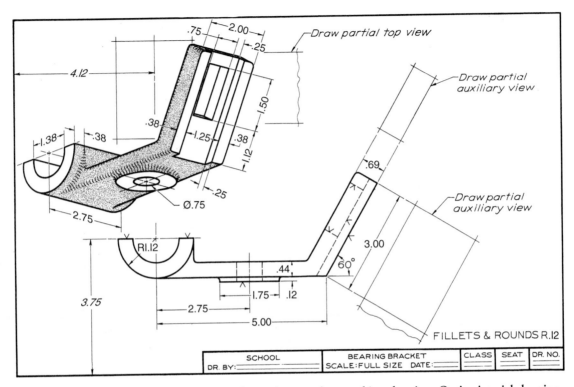

Fig. 13-35. Using Layout D in the Appendix, make complete working drawing. Omit pictorial drawing, instructional notes, and spacing dimensions. Move dimensions from pictorial to views.

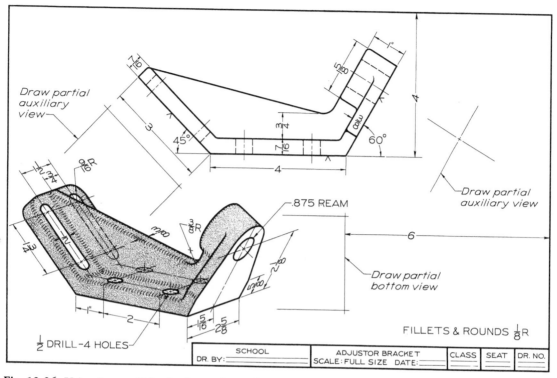

Fig. 13-36. Using Layout D in the Appendix, make complete mechanical drawing. Omit pictorial drawing, instructional notes, and spacing dimensions. Move dimensions from pictorial to views.

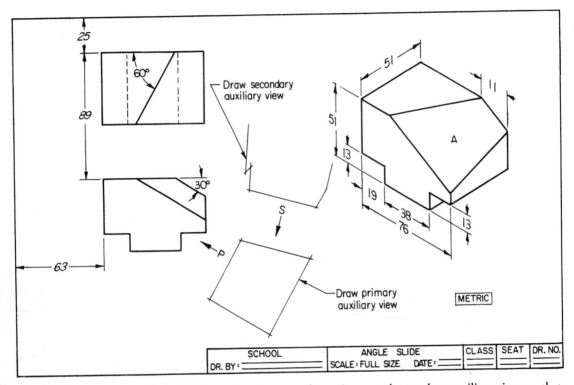

Fig. 13-37. Using Layout D in the Appendix, draw complete primary and secondary auxiliary views so that the latter shows the true shape of oblique surface A. Omit pictorial drawing, instructional notes, and all dimensions.

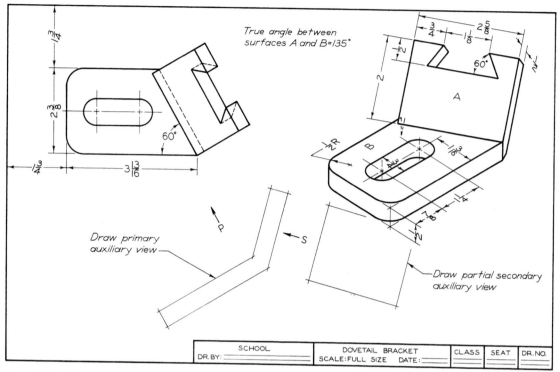

True angle between
surfaces A and B=135°

A

B

Draw primary
auxiliary view

Draw partial secondary
auxiliary view

P

S

60°

60°

SCHOOL	DOVETAIL BRACKET	CLASS	SEAT	DR.NO.
DR.BY:	SCALE:FULL SIZE DATE:			

Fig. 13-38. Using Layout D in the Appendix, draw a complete primary auxiliary view, and a partial secondary auxiliary view that shows the true shape of the oblique surface. Omit pictorial drawing, instructional notes, and all dimensions.

CHAPTER 14

IMPORTANT TERMS

axis of revolution   direction of
                      revolution

OBJECTIVES

After studying this chapter, you should be able to:

- Explain the relationship between an auxiliary view and a revolution.

- Prepare a revolution about an axis perpendicular to a front plane.

- Prepare a revolution about an axis perpendicular to a top plane.

- Prepare a revolution about an axis perpendicular to a side plane.

- Prepare successive revolutions.

- Find the true length of a line by means of revolution.

CHAPTER 14

Revolutions

A revolution presents an object from a different view. Revolutions are used to simplify a drawing, avoid confusion, or save time. They are important in all technical drawing careers. For example, naval architects may use revolved views to show the true size and shape of inclined surfaces in a ship's design.

Naval architects design boats, ships, and marine structures of all kinds. Civilian boats might include anything from small personal sailboats and fishing boats to vessels designed for commerce, such as oil barges, ocean-going freighters, and cruise ships. Military vessels of all types are also designed by the naval architect.

Naval architects must have thorough engineering training, with a background in structural materials, fluid dynamics, power plants, naval technology, and human engineering. They constantly work with drawings of all types, ranging from simple sketches to complex mechanical assembly drawings.

■ CAREER LINK ■

To learn more about careers in naval architecture, contact JETS-Guidance, 1420 King Street, Suite 405, Alexandria, VA 22314 or visit their web site at *www.asee.org/jets*

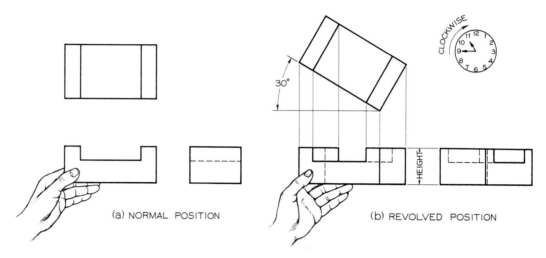

Fig. 14-l. Revolution.

14.1 REVOLUTION. The normal position of an object for its three regular views is shown in Fig. 14-1(a). If the object is revolved so as to bring the right end nearer to the observer in the front view, as shown at (b), the front and side views will change greatly in appearance. However, the *height* will remain the same. The top view will be the same size and shape. It will be changed in position only. The amount of rotation, in this case 30°, shows in the top view. The revolution shown is said to be *clockwise*. In the opposite direction, it would be *counterclockwise*.

14.2 AUXILIARY VIEWS AND REVOLUTION. As shown in the previous chapter, an auxiliary view is obtained when you *view the object at an angle*, Fig. 14-2(a). In other words, for an auxiliary view you shift your position with respect to the object until you can view it in

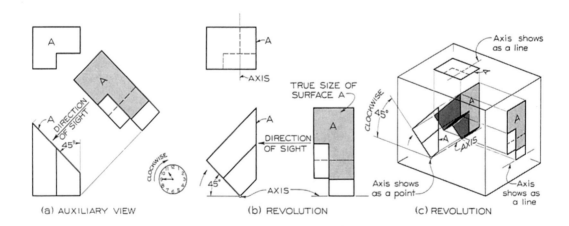

Fig. 14-2. Auxiliary views and revolution.

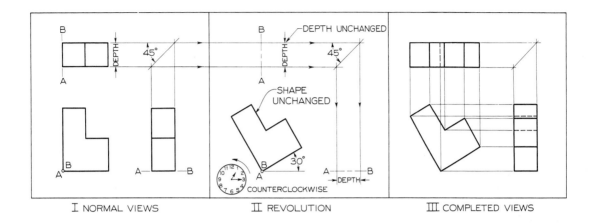

Fig. 14-3. Revolution about an axis perpendicular to front plane.

the direction you wish. In this case the purpose is to obtain a view that shows the true size and shape of surface A.

You can obtain exactly the same view by *revolving the object* until the side view shows the true size and shape of surface A, as shown at (b). In this case the *axis of revolution* is assumed to be perpendicular to the front plane of the imaginary glass box, (c). The axis of revolution shows as a point in the front

view, as shown at (b). The *direction of revolution,* in this case, is said to be *clockwise,* as shown by the small clock face. The axis may be assumed through any convenient point on the object.

14.3 REVOLUTION ABOUT AXIS PERPENDICULAR TO FRONT PLANE – FIG. 14-3. In this case, as shown at I, the axis A-B is taken perpendicular to the front plane so that the axis shows

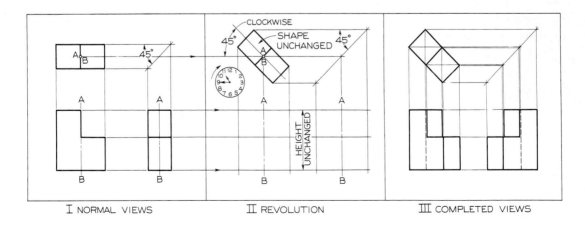

Fig. 14-4. Revolution about an axis perpendicular to top plane.

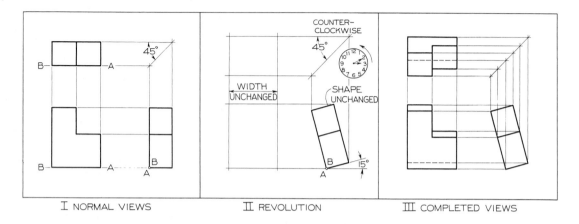

I NORMAL VIEWS   II REVOLUTION   III COMPLETED VIEWS

Fig. 14-5. Revolution about an axis perpendicular to side plane.

as a point in the front view. As shown at II, the front view is copied in the revolved position, in this case 30° *counterclockwise*. Notice that the front view is exactly the same size and shape as before, but in a revolved position. Remember this rule: *The view that is revolved is always the one where the axis is shown as a point. This view is not changed in shape and size.*

The top and side views will be changed in shape and size, but the depth dimension remains the same. Remember this rule: *In the views where the axis shows as a line, the dimension parallel to the axis remains unchanged.* Completed views are shown at III.

14.4 REVOLUTION ABOUT AXIS PERPENDICULAR TO TOP PLANE – FIG. 14-4. In this case, as shown at I, the axis A-B is taken perpendicular to the top plane so that the axis shows as a point in the top view. As shown at II, the top view is copied in the revolved position, in this case 45° *clockwise*. The top view is exactly the same size and shape as before, since the axis shows as a point in this view. However, the

view is revolved.

In the front and side views, the axis shows as a line. Therefore, the dimension of the object that is parallel to the axis — in this case the *height* — remains unchanged. It can be projected from I or transferred with dividers or scale. The completed views are shown at III.

14.5 REVOLUTION ABOUT AXIS PERPENDICULAR TO SIDE PLANE – FIG. 14-5. In this case, as shown at I, the axis A-B is taken perpendicular to the side plane so that the axis shows as a point in the side view. As shown at II, the side view is copied in the revolved position, in this case 15° *counterclockwise*. The side view is exactly the same size and shape as before, since the axis shows as a point in this view. However, the view is revolved. In the front and top views, the axis shows as a line. Therefore, the dimension of the object that is parallel to the axis — in this case the *width* — remains unchanged. It can be transferred from I with dividers or scale. The completed views are shown at III.

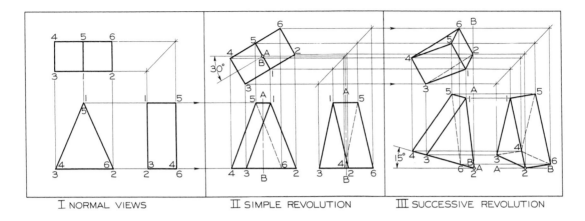

I NORMAL VIEWS   II SIMPLE REVOLUTION   III SUCCESSIVE REVOLUTION

Fig. 14-6. Successive revolution.

14.6 SUCCESSIVE REVOLUTIONS – FIG. 14-6.
After one revolution, the object can be revolved further through as many stages as desired. The given views in this case are shown at I. As shown at II, the object is first revolved 30° counterclockwise about an axis perpendicular to the top plane. Then, at III, the object is further revolved 15° clockwise about an axis perpendicular to the front plane. This can be continued indefinitely. Such drawings provide excellent practice in projection of views.

Important Rule: Lines that are parallel on the object will be parallel in any view. Therefore, in drawing the lines in Fig. 14-6 at II and III, make sure that the various sets of parallel lines are truly parallel. Use the triangle and T-square as shown in Fig. 4-20.

14.7 TRUE LENGTH OF LINE – FIG. 14-7. The
true length of any oblique line can be easily found by means of revolution. For example, in (a), the element of the cone 1-2 is revolved to the position 1-2′ where the line shows true length in the front view. At (b) the same method is applied to the oblique edge of a pyramid. At (c) the true length of a hip rafter 1-2 is found in the same manner.

The true length of a line can also be found by means of an auxiliary view, as shown previously in Fig. 13-14.

14.8 PRACTICAL APPLICATIONS OF
REVOLUTION. Revolution may sometimes be used instead of an auxiliary view to save time and simplify the drawing, as shown in Fig. 14-8(a). A common application of revolution often occurs where a part is to be bent after machining, as shown at (b).

In some cases revolution is helpful in clari-

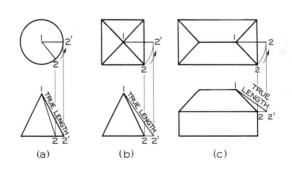

Fig. 14-7. True length of line.

301

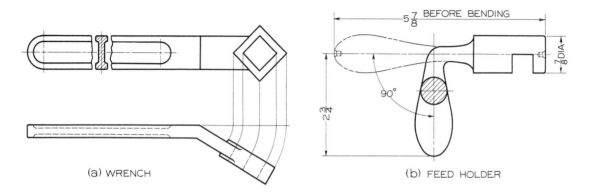

BEFORE BENDING

(a) WRENCH

(b) FEED HOLDER

Fig. 14-8. Practical applications of revolution.

fying a drawing that would otherwise be confusing and difficult to draw. For example, Fig. 14-9(a) shows a rib in an oblique position, which produces a confusing view, (b). Also, the holes in the base do not appear in their true distance from the rim. They are easily confused with other lines. If the rib and the hole are revolved, a clearer drawing results, as shown at (c). See also Sec. 12.10.

Another example is shown at (d) and (e), in which the slotted arm is foreshortened and distorted. In addition, the view at (e) appears

to be "amputated." Furthermore, the curves are difficult and time-consuming to draw. If the arm is revolved, (f), the resulting view is much simpler and more understandable.

14.9 COMPUTER GRAPHICS. Using computer graphics, objects can be quickly revolved around any desired axis. For example, CAD programs allow the user to move a part with an oblique surface through successive revolutions to show that surface in its true size and shape, Fig. 14-10.

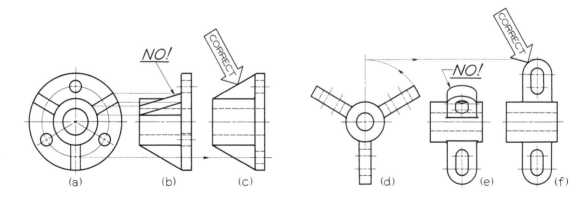

(a) (b) (c) (d) (e) (f)

Fig. 14-9. Practical applications of revolution.

Fig. 14-10. A revolution presents an object in a different view. In CAD, a 3D object can be revolved until the desired surface is shown. The CAD program can then be instructed to draw that view.

PROBLEMS

Simple revolution problems are given in Fig. 14-11. More difficult applications are given in Fig. 14-12. All are to be drawn with instruments. Those in Fig. 14-11 need not be dimensioned. However, dimensions should be shown on the problems in Fig. 14-12.

If assigned by the instructor, construct drawings using decimal-inch or metric scales by converting given fractional dimensions to decimal-inch or metric equivalents. Refer to Table 20 in the Appendix.

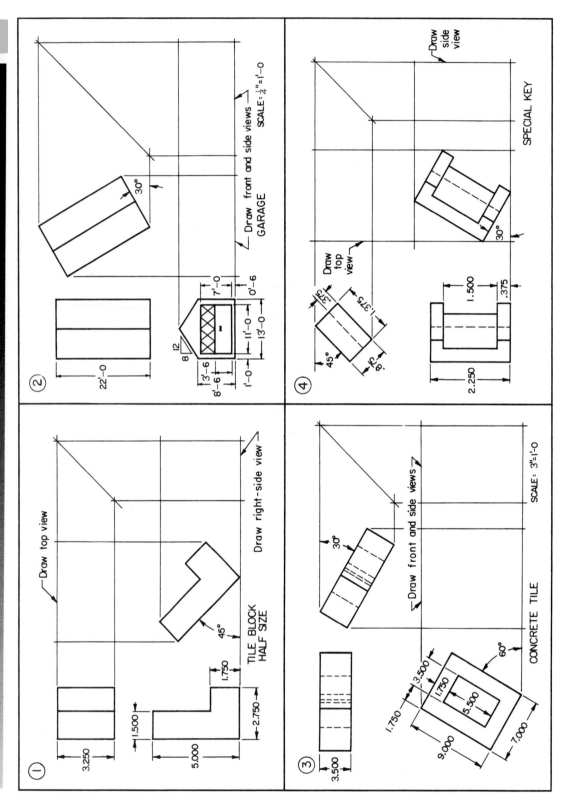

Fig. 14-11. Revolution problems. Using Layout C in the Appendix for each problem, draw given views of assigned problems as shown, and add missing views as indicated. Omit dimensions and instructional notes. Move titles and scales to title strip.

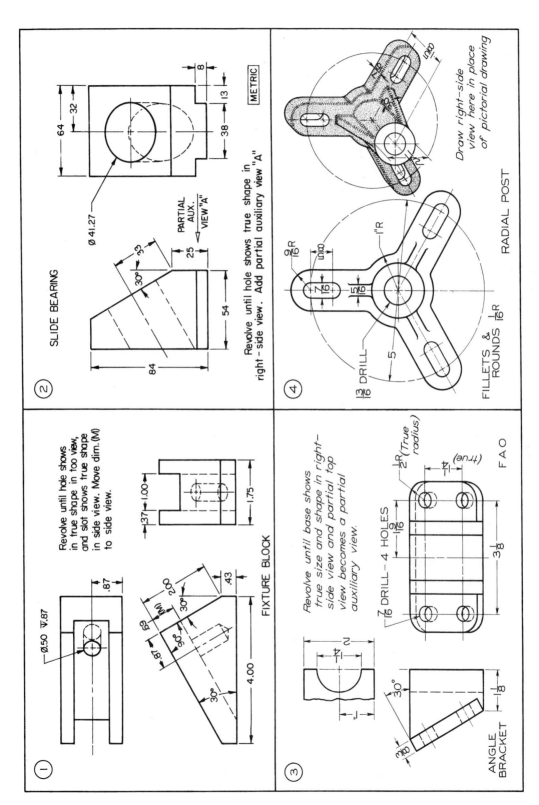

Fig. 14-12. Revolution problems. Using Layout C in the Appendix for each problem, draw problem assigned. Omit instructional notes. Dimension drawings completely. Move titles to title strip.

Arnold & Brown

CHAPTER 15

OBJECTIVES

After studying this chapter, you should be able to:

- Draw a helix.
- Identify three basic thread forms.
- List the steps in drawing a single RH external thread.
- Identify the three types of taps.
- Outline the main steps in drawing ANSI Standard bolts and nuts.

CHAPTER 15

Threads and Fasteners

The mechanical engineer, mechanical designer, and design technician constantly make or use drawings to show how items are designed and manufactured. Any item that is to be manufactured must first be drawn. Machinery, cars, equipment used in industry, and even household products such as toasters or doorbell buttons must all be drawn before they can be produced.

One important task of the mechanical engineer/designer is to carefully select the best fastener for a particular use. The wise engineer has a collection of catalogs showing standard screws, nuts, washers, etc. These fasteners are available in a wide variety of sizes and materials. The engineer is usually able to choose existing fasteners rather than spend time designing new ones.

Fasteners must be shown correctly on drawings so that the production team will know what is needed. This chapter describes fasteners and the standards that apply to them. You will learn how fasteners are drawn and how they are used.

CAREER LINK

To learn about careers in mechanical engineering, try the following sources:

The American Society of Mechanical Engineers, 345 E. 47th St., New York, NY 10017

American Society of Heating, Refrigerating, and Air-Conditioning Engineers, Inc., 1791 Tullie Circle NE, Atlanta, GA 30329. The Internet address is: //www.ashrae.org

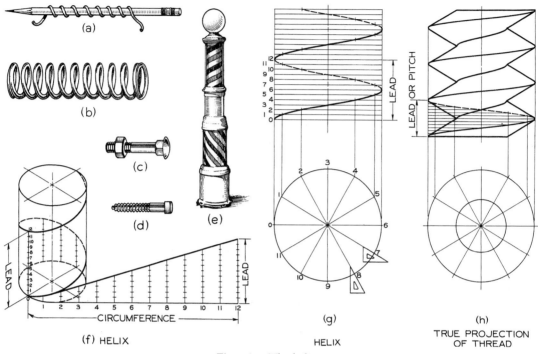

Fig. 15-1. The helix.

15.1 THE HELIX. All screw threads wind around a shaft in a curve called a *helix* (pronounced hē′ liks). If you wind a string around a pencil or any cylinder, the string will take the general form of a helix, Fig. 15-1(a). Other applications are found in the spring, (b), the threads on a bolt, (c), or on a screw, (d), and the stripes on a barber pole, (e). If a right triangle is wrapped around a cylinder, as at (f), the hypotenuse will form a helix on the cylinder. The *lead* (pronounced leed) of the helix is the distance, parallel to the axis, from a point on one turn of the helix to the corresponding point on the next turn, as shown.

To draw a helix, draw two views of the cylinder, (g). Divide the circle into any number of equal parts, say 12. Then, on the rectangular view, assume a lead. Divide it into the same number of equal parts. Then draw

parallel lines and number the divisions as shown. Project point 1 on the circle up to line 1, point 2 up to line 2, etc. Draw the helix through the points with the aid of the irregular curve, Sec. 4.30.

The use of the helix in drawing the true projections of a screw thread is shown at (h). The drafter should understand the helix and how to draw it. Nonetheless, a true helical thread drawing takes so much time that simpler approximations are used, as explained in the paragraphs that follow.

15.2 SCREW THREADS. Figure 15-2(a) shows a photograph of an *external thread* and of the corresponding *internal thread* in section, illustrating the various thread terms. At (b) is shown a drawing of the same threads in which straight lines are used to replace the difficult helical curves.

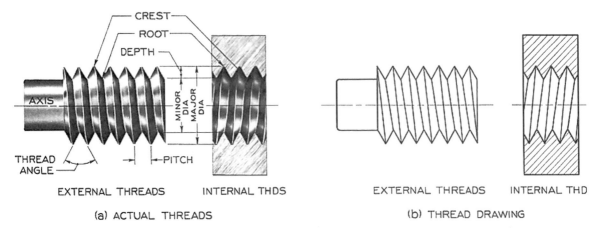

Fig. 15-2. Screw-thread terms.

Screw threads are used mainly to (1) *hold* parts together, as on a bolt or screw, (2) *adjust* parts with respect to each other, as on the adjusting screw on a bow compass, or (3) *transmit power*, as on a vise screw or a valve stem.

15.3 CUTTING THREADS ON A LATHE. Threads are often cut on a lathe, particularly the larger threads, Fig. 15-3. The threads are formed by cutting helical grooves around the shaft, leaving the threads standing as helical ridges. As the shaft rotates in the lathe, the cutting tool moves slowly to the left. Several cuts are made, each a little deeper, until the correct depth is reached. For a top view of this operation, see Fig. 15-4(a).

Chicago Public Schools

Fig. 15-3. Cutting external threads on a lathe.

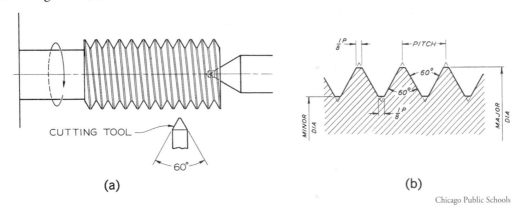

(a)

(b)

Chicago Public Schools

Fig. 15-4. Form of thread.

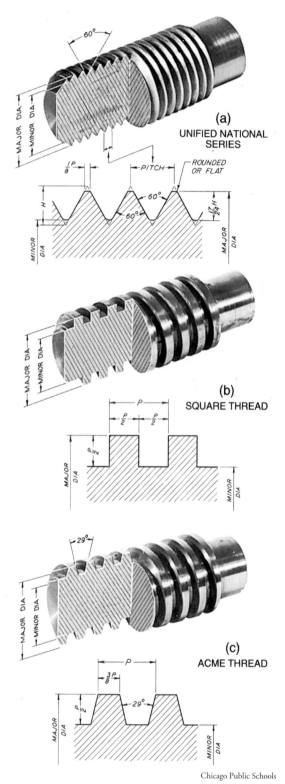

(a)
UNIFIED NATIONAL
SERIES

(b)
SQUARE THREAD

(c)
ACME THREAD

Fig. 15-5. Thread forms.

Large internal threads are cut in a manner similar to boring, Figs. 11-17 and 11-20(g). Small internal and external threads are usually cut with taps and dies, Sec. 15.12.

15.4 THREAD FORMS. The *form* of thread is determined by the shape of the cutting tool, Fig. 15-4. Some of the basic forms are shown in Fig. 15-5. The *square* and *Acme* threads are used (also in modified forms) for transmitting power, as on a vise.

Additional thread forms are shown in Fig. 15-6. The *sharp-V thread,* (a), is now used only to a limited extent where friction holding or adjusting is important. The Unified thread is now the new ANSI Standard thread. The *worm thread,* (c), is used on shafts to carry power to worm gears. The *knuckle thread,* (d), is usually rolled from sheet metal. However, it is sometimes cast. It is used in various modified forms on bottle tops, light sockets and globes, and such items. The *buttress thread,* (e), is used to transmit great power in one direction, as on jacks and on breech locks of large guns. The *metric thread* (f) has a profile similar to the Unified series, but the two are not interchangeable. For unusual requirements, there are many different special thread forms.

15.5 THREAD PITCH. The *pitch* of a thread is the distance, parallel to the axis, from a point on one thread to the corresponding point on the next adjoining thread, Figs. 15-2(a), 15-5, and 15-6. The pitch, P, is usually expressed in tables (see the Appendix) in terms of the *number of threads per inch,* which means the number of pitches per inch. To find the pitch, divide 1 by the number of threads per inch. Thus, if a thread has 4 threads per inch,

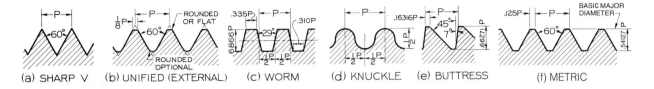

(a) SHARP V (b) UNIFIED (EXTERNAL) (c) WORM (d) KNUCKLE (e) BUTTRESS (f) METRIC

Fig. 15-6. Other thread forms.

Fig. 15-7(a), the pitch is $\frac{1}{4}''$, and the threads are relatively large. If a thread has a larger number of threads per inch, say 16, as at (b), the threads are relatively small. In measuring "threads per inch" the scale, (c), or a thread pitch gage, (d), can be used. For square or Acme threads, (e) and (f), each pitch includes a thread and a space.

15.6 RIGHT-HAND (RH) AND LEFT-HAND (LH) THREADS.

A right-hand, or RH, thread advances into a nut when turned clockwise, Fig. 15-8(a). A left-hand, or LH, thread advances into a nut when turned counterclockwise, (b). If your right or left hand is placed alongside the threads, the direction of the thumb indicates whether the thread is RH or LH, as shown. All threads are understood to be RH unless designated specifically LH in the thread note, Sec. 15.18.

15.7 MULTIPLE THREADS.

A *single thread* is composed of a single ridge around a shaft, with each turn next to the previous turn, Fig. 15-9(a). This can be likened to a single cord wound around a rod, each turn packed tightly against the previous one, as shown. *Multiple threads* are composed of two or more ridges side by side, (b) to (d). This may be likened to two or more cords wound around a rod. Note that the *slope* (or slant of the crest lines) of a single thread is determined by the offset measurement $\frac{1}{2}$P, a double thread by the offset measurement P, a triple thread by the offset measurement $1\frac{1}{2}$P.

The *lead* of a thread is the distance a threaded shaft advances into the nut in one revolution. As shown in Fig. 15-9, in a single thread the lead is equal to the pitch. In a double thread, it is twice the pitch. In a triple thread, it is three times the pitch.

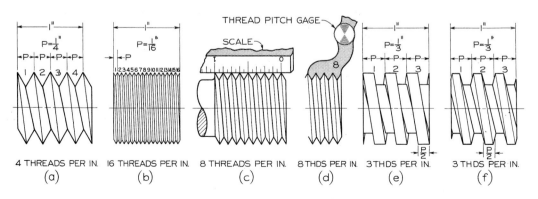

4 THREADS PER IN. (a) 16 THREADS PER IN. (b) 8 THREADS PER IN. (c) 8 THDS PER IN. (d) 3 THDS PER IN. (e) 3 THDS PER IN. (f)

Fig. 15-7. Threads per inch.

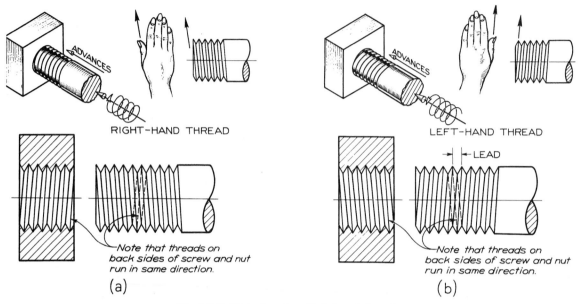

Fig. 15-8. Right-hand and left-hand threads.

Note that in a drawing of a single or a triple thread a crest is opposite a root. In the case of a double or a quadruple thread a crest is opposite a crest. This holds for all forms of threads.

Multiple threads are used wherever quick action but not great power is required. Such threads are used on valve stems, fountain pen caps, and toothpaste tube caps.

15.8 SYMBOLIC AND DETAILED THREADS. In drawing threads, represent them as simply as possible. The simplest method of drawing threads is the *symbolic method,* Fig. 15-10(a)

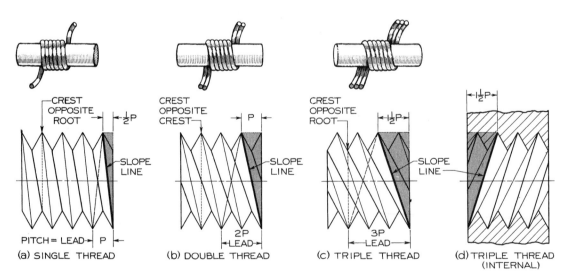

Fig. 15-9. Multiple threads.

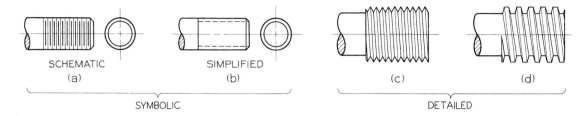

Fig. 15-10. Thread representation (external).

and (b). This method is used for small diameters (under approximately 1″ on the drawing) where more complete thread pictures would be difficult or impractical to draw. The symbolic method is used for *all forms of threads,* as American National, National series, Unified, square, or Acme.

The nearest approximation to the true thread picture is the *detailed* thread representation, Fig. 15-10(c) and (d), in which straight lines replace the helical curves. This method may be used for diameters of approximately 1″, or over, on the drawing. Thus, a $2\frac{1}{2}$″ diameter thread to half scale will be $1\frac{1}{4}$″diameter on the drawing. It may be drawn by the detailed method. However, many companies have adopted the schematic representation for all threads.

15.9 DETAILED SHARP-V AND UNIFIED NATIONAL THREADS. As shown in Figs. 15-4 to 15-6, the Sharp-V, and Unified National thread forms are the same except for rounds or flats on the crests and roots. These small features are ignored. Both threads are drawn in the same way by the detailed method, Fig. 15-11. Note how the external threads appear on the shaft, (a), and in the end view, (b). Note how the internal threads appear in section, (c), in elevation, (d), and in the end view, (e).

The steps in drawing a single RH external thread are shown in Fig. 15-12.

I. Lay out the major diameter and the thread length with light construction lines. Then find the pitch, P, for this diameter by using Table 2 in the Appendix. See also Fig. 15-7(a). For

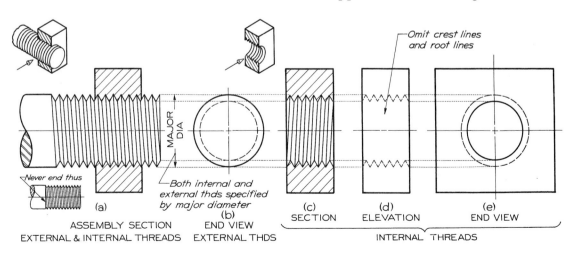

Fig. 15-11. Detailed Sharp-V or Unified threads.

313

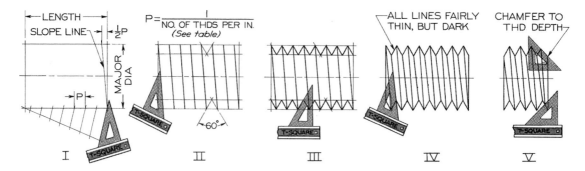

Fig. 15-12. Steps in drawing external American National, Unified, or Sharp-V threads.

example, if this is a 3″ diameter thread, there are 4 threads per inch, and $P = \frac{1}{2}$″. Establish the slope of the crest lines by the offset measurement $\frac{1}{2}P$, or $\frac{1}{8}$″. Then set off distances P along the lower line, as shown. In this case $P = \frac{1}{4}$″. It can easily be set off with the scale. If the pitches cannot be set off with the scale, use the parallel-line method, as shown (see Fig. 6-2 or 6-3).

II. Draw fine parallel crest lines, using triangle and T-square, as shown. These lines can be drawn finished weight at once. Draw two 60° V's. Then draw guide lines for the root diameter of the

thread, as shown.
III. Draw all V's finished weight.
IV. Draw root lines finished weight.
V. Construct chamfer, if required. Note that this will cause the last crest line to change position slightly.

In the final drawing, all lines should be fairly thin and dark. Note that the root lines will not be parallel to the crest lines.

15.10 DETAILED SQUARE THREADS. Detailed square threads are shown in Fig. 15-13. Note carefully the differences between the internal and external threads at (a) and (c), and the differences between the external thread alone

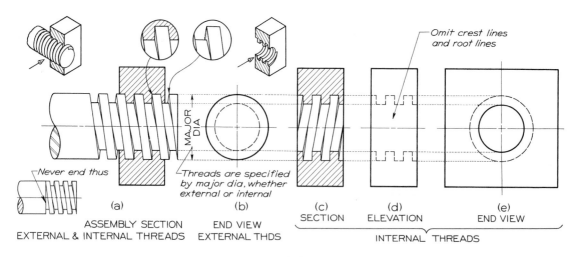

Fig. 15-13. Detailed square threads.

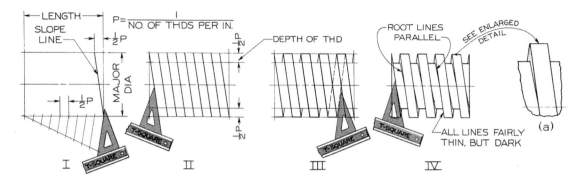

Fig. 15-14. Steps in drawing external square threads.

and when mated with the internal thread in section, (a), as shown in the small circles.

The steps in drawing an external square thread are shown in Fig. 15-14.

I. Lay out the major diameter and the thread length with light construction lines. Then find the pitch, P, for this diameter by using Table 9 in the Appendix. See Fig. 15-7(e). For example, if this is a $1\frac{1}{2}''$ diameter thread, there are 3 threads per inch, and $P = \frac{1}{3}''$. Along the lower line, set off distances $\frac{1}{2}P$ ($\frac{1}{6}''$). You can do this by using the parallel-line method, Fig. 6-2 or 6-3, as shown, or by dividing 1" into 6 parts with the bow dividers. Then establish the slope of the threads by the offset measurement $\frac{1}{2}P$ ($\frac{1}{6}''$).

II. Draw parallel crest lines, using the triangle and T-square. These can be drawn finished weight at once. Draw light guide lines for the depth of the thread, as shown, using measurements $\frac{1}{2}P$.

III. Heavy-in tops of threads. Draw visible back crest lines, finished weight.

IV. Draw parallel visible root lines finished weight. Heavy-in bottoms of thread spaces. See detail at (a).

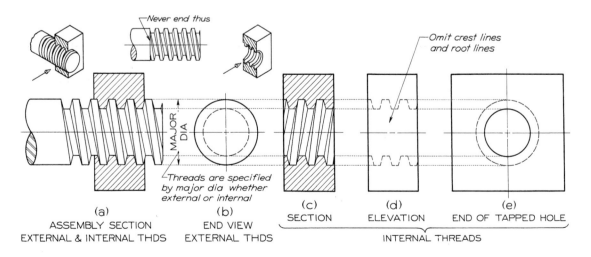

Fig. 15-15. Detailed Acme threads.

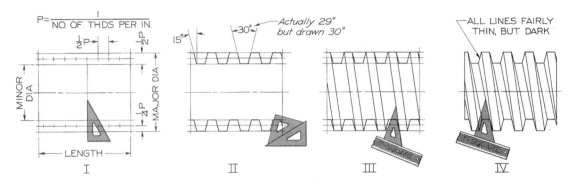

Fig. 15-16. Steps in drawing external Acme threads.

All final lines should be fairly thin and dark.

15.11 DETAILED ACME THREADS. Detailed Acme threads are shown in Fig. 15-15, and the steps in drawing an external Acme thread in Fig. 15-16.

I. Lay out the major diameter and the thread length with light construction lines. Then find the pitch, P, for this diameter by using Table 9 in the Appendix. See Fig. 15-7(f). For example, if this is a $1\frac{1}{2}''$ diameter thread,

there are 3 threads per inch, and $P = \frac{1}{3}''$. Draw root-diameter guide lines, making the thread depth $\frac{1}{2}P$ or $\frac{1}{6}''$. Divide $1''$ into sixths by trial with the bow dividers, or use the parallel-line method, Fig. 6-2 or 6-3. Then draw construction lines halfway between these lines and the outside lines. Set off $\frac{1}{2}P$ ($\frac{1}{6}''$) distances along these construction lines.

II. Draw sides of threads 15° with vertical through the $\frac{1}{2}P$ P points. These can be drawn finished weight at once. Then heavy-in tops of threads and bottoms of thread spaces.

III. Draw parallel crest lines, finished weight, using the triangle and T-square.

IV. Draw parallel root lines, finished weight, as shown. All final lines should be fairly thin and dark.

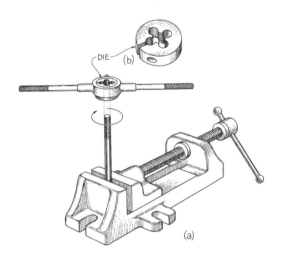

Fig. 15-17. Cutting a small external thread with a die.

15.12 CUTTING THREADS WITH TAPS AND DIES. Small external threads are often cut by hand with a *die*, Fig. 15-17. An enlarged view of a die is shown at (b). The die fits in a die holder. The thread is cut merely by turning the tool until the proper thread length is obtained. For production work, the die may be power driven on a lathe or in a drill press.

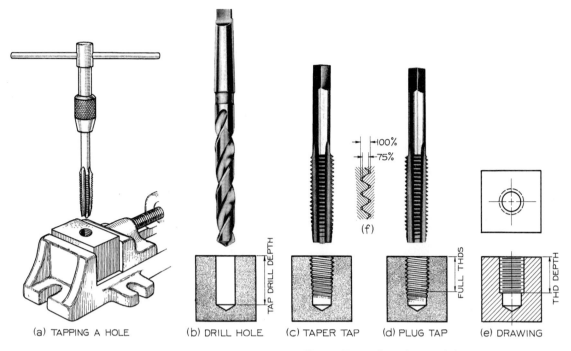

(b) National Twist Drill & Tool Co. (c) and (d) Morse Twist Drill & Machine Co.

Fig. 15-18. Tapping a small hole.

Small internal threads are usually cut with a *tap,* such holes being called *tapped holes,* Fig. 15-18(a). The tap is a small fluted cutting tool with cutting teeth shaped to cut the threads that are desired. First, the original hole is drilled, (b). The depth of the hole must be several thread pitches deeper than the thread length required, to allow space for the end of the tap and for the metal chips. The diameter of the drill is slightly greater than the root diameter of the thread so that

the actual thread engagement will be, for best results, about 75% of the full depth of the thread. Tap drill sizes for Unified threads are given in Table 2 of the Appendix. It is good practice to give tap drill information in the thread note, Fig. 15-22(I). However, this is often omitted. The shop person consults tap drill tables.

Taps are available in sets of three, the *taper tap,* the *plug tap,* and the *bottoming tap.* The taper tap, Fig. 15-18(c), is always used first. It

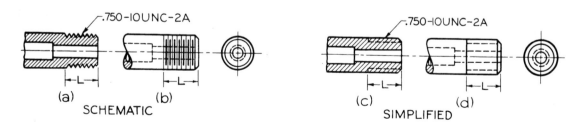

Fig. 15-19. Symbolic external threads.

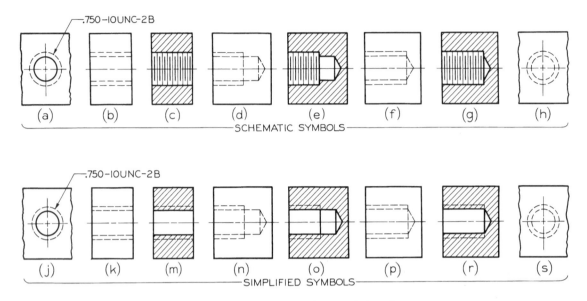

Fig. 15-20. Conventional internal threads.

has a considerable tapered portion at the end to guide the tap, but does not have full threads near the end. To extend the threads deeper in the hole, the plug tap, (d), is then used. Although this still leaves several imperfect threads in the bottom of the hole, this tap is usually the last one used. The tapped hole as it would appear on a drawing is shown at (e). If

necessary to tap practically to the bottom of the hole, the bottoming tap (not shown in Fig. 15-18) is used, producing a hole as shown in Fig. 15-20(f), (g), (p), and (r).

A "blind" tapped hole is one that does not go through a piece, Fig. 15-18(e). A "through" tapped hole goes entirely through. In general, blind tapped holes should be

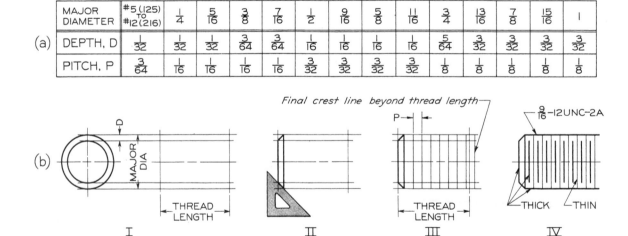

	MAJOR DIAMETER	#5 (125) TO #12 (216)	$\frac{1}{4}$	$\frac{5}{16}$	$\frac{3}{8}$	$\frac{7}{16}$	$\frac{1}{2}$	$\frac{9}{16}$	$\frac{5}{8}$	$\frac{11}{16}$	$\frac{3}{4}$	$\frac{13}{16}$	$\frac{7}{8}$	$\frac{15}{16}$	1
(a)	DEPTH, D	$\frac{1}{32}$	$\frac{1}{32}$	$\frac{1}{32}$	$\frac{3}{64}$	$\frac{3}{64}$	$\frac{1}{16}$	$\frac{1}{16}$	$\frac{1}{16}$	$\frac{1}{16}$	$\frac{5}{64}$	$\frac{3}{32}$	$\frac{3}{32}$	$\frac{3}{32}$	$\frac{3}{32}$
	PITCH, P	$\frac{3}{64}$	$\frac{1}{16}$	$\frac{1}{16}$	$\frac{1}{16}$	$\frac{1}{16}$	$\frac{3}{32}$	$\frac{3}{32}$	$\frac{3}{32}$	$\frac{3}{32}$	$\frac{1}{8}$	$\frac{1}{8}$	$\frac{1}{8}$	$\frac{1}{8}$	$\frac{1}{8}$

Fig. 15-21. Schematic external threads (full size).

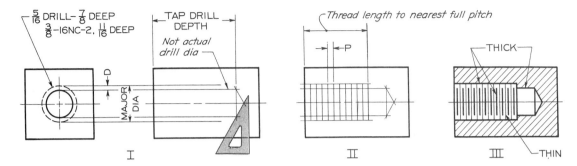

Fig. 15-22. Schematic internal threads (full size).

avoided where possible. Chips in the bottom may cause jamming.

For production work, tapping is done in a lathe or drill press, in which a tapholder is used to hold and drive the tap.

15.13 SYMBOLIC THREADS. *Symbolic thread symbols* are used for small diameters, under approximately 1″ on the drawing, to represent *all forms of threads*. Two forms are approved as ANSI Standard, the *schematic* and the *simplified*. External thread symbols are shown in Fig. 15-19. Note that when the schematic form is drawn in section, (a), the actual thread forms are drawn. Otherwise, the presence of threads would not be evident. In elevation, (b), the crest lines are represented by long thin lines and the roots by short thick lines, both at right angles to the shaft. At (c) and (d) are shown the simplified forms in which hidden lines are drawn parallel to the axis at the approximate depth of the thread.

Symbolic internal threads are shown in Fig. 15-20. Note that schematic and simplified symbols are alike except in sectional views.

15.14 TO DRAW SCHEMATIC EXTERNAL SYMBOLIC THREADS. The steps in drawing external symbolic threads are shown in Fig. 15-21. Since the representation is strictly

symbolic, no attempt is made to use actual thread depth and pitch. Values for depth and pitch can be easily laid off with the scale. They result in a pleasing approximation of the threads. Such values are shown in the table, Fig. 15-21(a).

I. Lay out the major diameter (in this case $\frac{9}{16}$″) and the desired thread length with light construction lines. In the table find the depth D for a $\frac{9}{16}$″ diameter thread. This is $\frac{1}{16}$″. Draw the inner circle in the end view $\frac{1}{16}$″ smaller in radius than the large circle. Draw guide lines for the depth of the thread, as shown.

II. Draw 45° chamfer, finished weight.

III. Draw thin dark crest lines, finished weight, spaced a distance P apart ($\frac{3}{32}$″, as shown in table). Note that the final crest line on the right may fall slightly beyond the actual thread length, which is satisfactory.

IV. Draw dark finished-weight root lines by eye midway between the crest lines. Note the strong contrast between the thin crest lines and the thick lines.

If a thread is drawn to a reduced scale, use the values in the table in Fig. 15-21, which correspond to the diameter as actually drawn. Thus, for a 1″ diameter thread drawn to half scale, use the values for $\frac{1}{2}$″ diameter.

15.15 TO DRAW SCHEMATIC INTERNAL SYMBOLIC THREADS. The steps in drawing "blind" tapped holes are shown in Fig. 15-22. Values for depth and pitch are shown in the table in Fig. 15-21.

I. Draw hidden circle equal to the major diameter (say $\frac{3}{8}$″) and solid circle a distance D ($\frac{3}{64}$″) less in radius than the hidden circle. Lay out rectangular view with light construction lines. Make the tap drill depth several pitches greater than the thread length if the tap drill depth has not been specified in the note. In this case the tap drill depth has been given as $\frac{7}{8}$″. Note that the drill depth does not include the conical point of the hole.

II. Draw finished-weight thin dark crest lines spaced a distance P apart, or $\frac{1}{16}$″.

III. Heavy-in all final lines as necessary to complete the drawing, with a strong contrast between the thin lines and the thick lines.

15.16 AMERICAN NATIONAL SCREW THREADS. The American National thread *form* is shown in Fig. 15-4(b). The old American National thread tables list five *series* of numbers of threads per inch for different diameters: the *Coarse Thread Series* (NC), the *Fine Thread Series* (NF), and the *8-Pitch Series* (8N), *12-Pitch Series* (12N), and *16-Pitch Series* (16N). The 8-, 12-, and 16-Pitch series all have the same number of threads per inch for all diameters. The *Coarse* and *Fine* series are given in Table 2 of the Appendix.

Also standardized are four *classes of fit*. The term *fit* refers to how closely the screw fits in the threaded hole — that is, to the amount of "play" between the two parts. The four fits are the Loose Fit (Class 1), used for rough work; the Free Fit (Class 2), used for the great bulk of screw thread work; the Medium Fit (Class 3), used for the better grades of work, as in automobile engines; and the Close Fit (Class 4), used where a very snug fit is required, as in certain aircraft engine parts.

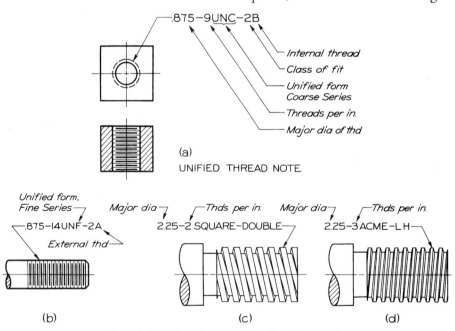

Fig. 15-23. Thread notes. (See also Fig. 11-2.)

15.17 UNIFIED THREADS. The Unified National Standard *form* is shown in Fig. 15-6(b). The ANSI Standard "Unified Screw Threads" (ANSI/ASME B1.1) lists six different *series* of numbers of threads per inch for different diameters, and selected combinations of special diameter and special pitch. The six series are the *Coarse Thread Series* (UNC or NC) recommended for general use, the *Fine Thread Series* (UNF or NF) for automotive and aircraft work, the *Extra Fine Thread Series* (UNEF or NEF) for aircraft work and other uses where a very fine thread is required, and the *8-Thread Series* (8N), *12-Thread Series* (12UN or 12N), and *16-Thread Series* (16UN or 16N). The 8-, 12-, and 16-Thread series all have the same number of threads per inch for all diameters. For numbers of threads per inch and tap drill sizes, see Table 2 in the Appendix.

Exact fits are controlled by dimensions given in tables. The drafter simply indicates by a symbol the fit desired. In these symbols the letter A refers to external threads, and B refers to internal threads. See Fig. 15-23(a) and (b). Classes 1A and 1B take the place of and are similar to Class 1 of the old American National. Classes 2A and 2B are intended for the normal production of screws, bolts, and nuts and for a variety of other uses. Classes 3A and 3B provide for highly accurate and close fitting requirements. Classes 2 and 3 have been retained from the old American National.

15.18 THREAD NOTES. Typical thread notes are shown in Fig. 15-23. Notes for internal threads are attached to the circular views, (a). Notes for external threads are attached to the "side" views of the threads, (b). Threads are always understood to be single threads, Sec. 15.7, unless the note indicates a double thread, (c), or some other multiple. Also, threads are understood to be RH, Sec. 15.6, unless LH is given in the note, as at (d).

15.19 ANSI STANDARD PIPE THREADS. ANSI Standard pipe threads (ANSI/ASME B1.20.1) are either *tapered* or *straight* and have a 60° angle of thread. As shown in Fig. 15-24, pipe threads are drawn by the symbolic method, the detailed method seldom being used. The actual taper of a taper pipe thread is $\frac{1}{16}''$ per inch on *diameter*. This is exaggerated to $\frac{1}{16}''$ per inch *on radius* to make the taper show up more clearly, as shown at (a). Straight pipe threads are drawn as in Fig. 15-24, but without taper.

The *nominal size* of pipe does not correspond to the outside diameter (OD) except in

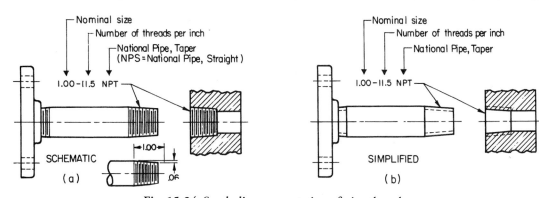

Fig. 15-24. Symbolic representation of pipe threads.

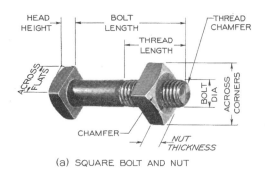

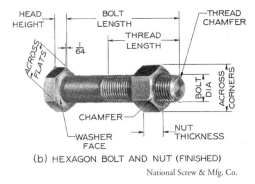

National Screw & Mfg. Co.

Fig. 15-25. Bolt and nut terms.

sizes 14″ and larger. For smaller sizes, the outside diameter is larger than the nominal size. Thus, a 1″ pipe has an outside diameter of 1.315″.

To draw the threads as in Fig. 15-24, lay out the *outside* diameter of the pipe as given in the table. For drawing purposes, make the thread length approximately equal to the outside diameter. Space crest and root lines as in Fig. 15-21. Use the outside diameter of the pipe as the "major diameter" in the table.

For further information on piping drawings, see Sec. 16.14.

15.20 ANSI STANDARD BOLTS AND NUTS.

Bolts and nuts, Fig. 15-25, are used to hold parts together, usually so that they can be taken apart later for repair or replacement. Two standard forms are used, the *hexagon* and the *square*. Square heads and nuts are chamfered at 25°. Hexagon heads and nuts are chamfered at 30°. Both are drawn at 30° for simplicity.

Regular Series bolts and nuts are for general use. Bolts and nuts in the *Heavy Series* are slightly larger and intended for heavier use. Square head bolts appear only in the Regular Series. Hexagon-head bolts, hexagon nuts, and square nuts appear in both series.

Square heads and nuts are rough, or *unfinished*. Hexagon heads and nuts may be *unfinished* or *finished*. A *washer face* is machined on finished hexagon bolts and nuts. This is actually $\frac{1}{64}$″ thick, but is drawn $\frac{1}{32}$″ to show up more clearly. The threaded end of a bolt may be rounded or chamfered. It is usually drawn with a 45° chamfer to the thread depth.

The proportions of bolts and nuts, based on the diameter D of the bolt, which are either exact sizes or close approximations for drawing purposes, are as follows, where W = width across flats, H = height of head, and T = thickness of nut:

Regular Hexagon and Square Bolts and Nuts

$$W = 1\tfrac{1}{2}D \qquad H = \tfrac{2}{3}D \qquad T = \tfrac{7}{8}D$$

Heavy Hexagon Bolts and Nuts and Heavy Square Nuts (There are no heavy square bolts.)

$$W = 1\tfrac{1}{2}D + \tfrac{1}{2} \qquad H = \tfrac{2}{3}D \qquad T = D$$

Unfinished bolts have coarse threads, Class 2A, while finished bolts have coarse, fine, or 8-pitch threads, Class 2A. Unfinished nuts have coarse threads, Class 2B, while finished nuts have coarse, fine, or 8-pitch threads, Class 2B.

Minimum thread lengths are 2D plus $\frac{1}{4}''$ for bolts up to 6″ in length, and 2D plus $\frac{1}{2}''$ for bolts over 6″ in length. Bolts too short for these formulas are threaded up to the head.

Bolt lengths have not been standardized because of the endless variety required by industry. The following increments (steps in lengths) are compiled from manufacturer's catalogs:

Square Head Bolts
 Lengths $\frac{3}{4}''$ to $1\frac{1}{2}''$: $\frac{1}{4}''$ increments
 Lengths 2″ to 10″: $\frac{1}{2}''$ increments
 Lengths 11″ to 30″: 1″ increments
Hexagon Head Bolts
 Lengths $\frac{3}{4}''$ to 8″: $\frac{1}{4}''$ increments
 Lengths $8\frac{1}{2}''$ to 20″: $\frac{1}{2}''$ increments
 Lengths 21″ to 30″: 1″ increments

15.21 To Draw ANSI Standard Bolts and Nuts. Steps in drawing ANSI Standard

square and hexagon bolts and nuts are shown in Figs. 15-26 and 15-27. Proportions are based on diameter D. They closely conform to the actual dimensions. Occasionally, where unusual accuracy is required, bolts and nuts are drawn to actual dimensions, as given in Table 5 of the Appendix. In Figs. 15-26 and 15-27:

 I. Lay out bolt diameter D, the bolt length (under side of head to tip end), the height of the head and thickness of the nut, and the chamfer circle. For finished hexagon bolts and nuts, draw a washer face (actually $\frac{1}{64}''$, but drawn $\frac{1}{32}''$ thick), Fig. 15-27, *included in* the head height and nut thickness. Note that square bolts and nuts are always unfinished. Hence, they have no washer faces.

An easy way to find the various proportions based on D is to use the scale as shown in both

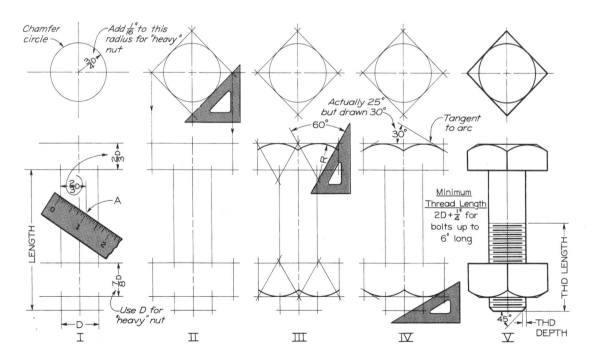

Fig. 15-26. Regular square bolt and nut (unfinished).

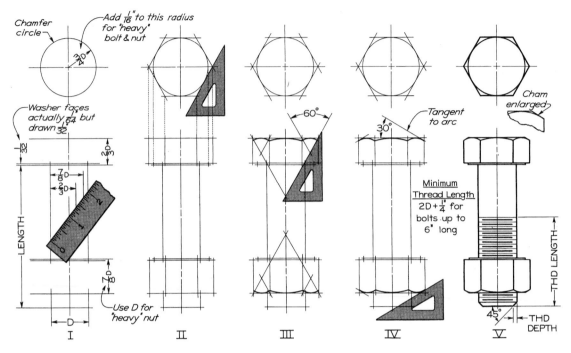

Fig. 15-27. Regular hexagon bolt and nut (finished). Note: Unfinished hexagon bolt has this same head height and nut height, but washer face is omitted.

figures. For example, Fig. 15-26(I), make a small dot at A. Then set the dividers between this dot and the left side of the bolt. Transfer the $\frac{2}{3}$D distance to set off the height of the head.

 II. Draw a square or a hexagon about the chamfer circle. Project down to the head and nut to establish the corners.

 III. Locate centers for arcs with 30° × 60°

triangle as shown. Draw final dark arcs.

 IV. Draw 30° chamfers tangent to arcs, as shown.

 V. Lay off thread length (see Sec. 15.20) from tip end of bolt, and draw threads. Construct 45° chamfer on end of bolt to thread depth, as shown. Darken all lines as necessary to complete the drawing.

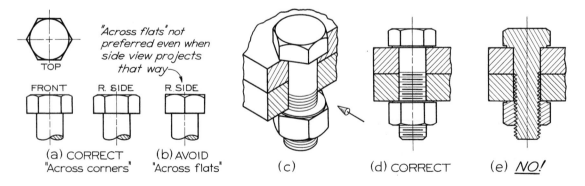

Fig. 15-28. Drawing bolts and nuts.

Bolts and nuts are preferably drawn "across corners" so as to show the maximum number of faces in each view, as shown in Figs. 15-26 and 15-27. In Fig. 15-28(a) the top view is correctly placed to show the hexagon bolt head "across corners" in the front view. However, the true projection of the side view would show "across flats," (b). This is not desirable because it looks like a square head "across corners." The conventional and correct method is to draw the head "across corners" in the side view, as shown at (a). Occasionally, a head or a nut may be in a restricted position in an assembly drawing. Then it must be represented "across flats." In such cases, use the proportions shown in Fig. 15-29.

It is customary not to section bolts and nuts when drawn in assembly, Fig. 15-28(c) and (d), because they do not in themselves require sections for clearness. Incorrect sectioning is shown at (e). The same holds true for all screws and similar parts. See Sec. 16.13.

It is a good rule to make holes $\frac{1}{32}''$ larger than the bolts up to $\frac{3}{8}''$ diameter, and $\frac{1}{16}''$ larger above $\frac{3}{8}''$ diameter. However, clearance is not shown on the drawing unless necessary to make clear that the bolt threads are not engaged with internal threads in the hole.

15.22 SPECIFICATIONS FOR BOLTS AND NUTS.
Bolts and nuts are often listed in parts lists,

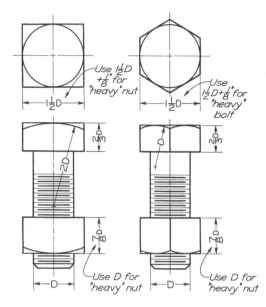

Fig. 15-29. Bolts "across flats."

on assembly drawings, and other records. Standard specifications are given, of which the following are typical:

Example (Complete):
$\frac{7}{16}$-14UNC-2A × $2\frac{1}{4}$ FIN HEX BOLT
Example (Abbreviated):
$\frac{7}{16}$ × $2\frac{1}{4}$ FIN HEX BOLT
Example (Complete):
$\frac{3}{4}$-10UNF-2A HEAVY SQ NUT
Example (Abbreviated):
$\frac{3}{4}$ HEAVY SQ NUT

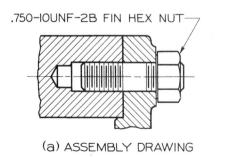
(a) ASSEMBLY DRAWING

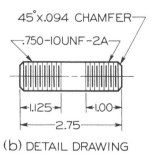
(b) DETAIL DRAWING

Fig. 15-30. Stud application and dimensioning.

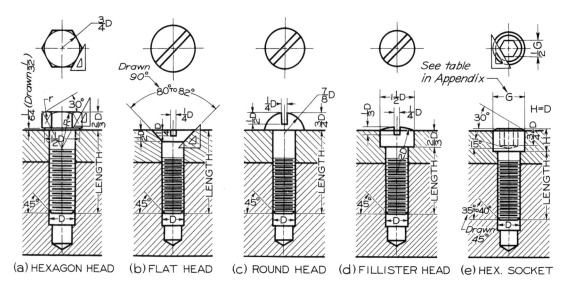

Fig. 15-31. ANSI standard cap screws.

15.23 STUDS. A stud is a rod threaded on both ends, one end of which is screwed tightly in a permanent position into a main casting. The other end passes through a clearance hole in the part to be held down, such as a cylinder head, and a nut is screwed on the free end, Fig. 15-30(a). Studs are not standardized. Hence, they must be drawn and dimensioned on detail drawings. See Fig. 15-30(b).

15.24 ANSI STANDARD CAP SCREWS. Cap screws ordinarily pass through a clearance hole in one member and screw into the other member. The clearance hole is usually drilled $\frac{1}{32}$" larger than the screw, up to $\frac{3}{8}$" diameter, and $\frac{1}{16}$" larger for diameters over $\frac{3}{8}$". However, the clearance is not shown on the drawing unless necessary for clearness, in which case it is drawn $\frac{1}{16}$" larger in diameter. Similarly, the counterbores for the fillister heads and socket heads are usually made $\frac{1}{32}$" or $\frac{1}{16}$" larger in diameter than the heads. This clearance is usually not shown on the drawing. Cap screws are

regularly produced with finished heads and chamfered points. They are used on machines where accuracy and appearance are important.

There are five types of heads, as shown in Fig. 15-31. Exact dimensions for the hexagon head cap screw are given in Table 5 of the Appendix. Exact dimensions for the slotted and socket head types are given in Table 6 of the Appendix. These dimensions may be used where the drawing requires exact sizes, but this seldom occurs. In Fig. 15-31 proportions, given in terms of diameter D, closely conform to the actual dimensions. They produce almost exact drawings with relatively little effort. The hexagon-head cap screw, (a), is drawn in the same manner as the finished hexagon bolts, Fig. 15-27. The minimum length of thread on hexagon head cap screws (ANSI B18.2.3.1M) is 2D + $\frac{1}{4}$" for lengths up to 6", and 2D + $\frac{1}{2}$" for lengths over 6". Hexagon head cap screws too short for these formulas are threaded to within $2\frac{1}{2}$ threads of the head for lengths up to 1", and

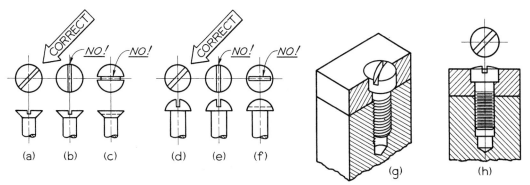

Fig. 15-32. Drawing cap screws.

$3\frac{1}{2}$ threads for lengths greater than 1″. The minimum length of thread on slotted cap screws (ANSI/ASME B18.6.2) is 2D $+\frac{1}{4}$″. When the screws are too short for this formula, they are threaded to within $2\frac{1}{2}$ threads of the head. See tables in ANSI B18.2.3.1M for thread lengths of hexagon socket cap screws. Lengths of screws and thread data are given in notes with Table 6 of the Appendix.

In drawing tapped holes for cap screws, allow one or two threads beyond the end of the screw for clearance. Allow a drill depth beyond the threads equal to several more threads, Fig. 15-31. See also Secs. 15.12 and 15.15. Note that in the bottom of all the tapped holes in Fig. 15-31 the threads are omitted so as to show the ends of the screws more clearly. For information on drilled, countersunk, or counterbored holes, see Secs. 10.24, 11.11, and 11.12.

Correct and incorrect methods of drawing slotted screw heads are shown in Fig. 15-32(a) to (f). It is customary not to section screws when drawing them in assembly, (g) and (h). They do not in themselves require sectioning for clarity. See also Fig. 15-28 for similar treatment of bolts.

Typical cap screw notes are as follows:

Example (Complete):
.437-14UNC-3 × 2.25 HEXAGON CAP SCREW
Example (Abbreviated):
.437 × 2.25 HEX CAP SCR
Example (Complete):
.500-13NF-3 × 2.50 FILLISTER HEAD CAP SCREW
Example (Abbreviated):
.500 × 2.50 FILL HD CAP SCR

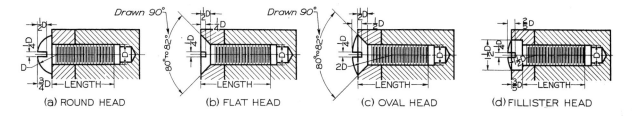

Fig. 15-33. ANSI standard machine screws.

15.25 ANSI STANDARD MACHINE SCREWS.

Machine screws are in general smaller than cap screws. They are available in a greater variety of heads. Eight styles of head are standardized. Four of the most common are shown in Fig. 15-33. Machine screws are particularly adapted to screwing into thin materials. On all lengths under 2″ they are threaded to the head (ANSI B18.6.3). Machine screws over 2″ in length have a minimum thread length of $1\frac{3}{4}$″. Machine screws are used extensively in firearms, jigs, fixtures, and other small mechanisms.

In Fig. 15-33 proportions, given in terms of diameter D, conform closely to the actual dimensions. They produce almost exact drawings. Actual dimensions are given in Table 7 of the Appendix. Threads are National Coarse or National Fine, Class 2 fit. Clearance holes and counterbores should be made larger than the screws, as explained for cap screws in Sec. 15.24.

Typical machine screw notes are as follows:

Example (Complete):

NO. 5 (.125)-4UNC-2 × 1.50
FILLISTER HEAD MACHINE SCREW

Example (Abbreviated):

NO. 5 (.125) × 1.50 FILL HD MACH SCR

Example (Complete):

.312-24UNF-2 × 2.25 OVAL HEAD
MACHINE SCREW

Example (Abbreviated):

.312 × 2.25 OVAL HD MACH SCR

15.26 ANSI STANDARD SETSCREWS.

Setscrews are used to prevent relative motion, usually rotary, between two parts, such as the motion of a pulley hub on a shaft. A setscrew is screwed into one part so that its point bears firmly against another part, Fig. 15-34(a). If the point of the setscrew is cupped, (e), or if a flat is milled on the shaft, (a), the screw will hold more firmly. There are four types of head, (a) to (d), and six styles of points, (e) to (k). The headless setscrews are coming into greater use. They have no projecting heads to catch clothing and cause accidents.

Most of the dimensions in Fig. 15-34 are ANSI Standard (ANSI/ASME B18.6.2) for-

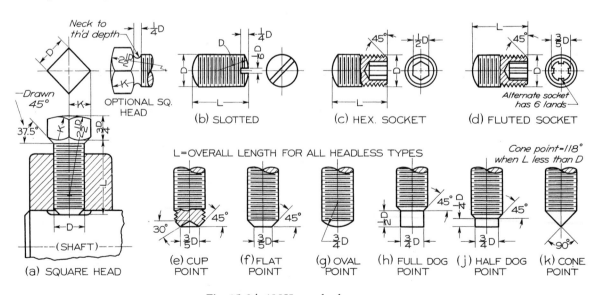

Fig. 15-34. ANSI standard setscrews.

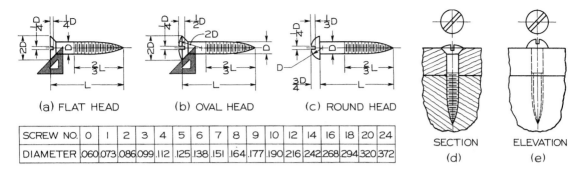

SCREW NO.	0	1	2	3	4	5	6	7	8	9	10	12	14	16	18	20	24
DIAMETER	.060	.073	.086	.099	.112	.125	.138	.151	.164	.177	.190	.216	.242	.268	.294	.320	.372

Fig. 15-35. ANSI standard wood screws.

mula dimensions. The resulting drawings are almost exact representations. Length L is the overall length for headless setscrews. It is the distance from the underside of the head to the point for square head setscrews.

Setscrews are usually made of steel and are casehardened. Headless setscrews are threaded with either National Coarse or National Fine threads, Class 2. Square-head setscrews may have Coarse, Fine, or 8-Pitch threads, Class 2A. The Coarse thread is usually applied to the square head setscrew, which is generally used in the rougher grades of work. Typical setscrew notes are as follows:

Example (Complete):
.500-13UNC-2A × 1.50 SQUARE HEAD
FLAT POINT SETSCREW

Examples (Abbreviated):
.500 × 1.50 SQ HD FLAT PT SET SCR
.375 × 1.25 HEX SOCK CUP PT SET SCR

15.27 ANSI STANDARD WOOD SCREWS.
Wood screws with three types of heads have been standardized, (ANSI B18.6.1), Fig. 15-35. Any of them may be plain-slotted or cross-recessed. The proportions, based on diameter D, closely conform to actual dimensions. They are more than sufficiently accurate for use on drawings. Applications are shown at (d) and (e). Note that the threads are drawn by the symbolic method, Sec. 15.13. Crest lines and root lines are spaced by eye to appear approximately as in Fig. 15-35.

A typical wood screw note is:

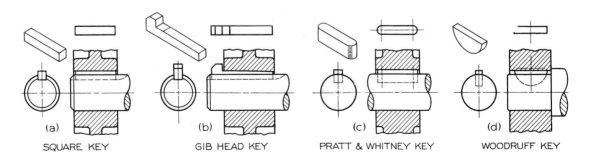

Fig. 15-36. Keys.

NO 5 × 1¼ FILL HD WOOD SCR

15.28 KEYS. *Keys* are used to prevent relative motion between shafts and wheels, couplings, cranks, and other machine parts. A square key is shown in Fig. 15-36(a). A flat key is the same, except that it is not as high as the square key. Either of these may have the top surface tapered ⅛″ in 12″, in which case they become square taper or flat taper keys. The width of square and flat keys is generally about one fourth the shaft diameter. One-half the height of the key is sunk into the shaft. See Table 8 in the Appendix.

A gib head key is shown at (b). It is exactly the same as the square taper or flat taper key, except that a gib head is added. The head provides for easy removal. See Table 8 in the Appendix.

The Pratt & Whitney key, (c), has rounded ends and is seated in a keyseat of the same shape. Two-thirds of a P & W key is sunk into the shaft.

The Woodruff key, (d), is semicircular in shape. It fits in a semicircular keyseat in the shaft. The top of the key fits into a plain rectangular keyway. Sizes of keys for given shafts are not standardized. A good rule is to select a key whose diameter is about equal to the shaft diameter. See Tables 10 and 11 in the Appendix.

Typical notes for keys are as follows:

$$\tfrac{1}{4} \times 1\tfrac{1}{2} \text{ SQ KEY}$$
$$\tfrac{1}{4} \times \tfrac{3}{16} \times 1\tfrac{1}{2} \text{ FLAT KEY}$$
NO. 204 WOODRUFF KEY
NO. 10 PRATT & WHITNEY KEY

A *keyseat* is in the shaft. A *keyway* is in the hub or surrounding part. Typical notes for keyseats and keyways are shown in Fig. 10-30(C), (D), (J), and (N).

In keyway notes, the width dimension is given first and then the depth. The depth dimension is measured along the side of the keyway and not at the center. Typical notes for keyways are as follows:

$$\tfrac{3}{8} \text{ WIDE} \times \tfrac{3}{16} \text{ DEEP KEYWAY}$$
or
$$\tfrac{3}{8} \times \tfrac{3}{16} \text{ KEYWAY}$$

15.29 LOCKING DEVICES. Many types of devices to prevent nuts from unscrewing too easily are available. The most common of them are shown in Fig. 15-37. ANSI Standard *jam nuts,* (a) and (g), are somewhat thinner than the regular nuts (see ANSI B18.2.4.5M). The finished and semifinished

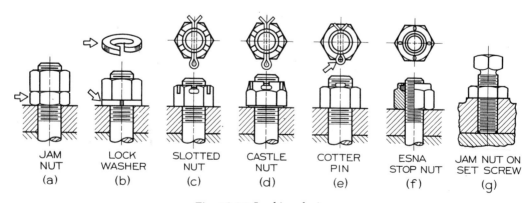

Fig. 15-37. Locking devices.

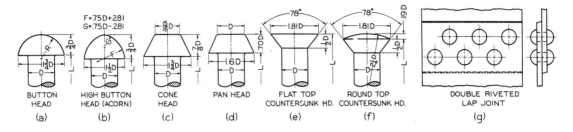

Fig. 15-38. ANSI standard large rivets (adapted from ANSI/ASME B18.1.2).

nuts have a $\frac{1}{64}''$ washer face or are chamfered on both sides. Jam nuts are drawn in the same manner as ordinary hexagon nuts, Fig. 15-27, but with a thickness of $\frac{1}{2}D$ (approximate). The ANSI Standard lock washer, (b), has a slot whose edges tend to "dig in" when the nut is loosened. Cotter pins are used with slotted nuts, (c), castle nuts, (d), or with ordinary nuts, (e). There are many patented nuts of various kinds, of which the Esna stop nut, (f), is representative. A setscrew may be kept from loosening by means of a jam nut, as shown at (g).

15-30 RIVETS. *Rivets* are permanent fastenings, as distinguished from removable fastenings such as bolts or screws. They are generally used to hold sheet metal or rolled steel shapes together. They are made of wrought iron, soft steel, copper, or other materials. Although seldom used in new construction, rivets will appear in drawings of existing structures, and it is useful to be aware of how they are drawn when modifications are being made.

ANSI Standard large rivets, Fig. 15-38, were once used in the structural work of bridges, buildings, and in ship and boiler construction. The button head and cone head rivets were used in tanks and boilers.

A typical riveted joint is shown at (g). Note that the side view of each rivet shows the shank of the rivet with both heads made with circular arcs. The circular view of each rivet is represented by only a visible circle for the head.

15-31 COMPUTER GRAPHICS. By using symbol libraries available in CAD programs, the drafter can select standard detailed or schematic symbols for a variety of screws and other fasteners. This makes it easy to quickly add or change the number, size, or location of fasteners shown on a drawing.

PROBLEMS

The problems in Figs. 15-39 and 15-40 are given to provide practice in drawing detailed threads, symbolic threads, and some of the more common fasteners. These problems are all designed to be drawn with instruments on Layout C in the Appendix.

A large number of practical problems involving threads and fasteners are given at the end of the following chapter.

If assigned by the instructor, construct drawings using decimal-inch or metric scales by converting given fractional dimensions to decimal-inch or metric equivalents. Refer to Table 20 in the Appendix.

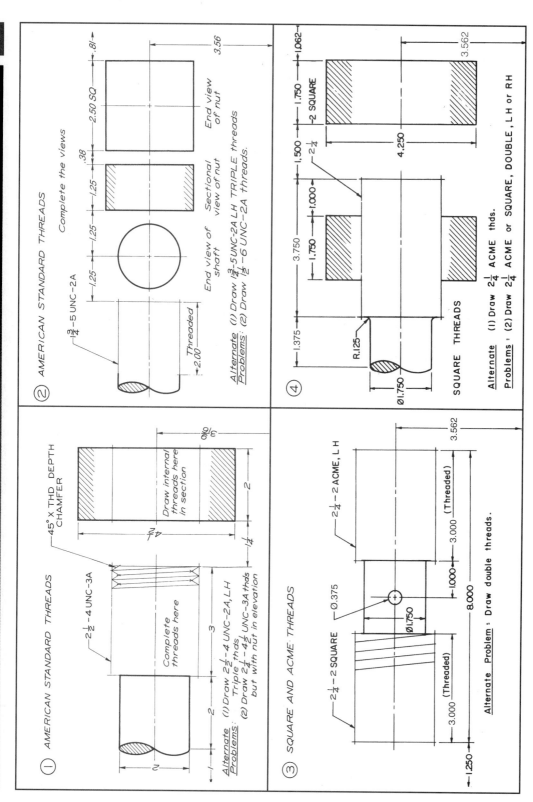

Fig. 15-39. Detailed threads. Using Layout C in the Appendix, draw problems assigned. Omit all inclined lettering. Transfer titles to title strip.

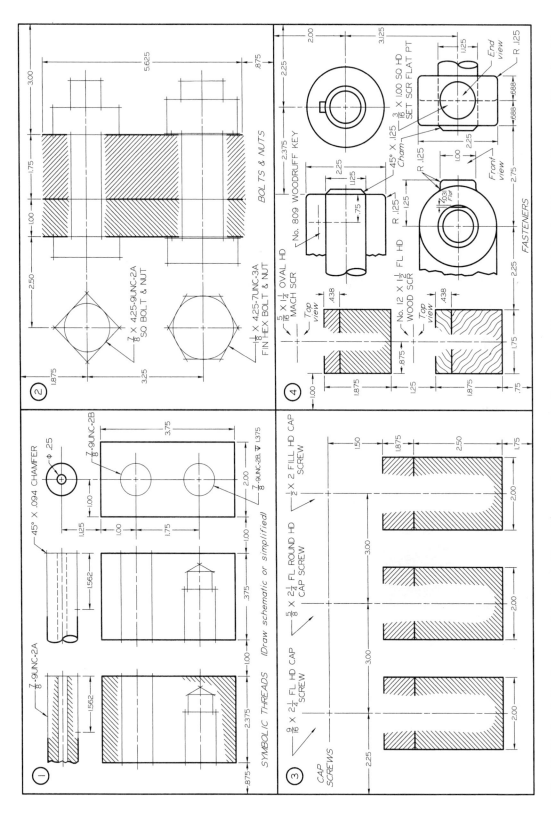

Fig. 15-40. Symbolic threads and fasteners. Using Layout C in the Appendix, draw problem assigned. Omit all inclined lettering. Transfer titles to title strip.

CHAPTER 16

IMPORTANT TERMS

layout drawing

assembly drawing

title strip

change strip

outline assembly

working-drawing assembly

general assembly

sub-assembly

OBJECTIVES

After studying this chapter, you should be able to:

- Identify the types of drawings that are prepared as a design is routed from one department to another.

- Specify the general methods used to group details on sheets.

- Explain the purpose of a change strip, or revision strip.

- Identify the various parts of the drawing that need to be checked for correctness.

- Identify the various types of assembly drawings.

334

CHAPTER **16**

Working Drawings

Working drawings, as the name suggests, are used during the work of making a product or structure. They provide the information needed to make the parts and assemble the final product. Chapter 16 discusses production drawings, assembly drawings, the outline assembly, the working-drawing assembly, the general assembly, and the sub-assembly.

Working drawings tell how a design is to be manufactured and assembled. Setting up the production line to carry out these instructions is the job of a systems engineer. The systems engineer devises a plan to combine tooling and assembly, and, using robotic techniques and automation, to achieve an efficient, interactive production environment. The machinery, control systems (including computers), and various manufacturing processes used in any particular assembly line operation must all be considered. It will be necessary to make frequent sketches and flow chart diagrams to display all the information collected about a project before a final, most efficient solution can be agreed upon.

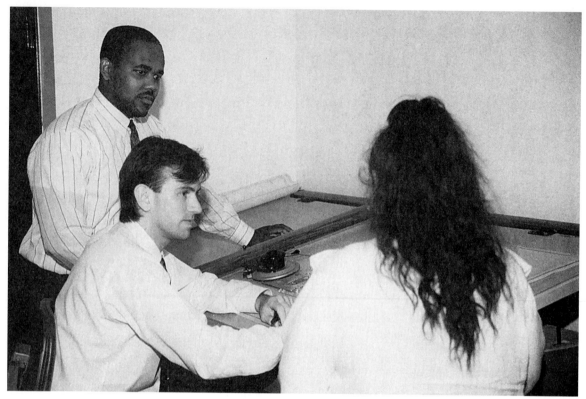

Fig. 16-1. Industrial drafting room.

16.1 INDUSTRIAL DRAFTING. A typical drafting room in a large manufacturing company is shown in Fig. 16-1. Engineers, designers, and drafters work here. They each do their parts in planning the products to be manufactured. When you look at a complicated structure, such as a supersonic plane, a full-color printing press, or a numerically controlled milling machine, do not be discouraged. Not one, but hundreds of persons have contributed their time, their knowledge, and their originality to these modern wonders. Each person is more or less a specialist who has learned how to handle problems in a relatively limited field.

Figure 16-2 shows how drawings are routed from the engineering department to the drafting department and on to all the main divisions of a manufacturing plant. For a discussion of the various shops, see Chapter 11. Usually a design engineer conceives the basic idea or an idea for improvement. Often, however the engineering department gets valuable ideas from drafters, mechanics, or salespersons. In any case, it is only natural at this point for the designer to set down the new idea in the form of a freehand sketch. For example, at the Brown & Sharpe Manufacturing Company, a design engineer had an idea for improving the adjustment of the arm of a milling machine. The designer set this down rapidly in a freehand sketch, Fig. 16-3. The device that the designer had in mind, which was eventually developed, was an "Arm Adjusting Mechanism." This is shown on the machine in Fig. 16-4.

Next, the designer who made the sketch, or someone equally experienced, made a *layout drawing* or *design assembly*. This was drawn accurately to scale in pencil and without dimensions, Fig. 16-5. This is an assembly drawing showing the mechanism in position in the milling machine. It includes the views necessary to show the size and shape of each part of the mechanism, but dimensions are omitted. Layouts are drawn full size, if possible. These enable the designer to visualize more clearly the actual sizes of the parts while drawing them. When drawing to half size or less, the designer may tend to design large, heavy, or clumsy parts. On the other hand, in drawing double size or larger, the designer may tend to get an exaggerated idea of sizes and draw the parts too small.

Special attention is given to clearances of moving parts so that they will not interfere with other parts, to ease of assembly, and to serviceability. Designers make use of all the

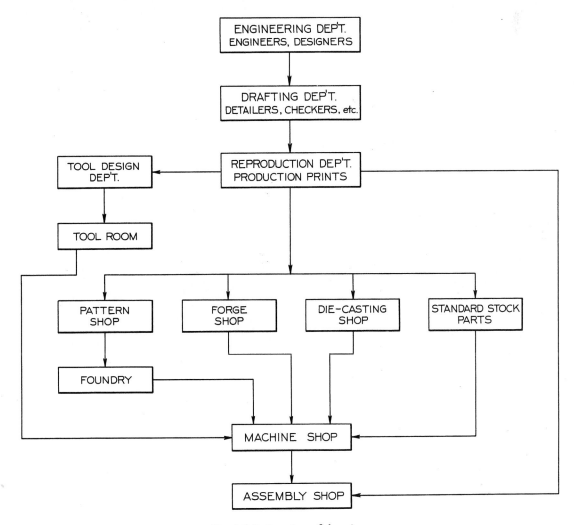

Fig. 16-2. Routing of drawings.

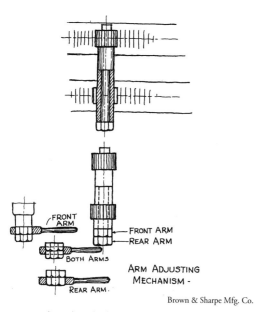

Brown & Sharpe Mfg. Co.

Fig. 16-3. Idea sketch for improvement on milling machine.

Brown & Sharpe Mfg. Co.

Fig. 16-4. "Arm adjusting mechanism" on the milling machine.

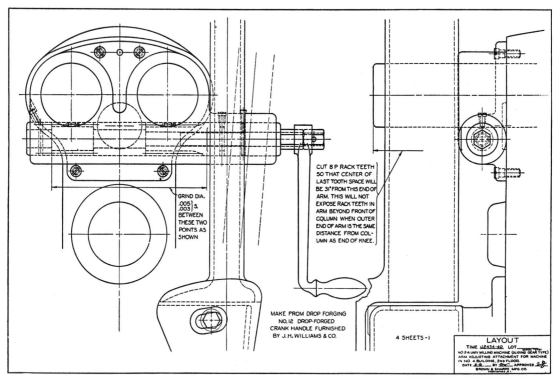

Brown & Sharpe Mfg. Co.

Fig. 16-5. Design layout of "arm adjusting mechanism."

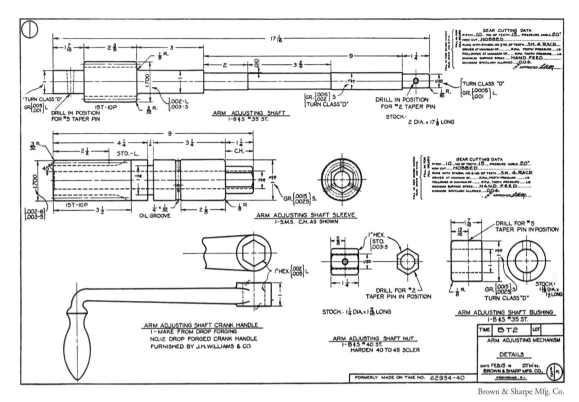

Brown & Sharpe Mfg. Co.

Fig. 16-6. Details of "arm adjusting mechanism."

sources of information available to them. These sources include physics, chemistry, mathematics, mechanics, strength of materials, and other subjects that the engineering student studies in college. Designers adopt the conclusions and recommendations obtained from experimental tests and laboratory studies. They also use the sound experience that they and their company have obtained from the success and failure of past machines. They carefully follow the performance and maintenance records of machines in use. Much design information is compiled in handbooks, to which they often refer.

Some mechanisms are relatively simple in form and operation. Thus, most of the design effort consists in determining functional shapes and arrangement. However, the parts may be subject to heavy loads, in which event careful computation of strength is necessary in order that the parts may be designed to correct sizes. A good example of this type of design problem would be a simple hoist. Other mechanisms, such as typewriters and adding machines, are not subject to large stresses. With them, the chief problem is one of the arrangement and shape of parts for effective operation and low-cost production.

After the layout has been approved by the chief engineer or delegated assistants, it is turned over to the drafter to make the *production drawings*. In relatively small firms, the engineering and drafting may be concentrated in a single department. In larger firms they are separated, as shown in Fig. 16-2. In any case, the *drafters* or *detailers* use the layout as a guide and make the detail drawings, Fig. 16-6. The drafter "picks off" the sizes directly from the layout with the scale or the dividers. The detail drawings show the necessary views, dimensions, and notes required in the shop to make the parts without additional instructions. The details may all be drawn on one sheet, as in Fig. 16-6, or each part on a separate sheet.

Finally, an *assembly drawing,* Fig. 16-7, will be needed to guide workers in assembling parts properly and for general reference throughout the shops. Since the original layout, Fig. 16-5, is an assembly drawing, it can often be traced, omitting unnecessary adjacent parts and making any other desirable changes to suit the purpose.

Before blueprints or copies are made, the production drawings must be carefully checked by a *checker,* Sec. 16.6, to make sure they are correct before releasing prints to the shops.

If the mechanism contains new ideas, a *patent drawing* is made, Fig. 16-8. It is sent to the United States Patent Office. Patent drawings must follow the rigid rules of the Patent Office.

16.2 NUMBER OF DETAILS PER SHEET. Two general methods are used regarding grouping of details on sheets. If the structure or mechanism is small or composed of few parts, all the details may be drawn on one sheet, as in Fig. 16-6. This is an older method. It is common practice in drawing details of jigs, fixtures, valves, and similar mechanisms. The assembly may also be shown with the details, but usually it is drawn on a separate sheet.

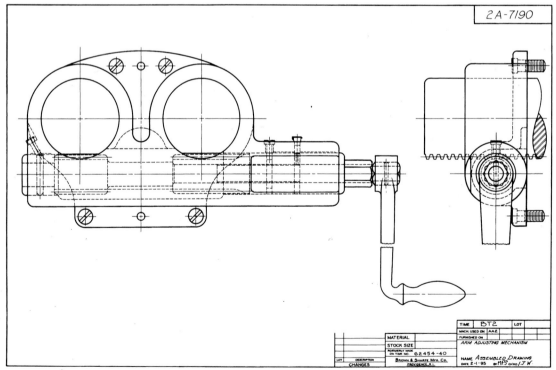

Brown & Sharpe Mfg. Co.

Fig. 16-7. Assembly of "arm adjusting mechanism."

When several details are drawn on one sheet, attention should first be given to spacing. Before beginning to draw any one view or part, block in all details with construction lines, as shown in Fig. 16-9. Ample spacing should be allowed for dimensions and notes. A good method to determine spacing is to cut out rectangular scraps of paper roughly equal to the sizes of the views. Place these on the sheet in position. Mark the locations lightly on the sheet. Then remove the scraps of paper.

If possible, the same scale should be used for all details. Otherwise, the different scales should be clearly noted under each detail. Each detail should be represented by the regular views, sections, or auxiliary views needed to describe the part clearly, as shown in Chapters 7, 8, 12, and 13. It should have all necessary dimensions and notes, as described in Chapter 10.

One common method of identifying the parts is to letter a title note under each detail, Fig. 16-10. The detail number may be encircled, and the title underlined for emphasis. Under the title the material and the number of pieces required in the assembly are indicated, together with any other general information regarding the piece.

A method used today to identify parts is to give a *parts list,* or bill of materials, Fig. 16-11. Such a parts list takes the place of the title notes shown in Fig. 16-10, as regards part number, title, material, and number required. It may include other information, such as pattern numbers, stock sizes, and weights. The parts list may be located above the title block, *reading upward,* Fig. 16-11, or in the upper right corner of the sheet, *reading downward.*

Parts should be listed in general order of

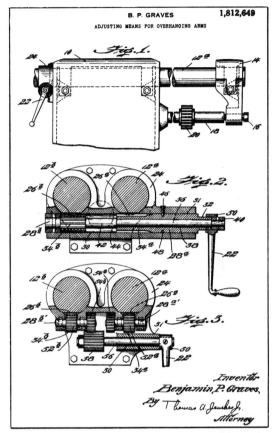

Brown & Sharpe Mfg. Co.

Fig. 16-8. Patent Office drawing for "arm adjusting mechanism."

size or importance of the details. In general, main castings or large parts are listed first. The standard parts or small details are listed last. Standard parts are listed but not drawn unless they are to be altered in some way.

Many companies have adopted the practice, for convenience in filing, of drawing each detail, no matter how small, on a separate sheet, Fig. 16-12. Many have adopted the ANSI Standard sheet sizes, based on 8.5″ × 11.0″, or 9.0″ × 12.0″, and multiples thereof, Sec. 4.5. (Sheet sizes are also available from suppliers in fractional inch or metric equivalents.) The sizes based on 8.5″ × 11.0″ are most common because of the convenience in

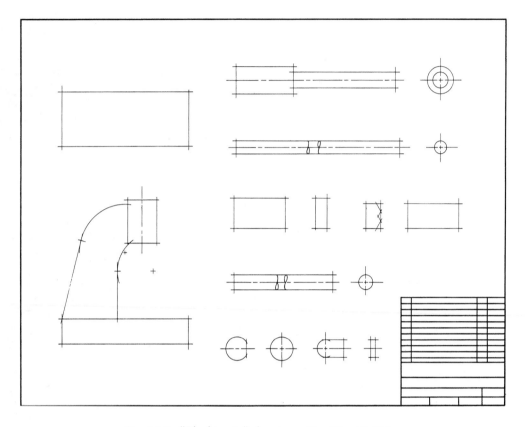

Fig. 16-9. "Blocking in" the views. (See Fig. 16-45.)

correspondence and filing in standard letter file drawers.

16.3 TITLE STRIPS. Every drawing should have a *title strip* or title block, Figs. 16-10 to 16-12. The purpose of this is to show in an organized way all necessary information not shown on the drawing itself. In industry it is almost universal practice to use printed sheets on which the border and title strip are already printed. The drafter merely fills in the blanks.

16.4 CHANGE RECORDS. After a drawing has been released and used, it is often necessary to make changes. It is important to keep a complete and accurate record of every change. For this purpose a *change strip*, or *revision strip*, is

included at some convenient place on the drawing, as shown at the lower left of Fig. 16-12. An encircled letter is placed on the drawing near the place where the change was made. The same letter is given in the change strip. Some companies use triangles or squares around the letters. Others use numbers instead of letters. In the change strip, the change is briefly described, usually by stating what the affected part of the drawing *was* before the change. In addition, the drafter gives his or her initials and the date.

16.5 DRAWING NUMBERS. Every drawing should be numbered. This may be a simple serial number, such as 30418. It may have a letter after it or before it to indicate the sheet size,

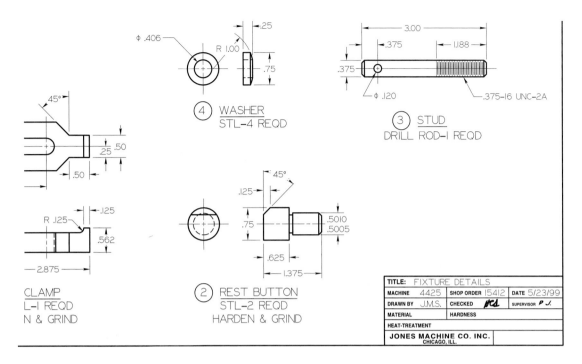

Fig. 16-10. Parts notes.

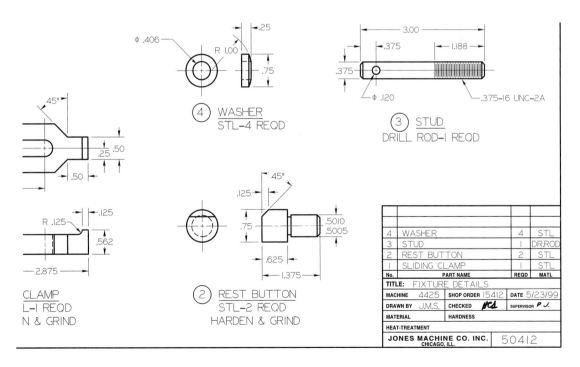

Fig. 16-11. Parts list.

343

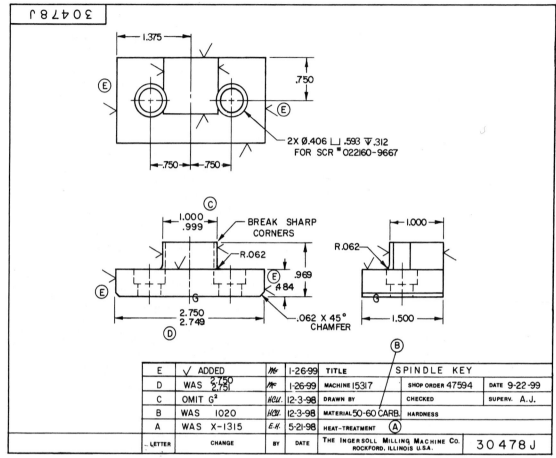

Fig. 16-12. Detail drawing.

such as 30418A or A30418. Thus, a size A sheet may be 8.5″ × 11.0″. A size B sheet may be 11.0″ × 17.0″. See Sec. 4.5. All sorts of numbering schemes are in use. The various parts of the number may be used to indicate such things as the model number of the machine, the general nature of the part, the use of the part, and so on. In general, it is advisable to use simple serial numbers. Avoid using the drawing number to convey other information. Such practice can lead to confusion.

The drawing number is also the number of the part itself. It should be lettered very bold-ly, at least $\frac{1}{4}$″ high, in the lower right corner and also in the upper left corner of the sheet, Fig. 16-12. For this purpose a technical fountain pen is recommended.

16.6 CHECKING. Young drafters in industry soon find that *correctness* is very important. Their initials on drawings identify them. When a drawing is completed, it is turned over to another person to check it. In small offices, checking may be done by the original designer or by an experienced drafter. In large offices, drawings are inspected by a *checker*.

This checker follows a rigid procedure to be sure that nothing is overlooked. In general, the checker covers the following:

1. Soundness of the Design — function, strength, economy, manufacturability, serviceability, ease of assembly and disassembly, lubrication, repairs, and like considerations.
2. Views — scale, sections, partial views, auxiliary views, fillets and rounds.
3. Dimensions and Notes — errors, omissions, duplications, tolerances, shop operations, finish marks.
4. Legibility — linework and lettering for clearest reproduction.
5. Clearances — no interference of moving parts.
6. Materials — best choice, stock sizes, heat treatment, other considerations.
7. Standard Parts — specified wherever possible.
8. Title Block Information.

Throughout this procedure, the checker follows carefully the company engineering and drafting standards, Sec. 16.7. Checking is usually done on a print so as not to mark up the original. Ordinary lead pencils may be used on Ozalid prints or others that have white backgrounds. On blueprints, colored pencils are most satisfactory. Often, a "check-assembly" is drawn to make sure that all parts fit and function in the assembly in a proper manner.

16.7 STANDARDS. Each firm has its own engineering standards, made available to all engineers and drafters in loose-leaf form. These include, in addition to engineering data, definite standards that all drafters are expected to follow regarding the items listed in Sec. 16.6. Various groups of industries also have standards that sum up the recommended practices in each type of industry — as, for example, the standards of the Society of Automotive Engineers covering the entire automotive industry. Above these are the many publications of the American National Standards Institute, including the ANSI *Standard Drafting Manual,* Y-14. It is the "last word" in drafting standards in this country. All drawings in this book are drawn in conformity with it.

16.8 ASSEMBLY DRAWINGS. An *assembly drawing* shows the entire assembled machine or structure, or the assembly of some unit such as the carburetor or the transmission of an automobile. Assemblies vary in character according to their uses. In general, they are: (1) *design assemblies* or *layouts,* (2) *outline* or *installation assemblies,* (3) *working-drawing assemblies,* and (4) *general assemblies.*

16.9 OUTLINE ASSEMBLY. An *outline assembly,* or *installation assembly,* Fig. 16-13, shows one or more views of an assembly "in outline." The purpose is to give general information regarding the character and size of the unit and how it fits in its environment. Little or no sectioning is generally needed, but it may be used if necessary. Small, relatively unimportant details are omitted — as, for example, many screw heads in Fig. 16-13. The dimensions given are only the principal overall and center-to-center distances needed to clarify questions of installation. The outline assembly is widely used in catalogs and other sales literature. When there are several sizes of one machine, dimensions are usually indicated by capital letters. Values for each letter are given in a table.

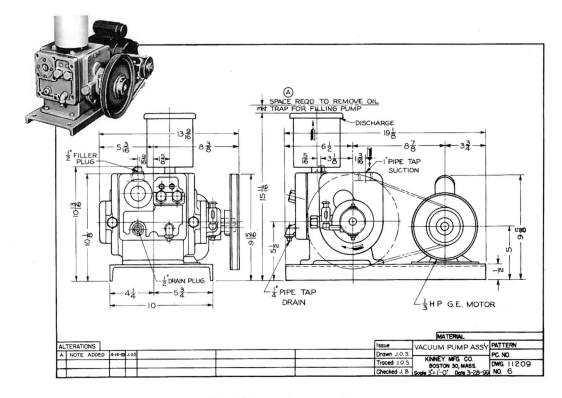

Fig. 16-13. Outline assembly.

16.10 WORKING-DRAWING ASSEMBLY. A *working-drawing assembly,* Fig. 16-14, is a combined detail and assembly drawing giving complete dimensions and notes for all parts. This method is used in place of separate detail and assembly drawings when the mechanism is relatively simple and all parts can be adequately represented in a single assembly drawing. It often happens that, while most parts can be clearly shown and dimensioned in the assembly, some cannot. In this case these parts are detailed separately on the same sheet, and the drawing becomes a combination of working-drawing assembly and detail drawing.

Since working-drawing assemblies eliminate the cost of separate detail drawings, they are widely used in classes of work where this is possible. Examples are drawings of jigs, fix-

tures, valves, aircraft sub-assemblies, and certain work not requiring the most complete manufacturing information for each detail.

16.11 GENERAL ASSEMBLY. A *general assembly,* Fig. 16-15, shows *how the parts fit together and how the assembly functions.* Its chief use is the assembly shop where all the finished parts are received and put together. Assembly workers do not need to learn the shape of any part from the assembly drawing, as they have all the actual parts, ready to be assembled. Consequently, the views used on an assembly drawing are simply those that show clearly how all the parts fit together. These views may be one or more regular views, sections of any kind, auxiliary views, and partial views. Sectioning in assembly drawing is discussed in Sec. 16.13.

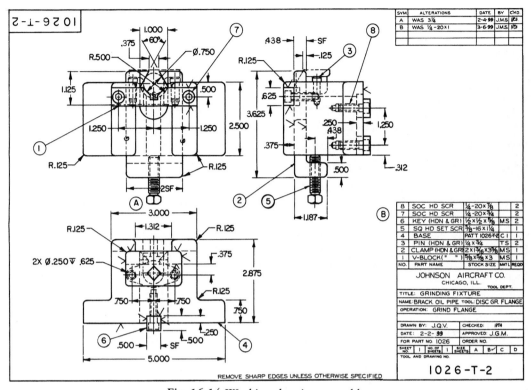

Fig. 16-14. Working-drawing assembly.

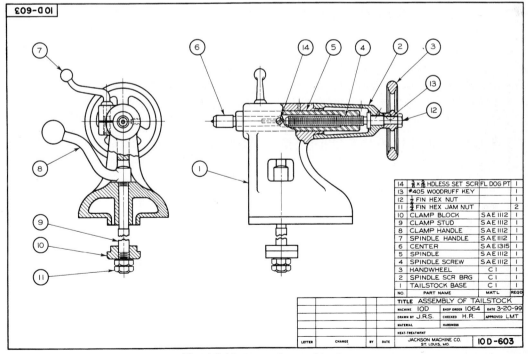

Fig. 16-15. General assembly drawing.

As a rule, no dimensions are given on a general assembly drawing. Occasionally some special dimension is given that is related to the function of the entire assembly. For example, a drawing of a jack may have a dimension showing the maximum height when open and the minimum height when closed. A drawing of a vise may have a dimension showing the maximum opening of the jaws.

Frequently, with large or complicated machines it is not possible to show all parts in one assembly. In such situations a separate drawing is made. It shows a group of related parts that form a unit of the whole machine and make what is called a *subassembly* or *unit assembly*. See Fig. 16-7.

16.12 TITLE STRIPS AND PARTS LISTS ON ASSEMBLIES. Title strips on assembly drawings are the same as on detail drawings, Sec. 16.3, except that the title includes the word "assembly," as in the following examples:

CONNECTING ROD ASSEMBLY
ASSEMBLY OF GRINDER VISE

Parts lists are similar to those on detail drawings where a number of details are shown on the same sheet, Fig. 16-11. The parts list may be placed in any convenient open corner on the drawing. Preferably, it is located to read up from the title block, Fig.

16-11, or down from the upper right-hand corner of the sheet. Ordinarily the information given for each part includes the part number, part name, material, and number of pieces required in the assembly. Part No. 1 is usually the main base or casting. It is followed by the other parts in general order of decreasing size, and finally by the standard parts, such as bolts, screws, bearings, and pins.

To save drafting time in lettering and to facilitate filing and record-keeping, it is quite common, particularly in large plants, to give parts lists in typewritten form on separate parts list sheets.

On the assembly drawing, each part usually is identified with its description in the parts list. This is done by lettering the part numbers in $\frac{7}{16}''$ or $\frac{1}{2}''$ diameter circles near the assembly, and drawing leaders to each part where it is most clearly shown, Figs. 16-14 and 16-15. The circles should be arranged in groups in vertical or horizontal rows. These should not be scattered in disorder on the sheet. Leaders should not cross, and none should be drawn vertically or horizontally on the sheet.

16.13 ASSEMBLY SECTIONS. In assembly drawings where several adjacent parts are sectioned, it is necessary to draw the section

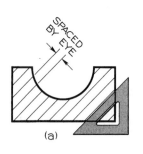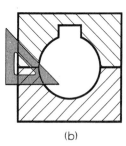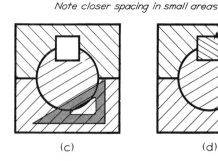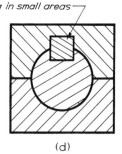

Fig. 16-16. Section lining (full size).

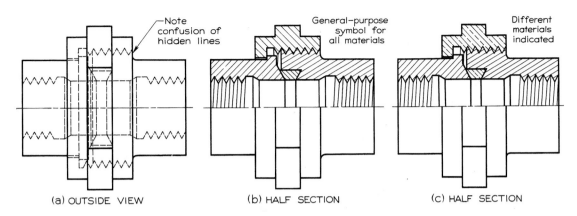

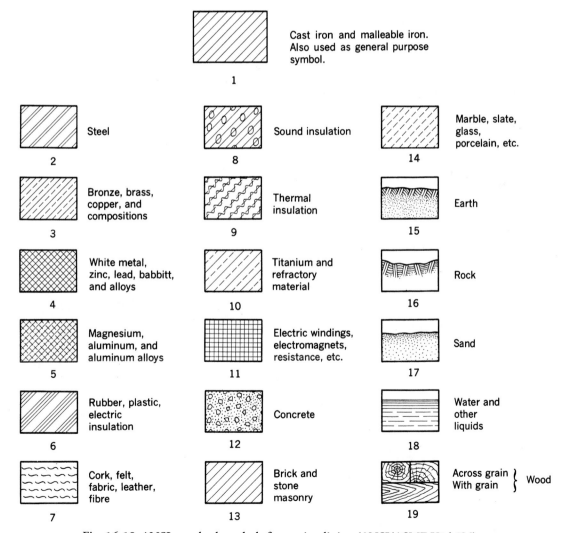

Fig. 16-17. Assembly sectioning.

1 — Cast iron and malleable iron. Also used as general purpose symbol.

2 — Steel

3 — Bronze, brass, copper, and compositions

4 — White metal, zinc, lead, babbitt, and alloys

5 — Magnesium, aluminum, and aluminum alloys

6 — Rubber, plastic, electric insulation

7 — Cork, felt, fabric, leather, fibre

8 — Sound insulation

9 — Thermal insulation

10 — Titanium and refractory material

11 — Electric windings, electromagnets, resistance, etc.

12 — Concrete

13 — Brick and stone masonry

14 — Marble, slate, glass, porcelain, etc.

15 — Earth

16 — Rock

17 — Sand

18 — Water and other liquids

19 — Across grain / With grain } Wood

Fig. 16-18. ANSI standard symbols for section lining (ANSI/ASME Y14.2M).

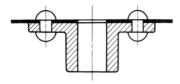

Fig. 16-19. Sectioning thin parts.

lines in different directions to distinguish the pieces clearly, as shown in Fig. 16-16. The first large area, (a), is section-lined at 45°. The next large area, (b), is then section-lined at 45° in the opposite direction. Additional areas are section-lined at 30° or 60° with horizontal, (c) and (d). If necessary, to make any area contrast with the others, any other angle may be used. Note that section lines do not meet at the visible lines separating the areas. Note that for small areas the lines are drawn closer together.

In Fig. 16-17(a) is shown a regular outside view of an assembled unit, illustrating how interior shapes are not made clear by means of hidden lines. The separate pieces overlap. It is impossible to distinguish between them.

A half section of the same unit is shown at (b) in which the four pieces are clearly shown by section lines in opposite directions.

It is sometimes desirable on assembly sections to indicate the different materials by means of symbolic section lining, as shown at (c). The ANSI Standard symbols for section lining are shown in Fig. 16-18. On detail drawings of individual parts, it is recommended that the symbol for cast iron be used for all materials and that the exact specification of materials be given in a note under the views or in the title strip.

In sectioning very thin parts, such as sheet metal, gaskets, and like pieces, when there is not enough space for section-lining, the sectioned parts may be shown solid, as in Fig. 16-19.

In assembly sections, there may lie in the path of the cutting plane some solid parts that themselves do not require sectioning and where the use of sectioning would make the drawing less clear and would lengthen the time required to draw it. In such cases it is

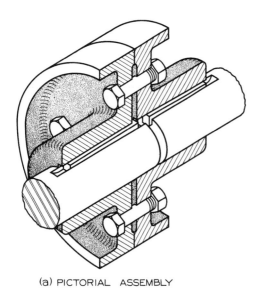

(a) PICTORIAL ASSEMBLY

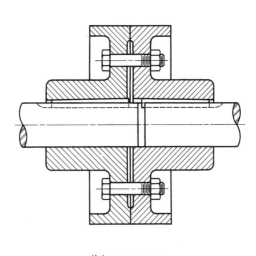

(b) ASSEMBLY

Fig. 16-20. Solid parts in assembly sections.

customary *not to section* such solid parts. A pictorial drawing of a flange coupling is shown in Fig. 16-20(a). Here the bolts, nuts, shafts, and keys are not sectioned but are left in full form. The corresponding assembly drawing is shown at (b). Other parts that are usually not sectioned are ribs, gear teeth, spokes, screws, nails, ball and roller bearings, and pins. See Figs. 15-28 and 15-32.

16.14 PIPING DRAWINGS. Piping drawings are actually assembly drawings. In drawings for power plants, pumping plants, heating systems, and plumbing systems, piping is represented by *double-line drawings,* Fig. 16-21(a), or *single-line drawings,* (b), depending upon how complete a picture is necessary. For the

most part, the single-line drawings are used because of the saving in drafting time. Piping usually runs in many directions. Thus, ordinary top, front, and side views may be very confusing, if not impossible to read, because of many overlapping pipes. For this reason it is common to draw piping systems pictorially, either in isometric, Fig. 16-22(a), or in oblique, (b). For complete information on pictorial drawing methods, refer to Chapter 17.

16.15 USING CAD FOR WORKING DRAWINGS. The working drawing is the culmination and final graphical report of the design process. The CAD system simplifies the production of working drawings. It incorporates two-dimensional as well as three-dimensional drafting. All parts

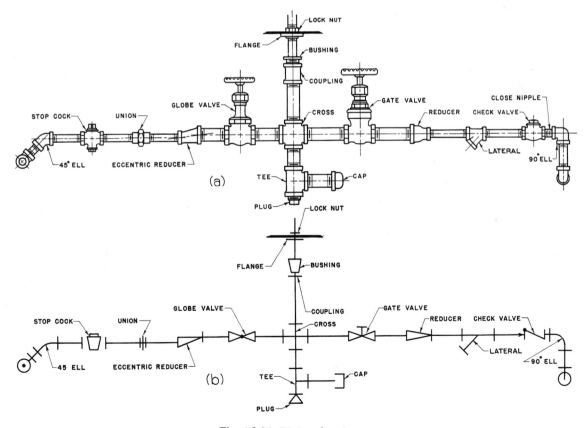

Fig. 16-21. Piping drawings.

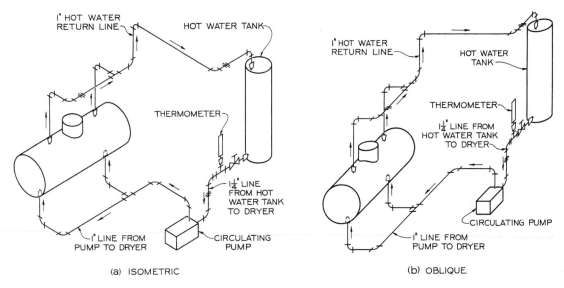

Fig. 16-22. Pictorial piping drawings.

of the design are incorporated into the final set of drawings prior to construction of the project. Figure 16-23 shows the use of all phases of CAD to develop working drawings.

Using CAD, the drafter pulls together all the parts needed to develop the finished design. Existing details, pictorials, assemblies, sections, welding, and any required form of design are easily inserted into the drawing sheet. The saved views from three-dimensional designs can be placed into assembly drawings and details.

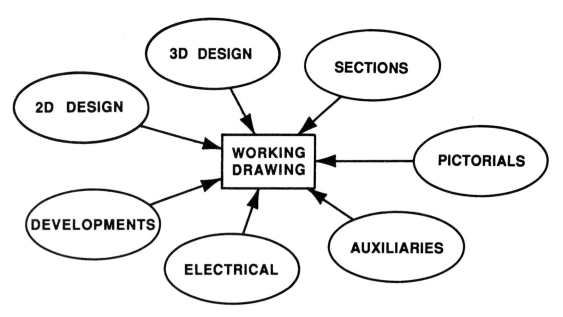

Fig. 16-23. How the various aspects of CAD are used to develop working drawings.

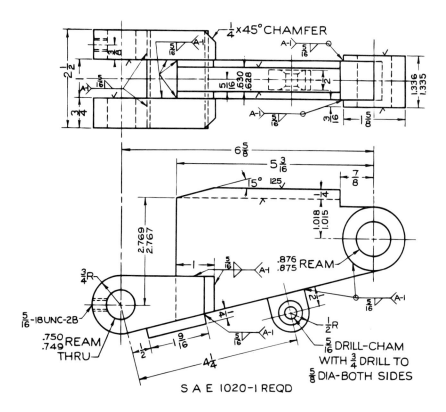

Fig. 16-24. Welding drawing.

Notes and other written information can be produced on a word processor or typed directly in the CAD program. Spelling is checked and the completed document is inserted into the drawing. The inserted document converts to the style and size of the existing text. Bills of materials can be extracted from attribute database files. They can be inserted into the drawing without the need to retype the information. The information is merely inserted where required.

16.16 WELDING DRAWINGS. A welding drawing is, in one sense, an assembly drawing. It shows a number of individual pieces joined together by welding to form a single unit.

In earlier days, it was common practice to indicate welding information on a drawing by simply lettering a note such as "To be completely welded" or other similar instructions. This practice resulted in costly shop mistakes, with not enough or too many welds being made. This was due primarily to a lack of specific welding information on the drawing. Since it would be impractical to show the welds themselves on a drawing, graphical welding symbols were developed by the American Welding Society. They are completely described in the publication ANSI/AWS A2.4, "Symbols for Welding, Brazing, and Nondestructive Examination." These symbols have also been adopted by the American National Standards Institute. Standard welding symbols are shown in Table 14 of the Appendix. Figure 16-24 shows a completely dimensioned welding drawing.

The complete welding symbol, shown in Table 14 of the Appendix, consists of eight elements, or as many of these elements as are necessary. These are (1) reference line, (2) arrow, (3) basic weld symbols, (4) dimensions and other data, (5) supplementary symbols, (6) finish symbols, (7) tail, and (8) specification, process, or other references.

A welding symbol may be drawn mechanically or freehand. It is important, however, that the elements of a symbol maintain standard locations with respect to each other. For complete information and procedures regarding the specification and application of welding symbols, the ANSI/AWS Standards should be consulted.

16.17 AEROSPACE DRAFTING. No other industry in modern times has experienced such a rapid growth as the aerospace industry. This is a vast industrial complex that designs and builds aircraft, missiles, and space vehicles. The name aerospace is used because it generally is understood to include all means of both air and space travel. What a generation ago was considered science fiction is today scientific fact. Through cooperation with government agencies such as the National Aeronautics and Space Administration (NASA), it has conducted successful space programs that have resulted in astronauts being orbited into space and rockets sent to the moon.

Aerospace drafting, Fig. 16-25, is a composite of many specialized types of drafting such as mechanical, electrical, structural, and sheet-metal drafting. It includes many different kinds of drawings such as layout draw-

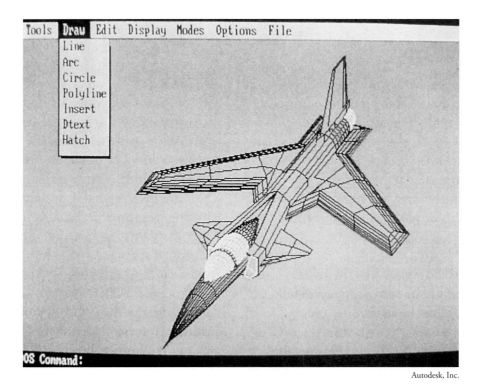

Autodesk, Inc.

Fig. 16-25. Typical aerospace drawing.

ings, working drawings, detail drawings, assembly drawings, pictorial drawings, installation drawings, and many others. A competent aerospace drafter will possess a working knowledge of the principles and techniques of technical drawing. He or she will have a basic understanding of the terminology, methods, and procedures used in the aerospace industry. The activities in the aerospace industry

are complex and varied. It would be impossible to cover the subject in any detail in a few short paragraphs. Standards have been developed, however, with which the drafters should be familiar and which will serve as a source of reference. Among these are the ANSI Y14 *Drafting Manual,* government standards and publications, and individual company drafting standards.

PROBLEMS

The problems on the following pages have been taken from plans of actual machines or devices. These range from comparatively simple problems to the more complicated problems that may be assigned to superior students. Then, with the advice of their instructor, students are to select a sheet size (see the Appendix) for each problem assigned. The first step should be to make "thumbnail sketches" of the views of each part, and obtain the instructor's approval. After that, the instructor may assign more complete sketches, fully dimensioned, followed by the mechanically drawn detail and assembly drawings.

The problems should be fully dimensioned using decimals and current ANSI drafting standards.

If assigned by the instructor, construct drawings using decimal-inch or metric scales by converting given fractional dimensions to decimal-inch or metric equivalents. Refer to Table 20 in the Appendix.

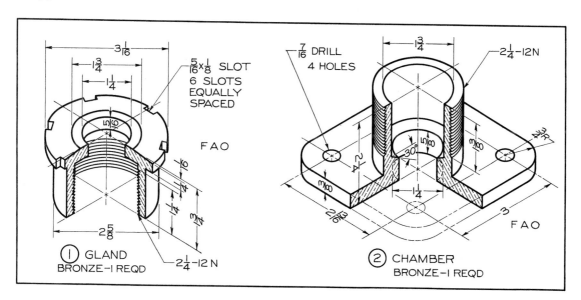

Fig. 16-26. Stuffing box.

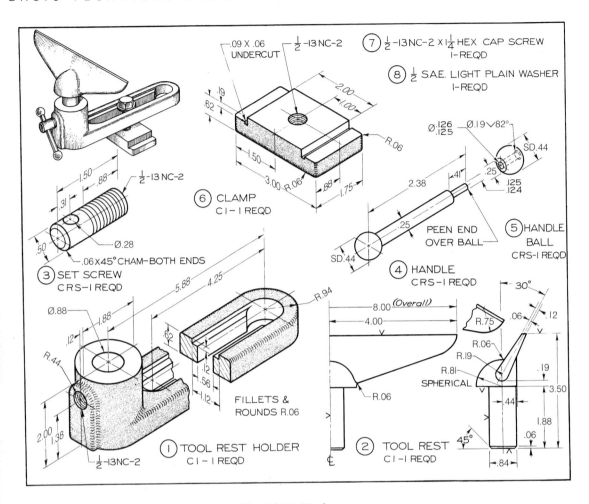

Fig. 16-27. Tool post.

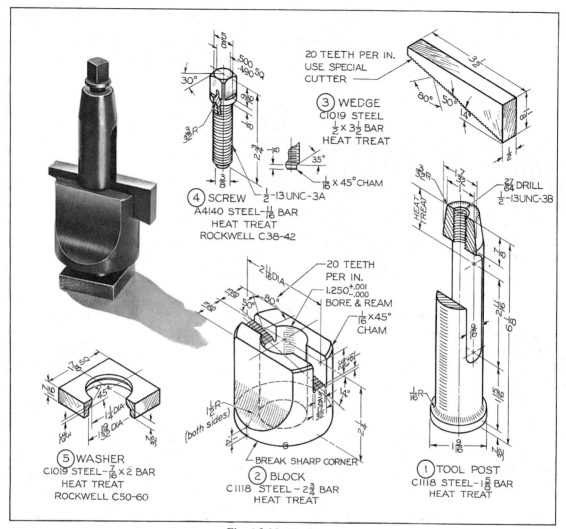

Fig. 16-28. Tool post.

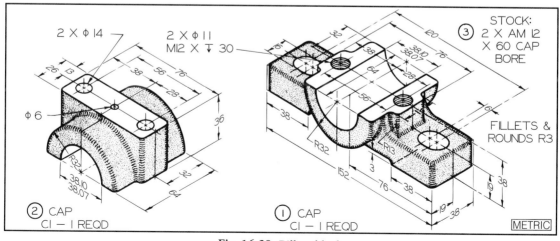

Fig. 16-29. Pillow block.

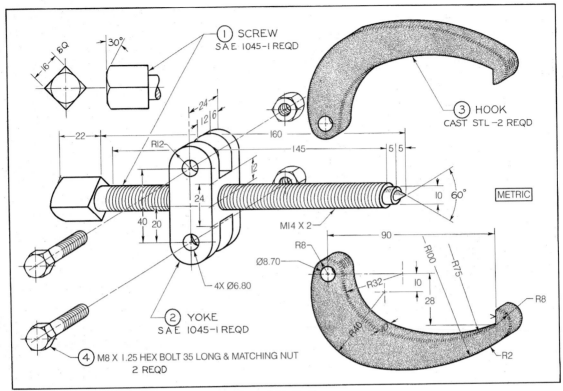

Fig. 16-30. Puller.

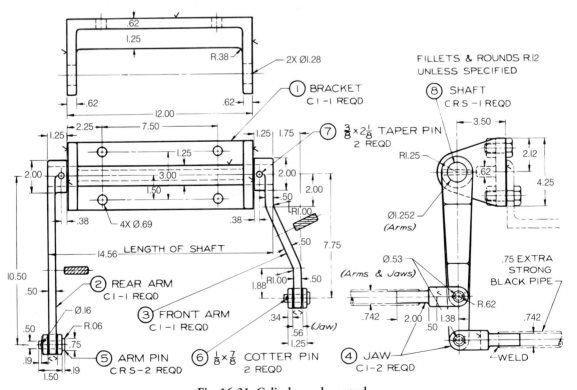

Fig. 16-31. Cylinder cock controls.

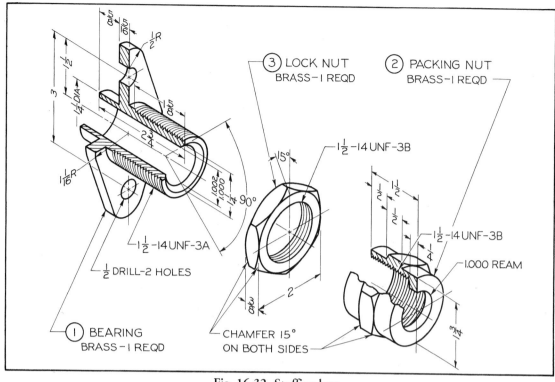

Fig. 16-32. Stuffing box.

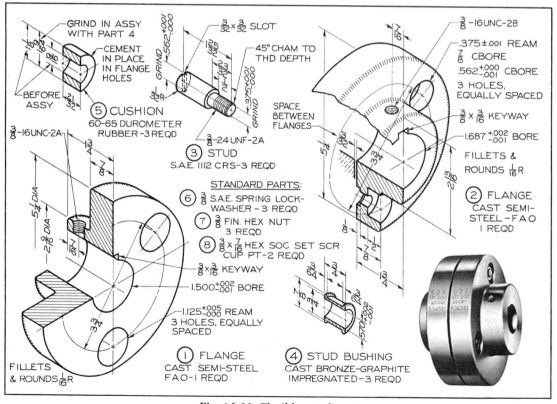

Fig. 16-33. Flexible coupling.

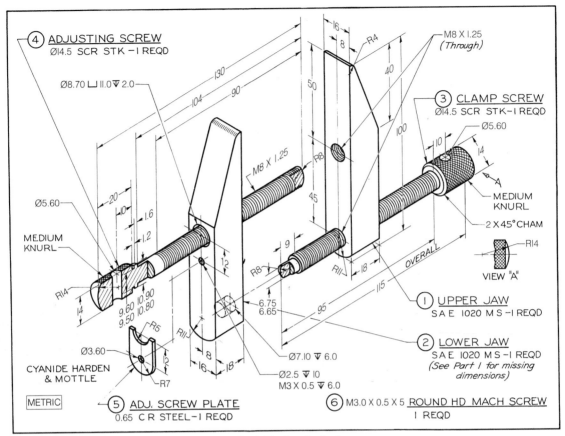

Fig. 16-34. Toolmaker's clamp.

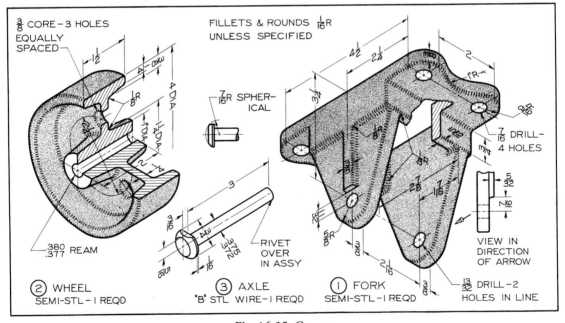

Fig. 16-35. Caster.

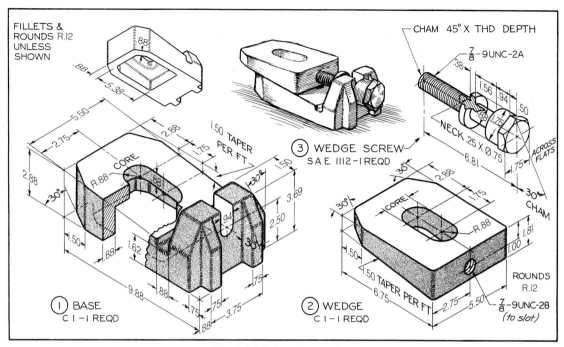

Fig. 16-36. Leveling wedge.

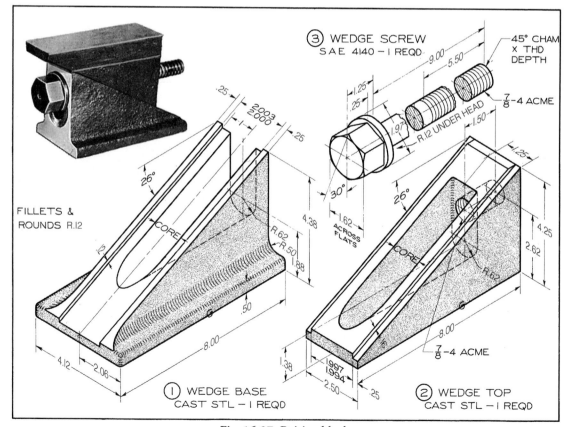

Fig. 16-37. Raising block.

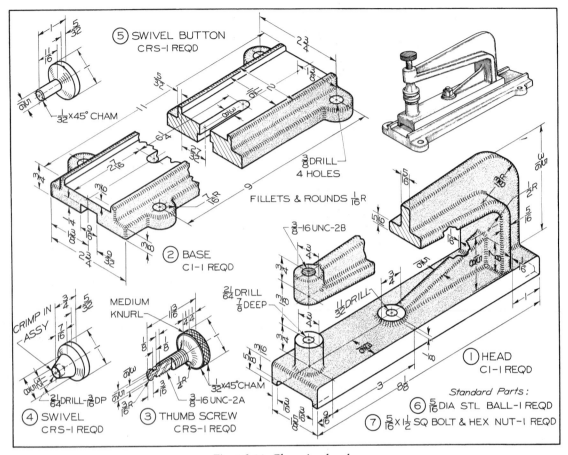

Fig. 16-38. Clamping head.

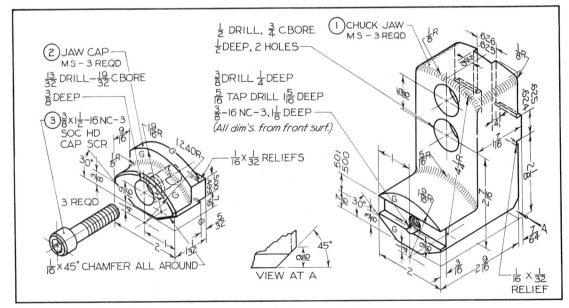

Fig. 16-39. Chuck jaw for lathe.

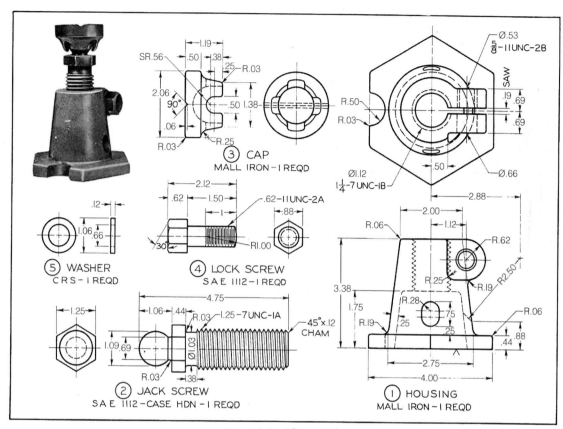

Fig. 16-40. Planer jack.

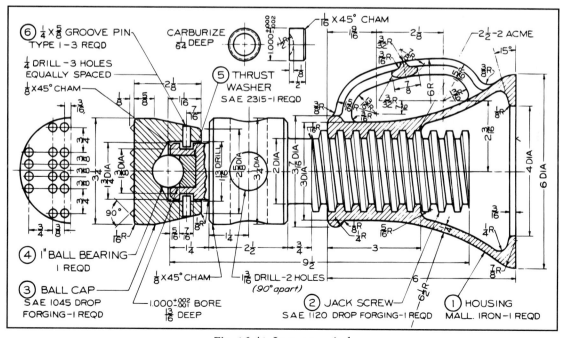

Fig. 16-41. Loco screw jack.

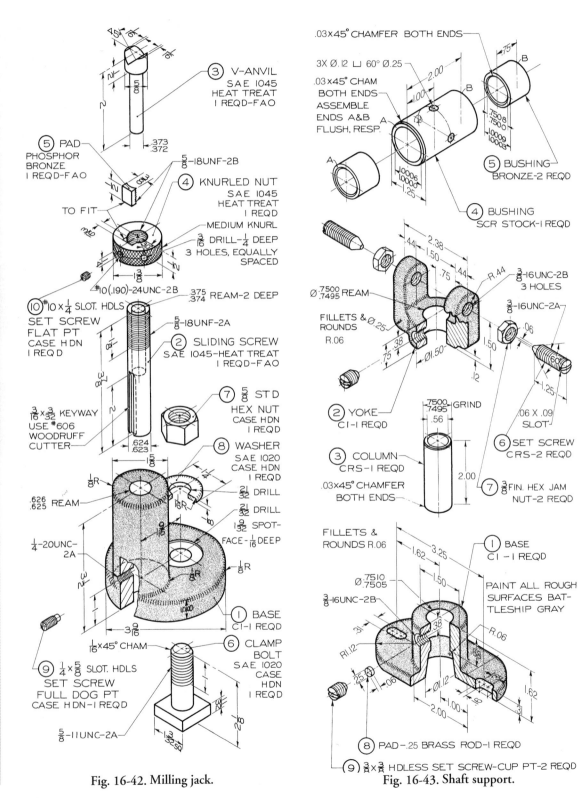

Fig. 16-42. Milling jack.

Fig. 16-43. Shaft support.

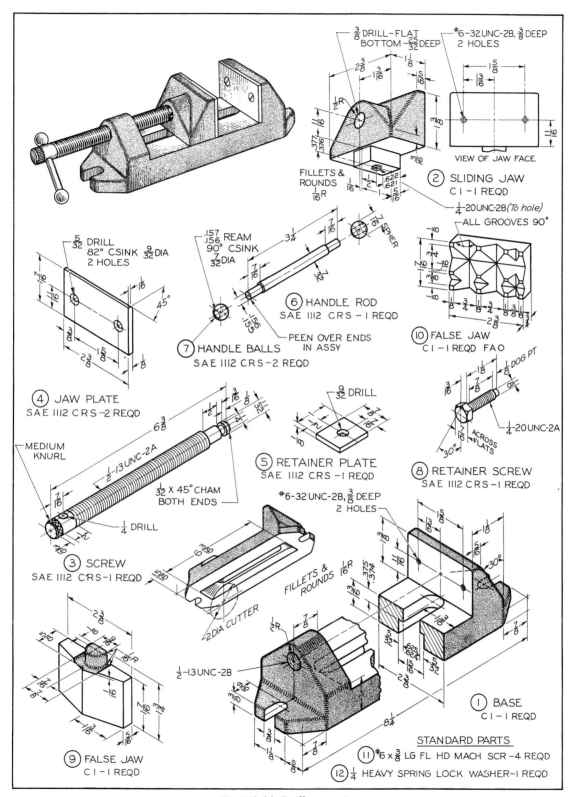

Fig. 16-44. Drill press vise.

365

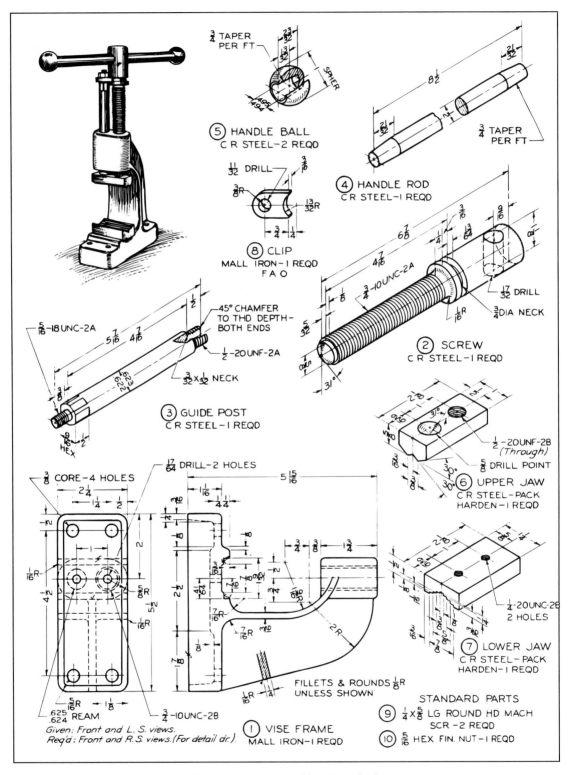

Fig. 16-45. Pipe vise. (See Fig. 16-9.)

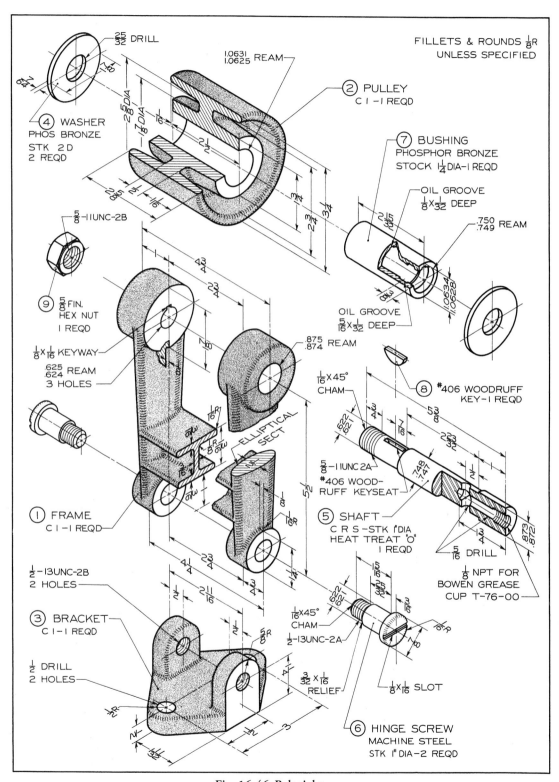

Fig. 16-46. Belt tightener.

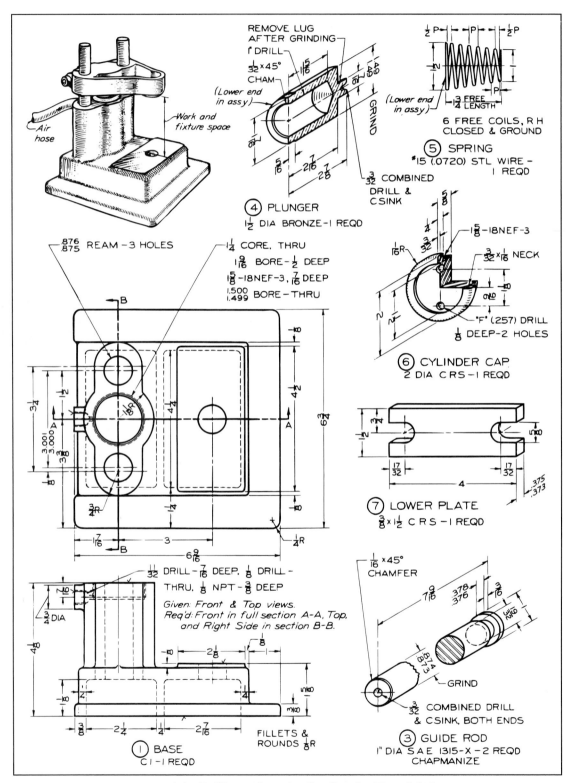

Fig. 16-47. Drilling and tapping jig. (See also Fig. 16-48.)

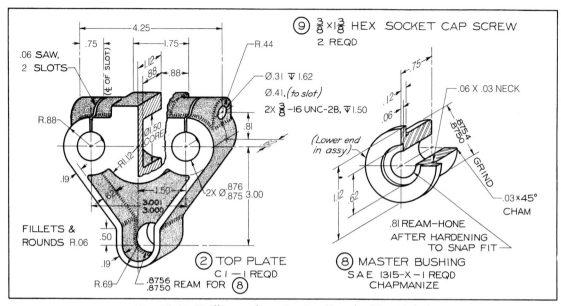

Fig. 16-48. Drilling and tapping jig. (See also Fig. 16-47.)

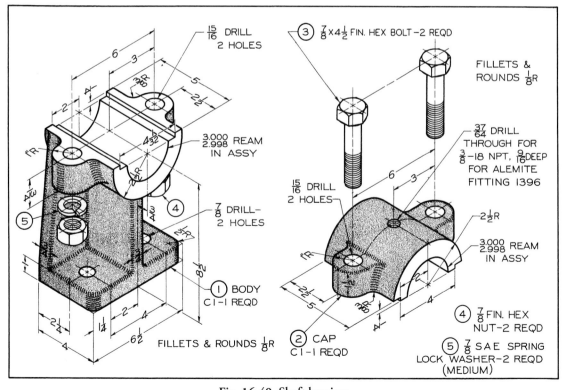

Fig. 16-49. Shaft bearing.

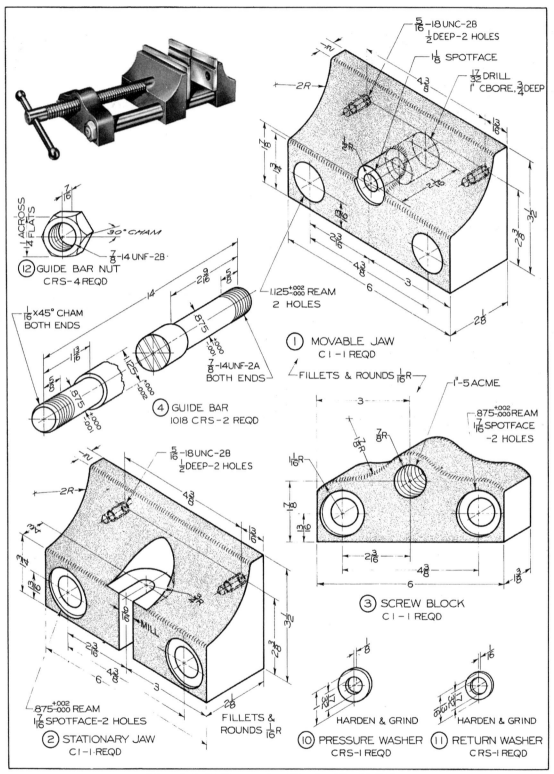

Fig. 16-50. Drill press vise. (See also Fig. 16-51.)

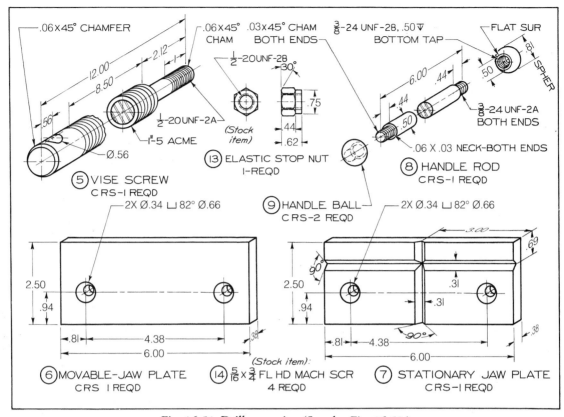

Fig. 16-51. Drill press vise. (See also Fig. 16-50.)

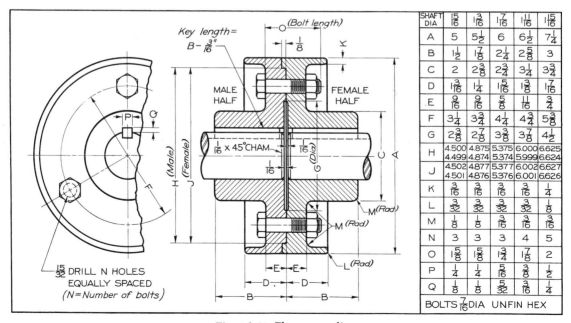

SHAFT DIA	$\frac{15}{16}$	$1\frac{3}{16}$	$1\frac{7}{16}$	$1\frac{11}{16}$	$1\frac{15}{16}$
A	5	$5\frac{1}{2}$	6	$6\frac{1}{2}$	$7\frac{1}{4}$
B	$1\frac{1}{2}$	$1\frac{7}{8}$	$2\frac{1}{4}$	$2\frac{5}{8}$	3
C	2	$2\frac{3}{8}$	$2\frac{3}{4}$	$3\frac{1}{4}$	$3\frac{3}{4}$
D	$1\frac{3}{16}$	$1\frac{1}{4}$	$1\frac{5}{16}$	$1\frac{3}{8}$	$1\frac{7}{16}$
E	$\frac{9}{16}$	$\frac{9}{16}$	$\frac{5}{8}$	$\frac{11}{16}$	$\frac{3}{4}$
F	$3\frac{1}{4}$	$3\frac{3}{4}$	$4\frac{1}{4}$	$4\frac{3}{8}$	$5\frac{5}{8}$
G	$2\frac{3}{8}$	$2\frac{7}{8}$	$3\frac{3}{8}$	$3\frac{7}{8}$	$4\frac{1}{2}$
H	4.500 4.499	4.875 4.874	5.375 5.374	6.000 5.999	6.625 6.624
J	4.502 4.501	4.877 4.876	5.377 5.376	6.002 6.001	6.627 6.626
K	$\frac{3}{16}$	$\frac{3}{16}$	$\frac{3}{16}$	$\frac{3}{16}$	$\frac{1}{4}$
L	$\frac{3}{32}$	$\frac{3}{32}$	$\frac{3}{32}$	$\frac{3}{32}$	$\frac{1}{8}$
M	$\frac{1}{8}$	$\frac{1}{8}$	$\frac{3}{16}$	$\frac{3}{16}$	$\frac{3}{16}$
N	3	3	3	4	5
O	$1\frac{5}{8}$	$1\frac{5}{8}$	$1\frac{3}{4}$	$1\frac{7}{8}$	2
P	$\frac{1}{4}$	$\frac{1}{4}$	$\frac{5}{16}$	$\frac{3}{8}$	$\frac{1}{2}$
Q	$\frac{1}{8}$	$\frac{1}{8}$	$\frac{5}{32}$	$\frac{3}{16}$	$\frac{1}{4}$
BOLTS $\frac{7}{16}$ DIA UNFIN HEX					

Fig. 16-52. Flange coupling.

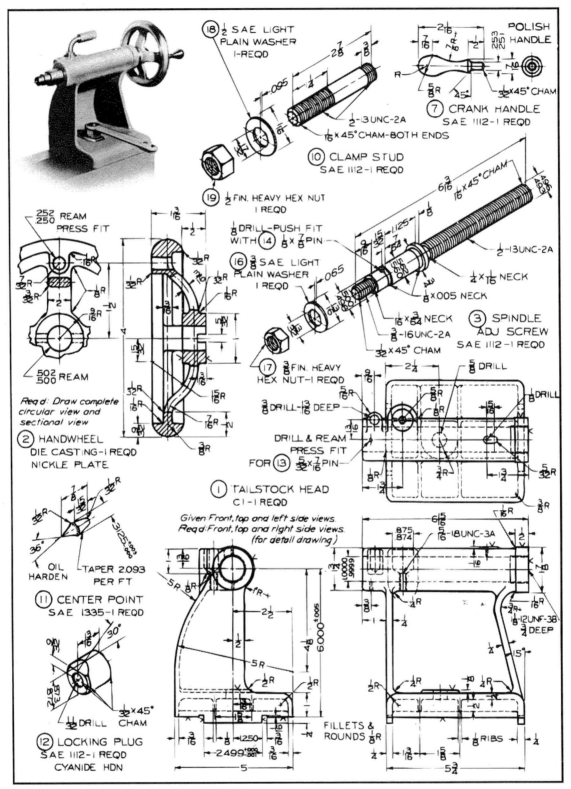

Fig. 16-53. Lathe tailstock. (See also Fig. 16-54.)

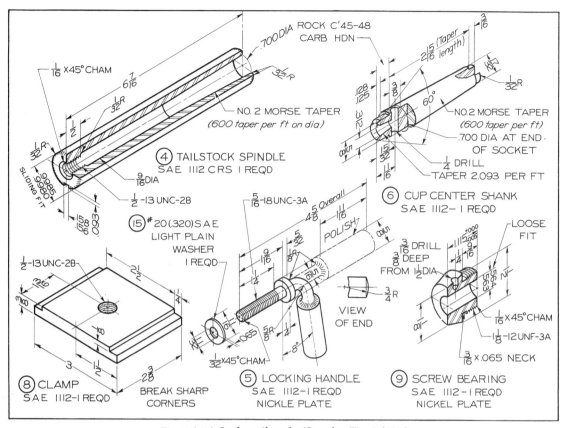

Fig. 16-54. Lathe tailstock. (See also Fig. 16-53.)

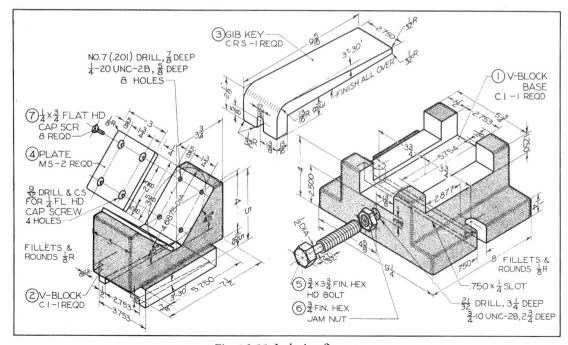

Fig. 16-55. Indexing fixture.

Ken Chernus/FPG International

CHAPTER 17

OBJECTIVES

After studying this chapter, you should be able to:

- Sketch an object in isometric.
- Construct angles in isometric.
- Draw circles and arcs in isometric.
- Draw a true isometric ellipse.
- Draw a curve in isometric.
- Dimension an isometric drawing.
- Draw angles in oblique.
- Draw arcs and circles in oblique.
- Draw a one-point perspective.
- Draw a two-point perspective.

CHAPTER 17

Pictorial Drawings

Pictorial drawings resemble a picture. Unlike multiview drawings, they show more than one face of an object in one view. Pictorial drawings are useful for showing the three-dimensional appearance of a design. They help people visualize the finished product.

Interior designers, for example, frequently use pictorial drawings to show designs to their clients. An interior designer designs the interior spaces of buildings. He or she is responsible for making sure the building's interior is functional, meets human needs, and makes efficient use of the space. Room layout, lighting, furniture, draperies, and carpeting as well as the color, style, and choice of accessories are all considerations of the interior designer. Many sketches will be made before a final design is agreed upon, and layouts will include floor plans as well as pictorial renderings such as one-point perspective drawings.

■ C A R E E R   L I N K ■

To learn more about careers in interior design, contact the American Society for Interior Designers, 608 Massachusetts Avenue NE, Washington, DC 20002. The web site is *www.asid.org.*

Mark Richards/PhotoEdit

Fig. 17-1. Automobile stylist at work.

17.1 PICTORIAL DRAWING. A *pictorial drawing* is one in which the object is viewed in such a position that several faces appear in a single view. Pictorial drawings are excellent for showing the *appearance* of objects. They are often used to supplement multiview drawings. They are used especially to show untrained people the appearance of objects that they cannot visualize from the views on blueprints.

Pictorial drawings are used in catalogs, sales literature, and technical books. They are used also in Patent Office drawings, in piping and wiring diagrams, and in machine, structural, and architectural drawing, Fig. 17-1.

17.2 TYPES OF PICTORIAL DRAWINGS. The three most common types of pictorial drawings are shown in Fig. 17-2. *Perspective,* (a), is the most natural representation, being geometrically the same as a photograph. However, perspectives are comparatively difficult to draw. Other methods, which are usually quite satisfactory and much easier to draw, are *isometric,* (b), and *oblique,* (c).

Isometric drawings are one type of *axonometric* drawing. The other two types are *dimetric* and *trimetric.* In isometric drawings, the three axes are equally spaced, 120° apart, and the scales along all three axes are the same. In dimetric, two of the axes make the same angle with the projection plane, and the third is either smaller or larger than the other two. For example, two angles could be 105° and the third 150°. The scale along the two axes making equal angles is the same, while the third is different. In trimetric drawing, no

376

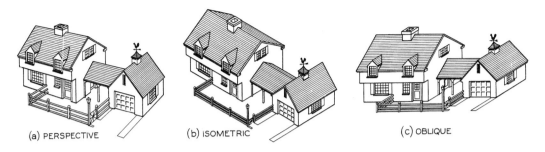

Fig. 17-2. Types of pictorial drawings.

two angles are the same (example: 105°, 120°, 135°), and all three axes have different scale ratios. Since they require special scales, dimetric and trimetric drawings are seldom used.

17.3 ISOMETRIC SKETCHING. One of the most effective ways to sketch an object pictorially is to sketch it in *isometric.* Take the object in your hand and tilt it toward you approximately as shown in Fig. 17-3(a). The front edge (and those edges parallel to it) will appear vertical. The two lower edges (and those parallel to them) will appear about 30° with horizontal.

    I. Start by sketching the enclosing box, sloping lines AC and AD at about 30° with horizontal. Make the height AB

equal to the height of the block. Make the width AD equal to the width of the block. Make the depth AC equal to the depth of the block.

    II. Block in the right-angled notch.

    III. Darken all final lines.

    This same procedure is used in sketching any rectangular object, such as the cabinet in Fig. 17-4.

17.4 SKETCHING ISOMETRIC ELLIPSES. In isometric, circles appear as ellipses. The steps in sketching a cylinder are shown in Fig. 17-5.

    I. Sketch the enclosing box.

    II. Sketch diagonals and center lines of the ends.

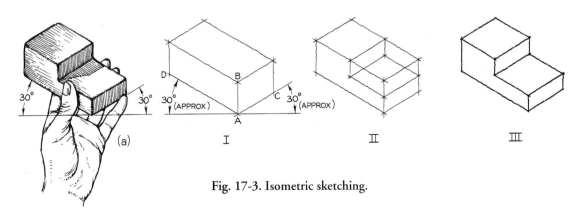

Fig. 17-3. Isometric sketching.

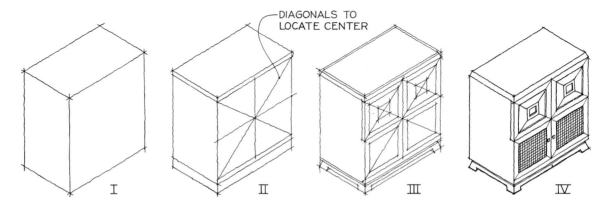

Fig. 17-4. Isometric sketch of cabinet.

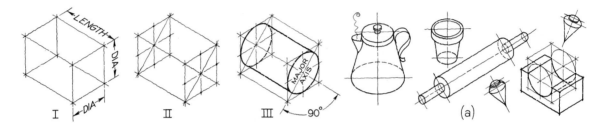

Fig. 17-5. Isometric ellipses.

III. Sketch the ellipses and complete the cylinder. Note that the major axes of the ellipses are at right angles to the center line of the cylinder. Isometric sketches of other common objects showing the major axes of ellipses at right angles to center lines are shown at (a).

17.5 SKETCHING ON ISOMETRIC CROSS-SECTION PAPER. One of the best aids in learning to sketch in isometric is cross-section paper, as shown in Fig. 17-6. Two given views are shown at (a).

I. Sketch the enclosing box. Count the isometric grid spaces equal to the corre-

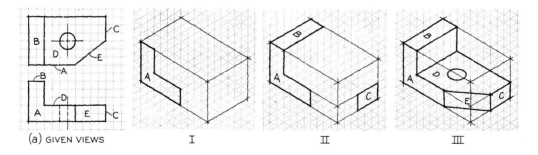

Fig. 17-6. Sketching isometric from given views.

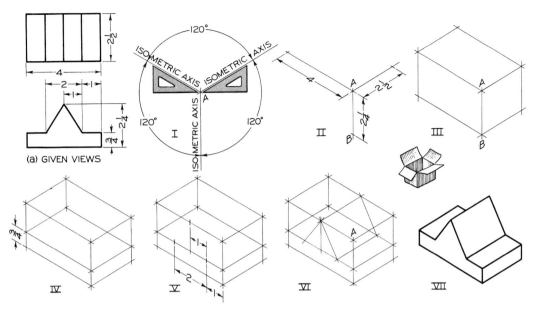

Fig. 17-7. Steps in isometric mechanical drawing.

sponding squares on the given views. Sketch surface A in isometric, as shown.

II. Sketch surfaces B and C in isometric.

III. Sketch surfaces D and E and the ellipse to complete the isometric sketch.

17.6 ISOMETRIC MECHANICAL DRAWING.

The steps in making an isometric mechanical drawing are shown in Fig. 17-7.

I. Draw the *isometric axes* 120° apart.

II. Set off along the axes the height ($2\frac{1}{4}''$), the width (4''), and the depth ($2\frac{1}{2}''$).

III. Complete the enclosing construction box, drawing lines parallel to the axes.

IV. Construct the base $\frac{3}{4}''$ high.

V. Locate top and bottom edges of inclined surfaces by measurements parallel to the axes.

VI. Complete construction of the wedge.

VII. Darken final lines.

Note, II and III, that the drawing could have been started at corner B instead of A.

17.7 ANGLES IN ISOMETRIC.

As shown in Fig. 8-15, angles may appear either larger or smaller than true size, depending upon the direction in which they are viewed. Therefore, in isometric, angles cannot be set off directly with the protractor.

Angles are constructed by locating the endpoints of inclined lines by measurements parallel to the axes, Fig. 17-7(V). Lines that are not parallel to the axes are called *non-isometric lines.* They are not true length. In isometric, all measurements must be made along *isometric lines* — that is, parallel to the axes.

If an angle is given *in degrees,* as in Fig. 17-8(a), it is necessary to convert it into linear measurements.

I. Draw the construction box. Note that the actual 90° angles at all corners of the object are in no case 90° in the isometric drawing.

II. The 30° angle cannot be set off in degrees. Thus, it is necessary to draw the

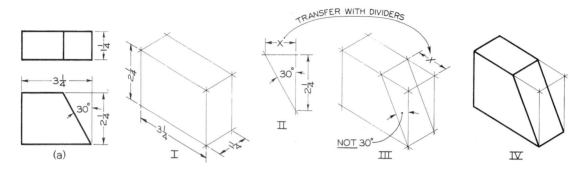

Fig. 17-8. Angles in isometric.

triangle full size to find the needed linear measurement X.

III. Transfer this dimension with dividers to the isometric drawing. Draw the parallel inclined lines. Remember: *Lines that are parallel on the object itself will be parallel in isometric.*

IV. Heavy-in all final lines.

17.8 OFFSET MEASUREMENTS. Figure 17-9(a) shows a triangular pyramid in which all surfaces are oblique except the base. Such an object is drawn by "box construction," I and II, in which all corners are located by offset measurements along isometric lines. Such objects are also drawn by "skeleton construc-

tion," III and IV, in which the base is drawn and then the vertex is located on the vertical center line.

17.9 OTHER POSITIONS OF THE ISOMETRIC AXES. The isometric axes may be drawn in any desired position provided that the angle between them is held at 120°. Objects customarily viewed from below, Fig. 17-10(a), may be drawn with *reversed axes* — that is, with two axes sloping downward. Also, reversed axes may be used to give a better view of an object, as shown at (b) and (c).

Long objects may be effectively drawn with the long axis horizontal, (d).

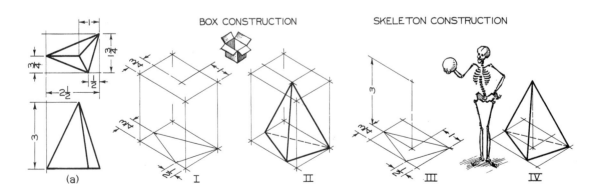

Fig. 17-9. Offset measurements.

17.10 CIRCLES IN ISOMETRIC. As shown in Secs. 2.9 and 17.4, circles appear as ellipses in isometric. An approximate ellipse, which can be easily drawn with the compass from four centers, is sufficiently accurate for nearly all isometric drawings. The steps in drawing this *four-center ellipse* are shown in Fig. 17-11.

Four-center ellipses, as they would be constructed on the four sides of a cube, are shown in Fig. 17-12(a). All diagonals are hor-

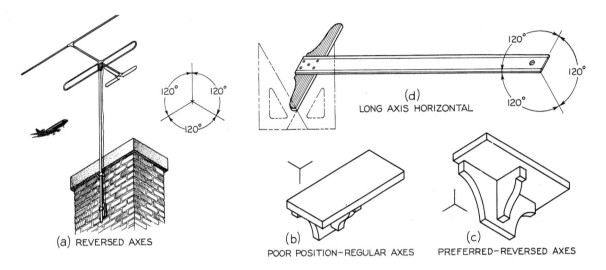

Fig. 17-10. Other positions of axes.

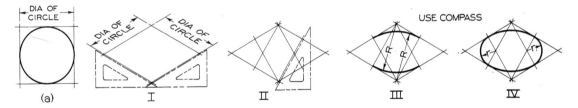

Fig. 17-11. Steps in drawing four-center ellipse.

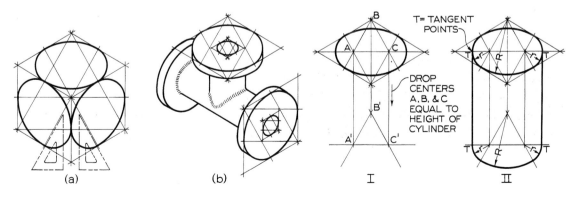

Fig. 17-12. Ellipse construction.

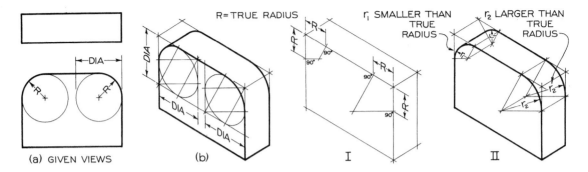

Fig. 17-13. Arcs in isometric.

izontal or 60° with horizontal. Thus, the entire construction can be made with the T-square and the 30° × 60° triangle. Note also that *all diagonals are perpendicular bisectors of the sides of the parallelograms.*

An application of ellipses in drawing a pipe fitting is shown at (b). Note that the smaller ellipses require their own construction. The same centers cannot be used for two or more concentric ellipses.

To make an isometric drawing of a cylinder with its axis vertical, refer to Fig. 17-12:

I. Draw the isometric ellipse for the upper end. Then drop centers A, B, and C down a distance equal to the height of the cylinder, as shown. Draw horizontal line A'C' and the lines B'A' and B'C' at

60° with horizontal.

II. Complete the cylinder by drawing two small arcs and one large arc corresponding to those in the upper ellipse.

17.11 ARCS IN ISOMETRIC. Two views of an object with rounded corners are shown in Fig. 17-13(a). The rounded corners could be drawn in isometric as shown at (b), in which the complete ellipses are constructed. However, only one arc is actually required at each corner. Thus, only a part of the construction is needed. It is only necessary, as shown at I, to set off the radius R from each corner and to draw perpendiculars that intersect at the required centers, as shown. Note that the compass arcs are not the same radius

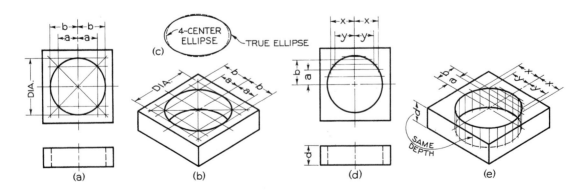

Fig. 17-14. True isometric ellipse construction.

at both ends of the block, II, or equal to the actual radius, R.

17.12 TRUE ISOMETRIC ELLIPSE.

In cases where the four-center ellipse is not accurate enough, the true isometric ellipse can be drawn by plotting points and using the irregular curve, Fig. 17-14.

Around the given circle, (a), draw a square and diagonals, as shown. Where the diagonals cut the circle, draw lines parallel to the sides of the square. Draw this set of lines in isometric, as shown at (b), transferring distances *a* and *b* with dividers. This method provides eight points on the ellipse. The ellipse is then drawn through the points with the irregular curve, Sec. 4.30. Use the curve as shown in Fig. 4.40.

A comparison of the true ellipse with the four-center ellipse is shown at (c). The four-center ellipse is slightly shorter and "fatter" than the true ellipse.

When more than eight points are needed for greater accuracy, draw as many parallel lines, spaced at random, across the given circle as desired, as shown at (d). Draw these lines in the isometric, (e). Transfer distances *a, b, x,* and *y* with dividers.

To locate points on the bottom ellipse, drop points of the upper ellipse down a distance equal to the height *d* of the block. Draw the ellipse, part of which will be hidden, through these points.

17.13 ELLIPSE GUIDES.

One of the chief difficulties in pictorial drawing is the frequent necessity for drawing ellipses. To allow the drafter to draw true ellipses in less time than required even for the four-center ellipse, many types of ellipse guides, or templates, are available. These are plastic sheets with various sizes of elliptical openings. Some of these, such as the Instrumaster Isometric Template, also provide the angles needed to construct isometric drawings, as well as scales printed along the edges, Fig. 17-15(a). The position for drawing an ellipse on top of a cube is shown at (b).

For ink work, technical fountain pens, Sec. 5.18, are recommended. They are available in a variety of sizes. Insert triangles under the stencil to separate it from the paper and prevent ink from running underneath, (c).

17.14 CURVES IN ISOMETRIC.

Curves are drawn in isometric by plotting a series of points on

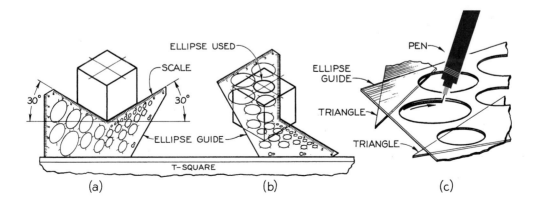

Fig. 17-15. Use of Instrumaster Isometric Template.

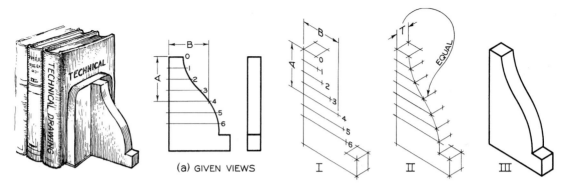

Fig. 17-16. Isometric curves.

the curve. Figure 17-16(a) shows the given views of a book end to be drawn in isometric. Draw a series of parallel construction lines across the view, as shown. It is best to space these equally with the scale or dividers so that they can be easily transferred to the isometric, I. Lines 1, 2, 3, 4, etc., are drawn the same length in isometric as in the given view.

To draw the back curve, draw parallel construction lines equal in length to the thick-

ness of the block, as shown at II. Finally, darken all required lines, III, using the irregular curve, Sec. 4.30.

17.15 INTERSECTIONS. To draw the curve of intersection between a cylindrical hole and an oblique plane, Fig. 17-17, first draw the ellipse, representing the hole in isometric, in the top plane of the enclosing isometric construction box. A series of imaginary parallel cutting planes are then drawn in the isometric. These lines correspond to those that were used to obtain the elliptical intersection in the regular views. Points are then projected down from the top plane of the construction box to the oblique plane to obtain the desired curve of intersection, as shown at (b).

To draw the curve of intersection between two cylinders, Fig. 17-18, pass a series of imaginary cutting planes through both cylinders and parallel to their axes, as shown. Each plane will cut lines (elements) from both cylinders that intersect at points common to both cylinders. More specifically, it will locate points on the curve of intersection, as shown at (b). In all problems involving a curve of intersection, as many points as are necessary should be plotted to assure a smooth curve.

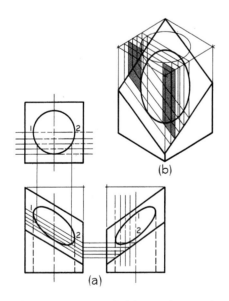

Fig. 17-17. Intersection of oblique plane and cylinder.

The curves of intersection may be drawn with ellipse guides or an irregular curve.

17.16 ISOMETRIC SECTIONING. Interior shapes are exposed in isometric, as in multiview drawing, by means of sections. A full section is shown in Fig. 17-19(a) and (b), with the steps in construction shown above. In drawing full sections in isometric, it is best to draw the cut surface first, and then add the remaining lines in the back half.

A half section is shown at (c) and (d), with the steps in construction shown above. In this case, it is best to block in the entire object and then cut out the section.

Avoid drawing section lines parallel or perpendicular to any principal lines of the drawing. Generally, section lines at 60° with horizontal produce the best effect, but other angles are permissible. In a half section, slope the lines in opposite directions, (d).

17.17 ISOMETRIC DIMENSIONING. Isometric drawings may be dimensioned, if desired. In general, dimensions should be made to lie in the isometric planes (extended) of the object. Many examples are shown in the problems in

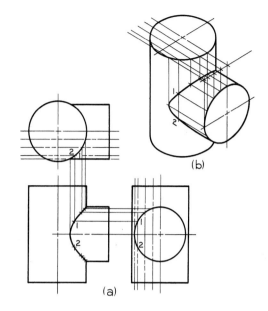

Fig. 17-18. Intersection of cylinders.

this book. See, for example, the problems in Figs. 8-56 through 8-61.

17.18 OBLIQUE SKETCHING. Another simple way to sketch a rectangular object pictorially is in oblique, Fig. 17-20.

I. Sketch the front of the object in true size and shape.

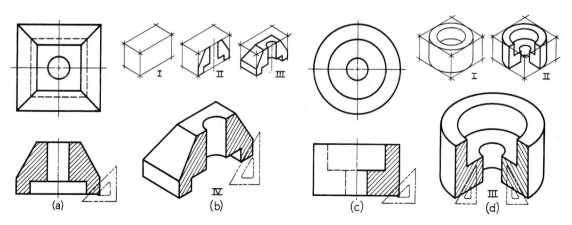

Fig. 17-19. Isometric sections.

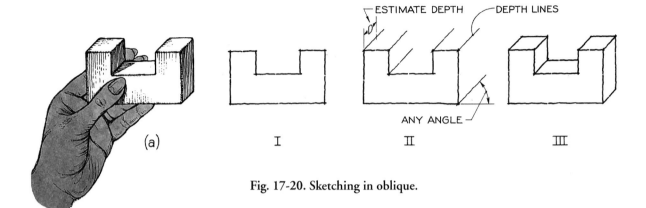

Fig. 17-20. Sketching in oblique.

II. Sketch the *depth lines* parallel to each other and at any convenient angle, say 30° or 45° with horizontal. Cut off the depth lines in such a way that the object will look natural. For simplicity in sketching, this may be full depth (actual depth of block). However, more natural results are obtained if this depth is three-quarter or half size. If the sketch is at half depth, it is called a *cabinet sketch.*

III. Sketch remaining lines to complete the drawing.

17.19 OBLIQUE SKETCHING ON CROSS-SECTION PAPER.
Oblique sketches may be made easily on ordinary cross-section paper. In Fig. 17-21(a), two views are given. Their dimensions can be determined by counting the $\frac{1}{4}$ " squares.

I. Sketch the given object $2\frac{1}{2}$ " wide (10 squares) and $1\frac{1}{2}$ " high (6 squares). Sketch the depth lines at 45° diagonally through the squares as shown. Excellent results are obtained if the depth lines pass diagonally through half as many squares as the actual number given — in this case, two as compared to four.

II. Sketch the remaining features.

III. Darken all final lines.

17.20 OBLIQUE MECHANICAL DRAWING.
The steps in making an oblique mechanical drawing are shown in Fig. 17-22.

I. Construct the *oblique axes* lightly. Draw the *depth axis* at any desired angle, usually 30°, 45°, or 60° with horizontal.

II. Set off the width, height, and depth on

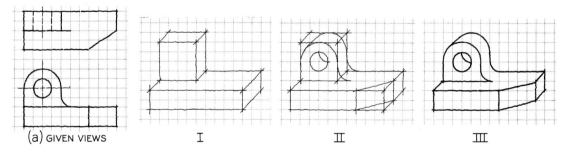

Fig. 17-21. Sketching in oblique on cross-section paper.

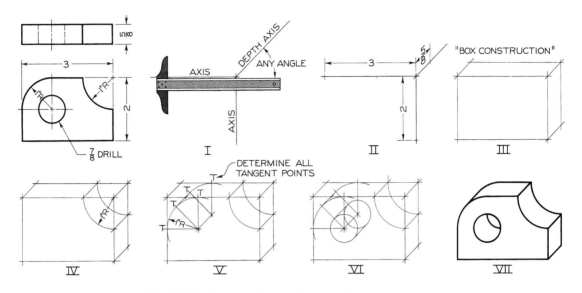

Fig. 17-22. Steps in oblique drawing—box construction.

the axes. In this case, the depth is drawn to full scale.

III. Draw construction box.

IV. to VI. Add arcs and circles, giving particular attention to points of tangency, T.

VII. Darken all required lines.

Note that all shapes lying in the front face of the object, or parallel to it, are shown in true size and shape. Hence, objects with circular shapes can be easily drawn directly with the compass. This makes oblique drawing much simpler for such shapes than isometric,

in which ellipses must be constructed.

If an object is essentially rectangular in shape, it is best drawn by "box construction," Fig. 17-22. Other objects lend themselves to "skeleton construction," Fig. 17-23. Note the points of tangency, III.

17.21 ANGLE OF DEPTH AXIS. As shown in Fig. 17-22(I), the depth axis may be drawn at any angle. However, the angles usually chosen are 30°, 45°, or 60° with horizontal. These can be readily drawn with the triangles. The

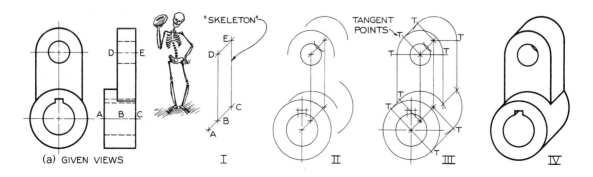

Fig. 17-23. Steps in oblique drawing—skeleton construction.

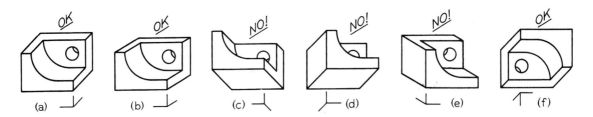

Fig. 17-24. Various angles of depth axis.

drafter must decide in each case which angle is most suitable. In Fig. 17-24, the drawings at (a) and (b) are suitable for this particular object, while those at (c) to (e) are not. However, if the object were turned over, as at (f), it is best shown with reversed axes.

17.22 SCALE OF DEPTH AXIS. The depth axis may be drawn full size or reduced. Figure 17-25(a) is an oblique drawing of a cube in which the depth axis is drawn to full scale. The cube appears to be too deep. Also, the depth lines appear to spread apart as they recede. When the depth axis is drawn to full scale, the oblique drawing is given the special name *cavalier drawing*. Cavalier drawing is perfectly satisfactory for representing many objects, Figs. 17-22 to 17-24. However, in others the distortion may be excessive.

If the depth axis is reduced to three-quarter size, (b), or half size, (c), the result is much more natural. These reductions can be easily set off with the architects scale. When the depth axis is reduced to half size, the name *cabinet drawing* is given because of its early use in the furniture industry.

A comparison between a cavalier drawing and a cabinet drawing of a bookcase is shown at (d) and (e). Note that if a cabinet drawing is drawn to half scale, the depth axis would be drawn to quarter scale.

17.23 CHOICE OF POSITION. The chief advantage of oblique drawing is the ease with which circular shapes can be drawn. For example, in Fig. 17-26(a), all circular shapes of the wheel are faced toward the front. Therefore these shapes are drawn easily with the compass. If the circular shapes are not faced toward the front, (b), the circles become ellipses that are distorted and tedious to draw. If one does not object to drawing ellipses, the

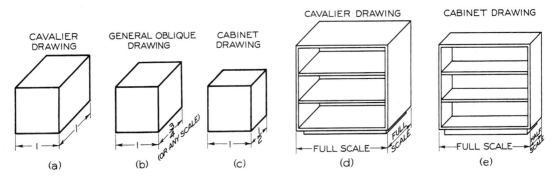

Fig. 17-25. Scale of depth axis.

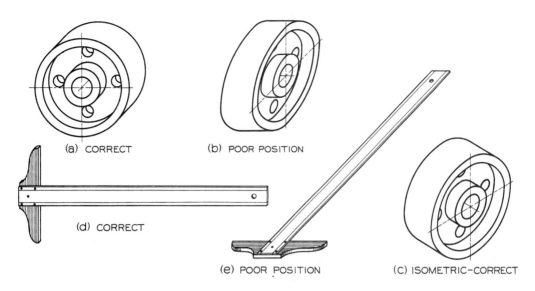

Fig. 17-26. Oblique positions.

wheel can be drawn in isometric, (c), with better results. *Rule: In oblique drawing, always face contours toward the front where they appear in true size and shape.*

The eye is accustomed to seeing parallel lines tend to converge as they recede into the distance. But in oblique drawing, the receding lines are drawn parallel. The result is sometimes very unnatural. A striking comparison between oblique drawing and perspective (the way the eye sees things) is shown in Fig. 17-27. Oblique drawing should not be used

Fig. 17-27. Unnatural appearance of oblique drawing.

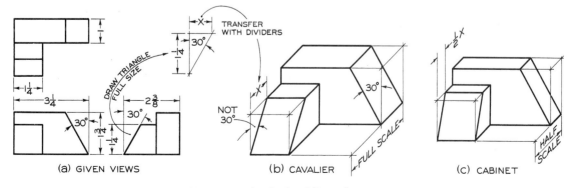

Fig. 17-28. Angles in oblique drawing.

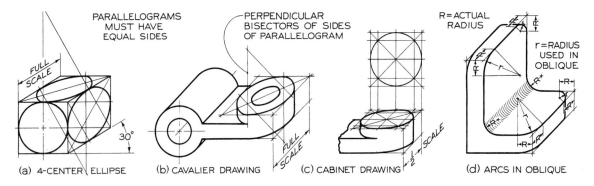

Fig. 17-29. Arcs and circles in oblique.

in such cases where distortion offends the eye. To minimize this distortion in oblique drawing, observe the following: *Draw long objects with the long dimension perpendicular to the line of sight,* Fig. 17-26(d).

17.24 ANGLES IN OBLIQUE. Figure 17-28(a) shows an object having two 30° angles. When drawn "in cavalier," (b), the angle that is faced toward the front is drawn true size. However, the angle that is in a receding plane will not be true size. It must be drawn by constructing the triangle full size, as shown, and transferring distance *x*.

If the object is drawn "in cabinet," every depth dimension, including distance x, must be drawn to half scale, (c).

17.25 ARCS AND CIRCLES IN OBLIQUE. Circles that are faced toward the front will appear as true-size circles. Those that are not faced toward the front will appear as ellipses, Fig. 17-29(a). The four-center ellipse requires an enclosing parallelogram with equal sides. Thus, it can be used only in cavalier drawing. The four centers are found by simply erecting perpendicular bisectors to the sides of the parallelograms, as shown.

An application of the four-center ellipse in a cavalier drawing is shown at (b). Note that the depth axis is full scale and that the sides of the parallelogram are therefore equal. If the depth axis is drawn to a reduced scale, such as half size (as in cabinet drawing), the four-center ellipse cannot be used. Instead, it is necessary to plot points on the ellipse, as shown at (c), and to draw the ellipse with the irregular curve, Sec. 4.30. If more points on the ellipse are needed, the method shown in Fig. 17-14(d) and (e) may be used.

The four-center method can be used to draw arcs in cavalier drawing only, Fig. 17-29(d). Simply set off given radius R from each corner and erect perpendiculars to locate centers, as shown.

17.26 OBLIQUE SECTIONS. Where necessary to expose interior shapes in oblique drawings, sections may be drawn, Fig. 17-30. Draw the section lines in opposite directions in a half section, as shown at (a). In general, avoid drawing section lines parallel or perpendicular to the visible lines bounding the sectional areas.

17.27 OBLIQUE DIMENSIONING. Oblique drawings may be dimensioned if desired, as

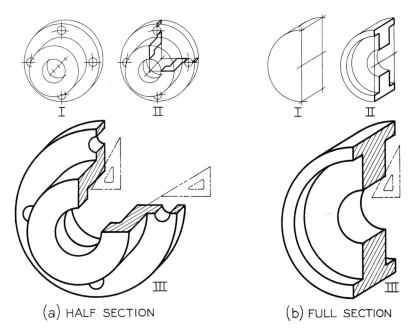

(a) HALF SECTION   (b) FULL SECTION

Fig. 17-30. **Oblique sections.**

shown in Figs. 16-30, 16-48, and 16-51. Dimension lines, extension lines, arrowheads, and dimension figures should be drawn to lie in the corresponding planes (extended) of the object. Notes should always be lettered horizontally and "in the plane of the paper."

17.28 PERSPECTIVE. Both the eye and the camera are constructed so that objects appear progressively smaller as they are farther away. For example, in Fig. 17-31, the spacing between the rails, the lengths of the ties, and the heights of the telephone poles appear to diminish with distance. Also, the rails and other lines parallel to them, such as the tops of the telephone poles, converge at a point on the horizon called the *vanishing point*. Remember this rule in perspective: *All parallel lines have*

Fig. 17-31. **One-point perspective.**

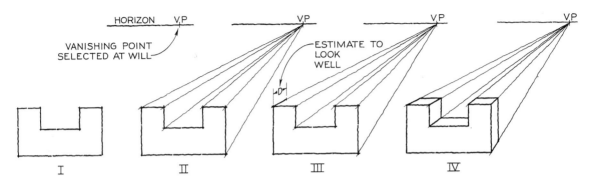

Fig. 17-32. One-point perspective sketch.

the same vanishing point. *If the lines are on or parallel to the ground, the vanishing point will be on the horizon.* Furthermore, the horizon will always appear to be at eye level. For example, if you should stand in the center track in the foreground in Fig. 17-31, your eye level would coincide with the horizon line.

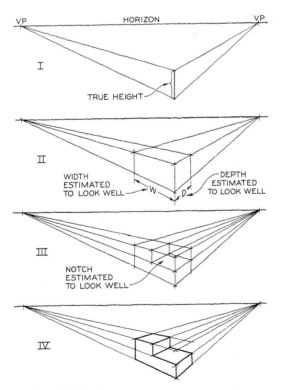

Fig. 17-33. Two-point perspective sketch.

17.29 PERSPECTIVE SKETCHING. The object sketched in oblique in Fig. 17-20 may be easily sketched in *one-point perspective* (one vanishing point), as shown in Fig. 17-32.

I. Sketch front face of object true size and shape. Select a vanishing point for the converging depth lines. Before deciding upon the location of the vanishing point, experiment with it in several different places.

II. Sketch depth lines toward vanishing point.

III. Estimate the depth by eye to make it look natural.

IV. Cut off all depth lines and complete the sketch. Note the similarity to the oblique sketch in Fig. 17-20.

The object sketched in isometric in Fig. 17-3 can be easily sketched in *two-point perspective* (two vanishing points), as shown in Fig. 17-33.

I. Sketch front corner of object true height. Locate two vanishing points by eye where you think they will produce the best picture. Both points must be on a horizontal "horizon" line.

II. Estimate width and depth. Sketch enclosing box.

III. Block in the right-angled notch.

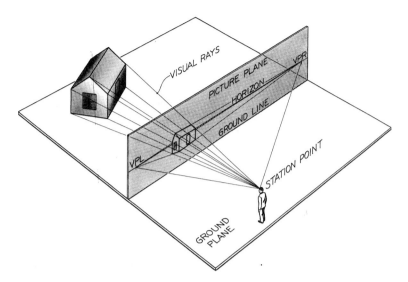

Fig. 17-34. Perspective projection.

IV. Darken all required lines. Note the similarity of this sketch to the isometric sketch in Fig. 17-3.

To gain further understanding, try sketching the same object with the horizon higher or lower than in Fig. 17-33. See what happens if the horizon is placed below the perspective. Also try out the effects of placing vanishing points closer together or farther apart.

17.30 THEORY OF TWO-POINT PERSPECTIVE.

To draw a correct perspective mechanically, it is first necessary to consider the theoretical method of projection. As shown in Fig. 17-34, a transparent *picture plane* (PP) is placed between the observer's eye or *station point* (SP) and the object. *Visual rays* extend from SP to all points on the object. Collectively, the piercing points of the visual rays in PP form the perspective or picture as seen by the observer. The *horizon line* on PP is drawn at eye level. The vanishing points VPL and VPR will be on this line.

17.31 TO DRAW TWO-POINT PERSPECTIVE.

The steps in drawing a two-point perspective are illustrated in Fig. 17-35.

I. The views of the object are given.

II. Draw the picture plane PP, horizon, and ground line GL. To simplify the construction, the front corner of the house (top view) is drawn touching PP, and at a convenient angle of 30° with PP. Draw the side view or front view resting on GL and to one side of the drawing. Height dimensions will be projected from this view across to the perspective. Locate the station point SP in front of the house, as shown.

III. Locate vanishing points by drawing lines from SP parallel to lines 1-2 and 1-3. From their intersections with PP, project down to the horizon to get VPL, the left-hand vanishing point, and VPR, the right-hand vanishing point. Notice that the closer SP is drawn to the top view the closer the vanishing points will be, and vice versa.

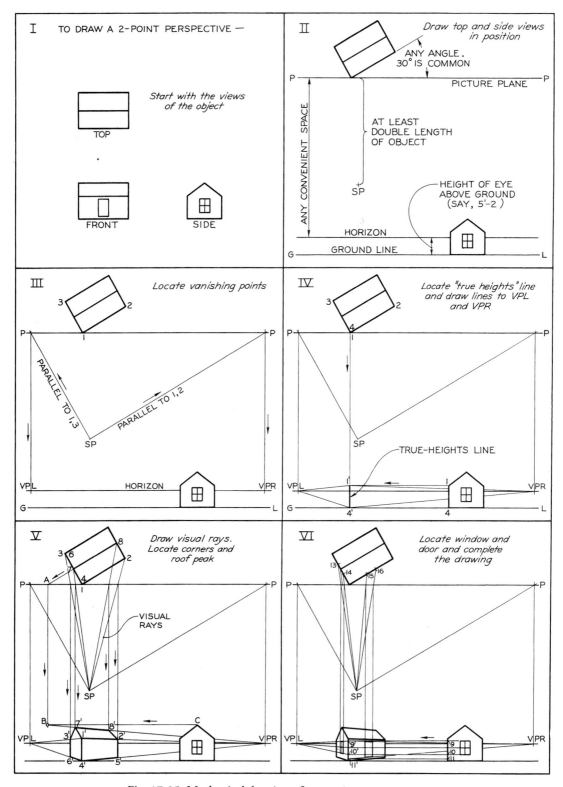

Fig. 17-35. Mechanical drawing of two-point perspective.

IV. Locate true-heights line by projecting down from corner 1-4 in PP to GL. Project across from points 1 and 4 in the side view to the true-heights line to establish 1'-4', the perspective of the front corner of the house. From 1' and 4', draw lines to VPL and VPR, as shown.

V. Draw visual rays from SP to the various points in the top view. From the intersections of these with PP, project down to locate corners 3'-6' and 2'-5'.

The method of finding the roof peak is a general method which is applied to finding the perspective of any horizontal line: First, extend line 7-8 (top view) until it intersects PP at A. Then project downward from A and across from the peak C to locate B, the piercing point of the line. Then draw line B-VPR. To determine the ends of the peak line 7'-8', draw visual rays SP-7 and SP-8. Where these intersect PP, project down to locate 7' and 8', as shown.

VI. Locate window and door. Project true heights 9, 10, 11 of window and door across from the side view to the true-

heights line at 9', 10', 11'. Then draw lines toward the two vanishing points, as shown. Draw visual rays from SP to 13, 14, 15, and 16 (sides of window and door in the top view). Project down from the intersections of these lines with PP to establish the widths of the window and door in the perspective.

17.32 TO DRAW ONE-POINT PERSPECTIVE.

Figure 17-36(a) shows a mechanical drawing in one-point perspective of the same object sketched in Fig. 17-32. The front face of the block is placed *in* PP so that it will be drawn in true size and shape. SP is located in front and to one side of the object, at a distance away from it equal to about twice its width. The horizon is placed well above the ground line. The single vanishing point is on the horizon directly above SP.

To determine the depth, construct the perspective of any convenient point on the back side of the object, as corner 2. Draw visual ray SP-2, intersecting PP at point A. Project down to point 2', which is the perspective of corner 2.

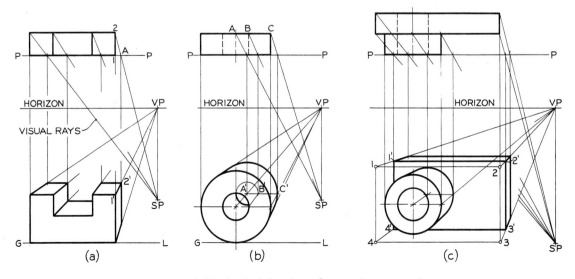

Fig. 17-36. Mechanical drawing of one-point perspective.

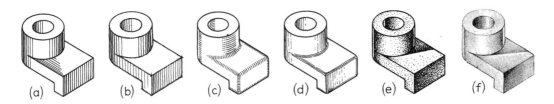

Fig. 17-37. Methods of shading.

A one-point perspective of a cylinder with a hole is shown at (b). Note that the radius AC of the back rim of the cylinder is reduced to A′C′ in the perspective. All circles are drawn with the compass, since they are parallel to PP.

In the one-point perspective shown at (c) it is necessary to construct the base in the picture plane at 1-2-3-4. You then locate it at the proper depth as shown at 1′-2′-3′-4′ by projecting down from the intersection in PP of the visual rays to the corners.

17.33 METHODS OF SHADING. The purpose of an industrial pictorial drawing is to show clearly the shape of the object and not necessarily to be artistic. Shading should, therefore, be simple and limited to producing a clear picture. Art training is required to produce professional results. However, the ordinary drafter can learn to do all the shading that is necessary.

Some of the most common types of shading are illustrated in Fig. 17-37. Pencil or ink lines are drawn mechanically at (a) or freehand at (b). Two methods of shading fillets and rounds are shown at (c) and (d). Shading produced by pen dots is shown at (e). Pencil "tone" shading is shown at (f). Pencil shading is often applied to pictorial drawings on tracing paper. Such drawings can be reproduced with good results by making diazo prints or other prints in which the background is white. In blueprinting, the darks and lights are reversed. Even then the results are quite satisfactory.

Examples of line shading on pictorial drawings in industrial sales literature are shown in

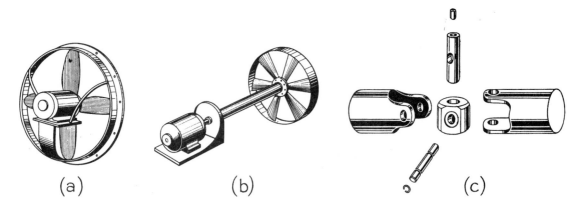

(a) and (b) Courtesy Power Fan Manufacturers Assn. (c) Courtesy Boston Gear Works

Fig. 17-38. Examples of mechanical line shading.

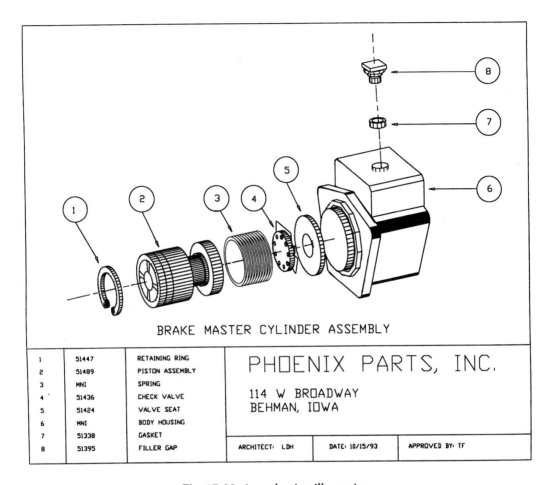

BRAKE MASTER CYLINDER ASSEMBLY

1	51447	RETAINING RING	PHŒNIX PARTS, INC.	
2	51489	PISTON ASSEMBLY		
3	MNI	SPRING		
4	51436	CHECK VALVE	114 W BROADWAY	
5	51424	VALVE SEAT	BEHMAN, IOWA	
6	MNI	BODY HOUSING		
7	51338	GASKET		
8	51395	FILLER GAP		

ARCHITECT: LDH	DATE: 10/15/93	APPROVED BY: TF

Fig. 17-39. A production illustration.

Fig. 17-38. An "exploded assembly" is shown at (c). Here, several parts are drawn in positions indicating how they are assembled.

17.34 PRODUCTION ILLUSTRATION. *Production illustration* is the term applied to a variety of pictorial drawings used in industry. In general, the need for pictorial illustrations arises from the fact that many manufactured items are becoming so complex that it is difficult to follow clearly all details from working drawings alone. This situation is particularly acute when a large number of workers cannot read complicated blueprints.

Production illustrations are used on the production lines, especially in assembling. They are used to show the way parts fit together and the sequence of operations to be performed.

Figure 17-39 is a computer-generated exploded assembly drawing. It shows how parts fit together. Production illustrations are also used extensively in parts catalogs and maintenance instruction handbooks. A typical parts catalog drawing is reproduced in Fig. 17-40. It is an exploded assembly showing a drill press.

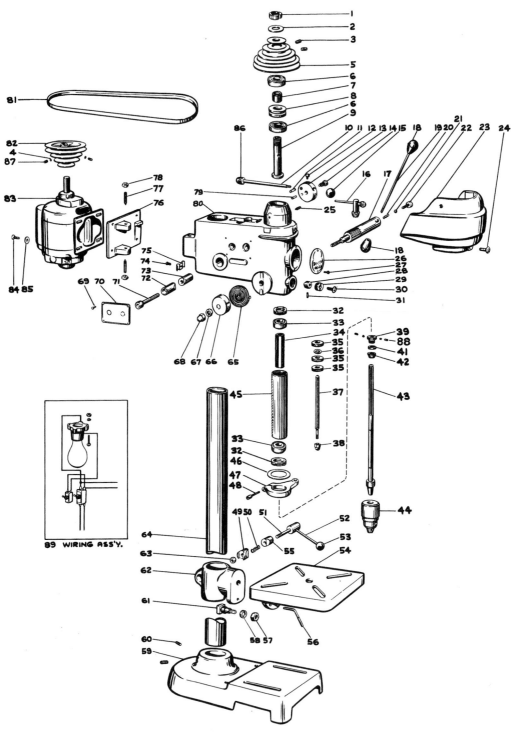

South Bend Lathe Works

Fig. 17-40. Parts catalog illustration.

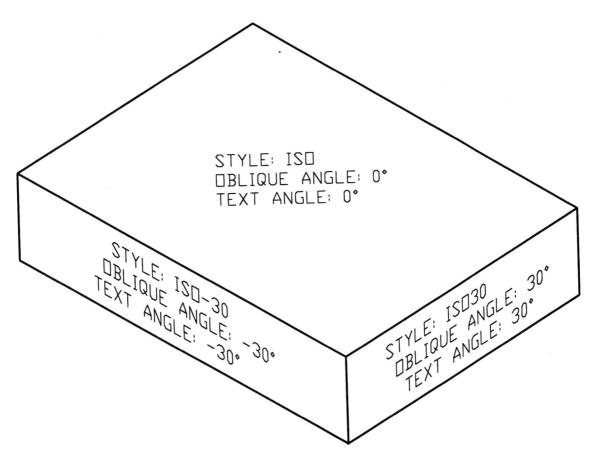

STYLE: ISO
OBLIQUE ANGLE: 0°
TEXT ANGLE: 0°

STYLE: ISO-30
OBLIQUE ANGLE: -30°
TEXT ANGLE: -30°

STYLE: ISO30
OBLIQUE ANGLE: 30°
TEXT ANGLE: 30°

Fig. 17-41. Text aligned with isometric planes.

17.35 COMPUTER GRAPHICS. Isometric drawings can be produced on a CAD system by activating the isometric mode, Fig. 17-41. The cursor and grids are automatically converted to an isometric plane. The drawing may be developed as in manual drafting. Isometric ellipses are converted to isometric style.

Using three-dimensional (3D) models simplifies the task of showing all types of pictorial views. A model can be viewed from any angle, including above or below or any place in between. The number of views is limited only by the imagination of the drafter. Once the desired view has been chosen, the display of the model can be converted to perspective or isometric. The view can also be saved and plotted to the desired scale. This procedure can be completed very quickly. It has the advantage of quickly producing as many views as are required. This saves design costs and allows companies to be more competitive in a world market.

PROBLEMS

Several sketching problems are given in Figs. 17-44 and 17-45. Additional problems may be assigned to be sketched in isometric or oblique from Figs. 7-29, 7-30, 8-26 to 8-28, and 10-45 to 10-47. Sketches may be made on cross-section paper or plain paper, as desired by the instructor.

In Figs. 17-47 to 17-52 are given problems to be drawn with instruments or with the use of a computer in isometric or oblique, as indicated. However, any of the isometric problems may be drawn in oblique, or oblique problems in isometric. Any problem may be assigned to be drawn freehand on cross-section paper or on plain paper. Many additional problems to be drawn mechanically or with the use of a computer may be assigned from Figs. 8-26 to 8-28, 10-45 to 10-47, and 13-19.

Isometric sectioning problems are given in the lower portion of Fig. 17-49. Oblique sectioning problems are given in the lower portion of Fig. 17-52. However, isometric sectioning problems may be assigned from the oblique group, or oblique sectioning problems from the isometric group. Many additional isometric or oblique sectioning problems may be assigned from Figs. 12-27 to 12-31.

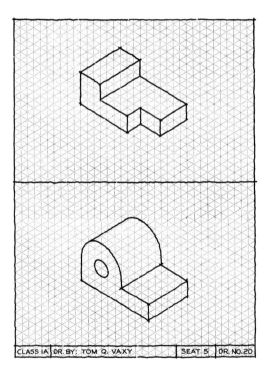

CLASS IA | DR. BY: TOM Q. VAXY | SEAT 5 | DR. NO. 20

Fig. 17-42. Isometric sketches

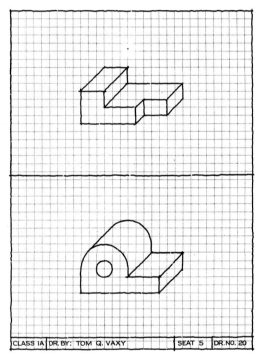

CLASS IA | DR. BY: TOM Q. VAXY | SEAT 5 | DR. NO. 20

Fig. 17-43. Oblique sketches.

Perspective problems are given in Figs. 17-53 and 17-54. Additional problems may be selected from those on preceding pages.

In Figs. 17-47 to 17-49 are given views of objects that are to be drawn in isometric with instruments. Use Layout C of the Appendix. In each problem the location of the starting corner A is given by two dimensions. The first is measured from the left border. The second is measured up from the top of the title strip. For example, in Fig. 17-46, point A is $6\frac{1}{4}''$ from the left border and $1''$ up from the title strip.

Figures 17-51 and 17-52 show views of objects to be drawn in oblique with instruments or with the use of a computer. Use Layout C in the Appendix. In each problem the starting corner A is located by two dimensions. The first is measured from the left border. The second is measured up from the top of the title strip. For example, in Fig. 17-50, point A is $4\frac{7}{8}''$ from the left border and $3\frac{1}{8}''$ up from the title strip. In all problems, assume the depth axis at full scale unless otherwise indicated.

If assigned by the instructor, construct drawings using decimal-inch or metric scales by converting given fractional dimensions to decimal-inch or metric equivalents. Refer to Table 20 in the Appendix.

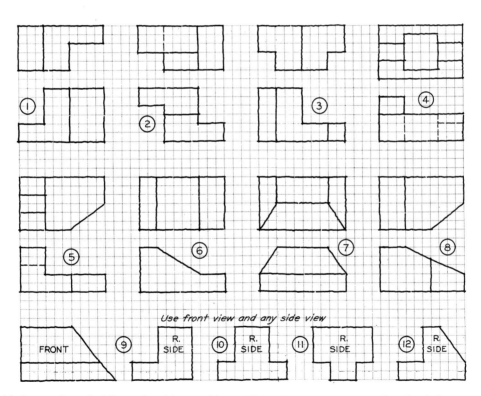

Fig. 17-44. Isometric and oblique sketching problems. Using Layout B in Appendix, divided into two parts as in Figs. 17-42 and 17-43, sketch problems in isometric or oblique as assigned. Each square = $\frac{1}{4}''$.

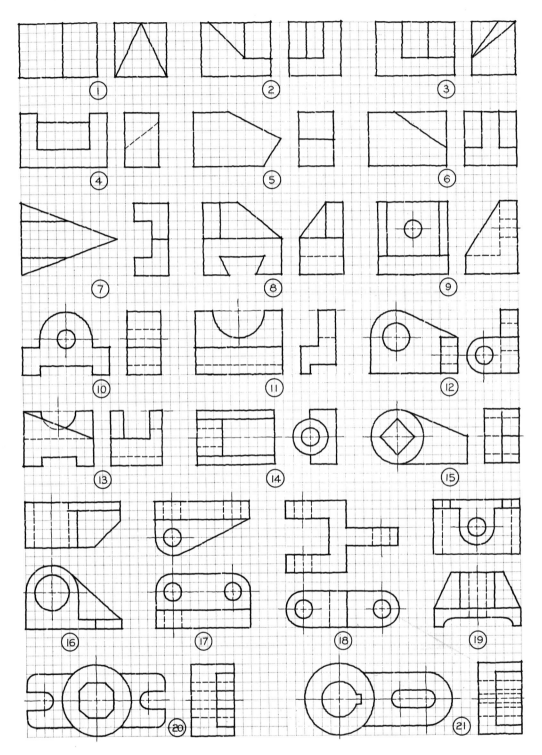

Fig. 17-45. Isometric and oblique sketching problems. Using Layout B in Appendix, divided into two parts as in Figs. 17-42 and 17-43, sketch problems in isometric or oblique as assigned. Each square = $\frac{1}{4}''$.

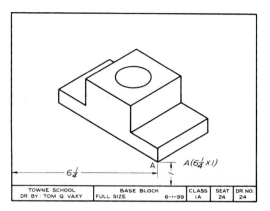

Fig. 17-46. Isometric problem.

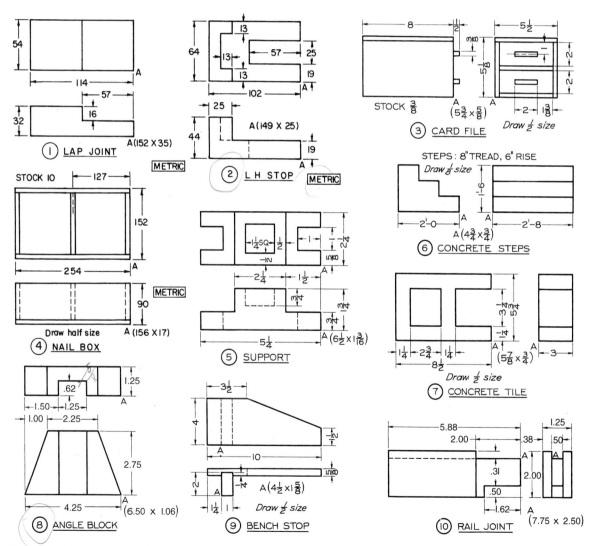

Fig. 17-47. Isometric problems. Locate starting corners A as explained at the beginning of this Problems section. Move titles to title strip and omit dimensions unless assigned.

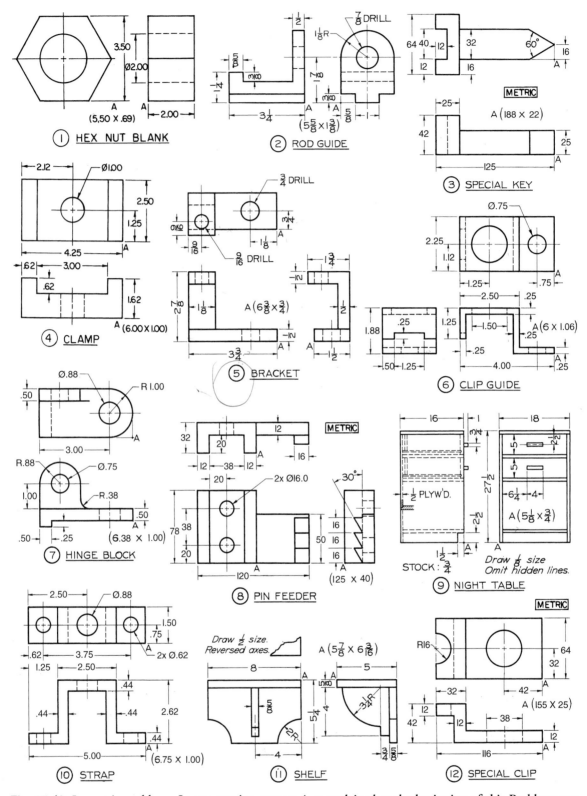

Fig. 17-48. Isometric problems. Locate starting corners A as explained at the beginning of this Problems section. Move titles to title strip and omit dimensions unless assigned.

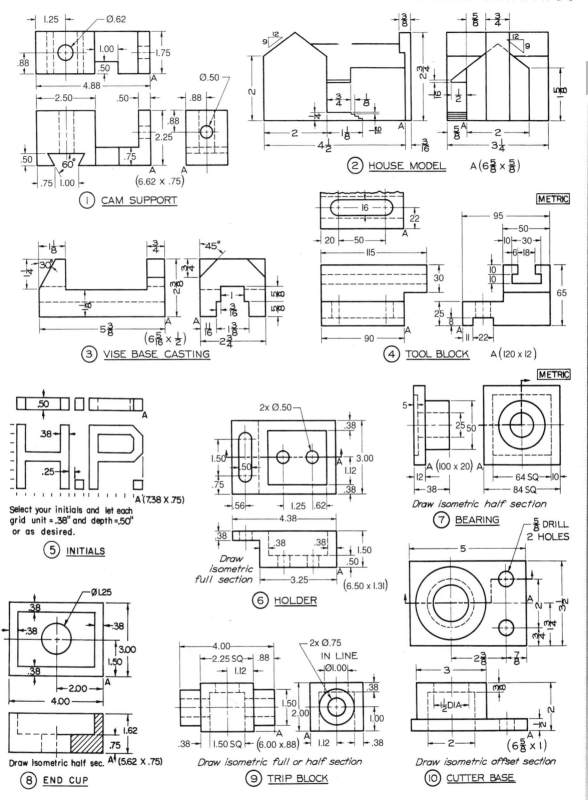

Fig. 17-49. Isometric problems. Locate starting corners A as explained at the beginning of this Problems section. Move titles to title strip and omit dimensions unless assigned.

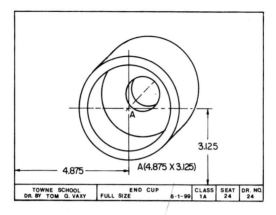

Fig. 17-50. Oblique problem.

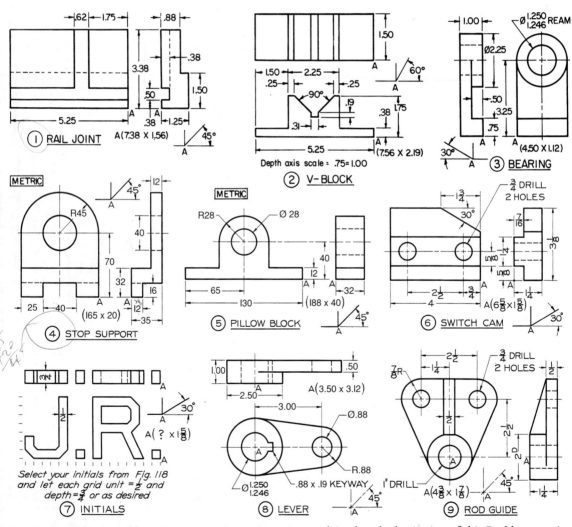

Fig. 17-51. Oblique problems. Locate starting corners A as explained at the beginning of this Problems section. Move titles to title strip and omit dimensions unless assigned.

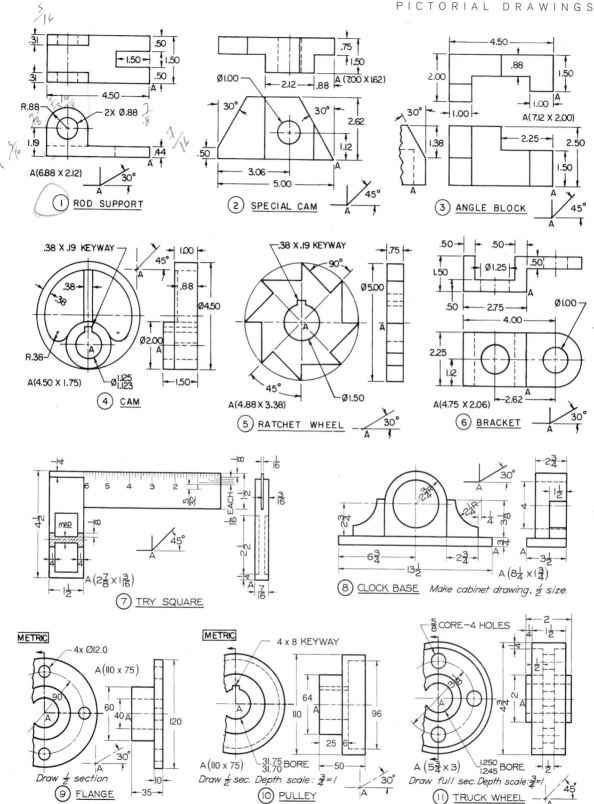

Fig. 17-52. Oblique problems. Locate starting corners A as explained at the beginning of this Problems section. Move titles to title strip and omit dimensions unless assigned.

407

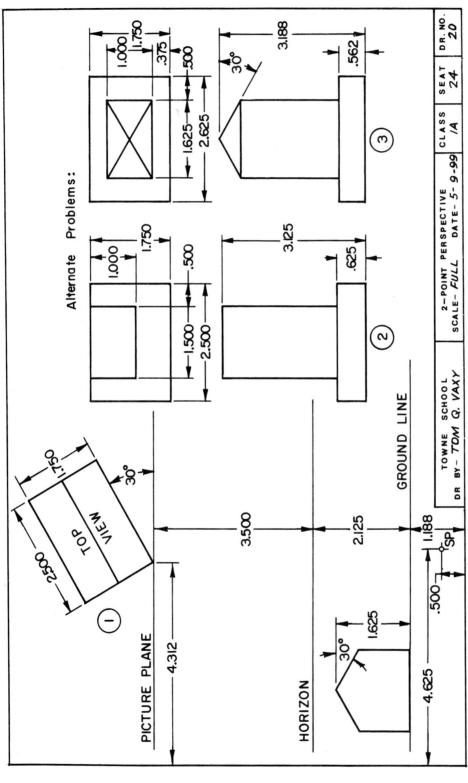

Fig. 17-53. Two-point perspective problems. Use Layout D in the Appendix. Omit all dimensions. Letter VPL, SP, etc.

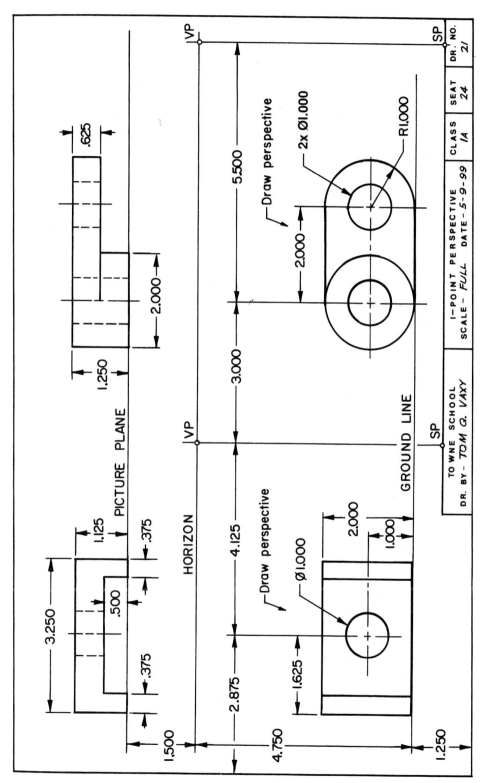

Fig. 17-54. One-point perspective problems. Use Layout D in the Appendix. Omit all dimensions. Letter VPL, SP, etc.

Felicia Martinez/PhotoEdit

CHAPTER 18

OBJECTIVES

After studying this chapter, you should be able to:

- Describe a parallel-line development.
- Prepare a pattern for a three-piece elbow.
- Prepare the pattern of a pyramid.
- Prepare the pattern of a cone.
- Explain an intersection.
- Explain the use of cutting planes in drawing intersections.
- Divide the base of a cone into a number of equal parts.

CHAPTER **18**

Developments and Intersections

An industrial designer is responsible for the shape and form of products, packaging, and displays of both consumer items and commercial systems. He or she must work with a variety of materials and create designs that are durable, economical, and easy to manufacture as well as attractive, functional, and easy to use.

Industrial designers need to have good three-dimensional visualization skills and the ability to rapidly lay out and sketch proposed solutions. They must be trained in production systems, materials, and human design factors. The ability to produce high-quality pictorial drawings of proposed designs is very important as a means of communicating ideas. Other kinds of drawings may also be required. Developments, for example, are useful in visualizing sheet metal and package designs.

CAREER LINK

To learn more about careers in industrial design, contact the Industrial Designers Society of America, 1142-E Walker Road, Great Falls, VA 22066 or visit their web site at *www.idsa.org*

411

18.1 DEVELOPMENTS. The *development* of an object is the surface of the object laid out on a plane. For example, if an ice cream carton is unfolded and laid out on a table, the result is a development. In the sheet-metal trade, a development is usually referred to as a *pattern* or a *stretchout*. Thousands of different manufactured objects are made by cutting out patterns and then folding them into shape, including pipes, air-conditioning ducts, heating ducts, pans, hoppers, bins, buckets, and even juice cartons and paper cups. A striking example of sheet-metal work in the oil industry is shown in Fig. 18-1.

The developments of the four most common solids are shown in Fig. 18-2. Other forms are usually more difficult and more expensive to make. They are avoided if possible. The patterns of the prism and pyramid are merely the sides and ends unfolded onto a plane surface. The patterns for the cylinder and cone are simply the surfaces and ends rolled out or unfolded onto a plane surface. Note that the prism and cylinder roll out into rectangular patterns, called *parallel-line developments*. The pyramid and cone roll out into pie-shaped patterns, called *radial-line developments*. The various geometrical solids are shown in Fig. 6-1.

Wyatt Metal & Boiler Works

Fig. 18-1. Catalyst collector or "cyclone" for petroleum refinery.

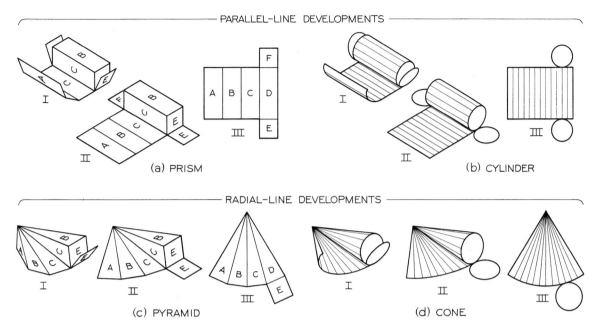

PARALLEL-LINE DEVELOPMENTS

(a) PRISM

(b) CYLINDER

RADIAL-LINE DEVELOPMENTS

(c) PYRAMID

(d) CONE

Fig. 18-2. Developments.

18.2 SHEET-METAL WORK. Patterns are made of paper, cardboard, plastic. They are made also of sheet metal, such as steel, brass, copper, and aluminum. After the metal is cut and folded or rolled into shape, the pieces are fastened together with solder, welds, rivets, or seams of various kinds. Where thickness of metal is a factor, some allowance must be made for stretching or crowding of metal at the bends. Also, extra material must be provided for laps and other kinds of joints, Fig. 18-4. In the following pages, these allowances for bends and seams will be disregarded.

It is customary to draw patterns so that the *inside* surfaces are up, as shown in Fig. 18-2. Thus, when the object is folded into shape, all fold lines will be on the inside.

After the pattern has been laid out from the drawing onto the metal, the metal is cut. It is cut by means of hand snips, chisels, circle cutters, ring and circular shears, or other tools or methods.

Sheet metal may be bent, folded, or rolled by hand in a number of ways. These in gener-

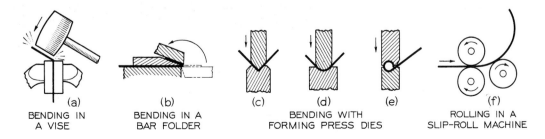

(a) BENDING IN A VISE

(b) BENDING IN A BAR FOLDER

(c) (d) BENDING WITH FORMING PRESS DIES

(e)

(f) ROLLING IN A SLIP-ROLL MACHINE

Fig. 18-3. Bending metal.

413

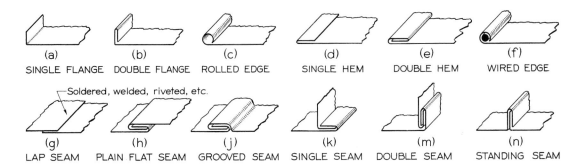

Fig. 18-4. Sheet metal edges and seams.

al consist of hammering the sheets over or around wood blocks or iron *stakes* (anvils of various shapes) with a mallet, Fig. 18-3(a). They may also be bent in a *hand seamer.* A machine used for making narrow bends is the *bar folder,* (b). For folding wide pieces, such as the sides of a box, a large machine called a *brake* is used.

Many types of bends are made on the *press brake* by means of various shapes of *forming press dies.* Some of these are shown at Fig. 18-3(c) to (e). Conical or cylindrical forms are made by hand-hammering over a rounded stake or by rolling on a *slip-roll forming machine,* as shown at Fig. 18-3(f).

Certain objects, such as automobile fenders and the warped skin of aircraft, are *nondevelopable.* (They cannot be laid out flat on a plane.) They are deformed into shape by pressing flat sheets into dies under heavy pressure.

Exposed edges, as on rims of pans or buckets, are usually flanged, hemmed, rolled, and so forth, as shown in Fig. 18-4(a) to (f). Various methods of joining metal at seams are shown at (g) to (n). Usually a pattern is fastened along the shortest edge to save labor and materials. However, the seams may be made along any convenient edge. Often the arrangement is simply the one that can be cut out most economically.

18.3 MODEL CONSTRUCTION. Paper or cardboard models should be made of at least a few of the earlier projects assigned. The final appearance and the fit of the pattern when folded and fastened together will depend upon the accuracy of your drawing.

For paper models, use any stiff paper, such as drawing paper. Draw tabs $\frac{1}{4}''$ wide along edges to be fastened, clipping the corners at 45°, as shown in Fig. 18-5(a). For curved

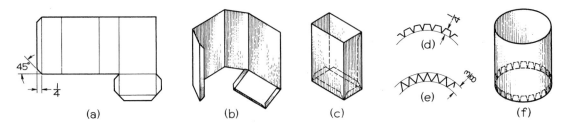

Fig. 18-5. Construction of a paper model.

edges, notch the tabs as shown at (d) or (e).

To cut out the pattern, use a razor blade or a sharp knife. Be sure to cut over a piece of heavy cardboard so as not to damage your drawing board or table top. *Never cut along a triangle or T-square.* A single nick will ruin it. To make the corners and the tabs fold smoothly, score along the fold lines by drawing the divider point along the lines. Use the triangle as a guide. Do not press hard. Keep the leg of the divider almost flat on the paper.

Fasten the seams of the model together. Use paste, glue, rubber cement, or cellulose tape. The model can be made more attractive by painting or spraying with colored lacquer, enamel, or acrylic paint.

▶PARALLEL-LINE DEVELOPMENTS
18.4 PATTERN OF TRUNCATED PRISM – FIG. 18-6.
Prisms have plane faces that intersect to form edges that are parallel, Fig. 6-1. The development of a square prism is shown in

Fig. 18-2(a), the side surfaces folding out into a simple rectangle.

If a prism is cut off at an angle, Fig. 18-6(a) and (b), it is said to be *truncated*. The top and front views of the truncated prism are shown at (c), together with an auxiliary view of the inclined surface. The lower end of the prism will develop into a straight line, 1-1. This is called the *stretchout line,* (d) and (e). The stretchout line is the perimeter of the base, or the total distance around the base, laid out in a straight line. The upper end of the prism will develop irregularly, as shown in the figure.

On the stretchout line, set off distances 1-2, 2-3, 3-4, and 4-1, taken from the top view. Through these points draw the edges, or fold lines, perpendicular to the stretchout line. These are parallel — hence the term "parallel line development." The upper ends A, B, C, etc., of these lines are found by projecting across from the front views of the corresponding points.

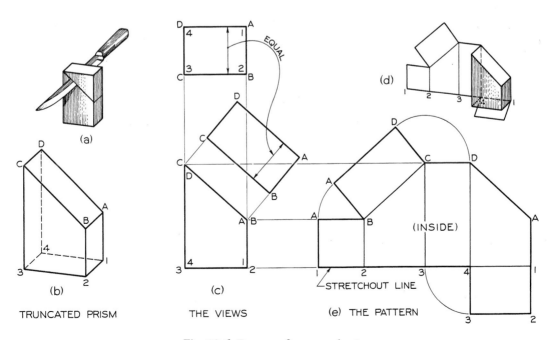

Fig. 18-6. Pattern of truncated prism.

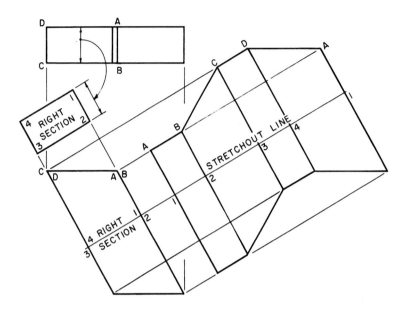

Fig. 18-7. Pattern of oblique prism.

The true size of the bottom is shown in the top view. The true size of the inclined surface is shown in the auxiliary view. These are transferred, if desired, to the pattern and attached along any convenient joining edges.

18.5 PATTERN OF OBLIQUE PRISM – FIG. 18-7.
If both ends of a prism are cut off at an angle other than 90°, Fig. 18-7, neither end will roll out into a straight line. However, if an imaginary sectioning plane is passed through the prism at right angles to the edges, a right section 1-2-3-4 is produced, as shown. This right section will roll out into a straight line. The true size of the right section is shown in the auxiliary view.

To draw the pattern, extend the stretchout line, as shown. Set off on the stretchout line distances 1-2, 2-3, 3-4, and 4-1, taken from the auxiliary view where they are shown true length. Through these points draw the edges perpendicular to the stretchout line. Establish

end-points A, B, C, etc., by projecting from corresponding points in the front view.

18.6 CYLINDERS.
A cylinder may be a *right cylinder* or an *oblique cylinder,* Fig. 6-1. Cylinders usually are circular. However, they may be elliptical or otherwise. A right cylinder, whose bases are perpendicular to its center line, will develop into a simple rectangle. A good way to demonstrate this is to take a paint roller and apply one revolution of the roller on a wall. The painted area will be a rectangle, Fig. 18-8(a).

18.7 CIRCUMFERENCE.
As shown in Fig. 6-1, the circumference of a circle is the distance around the circle. The circumference of a right circular cylinder is the distance around the base, as shown in the top view in Fig. 18-8(b). If the cylinder is rolled out on a plane, the length of the pattern will be the circumference of the cylinder. The height of the pat-

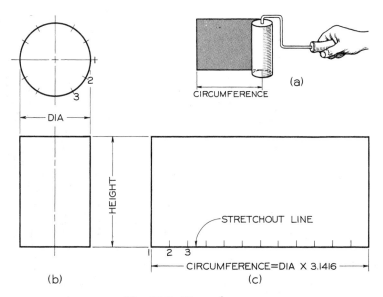

Fig. 18-8. Circumference.

tern will be the height of the cylinder, Fig. 18-8(c). The circumference may be laid out approximately by setting off a number of equal divisions 1-2, 2-3, etc., on the circle in the top view and then stepping off with the bow dividers the same number on the stretchout line on the pattern, as shown in the figure. However, the distances set off would be chords of the arcs, not the actual lengths of the arcs. The slight errors for each distance would add up to a sizable error on the total length.

The circumference of any circle divided by its diameter is 3.1416, or about $3\frac{1}{7}$. This number, 3.1416, is known to mathematicians as π. It is the Greek letter *pi* (pronounced *pie*). Thus, if you know the diameter of a circle, you can always get the circumference by multiplying the diameter by 3.1416. For example, if the diameter of the base of the cylinder in Fig. 18-8 is 2", the circumference of the base and the length of the pattern will be 2" × 3.1416 = 6.2832", or almost exactly $6\frac{9}{32}$". For most practical purposes, you can multiply by 3.14 instead of 3.1416, or even multiply by $3\frac{1}{7}$. Thus, if the diameter is 2", the circumfer-

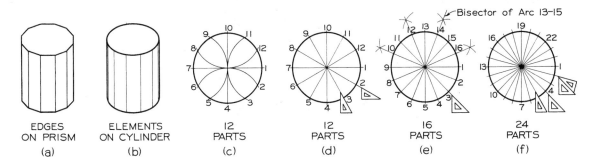

Fig. 18-9. Elements and divisions of a circle.

ence is $2'' \times 3\frac{1}{7}$, or $2'' \times \frac{22}{7}''$, or $\frac{44}{7}$, or $6.28''$, which again is $6\frac{9}{32}''$ (to the nearest $\frac{1}{64}''$).

18.8 ELEMENTS. An *element* of a cylinder is an imaginary straight line on the surface parallel to the axis. A cylinder may be thought of as a prism with an infinite number of edges. Even if the prism has as few as 12 sides, Fig. 18-9(a), the result is close to an actual cylinder. If lines are marked on the cylinder in the corresponding places, as shown at (b), the lines are *elements*. They are useful as will be shown in the following pages. Elements should be drawn as construction lines on the views and on the pattern. A sharp hard pencil should be used.

Practical methods used in dividing a circle into a number of equal parts are shown in Fig. 18-9(c) to (f). At (c) the compass is used, with centers at points 1, 4, 7, and 10, and

with a radius equal to the radius of the circle. At (d) the 30° × 60° triangle is used as in Fig. 4-14(d). At (e) the 45° triangle is used as in Fig. 4-14(b) to get 8 divisions. Then each of these is bisected with the compass, as in Fig. 6-6. At (f) the two triangles are used in combination, as seen in Fig. 4-18.

Elements of a cone are illustrated in Fig. 18-17.

18.9 PATTERN OF TRUNCATED CYLINDER – FIG. 18-10. If a cylinder is truncated, or cut off at an angle, the angled end will develop into a curved line, as shown in Fig. 18-10(a). The lower end will develop into a straight line 1-1, which will be the stretchout line.

To develop the cylinder, divide the top view, (b), into any convenient number of equal parts by one of the methods shown in Fig. 18-9. Project down to draw the elements

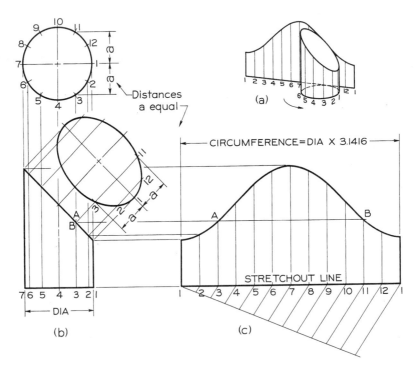

Fig. 18-10. Pattern of truncated cylinder.

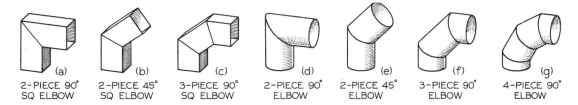

Fig. 18-11. Elbows.

in the front view. As shown at (c), draw the stretchout line 1-1, and set off the true circumference on it. Divide it into the same number of equal parts as in the top view, using the parallel line method, Fig. 6-2 or 6-3. At the division points, draw elements perpendicular to the stretchout line. Locate the top ends of the elements by projecting across from the top ends of the corresponding elements in the front view. Note that each time you project across, you can locate two points, as A and B, in the pattern. Sketch a light smooth curve through the points, and heavy-in the final curve with the aid of the irregular curve, Sec. 4.30.

If the bases are needed in the pattern, they can be cut out separately. The true size of the bottom is shown in the top view. The true size of the inclined surface is an ellipse and is shown true size in the auxiliary view.

18.10 ELBOWS. *Elbows,* Fig. 18-11, are common in sheet-metal work. Some are made up of prismatic shapes and others of cylindrical shapes. They may be composed of two, three, or more pieces. Each piece in the elbows at (a), (b), and (c) can be developed in a manner similar to that seen in Figs. 18-6 and 18-7. Each end piece in the elbows at (d) to (g) can be developed as shown in Fig. 18-10. The method of developing the center pieces at (f) and (g) is shown in Fig. 18-12.

A number of elbows in practical work are illustrated in Fig. 18-1.

18.11 PATTERN FOR THREE-PIECE ELBOW – FIG. 18-12. It is necessary first to draw the views of the elbow. The two end pieces are the same shape. Their patterns will be identical. The middle piece is double the size of an end piece, or the same as the two end pieces put together.

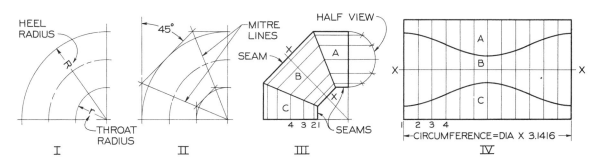

Fig. 18-12. Pattern of a three-piece elbow.

419

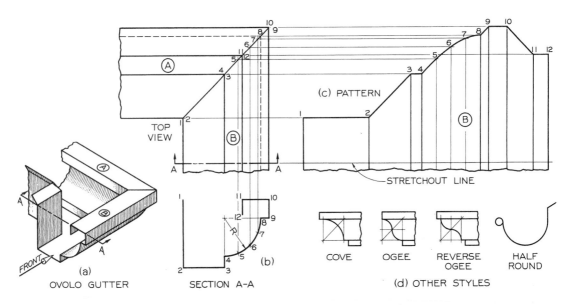

Fig. 18-13. Pattern of an ovolo gutter.

I. Draw *heel* and *throat radii,* and square off ends with light construction lines.

II. Draw vertical, horizontal, and 45° construction lines tangent to arcs as shown. Then, through the intersections, draw the *mitre lines* to the center of the elbow.

III. Draw a semicircular half view adjacent to either end piece. A full circle is unnecessary. No additional views of the elbow are needed. Divide the semicircle into equal parts, say 6, as shown in Fig. 18-9(d). From these points, draw elements on all three sections, as shown. Draw line X-X, the right section of the central piece.

IV. Draw the pattern. Pieces A and C are drawn in the way shown in Fig. 18-10. All elements are shown true length in the front view. Those in piece C are projected directly across from the front view. Those in pieces A and B are transferred with dividers. All three pieces can

be cut out of a rectangular piece, as shown. If desired, piece B could be developed in a manner shown for the oblique prism in Fig. 18-7.

18.12 PATTERN OF A GUTTER – FIG. 18-13.
Gutters are good examples of the extensive use of sheet metal work in building construction. Gutters are composed of various combinations of prisms and cylinders. The patterns are, therefore, parallel-line developments.

In Fig. 18-13(a) is shown a pictorial view of an *ovolo* gutter, sectioned at A-A. The sectional view, (b), shows the true right section of piece B. In the top view, the right section appears as a line A-A, which will roll out into a straight stretchout line. Piece B is imagined to be rolled to the right so that the inside of the pattern will be up. The plane surfaces are developed as for the truncated prism in Fig. 18-6, and the quarter-cylinder as for the truncated cylinder in Fig. 18-10. The stretchout for the quarter-cylinder will be one-fourth the circumference of the complete cylinder:

$$\frac{\text{Dia.} \times 3.1416}{4} \qquad \text{or} \qquad \frac{R \times 3.1416}{2}$$

The stretchout of the quarter-cylinder may be closely approximated by stepping off on the stretchout line the chord distances 4-5, 5-6, 6-7, and 7-8, taken from the sectional view at (b).

Other styles of gutters are shown in Fig. 18-13(d).

▶RADIAL-LINE DEVELOPMENTS

18.13 TRUE LENGTH OF LINE. In radial-line developments, the edges or elements "radiate" like spokes in a wheel from a point, instead of being parallel, Fig. 18-2(c) and (d). These lines usually do not show true length in the regular views. Hence the true lengths to be used in the patterns must be found. For example, in Fig. 18-14(a) the edge 1-3 does not appear true length in either view. As shown at (b), edge 1-3 is the hypotenuse of right triangle 1-2-3 (shaded). By constructing this right triangle true size, (c), we can get the

true length of the edge 1-3. The base 2-3 is taken from the top view at (a). The altitude is taken from the front view.

Another method, (d), is to revolve the triangle until 1-3 is horizontal in the top view. Then the front view of 1-3 will be true length. Note that the true size of the triangle (shaded) at (d) is exactly the same as at (c).

Stripped of all nonessentials, the true length can be found, (e), simply by revolving either view of the line (in this case the top view) until it is horizontal. The other view will then be true length.

For further information on true lengths, see Figs. 13-14 and 14-7.

18.14 PATTERN OF A PYRAMID – FIG. 18-15. *Pyramids* have flat triangular faces that intersect at a common point called the *vertex*. See Fig. 6-1. Bases of pyramids are polygons of three or more sides. In Fig. 18-15(a) is shown a *right rectangular pyramid* — "right" because its axis is perpendicular to the base, and "rectangular" because the base is a rectangle.

The top and front views of the pyramid are shown at (b). All inclined edges are the same length, but none are shown true length. To

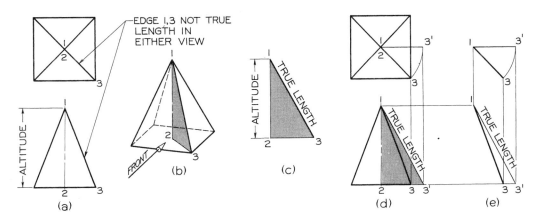

Fig. 18-14. True length of line.

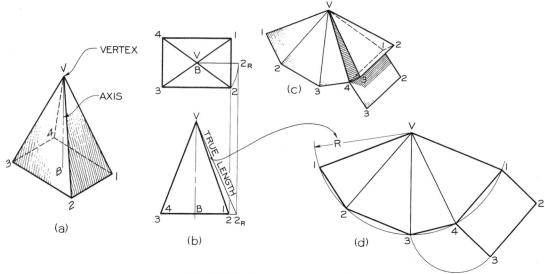

Fig. 18-15. Pattern of a pyramid.

get the true length, revolve the top view of edge V-2 until it is horizontal. The front view will then be true length, as shown. Use this true length as radius to draw the large arc in the pattern. Imagine the pyramid rolled about its vertex to the right as shown at (c). Then set off, on the large arc in the pattern, distances 1-2, 2-3, 3-4, and 4-1, taken from

the top view. Join points with straight lines, as shown, and add the base, taken from the top view.

18.15 PATTERN OF TRUNCATED PYRAMID – FIG. 18-16. A truncated pyramid is shown at (a). It is a right square pyramid because the base is square and the axis is perpendicular to

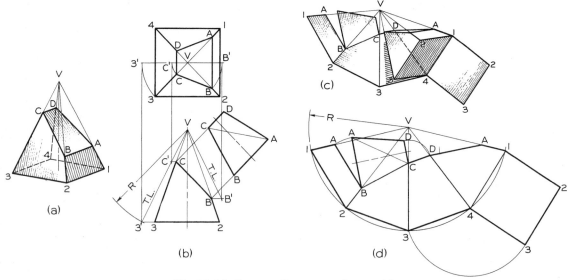

Fig. 18-16. Pattern of a truncated pyramid.

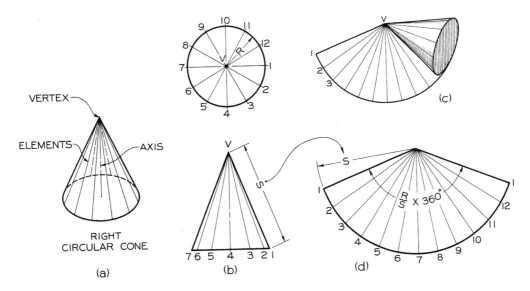

Fig. 18-17. Pattern of a cone.

the base. The top and front views are shown at (b), together with an auxiliary view of the cut surface.

To draw the pattern, (d), imagine the pyramid to be rolled to the right, as shown at (c). In the pattern, radius R is the true length of one of the inclined edges of the pyramid, as shown in the front view. Along the large arc in the pattern, set off distances 1-2, 2-3, 3-4, and 4-1, taken from the top view. Then join the points to each other and to the vertex with straight lines. In the front view, (b), the true lengths from vertex V down to points B and C on the cut surface are shown. Transfer the true lengths VB′ and VC′ to the pattern. Complete the pattern by adding the true size of the base, taken from the top view, and the true size of the cut surface, taken from the auxiliary view.

To transfer the auxiliary view ABCD to the pattern, draw diagonal CA so as to form two triangles CAD and CAB. Transfer the triangles by the method shown in Fig. 18-19.

18.16 PATTERN OF A CONE – FIG. 18-17. A *right circular cone* is shown at (a), with the top and front views at (b). When the cone is rolled out on a plane, the pattern will be a sector of a circle, or pie-shaped, (c). To draw the pattern, (d), draw an arc with radius S equal to the slant height of the cone, taken from the front view. The total angle included in the pattern is equal to

$$\frac{\text{Radius of base}}{\text{Slant height}} \times 360° \quad \text{or} \quad \frac{R}{S} \times 360°$$

Thus, if the cone has a 2″ radius base and a 5″ slant height, the formula would be $\frac{2}{5} \times 360°$, or $\frac{720°}{5}$, or 144°. Set off the angle in the pattern with the protractor, Fig. 4-19.

Another method is to divide the base (top view) into equal parts and draw elements as shown. Then set off chord distances 1-2, 2-3, etc., on the arc in the pattern. However, the chord of an arc is slightly shorter than the arc. When a number of chords are set off, there

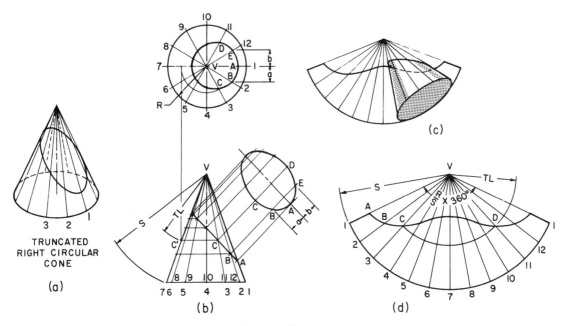

Fig. 18-18. Pattern of a truncated cone.

may be considerable cumulative error. If the bow dividers are used and set *very slightly* larger than the chord distance, the resulting error will be small. The method will be satisfactory in most cases.

The true size of the base is shown in the top view and may be added to the pattern or cut out separately.

18.17 PATTERN OF A TRUNCATED CONE – FIG. 18-18.

A truncated right circular cone is shown at (a), with the front, top, and auxiliary views at (b). When the cone is rolled out on a plane, (c), the lower end develops into a circular arc. The upper, or truncated, end develops into an irregular curve.

To develop the pattern, (d), draw the large arc with the radius S equal to the slant height of the cone. Then compute the included angle with the formula $\frac{R}{S} \times 360°$, as described in Sec. 18.16. Set it off with the protractor, Fig. 4-19. Divide the base in the

top view into equal parts. Draw the elements in both views. By trial, with the bow dividers, space off the same number of equal parts on the base arc in the pattern. Draw the elements. If care is exercised, you may use the spacing between divisions in the top view to determine the complete angle of the pattern instead of having to compute the angle. Then set off from V on the elements in the pattern the true lengths VA, VB, VC, etc., down to the inclined cut, taken from the front view. For example, in the front view, the true length of VC is VC'. This is transferred to the pattern to give points C and D. Each true length in the front view will give two points in the pattern in a similar manner. Note, at (b), that you obtain each true length merely by drawing a horizontal line from the point to an outside element of the cone. This is equivalent to revolving the element until it appears true length in the front view. When all points in the curve have been found, trace a smooth curve through them, using the irregular curve, Sec. 4.30.

The intersection of a plane and a cone, in this case, is a true ellipse. It will appear as a true ellipse in the top and auxiliary views. The true size of the ellipse is shown in the auxiliary view. In the auxiliary view, equal distances a and b, and others on each side of the center line, are transferred from corresponding points in the top view. The inclined face and the base may be cut out separately if needed in the pattern.

18.18 TRIANGULATION. *Triangulation* is the process of dividing a surface into a number of triangles and then transferring each of them in turn to the pattern. To transfer a triangle, say ABC in Fig. 18-19(a), draw side AB in the desired new location at (b). With the ends A and B as centers, and the lengths of the other sides of the given triangle as radii, strike two arcs to intersect at C. Then, as shown at (c), join point C to points A and B.

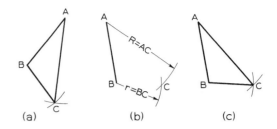

Fig. 18-19. Transferring a triangle.

18.19 PATTERN OF OBLIQUE CONE – FIG. 18-20. An *oblique cone* is shown at (a), with the top and front views at (b). Divide the base (top view) into equal parts. Draw elements to the vertex as shown. Only elements V-1 and V-7 are true length, as shown in the front view. The surface of the cone is thus divided into triangles by the elements. The pattern will be composed of these triangles laid out next to each other on a plane.

The simplest way to get true lengths is to construct a true-length diagram, as shown at (c). Revolve the top view of each element until it is horizontal. Then project down to

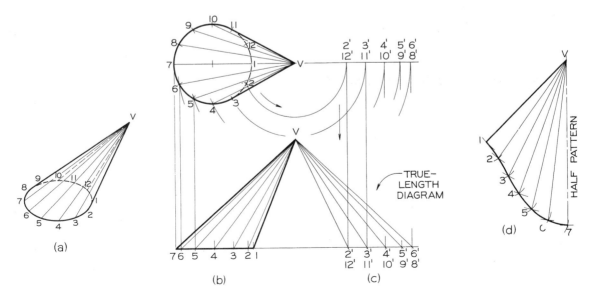

Fig. 18-20. Pattern of an oblique cone.

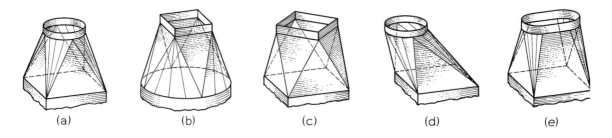

Fig. 18-21. Transition pieces.

the base line to get points 2', 3', etc., and connect with lines to the front view of the vertex V. These are true lengths to be used in the pattern. It is not necessary to find true lengths of elements V-1 and V-7. They are already shown true length in the front view.

If the pattern is divided on element V-1, the pattern is symmetrical. Only one half needs to be drawn. From V in the pattern, draw V-1 equal to V-1 in the front view. Then from V in the pattern, strike arc V-2 taken from the true-length diagram and arc 1-2 taken from 1-2 in the top view of the base of the cone. This triangle V-1-2 was transferred

in the same manner as was the triangle in Fig. 18-19. Complete the half-development by transferring the remaining triangles in the same way. Connect the points with a light freehand curve. Then heavy-in the curve with the aid of the irregular curve, Sec. 4.30.

18.20 TRANSITION PIECES – FIG. 18-21. A
transition piece is one that connects two different-shaped or skewed-position openings — as, for example, a round opening to a square opening. Transition pieces are widely used in air-conditioning, ventilating, heating, and similar installations. In most cases, transition

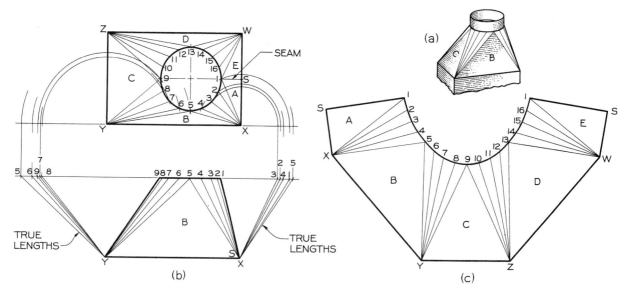

Fig. 18-22. Pattern of a transition piece.

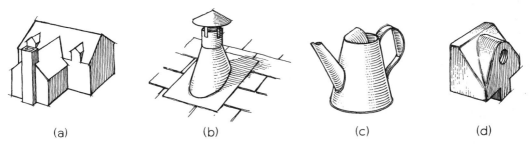

Fig. 18-23. Intersections.

pieces are composed of a combination of plane surfaces and conical surfaces. Therefore, the methods given above for pyramids and cones can be applied.

18.21 PATTERN OF TRANSITION PIECE – FIG. 18-22. A transition piece having a round opening at the top and a rectangular opening at the bottom is shown at (a). The top and front views are shown at (b). The surface is composed of four triangular plane surfaces and four conical surfaces. The conical surfaces are divided into narrow triangles so that they can be transferred to the pattern.

Assume the seam at 1-S (see top view). Triangle 1-S-X is a right triangle. It can be easily drawn in the pattern, with the true length of 1-S taken from the front view and the true length of SX taken from the top

view. Then triangle 1-X-2 and all others can be transferred by taking the small bases from the circle in the top view and the long sides from the true length diagrams. Transfer these in the manner of Figs. 18-19 and 18-20.

►INTERSECTIONS

18.22 INTERSECTIONS. A line intersects a surface in a point. Two surfaces intersect in a *line of intersection*. The complete intersection between two solids is called a *figure of intersection*. Such intersections are common in building construction, sheet-metal work, and machine construction, as illustrated in Fig. 18-23. The drafter or designer must know how to construct them.

Intersections are found by using one method over and over: finding the point where a line pierces a surface. In Fig. 18-24(a)

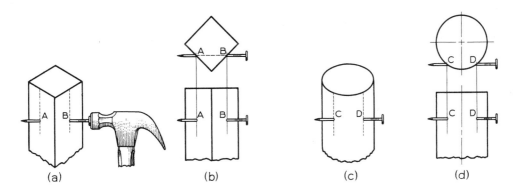

Fig. 18-24. Intersection of a line and a solid.

a nail is shown penetrating a prism, with two piercing points A and B. At (b) the top view shows clearly where the nail intersects the surfaces, because the surfaces appear edgewise. Project down to the front view of the nail to get piercing points A and B, as shown.

The same procedure applied to a cylinder is shown at (c) and (d).

18.23 INTERSECTION OF TWO SQUARE PRISMS – FIG. 18-25.

Two intersecting square prisms are shown pictorially at (a), with the three views at (b). The points in which edges W, X, Y, and Z of the horizontal prism pierce the vertical prism are points on the figure of intersection. Points 1 and 3 of the intersection are already evident in the front view. In the top view, the edges X and Z of the small prism intersect the large prism at points 2 and 4. Project down to get the front view of points 2 and 4. Join the points of the intersection with straight lines, as shown.

To draw the pattern of the small prism, (c), set off on the stretchout line W-W the widths of the faces WX, XY, etc., taken from the side view. Draw the edges through these points as shown. Set off from the stretchout line the lengths of the edges W-1, X-2, etc., taken from either the front or the top view. Join the points 1, 2, 3, etc., with straight lines.

To develop the pattern of the large prism, (d), set off on the stretchout line A-A the widths of the faces AB, BC, etc., taken from the top view. Draw the edges through these points as shown. Set off on the stretchout line distances BE and DF, taken from the top view. To locate points 1, 2, 3, and 4 in the pattern, project down from points E, C, and F, and across from points 1, 2, 3, and 4 in the front view. Join the points with straight lines.

18.24 INTERSECTION OF CYLINDERS – FIG. 18-26.

To obtain the intersection, divide the circle of the top view, (b), into a number of

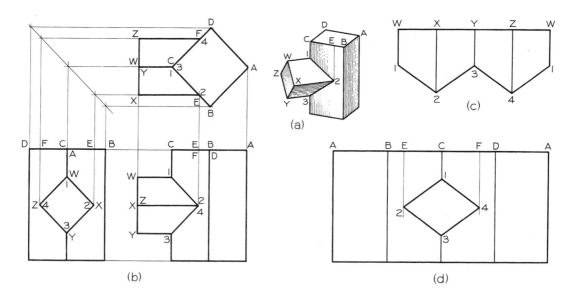

Fig. 18-25. Intersecting prisms.

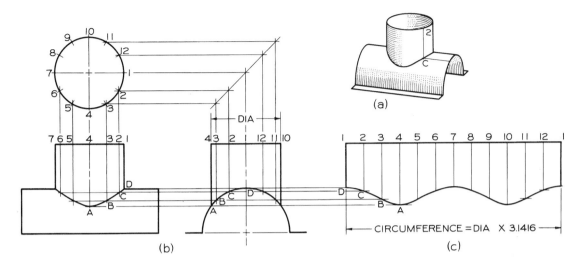

Fig. 18-26. Intersection of cylinders.

equal parts. Draw the elements in both the front and side views. Their points of intersection with the surface of the half-cylinder are shown in the side view at A, B, C, etc. They are located in the front view by projecting across to the corresponding elements in the front view. The accuracy of the curve depends, of course, on the number of points found. Connect the points smoothly with the aid of the irregular curve, Sec. 4.30.

The pattern of the vertical cylinder is shown at (c). The method used is the same as described in Sec. 18.9.

For other intersections of cylinders, see Fig. 8-20. Note especially at (d) that if the cylinders are the same size, the figure of intersection appears as straight lines.

18.25 CUTTING PLANES. Imaginary *cutting planes,* similar to those used in sectioning, Sec. 12.1, can be used to great advantage in drawing intersections. For example, consider the intersection between a prism and a cylinder in Fig. 18-27(a). If a cutting plane is passed parallel to the center lines or edges of the two solids, as shown at (b), straight lines are cut

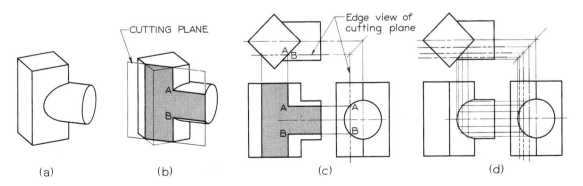

Fig. 18-27. Cutting planes.

429

from the prism. Elements are cut from the cylinder. These intersect at points A and B, which become points on the figure of intersection. Three views of this are shown at (c). The complete intersection is constructed by using several parallel cutting planes, as shown at (d).

18.26 OBLIQUE INTERSECTION OF CYLINDERS – FIG. 18-28. A pictorial view of an oblique intersection of cylinders is shown at (a). To obtain the intersection, (b), draw an auxiliary view of the inclined cylinder. Divide it into equal parts as shown. Draw cutting-plane lines through the divisions. Then draw the corresponding cutting-plane lines in the top view. Spacings between cutting-plane lines in the top and auxiliary views must be equal, as shown for distances *a* and *b*. Draw elements cut by the planes. Locate points where corresponding elements intersect. For example, element G of the vertical cylinder intersects elements 5 and 3 of the inclined cylinder at

points X and Y. This is shown pictorially at (c). Trace a smooth curve through the points to establish the figure of intersection, using the irregular curve, Sec. 4.30.

The true lengths of all elements of both cylinders are shown in the front view. The pattern of the inclined cylinder is symmetrical, and only half of the pattern is shown. The upper end develops into a straight stretchout line 1-7, the length being equal to the radius multiplied by 3.1416. The procedure is like that explained in Sec. 18.7.

The complete pattern of the vertical cylinder is shown at (d). The spacings BC, CD, etc., are chord-distances taken from the top view.

If the intersecting cylinders are the same size, the figure of intersection appears as straight lines, Fig. 8-20(d).

18.27 INTERSECTION OF PRISM AND PYRAMID – FIG. 18-29. The surfaces of the pyramid do not appear edgewise in any view. However,

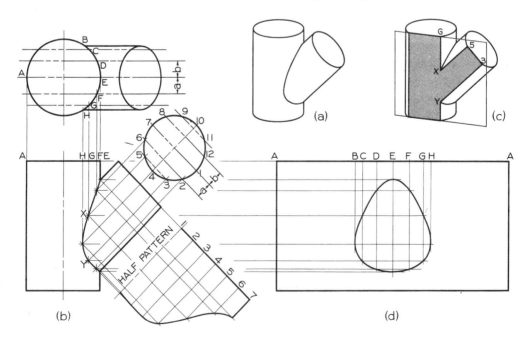

Fig. 18-28. Oblique intersection of cylinders.

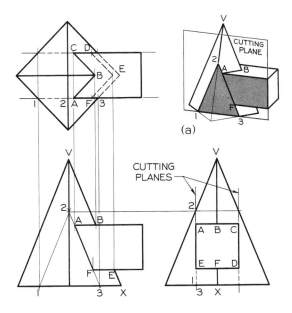

Fig. 18-29. Prism and pyramid.

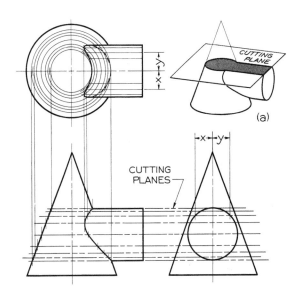

Fig. 18-30. Cylinder and cone.

the surfaces of the prism appear edgewise in the side view. Draw cutting planes *containing* these surfaces and cutting straight lines on the pyramid. These lines are lines of intersection between the surfaces of the pyramid and the surfaces of the prism. For example, as shown at (a), a cutting plane containing a vertical surface of the prism cuts lines 1-2 and 2-3 on the pyramid, and these intersect edges of the prism at points A and F. Edge VX of the pyramid is seen in the front view to intersect the top and bottom surfaces of the prism at points B and E. Locate the points of intersection in all views. Connect them with straight lines to complete the intersection as shown.

18.28 INTERSECTION OF CONE AND CYLINDER – FIG. 18-30. A convenient way to find the intersection of a cone and a cylinder is to use a series of cutting planes that cut circles on the cone and elements on the cylinder. The use of one such plane is shown at (a). For each plane, the intersections of elements cut on the

cylinder with the circle cut on the cone are shown in the top view. They can be projected down to the corresponding cutting-plane lines in the front view. This method is not convenient if the surfaces are to be developed.

18.29 INTERSECTION AND DEVELOPMENT OF CONE AND CYLINDER – FIG. 18-31. Divide the base of the cone into a number of equal parts, say 16, as shown at (a). Draw the elements of the cone in all three views. In the side view, all elements of the cylinder appear as points. Imagine each point where an element of the cone intersects the circle to be the end view of an element of the cylinder. Each element of the cylinder intersects an element of the cone, giving a point on the figure of intersection. Join all points with a smooth curve in the top and front views to complete the intersection.

Another method is to use cutting planes. Draw elements of the cone as before. In the side view, think of these elements as also the edge views of a series of cutting planes. Each

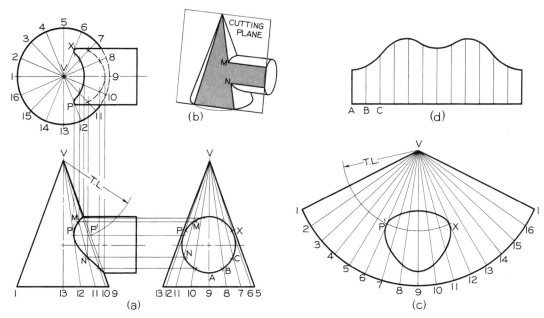

Fig. 18-31. Cone and cylinder.

plane will cut from each solid two elements. These intersect to give points on the figure of intersection. One such cutting plane, producing two points M and N, is illustrated at (b).

To draw the pattern of the cone, (c), draw a large arc with radius equal to the slant height of the cone. Set off on this arc the chord-distances 1-2, 2-3, 3-4, etc., taken from the top view, or compute the angle of the pattern as explained in Sec. 18.16. To locate points on the opening in the pattern, take true lengths in the front view from the vertex down to points on the figure of intersection. The true length (TL) of VP is shown at VP' in the front view and transferred to the pattern to get points P' and X.

To draw the pattern of the cylinder, divide the circle in the side view into an equal number of parts, and draw the elements. To simplify the illustration, these elements are omitted in the figure. Draw the stretchout line equal to the circumference of the cylinder

(dia. × 3.1416). Divide it into the same number of equal parts. The true lengths of the elements are shown in the front view. They can be transferred directly to the pattern.

18.30 COMPUTER GRAPHICS. The use of CAD in developments and intersections has simplified the design process. There are many software packages specifically designed for these functions. A few examples include HVAC (heating, ventilating, and air conditioning), piping, packaging, clothing pattern making, and surface modeling as required in the design of automobile bodies. These packages are varied in their approach. All accomplish the function of providing a detailed layout of the required part or assembly.

HVAC packages are designed to develop flat patterns that can be assembled into three-dimensional shapes. The user inputs the overall shape and dimensions of the part. The software then converts this information into a

432

flat pattern with each face in its true size and shape. Transition pieces are developed by choosing the sizes and shapes to be joined and specifying all required details. The software then develops the required transition shape, Fig. 18-32. If desired, the software will also determine the most economical method of cutting the pieces from standard sheets.

CAD software for the clothing industry develops the pattern to specific sizes. It then automatically scales the pattern up or down for other sizes.

The automotive industry uses surface modeling techniques to develop the design to the required shape. Each part of the design is then converted to a flat pattern to be cut for the shaping process. Sheet size, seam joining techniques, and shape forming details are included in the final layout design.

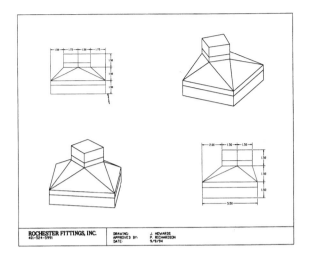

Fig. 18-32. A transition drawing prepared using CAD.

PROBLEMS

A very wide range of problems on developments and intersections is provided in Figs. 18-33 to 18-38. The first four groups, Figs. 18-33 to 18-36, consist of layouts (Layout D in the Appendix) in which there are many alternate problems to provide different assignments for students. Following these are a number of problems applying the principles of this chapter. These also are to be drawn on Layout D or 11.0″ × 17.0″ sheets.

In the illustrations of this chapter, cylinders and cones are usually divided into only 12 elements to keep the presentation as simple as possible. In the problems, the instructor may wish the student to use 16 or 24 divisions for greater accuracy.

Methods of constructing paper models are explained in Sec. 18.3. It is suggested that at least a few of the early problems in developments and later in intersections be actually cut out and formed into models.

In the problems that follow, disregard allowances of extra material for seams, rolled edges, and thickness of materials.

If assigned by the instructor, construct drawings using decimal-inch or metric scales by converting given fractional dimensions to decimal-inch or metric equivalents. Refer to Table 20 in the Appendix.

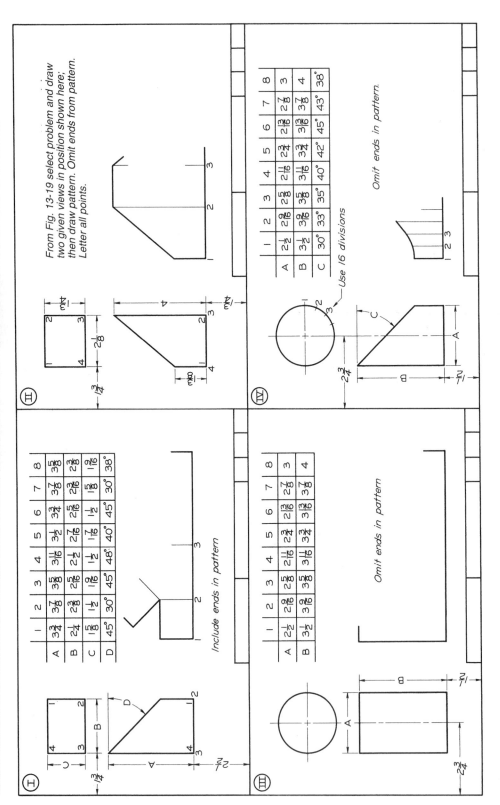

Fig. 18-33. Parallel-line development problems. Using Layout D in the Appendix, draw given views and pattern of problem assigned by instructor. Omit table and all spacing dimensions and instructional notes.

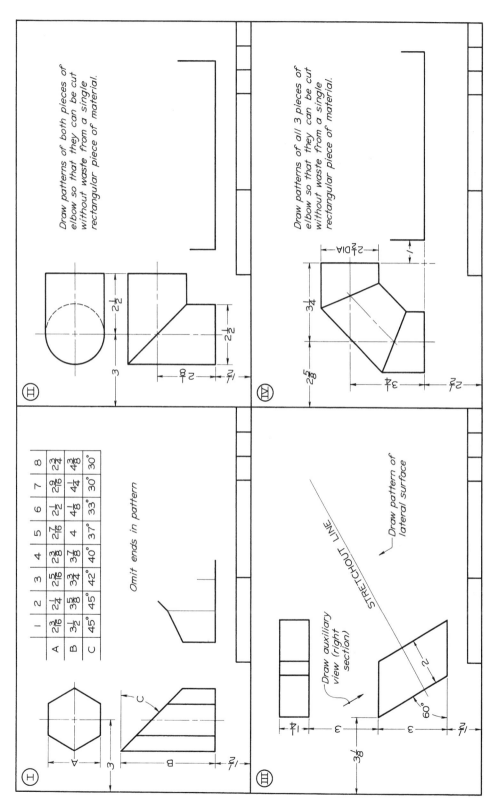

Fig. 18-34. Parallel-line development problems. Using Layout D in the Appendix, draw given views and pattern of problem assigned by instructor. Omit table and all spacing dimensions and instructional notes.

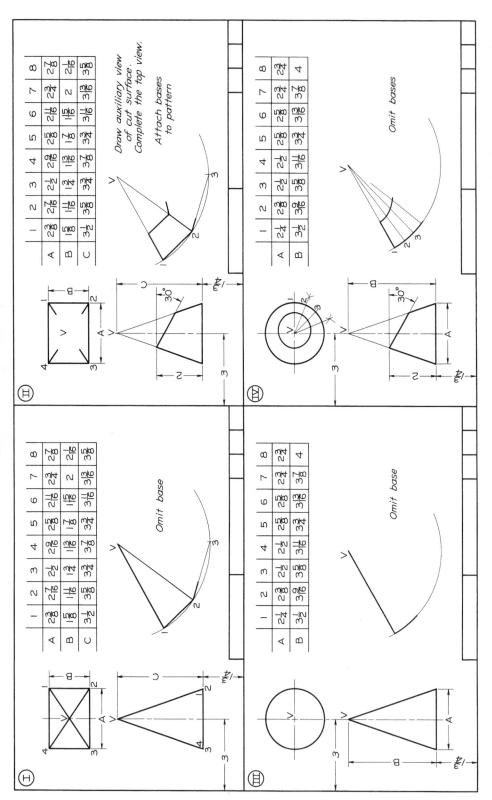

Fig. 18-35. Radial-line development problems. Parallel-line development problems. Using Layout D in the Appendix, draw given views and pattern of problem assigned by instructor. Omit all spacing dimensions, table dimensions, and instructional notes.

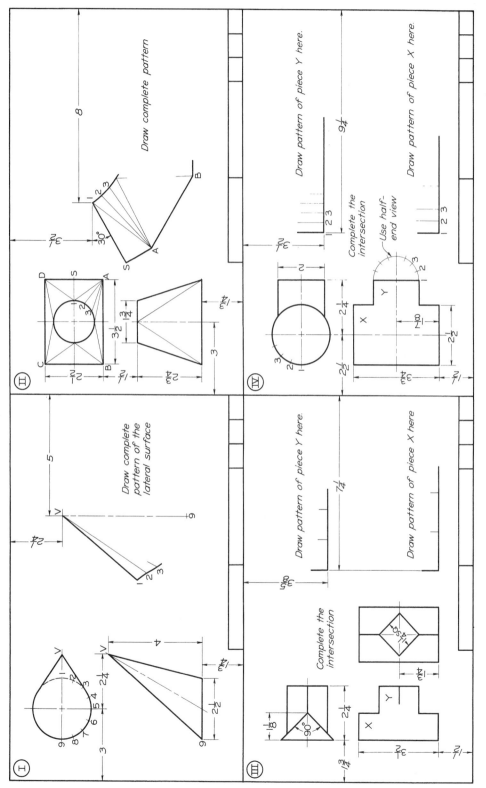

Fig. 18-36. Development and intersection problems. Using Layout D in the Appendix, draw views and pattern(s) of problem assigned by instructor. Omit all spacing and instructional notes.

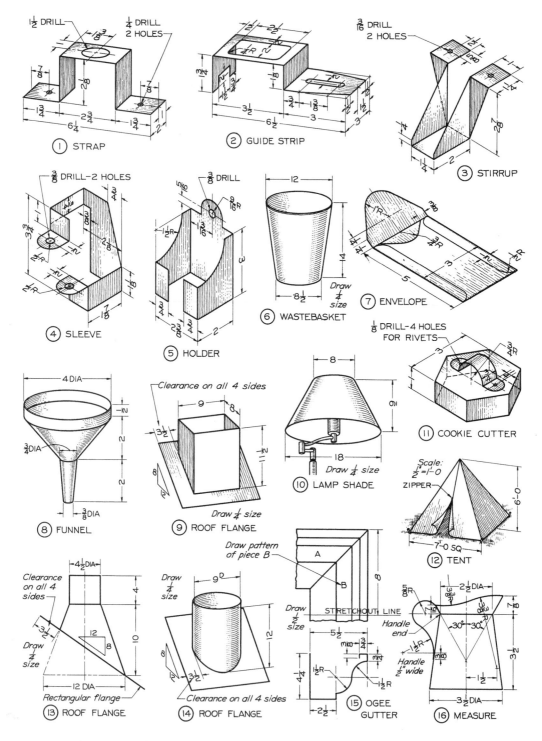

Fig. 18-37. Development problems. Using Layout D in the Appendix, draw pattern assigned, disregarding allowances for seams, rolled edges, and thickness of material.

438

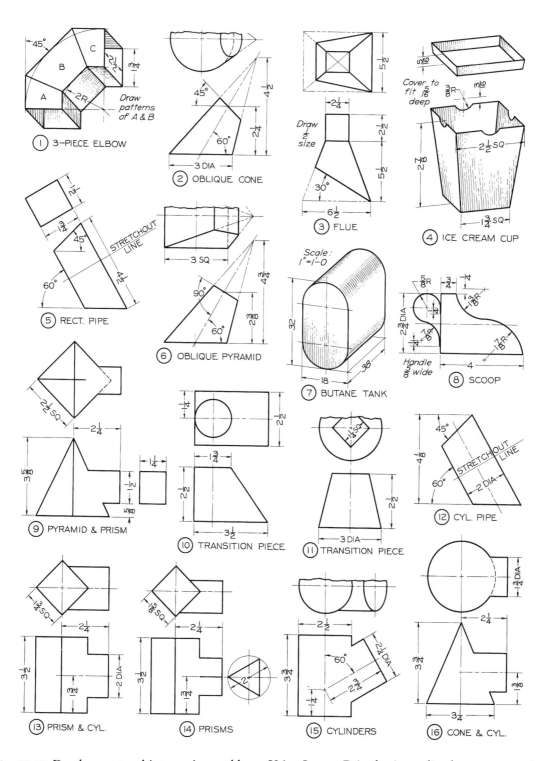

Fig. 18-38. Development and intersection problems. Using Layout D in the Appendix, draw pattern assigned. Disregard allowances for seams, rolled edges, and thickness of material.

CHAPTER 19

bar chart   pie chart

line chart   pictograph

OBJECTIVES

After studying this chapter, you should be able to:

- Prepare a simple bar chart.
- Prepare a simple line chart.
- Prepare a simple pie chart.
- Explain the use of pictographs in charts.

CHAPTER 19

Charts and Graphs

Charts and graphs can be used to present many kinds of information. One field that makes use of charts and graphs is bio-engineering. Bio-engineering applies engineering principles to biological or medical science. One aspect of bio-engineering is *ergonomics,* the science of adapting furniture, tools, appliances, and other objects to provide the most comfort and safety for the human body. A related area, *anthropometry,* is the study of human body measurements. Anthropometric data helps engineers design objects with the proper "fit" to the human body. For example, it is important to work at a computer keyboard that is at the correct height and provides a wrist support in order to prevent medical problems such as carpal tunnel syndrome or back and shoulder pain. Lighting, room temperature, humidity, and other environmental factors must also be considered.

Engineers working in this field also have a background in biological and psychological areas relating to humans and the environment, anatomy, physical form, and the nervous system. They may often find it convenient to work with data presented in various forms of charts and graphs, which makes it easy to compare the effects of several different procedures.

19.1 GRAPHICAL PRESENTATION.

We are constantly concerned with numbers, quantities, and comparisons of amounts. Statistics — facts expressed by numbers — are often regarded as "cold" or uninteresting because their meaning is not immediately apparent. The financier watches the Dow Jones stock averages. The sales manager keeps an eye on dollar volume of sales. The homemaker certainly is interested in the ups and downs of the cost of living.

If numerical facts are presented graphically — that is, by means of drawing — the information catches the eye immediately. Such drawings are called *charts, graphs,* or *diagrams.* There are many more kinds than can be discussed here. The most common are *bar charts, line charts,* and *pie charts.*

19.2 BAR CHARTS.

The bar chart is an effective device for showing a comparison between amounts. It is easily understood by everyone. It is used extensively in newspapers, magazines, and books.

The simplest form of bar chart is the *100-percent bar,* Fig. 19-1. This shows the percentage ratio of various parts to a given whole. The total length of the bar represents 100 percent. This distance is divided into segments that are proportional parts of the whole. The bars may be drawn horizontally or vertically. Appropriate shading or crosshatching is used to distinguish between the segments.

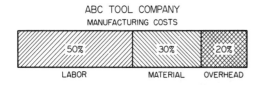

Fig. 19-1. A 100-percent bar chart.

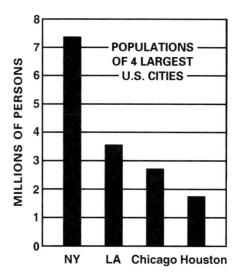

Fig. 19-2. Bar chart.

The bar chart in Fig. 19-2 shows very clearly the relative populations in four large cities. Bar charts are useful in comparing amounts either from large to small or the reverse. As a rule, the bars should be arranged in order of height and not at random.

A tabulation of points scored by three football teams in a given year is as follows:

School	Total Points
State College	78
Western University	246
Smith Institute	325

The steps in drawing a bar chart that gives this information in graphic form are shown in Fig. 19-3.

I. Draw a vertical line called the Y-axis or *ordinate* and a horizontal line called the X-axis or *abscissa.* The point of intersection will be zero on the chart.

II. Select a suitable scale for the points along the Y-axis. For example, using the architects scale, let $\frac{1}{8}'' = 10$ points; or using the

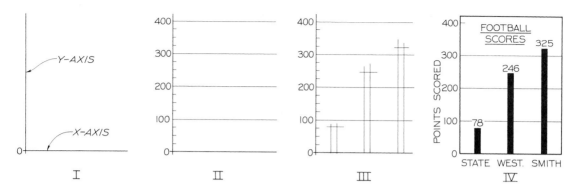

Fig. 19-3. Steps in drawing a bar chart.

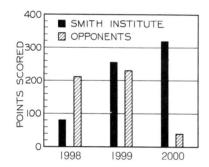

Fig. 19-4. Bar chart—scoring records.

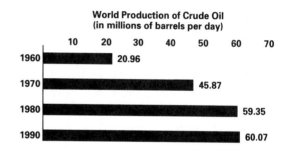

Fig. 19-5. World production of crude oil.

engineers scale, let 1″ = 100 points ($\frac{1}{10}$″ = 10 points). Letter the numbers at convenient intervals, say 100, 200, and 300, as shown. Draw horizontal lines across the chart at these locations.

III. Block in bars, spacing them far enough apart to permit lettering the names at the bottom. The width of the bars is determined simply on the basis of appearance.

IV. Fill in the bars solid, or shade with colored pencil, or use section-lining. Add necessary lettering, including a title.

Many variations of the bar chart are used. For example, suppose we want to show the scoring record of the Smith Institute in comparison with the scores made against Smith by all of its opponents. Such a chart is shown in Fig. 19-4. The bars are arranged in pairs. One set is filled in solid. The other is section-lined.

Bar charts may be drawn with the bars in a vertical position or in a horizontal position. A horizontal bar chart is shown in Fig. 19-5. In this case the quantities (millions) are expressed by means of bars and also in figures at the ends of the bars.

Bar charts may be drawn in a variety of ways. For example, the bars can be drawn pictorially in oblique, Fig. 19-6, or in isometric. They may be emphasized by means of shading or by colors.

19.3 CROSS-SECTION PAPER. Much time can be saved if the charts or graphs are construct-

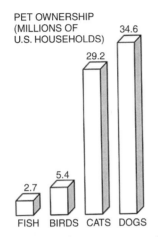

Fig. 19-6. Bar chart—pet ownership.

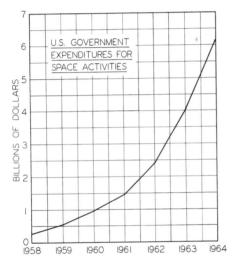

Fig. 19-7. Line chart.

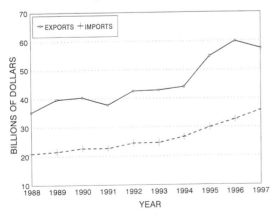

Fig. 19-8. Line chart comparing agricultural imports and exports.

ed upon cross-section paper such as we have already used for freehand sketching. Paper having $\frac{1}{8}''$ or $\frac{1}{4}''$ squares may be used. Generally, however, it is desirable to use paper with $\frac{1}{10}''$ or $\frac{1}{20}''$ squares.

19.4 LINE CHARTS. Line charts are used especially to show *trends* — as, for example, the ups and downs of the stock market.

The space program developed quickly in the early 1960s. The following figures indicate the expenditures, in billions of dollars, for space activities by the United States government in those early years:

Year	Billion $
1958	0.249
1959	0.521
1960	0.960
1961	1.468
1962	2.390
1963	4.077
1964	6.176

A line chart showing the trend of increasing expenditures for these activities is shown in Fig. 19-7. To make such a chart, draw the X- and Y-axes as in Fig. 19-3 for bar charts. Select suitable scales to be used along both axes. Draw horizontal and vertical grid lines. Then plot the numbers on the charts. For example, for 1962 the point for 2.390 is found on the vertical line of 1962 and above the horizontal marked 2. Actually, the amount above 2 is approximately $\frac{4}{10}$ of the distance between 2 and 3. Connect all plotted points with straight lines.

For the chart in Fig. 19-7, regular $\frac{1}{4}''$ cross-section paper was used. Along the vertical scale each two squares were taken to equal one billion dollars. Along the horizontal scale the years were spaced two squares apart.

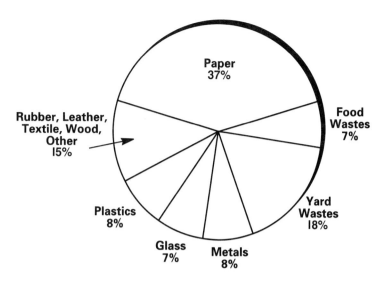

Fig. 19-9. This pie chart shows what materials make up solid waste.

Line charts are used in an endless variety of ways. For example, suppose we want to compare the agricultural exports and imports of the United States between 1988 and 1997. On Fig. 19-8, a solid line represents exports, and a broken line represents imports. This chart not only shows trends but also compares the dollar values of exports and imports for any given year.

19.5 PIE CHARTS. The idea for pie charts undoubtedly came from the custom of cutting pies into portions. For showing how the whole is split up into several unequal portions, the pie chart is excellent, Fig. 19-9. It is desirable to place the lettering horizontally in the sectors of the chart, as shown in the figure. Therefore, the pie chart is most suitable when there are no more than six divisions. If there are more divisions, so that each is too small to contain the lettering, leaders can be used.

To draw a pie chart, draw a circle large enough to contain the lettering without undue crowding. Then determine the angles

of the sectors by using the protractor, Fig. 4-19. To convert percentage into degrees, multiply the percentage by 360°. For example, for the 41 percent paper, $\frac{41}{100} \times 360° = 147.6°$.

POPULATION

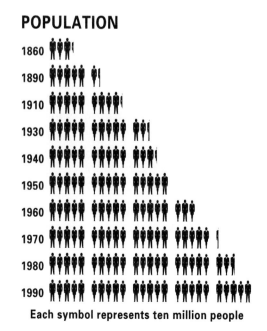

Each symbol represents ten million people

Adapted from graph by Twentieth Century Funds

Fig. 19-10. Pictograph—growth of U.S. population.

445

19.6 PICTOGRAPHS. A *pictograph* is a graph or chart in which pictures are used as symbols to represent units or quantities. For example, in Fig. 19-10 each symbol of a person represents 10 million people. These symbols are then repeated to show the various totals. The result is a form of bar chart. Another variation is to show one symbol opposite each year instead of several, but to vary the size of the symbol from small to large according to the number represented.

19.7 COMPUTER GRAPHICS. Presenting the progress of the design team using charts and graphs is an important part of the design process. A variety of CAD programs are available to produce complex, full-color charts and graphs. Using CAD, the drafter can easily experiment with different types of representations of the same data, such as the bar chart or pie chart. The drafter can then decide which is the most effective way to display the data.

PROBLEMS

It is suggested that before drawing any charts, each student bring to class at least one example of each type of chart discussed here. These can be found in newspapers, magazines, and in textbooks. The various charts can then be passed around and discussed by the class.

BAR CHARTS

PROBLEM 19-A. Assume that a typical personal budget dollar is spent as follows: Food — 18.4¢, Housing — 9¢, Transportation — 12.3¢, Household Operations — 11.7¢, Clothing — 7.4¢, Recreation — 7.6¢, Medical Care — 20¢, Personal Care — 2.3¢, Personal Business — 7¢, Education — 1.8¢, and Religious and Welfare Activities — 2.5¢. Make a vertical bar chart showing this information.

PROBLEM 19-B. Draw a bar chart showing, in terms of percentages, the final standings last year of baseball teams in either the National or the American League.

PROBLEM 19-C. Draw a bar chart showing the numbers of freshmen, sophomores, juniors, and seniors in your school.

PROBLEM 19-D. This is the same as Problem 19-C, except that the bar chart is to show a pair of bars for each grade, one bar for girls and one for boys.

LINE CHARTS

PROBLEM 19-E. Make a line chart showing how your grades have changed from month to month in a given subject throughout the school year.

PROBLEM 19-F. List the total points scored by a football team last year in each game. Make a line chart showing the progress of the team through the season.

PROBLEM 19-G. Paper discarded in the United States for the years indicated was as follows (in millions of tons):

1960 — 30; 1970 — 35; 1980 — 48; 1998 — 81.

Draw a line chart showing these facts. Let each one-quarter inch on the vertical scale equal one million tons of discarded paper.

PROBLEM 19-H. Make a line chart comparing the number of home runs hit by Mark McGwire and Sammy Sosa during the 1998 baseball season. Use different color or line types for each player's data.

	March	April	May	June	July	August	Sept.	Oct.
McGwire	1	10	16	10	8	10	15	0
Sosa	0	6	7	20	9	13	11	0

PIE CHARTS

PROBLEM 19-I. Assume that a nation's "budget dollar" is spent as follows: National Defense — 28¢, Space — 4¢, Fixed Interest Charges — 7¢, Veterans — 4¢, Social Security and Other Trust Funds — 26¢, and Other Costs — 31¢. Make a pie chart showing this distribution.

PROBLEM 19-J. Assume that employed people in the United States are employed as follows: Manufacturing — 14%, Agriculture — 4%, and All Others — 82%. Make a pie chart showing this information.

PROBLEM 19-K. A student has an allowance of $40 per month, and in addition earns $300 per month working for a grocery. The student's budget for expenditures is: School Lunches — $35, Supplies — $25, Amusement — $40, Clothing — $40, Savings — $200, Total — $340. Make a pie chart showing this distribution. To convert into percentages, divide the part by the whole, and multiply by 100. For example, for Clothing — $40, figure as follows:

$$\frac{40}{340} \times 100 = \frac{4000}{340} = 12\%$$

Arnold & Brown

CHAPTER 20

O B J E C T I V E S

After studying this chapter, you should be able to:

- Identify two types of cams.
- Identify the three types of cam followers.
- Present the main steps in laying out a cam.
- Identify three types of gears.
- Demonstrate how to approximate involute curves.

CHAPTER 20

Cams and Gears

The design of cams and gears is done by mechanical engineers, who generally begin with a basic design and modify it to meet specific needs. They may also try to improve an existing design by making the gear or cam operate more smoothly and quietly, with less wear and damage.

Computer modeling techniques make the designer's job much easier and faster. New materials and manufacturing processes are often investigated to make sure the cams or gears are being produced in the most efficient way. Graphical solutions of problems concerning cam and gear design can make the engineer's work faster and easier, especially when CAD is used to produce the drawings, which can then be easily changed.

▌ C A R E E R   L I N K ▌

To learn more about mechanical engineering, contact the American Society of Mechanical Engineers, 345 E. 47th Street, New York, NY 10017 or visit their web site at *www.asme.org*.

20.1 CAMS. A *cam* is a machine element used to obtain an irregular or unusual motion. Cams are manufactured in an endless variety of forms, some of which are shown in Fig. 20-1. *Plate cams,* or *disk cams,* are essentially flat with irregular edges, Fig. 20-2(a) to (c). A *cylindrical cam,* (d), is a cylinder with an irregular groove cut around it. The basic principle of a cam is illustrated at (a). A uniformly rotating *camshaft* has mounted upon it an irregularly shaped disk or plate, which is the cam. As the cam rotates counterclockwise, as shown by the arrow, the *follower* moves up gradually, then down more rapidly. Finally it remains "at rest" until the starting point is reached again.

The *roller* on the follower provides smooth operation. It is held in contact with the cam by gravity or a spring. The drafter or designer must design a cam that will produce a desired motion of the follower.

20.2 CAM FOLLOWERS. The type or shape of cam follower selected is determined by the requirements of the mechanism or the operation that is to be performed. The three most common types of cam followers are shown in Fig. 20-2(a) to (c). A *roller* follower is shown at (a), a *flat-faced* follower at (b), and a *pointed* follower at (c). The axis of the follower may be located on the vertical center line of the cam, as shown in (a) and (b). It may be offset as shown in (c). Many other special cam followers can also be designed or modified from the basic types to simplify a complicated cam profile or mechanism.

20.3 CAM MOTION. The desired motion of the follower of a cam can be shown in a *displacement diagram,* Fig. 20-3. The motion may be *uniform (constant velocity), harmonic,* or *uniformly accelerated* and *retarded (parabolic).* These motions may be used separately or in combination with one another. The horizon-

Fig. 20-1. A variety of cams.

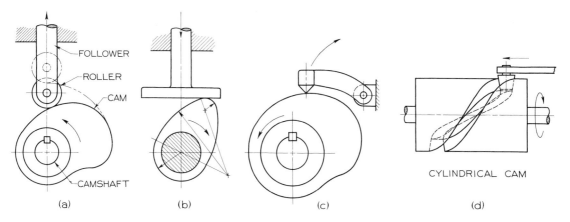

Fig. 20-2. Cam types and terminology.

tal base line (travel) on the diagram represents one revolution (360°) of the cam. Any convenient length can be used to represent this distance. The vertical distances (rise or fall) on the diagram are drawn to scale. They show the actual follower displacement.

The dashed line AB in Fig. 20-3 represents uniform motion in which both the rise and travel are divided into the same number of equal distances. This is the simplest of all motions. However, since it makes the follower start and stop suddenly it is also the least practical. This motion can be modified by an arc at each end of the motion, as shown by the heavy line.

Line BC represents a period of *dwell*, during which time the follower does not rise or fall.

To construct the curve CD, which gives harmonic motion to the follower, draw a semicircle whose diameter is equal to the desired rise. Divide the semicircle and the travel into the same number of equal parts. Points on the curve are located by drawing horizontal lines from the points on the semicircle to intersect the vertical lines drawn from the corresponding points on the travel, as shown. Sketch a smooth curve through the points. Heavy-in the curve with the aid of the irregular curve, Sec. 4.30.

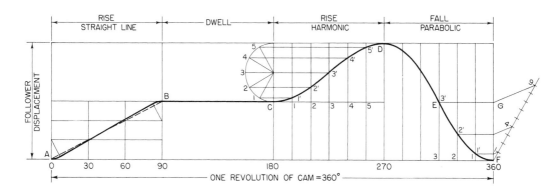

Fig. 20-3. Typical cam displacement diagram.

The parabolic curve DEF gives the follower uniformly accelerated and retarded motion. It consists of two halves, with the half of the curve from D to E being exactly the reverse of the curve from E to F. To construct the curve EF, divide the vertical height from G to F into parts proportional to 1, 3, 5, etc. The travel is divided into the same number of

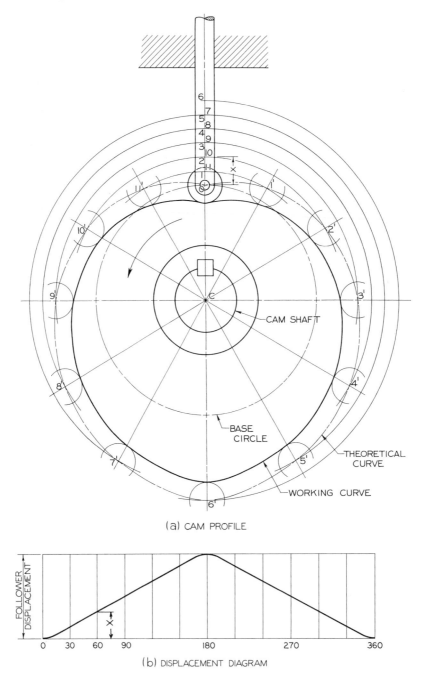

(a) CAM PROFILE

(b) DISPLACEMENT DIAGRAM

Fig. 20-4. Layout of a plate cam profile.

equal parts. Points on the curve are located by drawing horizontal lines from the points on line GF to intersect the vertical lines drawn from the corresponding points on the travel, as shown. Sketch a smooth curve through the points. Heavy-in the curve with the aid of the irregular curve, Sec. 4.30. The displacement diagram is usually constructed before the actual cam profile is drawn.

20.4 Layout of a Plate Cam Profile. The method of laying out a typical plate cam profile is illustrated in Fig. 20-4(a). In this case it is desired to move the follower a certain distance with uniform motion and then to return the follower to the starting point with uniform motion. This cam is to rotate counterclockwise, as shown by the arrow. The speed of rotation is constant. Therefore, *equal angles of rotation will be equivalent to equal units of time.* Thus, the follower will move the same distance for each angle of rotation of the cam.

The displacement diagram, (b), indicates the desired motion of the follower. The horizontal base line represents one revolution of the cam. It is divided into 12 equal parts corresponding to the 12 equal angular divisions of the base circle. The vertical distances show

the actual rise and fall of the follower. Note that the diagram has been modified by including arcs at each end of the follower motion, as shown in Fig. 20-3.

The steps in laying out the cam are:
1. Draw vertical and horizontal center lines. Then draw the *base circle*. The radius of the base circle is the distance from the center of the cam to the center of the roller when the two parts are closest together.
2. Divide the base circle into 12 equal angles of 30° each, as shown in Fig. 18-9(d). For greater accuracy, draw 24 divisions of 15° each, as shown at (f).
3. On the center line of the follower, from zero (center of roller), set off the desired "rise" of the follower. Divide this into 6 equal parts, as shown.
4. Imagine the cam to be stationary, and the roller moving *clockwise* around the cam. From cam center C strike construction arcs C-1, C-2, C-3, etc., to intersect 30° radial lines at 1′, 2′, 3′, etc. At each point, draw a construction arc representing the roller at that position.
5. Sketch a smooth curve tangent to the small arcs. Heavy-in the curve with the aid of the irregular curve, Sec. 4.30.

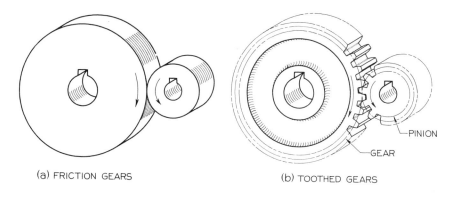

(a) FRICTION GEARS   (b) TOOTHED GEARS

Fig. 20-5. Friction gears and toothed gears.

453

Note: For a more detailed discussion of cams, see a book on mechanism or any standard text on the subject.

20.5 GEARS. If two round cylinders or *friction gears,* Fig. 20-5(a), are placed in contact, one, if rotated, will transmit motion to the other. Note that the two will turn in opposite directions. If they are the same size and one is turned around once, the other will revolve once. If one is twice the diameter of the other, it will turn halfway around for one revolution of the other. This points up one of the chief reasons for using gears — namely, to reduce or increase speeds.

However, friction gears are subject to slipping. Excessive pressure between them is required to obtain the necessary "traction." If teeth are provided, the resulting gears, Fig. 20-5(b), will transmit motion without slipping.

There are a great many kinds of gears, some of which are illustrated in Fig. 20-6. *Spur gears* connect parallel shafts. *Bevel gears* connect shafts whose axes intersect. *Worm gears* connect shafts whose axes ordinarily are at right angles to each other but are nonintersecting. The gears in Fig. 20-5(b) are spur gears, the smaller one being commonly called a *pinion* and the larger a *gear.* If teeth are cut on a flat piece, the part is called a *rack,* as shown at the lower right corner of Fig. 20-6.

Spur gear proportions, and the shape of the teeth, are well standardized. The terms used, Fig. 20-7, are common to practically all spur gears. The dimensions relating to tooth height are for "full-depth $14\frac{1}{2}°$ involute teeth," which is the most common type.

The *outside diameter* (OD) is the overall diameter of the gear. The *pitch diameter* (PD) is the diameter that corresponds to the fric-

Boston Gear Works

Fig. 20-6. A variety of gears.

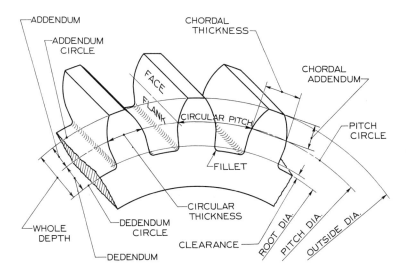

Fig. 20-7. Gear tooth terms.

tion gear without teeth. The *root diameter* (RD) is the diameter at the bottom of the tooth.

The *circular pitch* (CP) is the distance measured along the pitch circle from a point on one tooth to the corresponding point on the next tooth. The *diametral pitch* (DP), or simply *pitch,* is the number of teeth on the gear per inch of pitch diameter. The *addendum* (A) is the height of the tooth above the pitch circle. The *dedendum* (D) is the depth of the tooth below the pitch circle. The following formulas can be used to obtain various dimensions, where N equals the number of teeth:

Diametral pitch DP $= \dfrac{N}{PD}$

Circular pitch CP $= \dfrac{3.1416}{DP}$

Pitch diameter PD $= \dfrac{N}{DP}$

Outside diameter OD $= \dfrac{N + 2}{DP}$

Root diameter RD = PD - 2D

Addendum A $= \dfrac{1}{DP}$

Dedendum D $= \dfrac{1.157}{DP}$

Clearance C $= \dfrac{0.157}{DP}$

Whole depth WD = A + D

Circular thickness CT $= \dfrac{CP}{2}$

20.6 THE SHAPE OF THE TEETH. The so-called *involute system* is the most widely used gear tooth form in use today. An *involute* may be described as the path traced by the end of a cord as it is unwound, while kept taut, from a

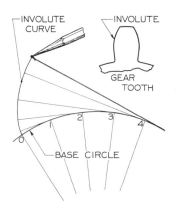

Fig. 20-8. Involute.

circle called the *base circle,* Fig. 20-8.

The exact form of a standard involute tooth, however, is seldom drawn. For display drawings, or whenever it is desired to show the teeth but not necessary to represent them exactly, the involute curves may be approximated by circular arcs as in Fig. 20-9.

Draw the addendum circle, pitch circle, and root circle, as shown. At any point P on the pitch circle, draw a tangent line. Then draw a second line through P at $14\frac{1}{2}°$ with the tangent line. This may be drawn 15° to simplify the construction. This is the so-called "pressure angle." Now draw the base circle tangent to the $14\frac{1}{2}°$ line. Along the

pitch circle, set off by trial, with the bow dividers, the spacing of the teeth. With radius one-eighth the pitch diameter, and with centers on the base circle, draw circular arcs as shown. The lower portions of the teeth are radial lines ending in small fillets.

The pressure angle of $14\frac{1}{2}°$ is most common. However, if the angle is increased to 20° and the height of the teeth reduced, the resulting teeth are *stub teeth.*

20.7 WORKING DRAWINGS OF SPUR GEARS. A typical working drawing of a spur gear is shown in Fig. 20-10. A sectional view is usually used. In many cases no circular view is needed. The gear teeth are not drawn, except as they appear (not section-lined) in the sectional view. In the circular view, if drawn, the addendum circle and the root circle are phantom lines. The pitch circle is a center line.

The dimensions are given for the "gear blank" in the views. The gear tooth information is given in a note or in a table. The gear-tooth data shown in the table in Fig. 20-10 are minimum, with details left to the gear manufacturer. If more complete gear tooth information on the drawing is desired, the circular-thickness, addendum, dedendum, tooth thickness, and other data must be provided.

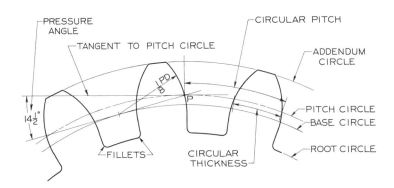

Fig. 20-9. Approximate representation of involute spur gear teeth.

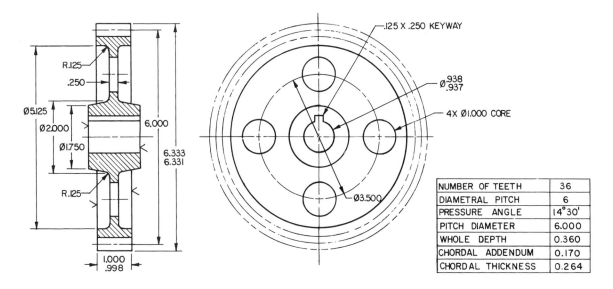

NUMBER OF TEETH	36
DIAMETRAL PITCH	6
PRESSURE ANGLE	14°30'
PITCH DIAMETER	6.000
WHOLE DEPTH	0.360
CHORDAL ADDENDUM	0.170
CHORDAL THICKNESS	0.264

Fig. 20-10. Working drawing of a spur gear.

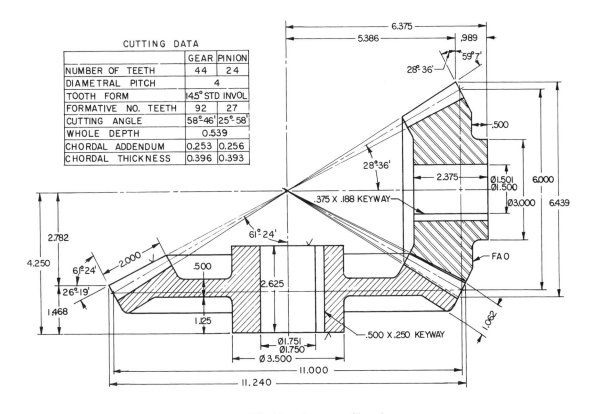

CUTTING DATA	GEAR	PINION
NUMBER OF TEETH	44	24
DIAMETRAL PITCH	4	
TOOTH FORM	14.5° STD	INVOL
FORMATIVE NO. TEETH	92	27
CUTTING ANGLE	58°46'	25°58'
WHOLE DEPTH	0.539	
CHORDAL ADDENDUM	0.253	0.256
CHORDAL THICKNESS	0.396	0.393

Fig. 20-11. Working drawing of bevel gears.

20.8 BEVEL GEARS. *Bevel gears,* Fig. 20-6 (second from upper left), transmit power between two shafts whose axes intersect. If the axes intersect at right angles, as is usually the case, the gears are often called *mitre gears.* The comparable friction wheels would be a pair of cones rolling against each other, with their vertexes at a common point. A technical treatment of bevel gears is outside the scope of this text. For complete information, see a standard text on mechanism.

20.9 WORKING DRAWINGS OF BEVEL GEARS. As in the case of spur gears, a working drawing of a bevel gear gives only the dimensions of the gear blank, while the data for cutting the teeth are given in a note or table, Fig. 20-11. Usually only a single sectional view is drawn, and if a second view is needed, only the gear blank or blanks are drawn, with the teeth omitted. It is common to draw a pair of mating gears together, as shown, although they are often drawn separately. The gear teeth are not section-lined. As in the case of spur gears, the larger is called the *gear,* and the smaller the *pinion.*

20.10 USING CAD TO DESIGN CAMS AND GEARS. Cams and gears can be designed and drawn using either conventional CAD software or special application CAD software. Many features in conventional software allow tedious functions such as the drawing of cam curves and profiles to be performed by specifying points and fitting the lines using spline curves. This requires the drafter to lay out the required points and angles in much the same method as in manual drafting, connecting the points, and invoking the spline command. Developing gears requires the layout of a single tooth of specific size on a base circle and then arraying (creating additional teeth) around the base circle. Details and cleanup are performed using standard commands.

Special application CAD software for cams and gears simplifies these tedious tasks. It also ensures adherence to ANSI standards. To develop a cam the drafter selects the type desired, the motion types, and degrees of motion. The cam and follower are automatically created to exact specifications. A gear is developed in much the same fashion. The drafter specifies the gear set design, including angles, number of teeth, size, etc. Software then develops multiple gear set solutions. The designer then chooses the best solution for the project. If required, both cam and gear designs are converted to three-dimensional models with properties that allow the designer to check for weight, fit, and all mass properties.

Designs developed by special application CAD software eliminate the need to redefine geometry for numerical control machining.

P R O B L E M S

CAMS

PROBLEM 20-A. Using Layout D in the Appendix, draw a displacement diagram, similar to that shown in Fig. 20-3, with the following follower motions: rise $1\frac{1}{4}''$ with modified uniform motion during 90°, dwell during 90°, rise $1\frac{1}{4}''$ with harmonic motion during 90°, and fall $2\frac{1}{2}''$ during the remaining 90°. Make the horizontal base line 10″ long and the follower displacement $2\frac{1}{2}''$ high.

PROBLEM 20-B. Using Layout B in the Appendix, design a plate cam with roller follower, similar to that shown in Fig. 20-4, with the following motions: rise $1\frac{1}{4}''$ with modified uniform motion during 180°, and fall $1\frac{1}{2}''$ with modified uniform motion during the remaining 180°. The diameter of the base circle is 3″, and the diameter of the roller is $\frac{1}{4}''$. Draw a displacement diagram for this cam.

PROBLEM 20-C. Same as Problem 20-B, except the follower has the following motions: rise $\frac{3}{4}$ ″ with modified uniform motion during 90°, dwell during 90°, rise $\frac{3}{4}''$ with modified uniform motion during 90°, and fall $1\frac{1}{2}''$ with modified uniform motion during the remaining 90°.

GEARS

PROBLEM 20-D. Using Layout D in the Appendix, make a full-size drawing of a spur gear, similar to that shown in Fig. 20-10, with 48 teeth and a diametral pitch of 8. Completely dimension the drawing and compute new values for the gear tooth data table appearing on the drawing.

PROBLEM 20-E. Same as Problem 20-D, except that the gear has 60 teeth and a diametral pitch of 10.

PROBLEM 20-F. Same as Problem 20-D, except that the gear has 72 teeth and a diametral pitch of 12.

If assigned by the instructor, construct drawings using decimal-inch or metric scales by converting given fractional dimensions to decimal-inch or metric equivalents. Refer to Table 20 in the Appendix.

CHAPTER 21

truss   design drawings

members   detail drawings

beams   erection
                     diagrams
layout drawings

foundation plans

OBJECTIVES

After studying this chapter, you should be able to:

- Identify the basic parts of a truss.
- Identify the plans and drawings in a complete set of drawings for a structure.

460

CHAPTER 21

Structural Drawing

The very first occupation that could be classified as engineering is *civil engineering*. George Washington was a land surveyor—a task that is still done by today's civil engineering crews. In addition, the modern civil engineer works with structural design and construction materials, soil and water resources, environmental concerns, and transportation system design.

Drawings are vitally important to the civil engineer, who must be able to accurately convey information about the construction of a new bridge or building, the revision of a highway intersection, or the design of a new commuter rail link. Modification of existing roads or structures requires the civil engineer to read, interpret, and make changes to old drawings. These old drawings can be scanned into a CAD system and easily redrawn to meet new requirements.

■ C A R E E R   L I N K ■

To learn more about civil engineering, contact the American Society of Civil Engineers, 1801 Alexander Bell Drive, Reston, VA 20191 or visit their web site at *www.asce.org*.

Another source of information is the Junior Engineering Technical Society, JETS-Guidance, 1420 King Street, Suite 405, Alexandria, VA 22314 or visit their web site at *www.asee.org/jets*

Fig. 21-1. Rib-type dome of the auditorium and ice rink at Brown University.

21.1 INTRODUCTION TO STRUCTURES. A spider's web is one of nature's fine examples of a structure. Such a web is made of many parts connected to form a unit strong enough to support the spider. Most of our structures, Fig. 21-1, also have many parts connected to form a framework strong enough to support loads. Figure 21-2(a) shows a simple bridge structure. The portion of the structure lettered ABCDE is a *truss*.

This truss has seven parts connected to each other at the *joints* A, B, C, D, and E. The parts are called *members*. All of the truss members are in the same vertical plane. AB is an *upper chord member*. CD and DE are *lower chord members*. AE, AD, BD, and BC are *inclined members*. The bridge is made up of

the truss ABCDE, the similar truss on the far side, the cross members connecting the two trusses at corresponding joints, and the roadway. Figure 21-2(b) shows the structure being used as a highway bridge. The width of the highway is the *span* of the bridge.

If the bridge were loaded with a truck, the weight of the truck would be transmitted by the bridge to the supports at the ends of the bridge. These supports are called *footings*. The truck would be supported by the roadway, the roadway by the lower cross members, the lower cross members by the trusses, and the trusses by the footings.

Each structural member performs a different function. The lower cross members have *bending loads*. These cross members are called

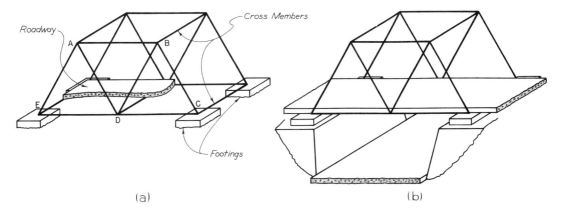

Fig. 21-2. A bridge structure.

beams. The roadway load acts perpendicular to the axes of these members and tends to bend them. The truss members have *axial loads.* The load on each truss member is in the direction of that member's axis. They will be either *tension members* or *compression members,* depending upon whether they have pulling or pushing loads.

A *loading diagram* for the lower cross member at the center of the bridge is shown in Fig. 21-3(a). The roadway load is called a *distributed load.* It is distributed over the length of the member. The forces designated R_1 and R_2 are called *reactions.* The cross-member reactions are the supporting forces supplied by the trusses. A loading diagram for one of the trusses is shown in Fig. 21-3(b).

The truss loads designated P_1, P_2, and P_3 are called *concentrated loads.* They are concentrated at the lower joints. Note that R_1 in (a) would be equal to P_2 in (b). They both represent the force between the center cross member and the truss. The truss reactions are the supporting forces supplied by the footings.

A multistory building is another example of a structure. Its members are *beams, girders,* and *columns.* The girders are large beams supporting other beams. The structural loads are transmitted through the members to the foundation of the building. The weight of a safe on one of the upper floors would be transmitted from the floor to the floor beams, from the floor beams to the girders, from the girders to the columns, and then from the columns to the foundation.

Drawings are necessary at all stages of structural work. A complete set of drawings for a large structure might include *layout drawings, foundation plans, design drawings, detail drawings,* and *erection diagrams. Layout drawings* show the overall dimensions of the structure and its location. *Foundation plans* show the footings or the location of piles to be driven. *Design drawings* indicate the design

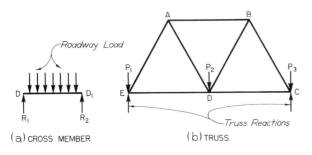

Fig. 21-3. Loading diagrams.

loads, showing the general arrangement of the structure, and giving specifications of the different members. *Detail drawings* give sufficient information about members for them to be made. *Erection diagrams* show piece markings and indicate the sequence to be followed in the final assembly of the structure.

A structural designer must know the strength and other properties of the materials in the structure. He or she also must know the different types of possible structural failure. A structural designer must consider *buckling, twisting, shear, vibrations,* and various combinations of loading. They also must consider tension, compression, and bending.

21.2 WOOD CONSTRUCTION. Wood was undoubtedly the first material used for structures. It is in common use for sills, studs, rafters, and roof trusses. It is often used in the construction of railway and highway trestles. In remote areas, wood may be the only available material for structures. The principal woods used in construction are white pine, yellow pine, fir, cypress, and oak.

Wood construction makes a satisfactory substitute for steel and reinforced concrete construction during shortages of steel. Wood is also used for formwork and supports in reinforced concrete construction.

21.3 STEEL CONSTRUCTION. The high strength and ductility of steel make it an ideal structural material. It has equal strength in either tension or compression. It can be rolled into shapes specifically designed for structural members. Its ductility facilitates fabrication and erection. It also enables a completed steel structure to withstand limited overloading without complete structural failure.

Cross-sections of the principal rolled structural shapes are shown in Fig. 21-4. The symbols beneath each section are used to specify particular shapes on drawings and in technical literature. Each shape is designed for some structural function. The standard "I"-beam and the wide flange beam are designed to resist bending with minimum cross-sectional area and minimum weight of the member. All of the shapes are available in different sizes and in lengths suitable for structural members. Steel *fabrication* is the process of *riveting, welding,* or *bolting* structural steel shapes into structures.

A complete listing of standard sizes for each structural shape is given in *Manual of Steel Construction* published by the American

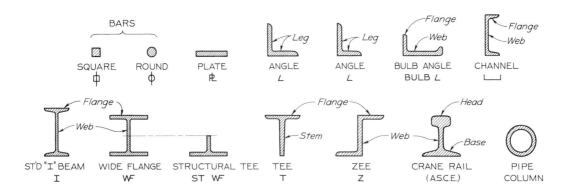

Fig. 21-4. Structural steel shapes.

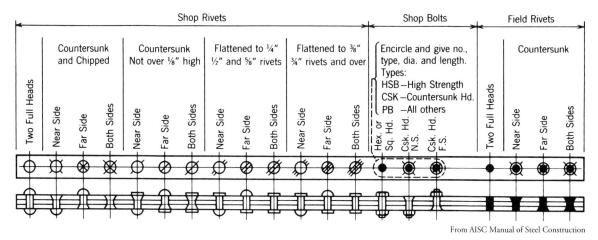

Fig. 21-5. Conventional symbols for rivets and bolts.

Institute of Steel Construction (AISC). This is the standard handbook in the field. It includes specifications and standards.

21.4 STRUCTURAL STEEL DRAWINGS. Detail (working) drawings for steel structures show the various parts as they are shipped. The shipping units for large steel structures are *subassemblies* of members with connecting parts attached. It is economical to do as much of the connecting as possible in the fabricating shop because of the favorable working conditions there. The drawings must specify the connections in detail. They must indicate which connections are to be made in the shop and which are to be made in the field during erection.

Connections are either *riveted, bolted,* or *welded.* Rivets are seldom used in new construction, but riveted connections are shown in existing drawings and will be of concern to a structural engineer dealing with inspection or renovation of older buildings or bridges. Almost all new structural work is joined by welding or bolting. Bolted connections are usually made with special high-strength struc-

tural bolts, nuts, and washers, which are available in both inch and metric sizes.

Figure 21-5 shows the AISC conventional symbols used on drawings to indicate structural rivets and bolts. *Shop rivets* are those applied to the structure in the fabricating shop before delivery to the site. *Field rivets* are those applied at the site. The center line of a line of rivets is the *gage line.* The uniform distance between centers of rivets is called the *pitch* of the rivets. These terms also apply to bolts.

Standard methods of making beam connections are shown in Fig. 21-6. The connections are between the I-beam shown and a column or girder perpendicular to the beam. In the framed-beam connection the short structural angles (called *clip angles*) have been shop-riveted to the web of the beam. They are shipped attached to the beam. In the seated-beam connection the clip angles have been shop-riveted to the web of the column or girder. The lower clip angle provides a seat for the beam during erection.

Figure 21-7 shows a detail drawing of a shipping unit for a large structure. The unit

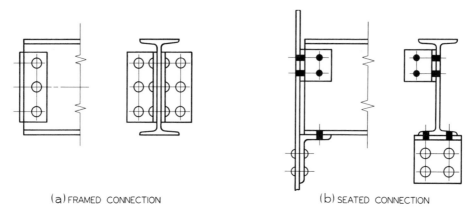

(a) FRAMED CONNECTION   (b) SEATED CONNECTION

Fig. 21-6. Beam connections.

consists of an I-beam with several short struc- tural angles shop-welded to its web. The structural angles at the left end are for con- nection to a column during erection. The intermediate angles are for seated connections to other beams framing into the beam shown. The right end of the beam frames into a gird- er. Note that the lower view is a section

viewed from the top. This is customary pro- cedure in structural drawing. In a machine drawing this view would be placed above the front view.

If a steel structure is small enough to be shipped and erected as a unit, both design and details may be shown on the same drawing. Figure 21-8 shows a partial drawing of such a

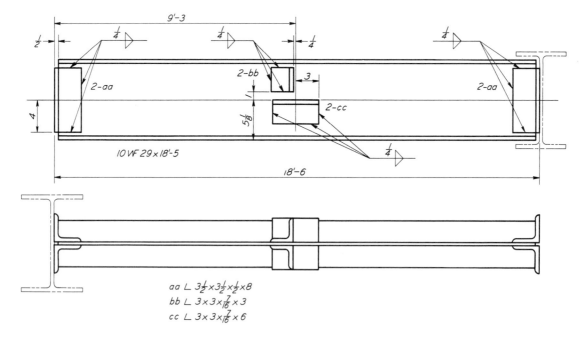

Fig. 21-7. Detail drawing of a welded shipping unit.

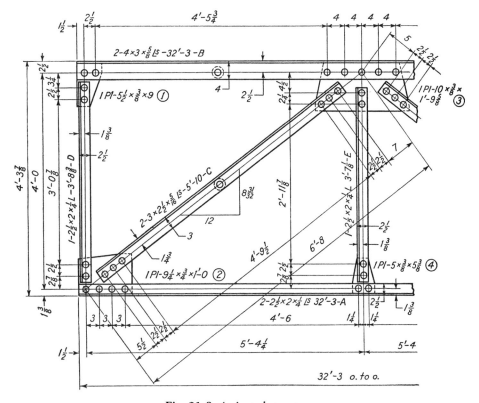

Fig. 21-8. A riveted structure.

structure. Note that the inch symbol is omitted and that the dimension figures are lettered above continuous dimension lines instead of being placed in a gap. Observe that the inclination of a structural member is shown by a right triangle, the lengths of whose sides are given. All of the rivet symbols indicate shop rivets. The plates designated 1, 2, 3, and 4 are called *gusset plates*. The members are structural angles and have the designations A, B, C, and D. The AISC publishes the authoritative, two-volume *Structural Shop Drafting*. This gives

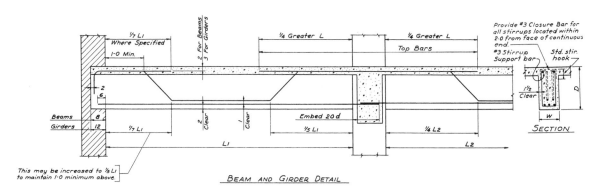

Fig. 21-9. Beam and girder detail.

further details of the standard methods of making structural steel drawings.

21.5 REINFORCED CONCRETE CONSTRUCTION.

The reinforcing of concrete with metal had its origin in France in about 1850 when Lambot built a boat of reinforced mortar. Today we find reinforced concrete structures everywhere.

The design of reinforced concrete structures is complicated. The designer must know the entire construction process to be able to design a structure that may be built economi-

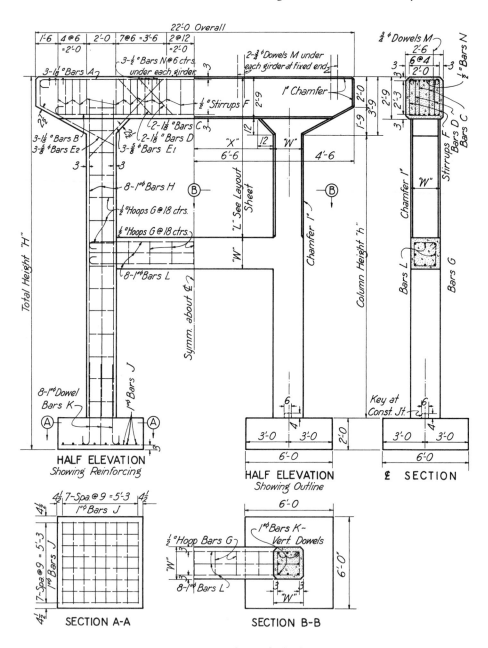

Fig. 21-10. An interior bent of a highway bridge.

cally. The process includes how the reinforcing bars are bent into shape and how they are tied together. It also includes how forms will be erected and supported, how the reinforcement is placed in the forms, how the concrete is to be poured, and how the forms are stripped. The designer should also know the methods of testing, curing, and inspection.

21.6 REINFORCED CONCRETE DRAWINGS. To assure uniformity in reinforced concrete drawing, the American Concrete Institute (ACI), publishes the approved *Manual of Standard Practice for Detailing Reinforced Concrete Structures.* This manual shows typical drawings, Fig. 21-9, for reinforced concrete buildings, bridges, piers, retaining walls, culverts, and arches.

A typical reinforced concrete structural drawing is shown in Fig. 21-10. Note that the right-side view is in full section. The solid dots in the sectioned areas are sections through the rods of reinforcing steel. In the front view, the left half is shown in section. The concrete is not indicated because to show it would take time and would not improve the clearness of the drawing. The long lines within the section represent the rods of reinforcing steel. Note that Sections A-A and B-B are both views "looking down." They would be placed above the front view in a machine drawing.

Concrete is often used in connection with structural steel, as seen in Fig. 21-11. In such buildings, the floor loads are carried by the steel skeleton. The concrete is used for floors and to cover structural members to obtain fireproofing and better appearance.

21.7 CAD AND STRUCTURAL DRAWING. Structural drawings made by using CAD are

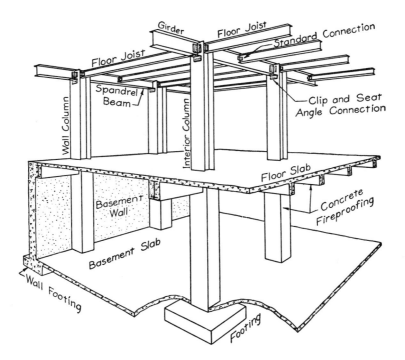

From *Theory of Modern Steel Structures* by L. E. Grinter (Macmillan)

Fig. 21-11. Fireproofed structure of a steel-frame building.

Autodesk, Inc.

Fig. 21-12. A CAD structural drawing.

prepared with specific software. Structural software has the symbols needed for designing a finished drawing. The CAD drafter need only call up the desired part and place it in the appropriate position on the drawing. Attributes (changeable information) are attached to each symbol. These allow the drafter to specify sizes and shapes. Standard construction details are included in the software. These details may be used as is or edited to conform to special construction applications. The drafter dimensions the pieces after they have been placed in position. Special symbols such as welding callouts are included as part of the dimensioning section of the software, Fig. 21-12.

PROBLEMS

PROBLEM 21-A. Make a detail drawing of diagonal member C in Fig. 21-8 with gusset plates 2 and 3 attached. (Assume that the structure is to be shipped "knocked down" instead of assembled.)

PROBLEM 21-B. Same as Problem 21-A, except make a detail drawing of the vertical member D in Fig. 21-8 with gusset plate attached.

PROBLEM 21-C. Same as Problem 21-A, except make a detail drawing of member E in Fig. 21-8 with gusset plate 4 attached.

PROBLEM 21-D. Assume that the structure shown in Fig. 21-8 is 4 ft. deep and that there are six panels similar to the one shown. Make an oblique drawing of the complete structure, showing very little detail. Make it similar to Fig. 21-2, with dimensions added.

PROBLEM 21-E. Make a bill of material for the complete structure in Fig. 21-8.

If assigned by the instructor, construct drawings using decimal-inch or metric scales by converting given fractional dimensions to decimal-inch or metric equivalents. Refer to Table 20 in the Appendix.

rightArnold & Brown

CHAPTER 22

After studying this chapter, you should be able to:

- Describe a topographic map.
- Explain how topographic features are represented on a map.
- Define the term *contour* as it applies to the earth's surface.
- Define the term *profile* as it applies to the earth.

map

atlas

plat

topographic map

contour

contour interval

elevations

profile

contour pen

CHAPTER 22
Map Drafting

Maps are drawn for many purposes. A map might show the location of a new housing subdivision being carved out of farmland, the route of a new highway, natural geographic features, or political boundaries of cities, counties, states, or countries.

All maps are based on accurate surveys, which may be made by a surveying crew working with laser-based instruments or derived from aerial photographs taken from a helicopter, airplane or satellite. Geographic positioning from satellite locational data is also used by the modern cartographer (mapmaker).

Drawings can be as simple as a set of rough sketches made on a job site or as complicated as a CAD-generated topographic map that shows how much earth must be moved to provide a level path for a new road.

CAREER LINK

To learn more about cartography, contact the American Congress on Surveying and Mapping, 5410 Grosvenor Lane, Suite 100, Bethesda, MD 20814. There also is an American Society for Photogrammetry and Remote Sensing at the same address, Suite 210. Their web address is *www.asprs.org*.

22.1 MAPS. A *map* is a drawing of the earth's surface or a part of it. When maps are bound together in book form, the result is an *atlas*. A map prepared primarily for navigation is generally known as a *chart*. A map of a small "parcel" of land is called a *plat* or a *plot*. A typical plot of farm land is shown in Fig. 22-1.

The *lines* of a plot are established by a surveyor with a device called a *transit*. Iron stakes are driven at corner points if no other fixed points are available. Actual distances are measured with a tape. The *bearings* of the lines, or the angles they make with due north or south, are also established by the surveyor. All

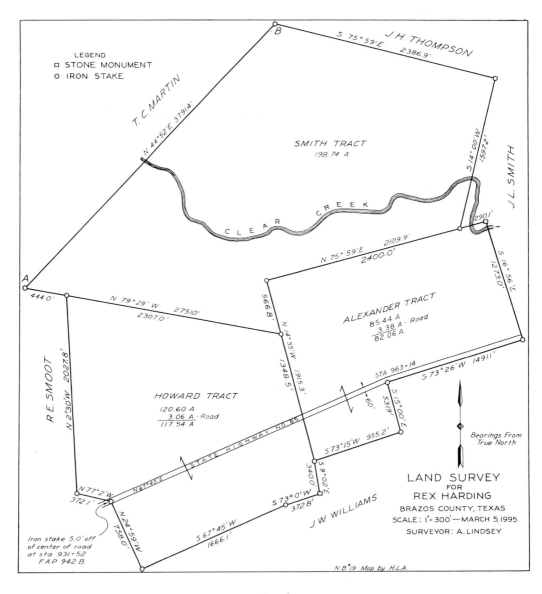

Fig. 22-1. Land survey.

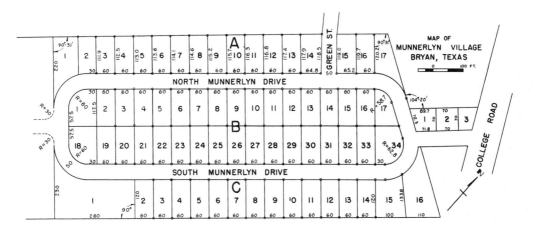

Fig. 22-2. A partial city map.

of this information is taken down in the form of *field notes.* The plot is drawn from these notes. Observe, in Fig. 22-1, that the bearing and the length are lettered along each line. For example, the line at the upper left near "T. C. Martin" is 44° 52' east of due north, and is 3791.4 ft. long. Note the arrow indicating true north. As a rule, a map is drawn with "north" toward the top.

Another common type of map is a *city map,* Fig. 22-2, showing streets and lots. Note the method of indicating the scale of the drawing.

A *topographic map* shows the physical features, such as lakes, streams, forests, mountains, and structures like bridges, dams, and buildings. Generally, the elevations of the various portions are represented by *contour lines.* Figure 22-3 is a topographic map of a country estate.

Keep in mind that the purpose for which a map is intended determines what features should be shown on it and how much detail should be included. Maps of large areas, such as towns, counties, or states, will not include all the details that may appear on a map of a

small area. The scale to which a map is drawn depends upon the size of the area to be represented and the amount of detail that must be shown. The scale for the map of a farm in Fig. 22-1 is 1″ = 300′, whereas the scale of a map by the United States Geological Survey, covering a large area, would often be 1:62,500. In inches, this would be approximately 1″ to a mile.

We represent topographic features with *symbols,* as shown in Fig. 22-3. For example, trees in general are always sketched in a certain way. Grasslands are indicated by groups of short marks that suggest grass. Streams and bodies of water are symbolized by freehand parallel lines along the water's edge. Refer to any standard text on topographic drawing for other symbols.

A *contour* is an imaginary line on the earth's surface, passing through points of equal elevation, or height above sea level. If a series of equally spaced planes are passed through a hill, Fig. 22-4, each plane will produce a contour line that is the intersection between the plane and the earth's surface. The distance between the planes is called the *con-*

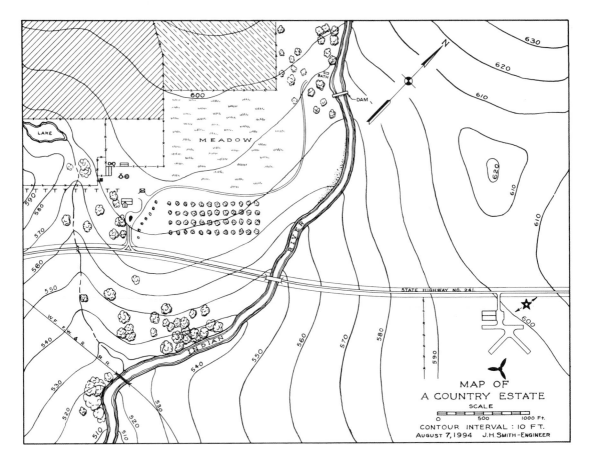

Fig. 22-3. A topographic map.

tour interval, in this case 10 ft. For a large rugged area, the contour interval may be as much as 100 ft. For a small area where the surface is relatively flat, such as a park, a garden, or a building site, the contour interval may be less than a foot. The *elevations* are lettered on each contour line, either along each line, as in Fig. 22-3, or in gaps left in each line, as in Fig. 22-4.

A *profile* is a sectional view through the earth produced by a vertical sectioning plane. In Fig. 22-4 the edge view of the imaginary plane is indicated in the top view by the line X-X. Note how each intersection of this plane

with a contour produces a point on the profile when projected down to the corresponding elevation. When a profile is drawn, a larger scale is often used for vertical distances in order to exaggerate surface irregularities so they will show up more clearly. This is quite common in profiles of a highway showing the grades of the roadway.

Contours are drawn freehand with an ordinary pen or with the aid of a *contour pen.* They can also be drawn using CAD. Note, in Fig. 22-3, that contour lines close together indicate a steep slope, whereas widely separated lines show a gentle slope. Also note in the same figure that contour lines tend to run upstream. Thus, even if no water is shown, a depression that at times may carry water is shown by the contour lines — as, for example, at the left of the figure. Anyone experienced in making or reading maps can easily visualize the shape of the earth's surface from a glance at the contour lines.

22.2 MAP DRAFTING USING COMPUTER SOFTWARE.

Specialized software packages are available to perform all tasks related to map and civil drafting. The complexity of the output is dictated by the information provided and by the sophistication of the software.

Layout for land surveys, site plans and subdivisions is easily accomplished using software packages. Readings from field survey equip-ment are transferred to the software, which converts the readings to accurate drawings. The CAD drafter adds the required dimensions and notes to develop the completed drawings. Existing maps can be scanned or digitized (traced) into the software. Details and notes can be added as required, Fig. 22-5.

Use of GIS (Geographic Information System) software is a more complex method of developing a plan to give added detail and information. The information for the software can be input from one or more of many systems. Field readings, satellite photography, and video capture are a few of the methods used to develop information. The GIS software will develop three-dimensional models of the site and develop plans in many formats, Fig. 22-6. Contours and profiles of the

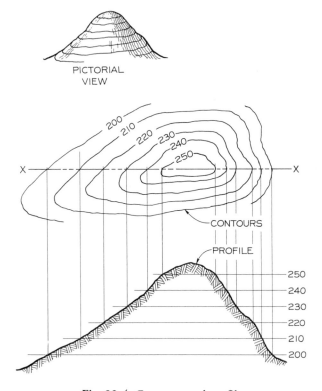

Fig. 22-4. Contours and profile.

site are developed and cut. Fill for roads and bridges is automatically generated. Site plans for subdivisions, parks, and industrial developments are easily generated.

Three-dimensional designs can be placed on the sites to compare compatibility of land use with traffic patterns and utility needs. The software database has the capability of storing information. This can include parcel numbers, street addresses, owners, taxes, utility services, and just about any required information. When the drawing of an area is on the screen, the information can be accessed by picking a point or spot on the plan. This method is used to compare densities for building purposes. It is also useful in determining the need for more services such as access roads, fire protection, and utilities.

Autodesk, Inc.

Fig. 22-5. A three-dimensional map prepared using CAD software.

Fig. 22-6. A map produced using GIS software.

Autodesk, Inc.

PROBLEMS

PROBLEM 22-A. Using Layout E in the Appendix, draw a plat of the survey shown in Fig. 22-1 to as large a scale as practicable. Use a protractor to set off bearings and an engineers scale to set off distances.

PROBLEM 22-B. Using Layout D in the Appendix, draw a city map, similar to that shown in Fig. 22-2, of your own town or neighborhood area. Select a suitable map scale.

PROBLEM 22-C. Using Layout D in the Appendix, make a double-size drawing of the topographic map shown in Fig. 22-3.

If assigned by the instructor, construct drawings using decimal-inch or metric scales by converting given fractional dimensions to decimal-inch or metric equivalents. Refer to Table 20 in the Appendix.

479

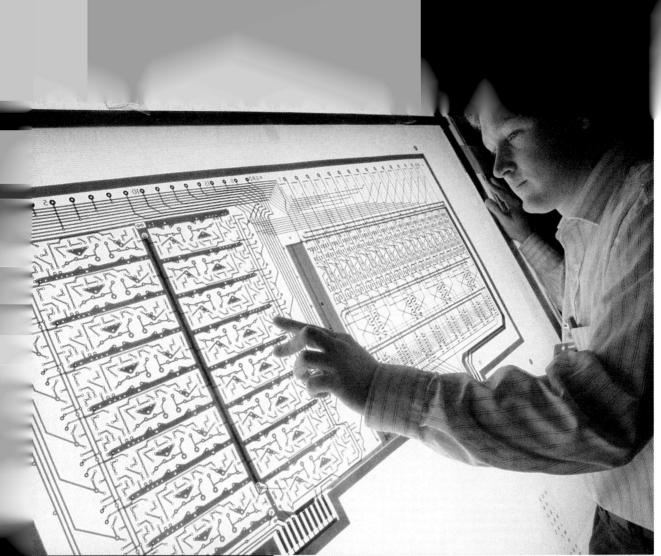

Richard Pasley/Stock Boston

CHAPTER 23

OBJECTIVES

After studying this chapter, you should be able to:

- Define electrical drafting.
- Identify three types of electrical diagrams.
- Identify the main steps in drawing a schematic diagram.

CHAPTER 23

Electrical and Electronic Drafting

The electrical engineer may be involved with development of electric power plants, devices, controls, and energy conversion systems. Beyond understanding the laws of physics and electrical principles, an electrical engineer must be able to place ideas and designs on paper using block and schematic diagrams, charts, and graphs.

Electrical devices must be mounted or housed in cabinets, with various plugs, jacks, and other connectors for input and output. All these electromechanical design factors require preliminary design sketches and final mechanical or CAD drawings.

23.1 ELECTRICAL AND ELECTRONIC DRAFTING.

Electrical or electronic drafting may be defined as the drawing of *electrical wiring diagrams, interconnection diagrams, block diagrams, layout diagrams,* and *schematics.* In addition, electronic drafting frequently involves the layout of printed circuit boards. Both electronic and electrical drafters will also be involved with mechanical layouts of control panels and housings. With the aid of drawing instruments or a computer and CAD software, a drafter will draw these diagrams. He or she uses standard graphical symbols to represent the various devices that are to be used. The completed electrical drawings provide the necessary information for the connection, installation, control, and maintenance of electrical or electronic devices.

The electrician responsible for the installation of the electrical work must have electrical drawings or wiring plans, Fig. 24-19. These plans must show the location of lights, switches, and receptacles, as well as the power source. The electronics technician, when assembling or repairing a computer or television set, requires a schematic diagram and circuit board layout to identify and trace the electronic circuits. In analyzing the various circuits, the electrician and technician will study the lines and symbols on the drawing prepared by the drafter. The ease with which these drawings can be interpreted depends on how well the wiring or schematic diagram is laid out and drawn by the drafter.

The drawings produced by the drafter are an important link between the electrical or electronic designer and engineer and the persons responsible for the construction of their ideas and designs. The primary purpose of the drafter is to relay this information from the designers and engineers to the electricians, technicians, and manufacturers in the electrical and electronic industries.

An electrical drafter must be well versed in the principles and techniques of technical drawing. He or she must also possess a working knowledge of electricity and electrical and electronic circuitry. A student will find that courses in physics and electronics are of considerable help in understanding and drawing electrical diagrams.

23.2 GRAPHICAL SYMBOLS AND TECHNIQUES.

The lines appearing on a wiring diagram indicate the connection of many different types of devices. Since it would be impractical and time-consuming to draw the electrical devices as they appear, *graphic symbols* are used to simplify representation. These symbols have been standardized by the American National Standards Institute. See ANSI Y14.15 and ANSI Y32.9. They are referred to by the drafter before drawing any electrical or electronic diagrams.

The symbols must be placed on the electrical diagram in the same place where the electrical device is to be indicated. The sizes of the symbols must be proportionate throughout the diagram. Symbols are not to be crowded. They should be drawn clearly and distinctly. If a symbol is identified with a letter and number prefix, no other symbol should be identified with the same letter and number. For example, a relay symbol may be drawn and identified as ⬭1CR . A second relay would be drawn and identified as ⬭2CR . Some of the more important symbols used on electronic diagrams are shown in Tables 17 and 18 in the Appendix.

Plastic *templates,* Fig. 23-1, are available for drawing electrical and electronic symbols. They are time-saving devices. However, their use may be limited if symbols smaller or larger than those appearing on the template are required.

In general, electrical diagrams need not be drawn to any specific scale. The smallest size standard layout compatible with the nature of the diagram should be selected. Line thicknesses and letter sizes should be selected on the basis of producing a legible drawing conforming to ANSI standards. A line thickness of *medium* weight is recommended for general use on electrical diagrams. *Thin* lines are used to represent brackets, leader lines, etc. A heavy line may occasionally be used to emphasize a particular portion of a circuit.

23.3 ELECTRIC CIRCUITS. An *electric circuit* may be defined as the path through which a current flows. *Current* is the time rate of flow of electricity. It is measured in *amperes*. However, we cannot have a current flowing in a circuit unless there is a closed circuit with a *voltage* and a *resistance*. Voltage is sometimes called *electromotive force* (emf) or *potential difference* (pd). Voltage is the difference in electrical pressure that causes the flow of electricity. It is measured in *volts*. The *resistance* is the opposition to the flow of electricity. It is measured in *ohms*.

The relationship among current, voltage, and resistance is known as *Ohm's Law*. It may be expressed by the following formula:

$$\text{Current (amperes)} = \frac{\text{Potential difference (volts)}}{\text{Resistance (ohms)}}$$

Expressed in symbols, it is:

$$I = \frac{E}{R}$$

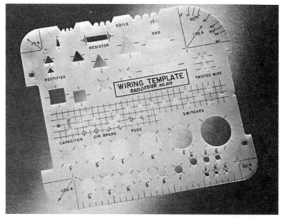

RapiDesign, Inc.

Fig. 23-1. Template guide for electrical symbols.

For example, suppose we want to find the amount of current passing through a resistance of 10 ohms connected to a 110-volt source.

$$\text{Solution: } I = \frac{E}{R} = \frac{110}{10} = 11 \text{ amperes}$$

Ohm's Law is one of the most important principles in electricity. It can be applied not only to an entire circuit, but also to any part of a circuit. While it applies only to direct current (dc) circuits, a special form of this law is used also in alternating current (ac) work.

Electric circuits may be connected in *series, parallel,* or a combination of both, Fig. 23-2. In a series circuit, (a), the various electrical devices are connected one after the other in a single line or path. In a series circuit the current is the same in every part of the circuit. The resistance is the sum of the separate resistances. The voltage is equal to the sum of the voltages across the separate parts.

In a parallel circuit, (b), the current may flow through two or more paths or branches. In a parallel circuit the voltage is the same

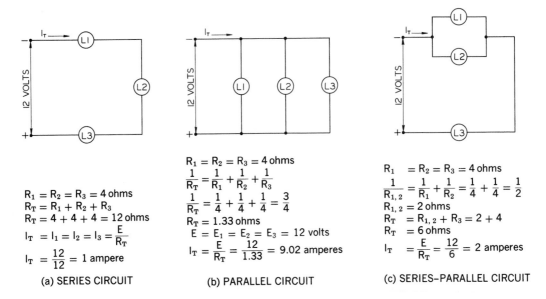

$R_1 = R_2 = R_3 = 4$ ohms
$R_T = R_1 + R_2 + R_3$
$R_T = 4 + 4 + 4 = 12$ ohms
$I_T = I_1 = I_2 = I_3 = \dfrac{E}{R_T}$
$I_T = \dfrac{12}{12} = 1$ ampere

(a) SERIES CIRCUIT

$R_1 = R_2 = R_3 = 4$ ohms
$\dfrac{1}{R_T} = \dfrac{1}{R_1} + \dfrac{1}{R_2} + \dfrac{1}{R_3}$
$\dfrac{1}{R_T} = \dfrac{1}{4} + \dfrac{1}{4} + \dfrac{1}{4} = \dfrac{3}{4}$
$R_T = 1.33$ ohms
$E = E_1 = E_2 = E_3 = 12$ volts
$I_T = \dfrac{E}{R_T} = \dfrac{12}{1.33} = 9.02$ amperes

(b) PARALLEL CIRCUIT

$R_1 = R_2 = R_3 = 4$ ohms
$\dfrac{1}{R_{1,2}} = \dfrac{1}{R_1} + \dfrac{1}{R_2} = \dfrac{1}{4} + \dfrac{1}{4} = \dfrac{1}{2}$
$R_{1,2} = 2$ ohms
$R_T = R_{1,2} + R_3 = 2 + 4$
$R_T = 6$ ohms
$I_T = \dfrac{E}{R_T} = \dfrac{12}{6} = 2$ amperes

(c) SERIES–PARALLEL CIRCUIT

Fig. 23-2. Series and parallel electric circuits.

across each branch of the circuit. The total current is the sum of the currents through the branches. The total resistance is equal to the voltage divided by the total current. In a parallel circuit the total resistance (R_T) may be determined by the following formula:

$$\frac{1}{R_T} = \frac{1}{R_1} + \frac{1}{R_2} + \frac{1}{R_3} + \text{etc.}$$

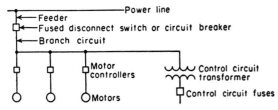

Reproduced with permission from JIC Electrical Standards for Industrial Equipment, 1957 Joint Industrial Conference, National Machine Tool Builders Association, 2139 Wisconsin Avenue, Washington, D.C. 20007.

Fig. 23-3. Single-line diagram of a typical branch circuit protective arrangement for a multi-motor machine.

A simple explanation of series and parallel circuits might best be illustrated by comparing two Christmas-tree light sets. One is wired in series and the other in parallel. In the series set, if one light burns out, all the remaining lights will also be out. In the parallel set, if one light burns out, the remaining lights will continue to operate.

An example of a combination series-parallel circuit is shown at (c).

23.4 ELECTRICAL DIAGRAMS. Several different types of electrical diagrams are ANSI Standard. See ANSI Y14.15.

The *single-line* or *one-line diagram*, Fig. 23-3, presents the various circuit information simply by means of single lines and standard graphical symbols.

The *schematic* or *elementary diagram*, Figs. 23-4 and 23-5, shows the electrical connections and functions of a circuit by means of standard graphical symbols. It shows them

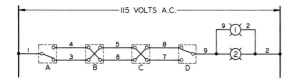

Fig. 23-4. Schematic diagram of three-way and four-way switches controlling two lamps.

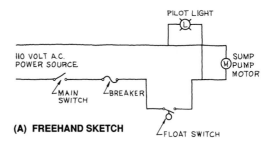

(A) FREEHAND SKETCH

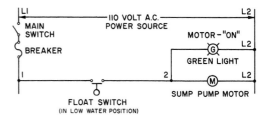

(B) MECHANICAL DRAWING

Fig. 23-5. Drawing the schematic diagram of a sump pump motor circuit.

without any regard to the actual physical size, shape, or location of the electrical devices or parts that make up the circuit.

The *connection* or *wiring diagram,* Fig. 23-6, shows the connections of an installation, or the electrical devices or parts that comprise the circuit. It includes the detail necessary to make or trace the internal and/or external connections involved. It usually shows the general physical arrangement of the component devices or parts.

The *interconnection diagram* shows only the external connections between unit assemblies or equipment. In this type of connection, or wiring diagram, the internal connections are usually omitted.

23.5 SCHEMATIC DIAGRAMS. A schematic diagram is one of the simplest forms of electrical or electronic diagrams. It is frequently used to illustrate an electrical circuit. In a schematic diagram, sometimes called an elementary diagram, no attempt is made to show the electrical devices in their actual positions. The electrical devices may be shown between vertical lines that represent the source of power. A schematic diagram of two three-way and two four-way switches controlling two lamps is shown in Fig. 23-4.

It is appropriate at this point to state several important principles of electricity. First, electricity flows from one terminal (negative) of the power source, through the electrical

devices, and around to the other side (positive) of the source. Next, a path must be provided for the flow of electricity. This path generally is a copper wire or path on a circuit board called a *conductor.* The nonmetallic parts of an electrical circuit are called *insulators.* Finally, a voltage is necessary to produce a flow of electricity through the electrical circuit.

23.6 DRAWING THE SCHEMATIC DIAGRAM. To draw a schematic diagram, the drafter will first make a freehand sketch, Fig. 23-5(a). Then, using the sketch as a guide, the drafter will locate the graphical symbols of the component devices and parts and determine the size of symbols to be used throughout the schematic, as shown at (b). The ability to do this with ease and speed can be acquired only by experience.

The size of the freehand sketch will also be used as a basis for determining the size of drawing paper required for the mechanical drawing. If larger than standard size, the rough sketch will have to be divided into sections.

The symbols on the schematic diagram can be positioned in approximately the same loca-tion as determined on the sketch. There must be an orderly layout of evenly spaced horizon-tal lines for each symbol from the top of the drawing to the bottom. The symbols represent-ing main components such as lights, coils,

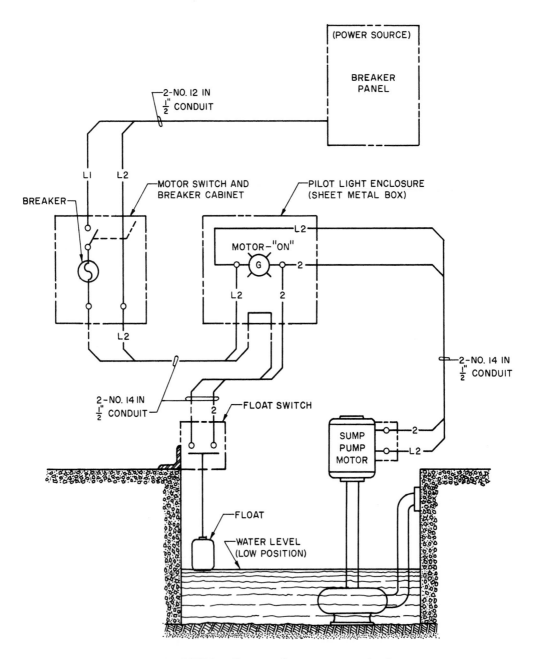

Fig. 23-6. Wiring diagram of a sump pump circuit.

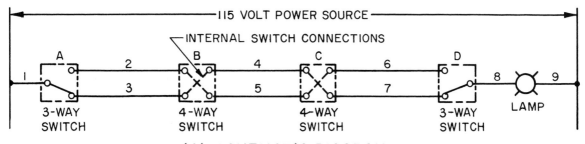

(A) SCHEMATIC DIAGRAM

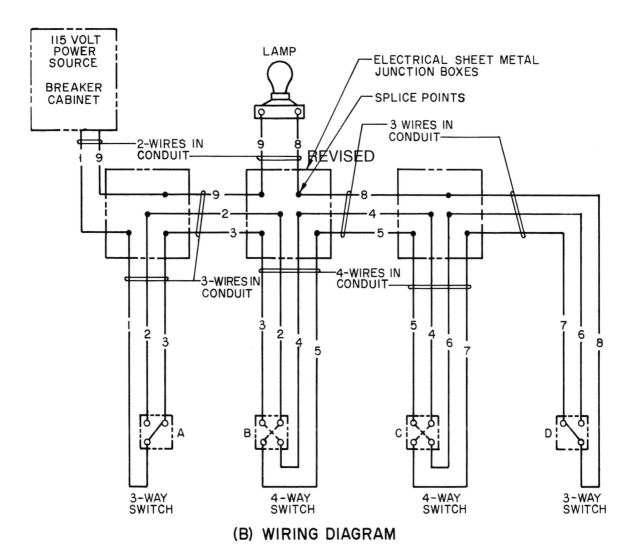

(B) WIRING DIAGRAM

Fig. 23-7. Schematic and wiring diagrams of four switches controlling one lamp.

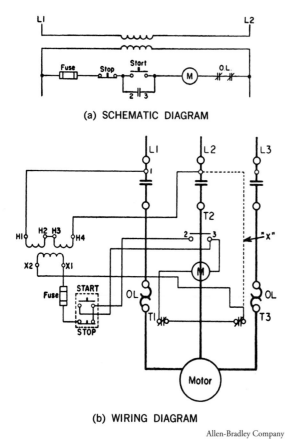

(a) SCHEMATIC DIAGRAM

(b) WIRING DIAGRAM

Allen-Bradley Company

Fig. 23-8. Schematic and wiring diagrams of a step-down transformer in a control circuit.

bells, and motors should be at the right of the drawing between the two vertical lines representing the power source. The symbols representing switches, or controlling elements, are drawn to the left in the same horizontal line as the associated main component located at the right of the diagram. The completed schematic diagram is shown in Fig. 23-5(b) on page 487.

23.7 WIRING DIAGRAMS. A wiring diagram, Fig. 23-6, is a type of electrical diagram in which the graphical electrical symbols are arranged in the same physical relationship as the actual equipment. The lines drawn connecting the symbols represent the actual wires

connecting the electrical devices. These wires may be in conduits or inside the enclosure of the electrical device. The connection lines show every individual connection made between components in a precise manner and indicate the shortest path.

A wiring diagram can be drawn by a drafter after all the symbols representing the electrical devices have been arranged on the drawing. The relative location of these symbols should approximate the actual physical and/or mechanical arrangement of the electrical devices. The schematic is the guide the drafter will follow to draw the wiring diagram and show the connection of electrical devices.

The development of a schematic into a wiring diagram, Fig. 23-7, is usually done by a senior or advanced electrical drafter. The conversion of a schematic to a wiring diagram is the basis of all electrical drafting. The beginning drafter should make every effort, through study and practice, to learn how to make wiring diagrams skillfully, correctly, and rapidly.

23.8 INDUSTRIAL ELECTRICAL DIAGRAMS. The electrical control circuit of a motor that operates an industrial machine is a very important and integral part of electrical diagrams. The method of controlling the electrical power to a motor can be indicated on a schematic diagram, Fig. 23-8(a). The actual wiring of the industrial machine is shown on a wiring diagram, (b).

Industrial electrical diagrams, as shown in *block diagram* form, Fig. 23-9, can be divided into two parts. One portion is the *power* circuit. The other portion is the *control* circuit. The power portion is drawn with heavy lines. The control portion is drawn with a thinner line to provide the desired contrast between the two portions. Generally both portions are

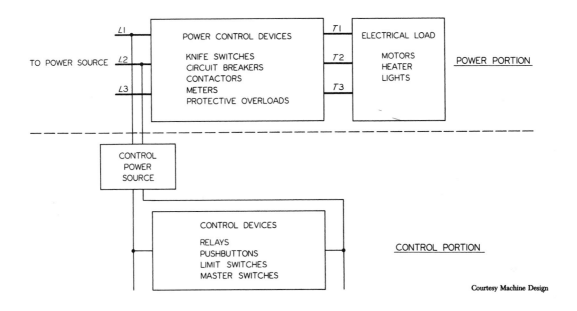

Fig. 23-9. **Block diagram of industrial electrical circuit.**

shown on one drawing. However, if necessary each portion may be shown on a separate drawing. The method of applying or regulating power to a motor is shown in the power portion. The devices in the power portion are controlled by a circuit shown in the control portion. Any required safety controls or motor-protective devices will be shown in the control-circuit portion, Fig. 23-9.

The power portion is drawn first. It will consist of heavy lines showing single-phase power (two wires to the motor) or three-phase power (three wires to the motor). The control portion will consist of the control devices such as relays, pushbuttons, and switches. This portion is shown with thinner lines.

The electrical circuit devices are shown by using standard graphical symbols and abbreviations of their common names. These should conform to NEMA (National Electrical Manufacturers Association) and ANSI standards.

A system of identifying the control devices is necessary to show the difference between similar control devices in a circuit wiring diagram or schematic diagram. One widely used system combines a number with the abbreviation of the common name of the control device. The number may precede or follow the abbreviation. For example, if two relays are used in a circuit, they may be identified as 1CR and 2CR or as CR1 and CR2. The contacts and coil of each relay must be identified by the same combination of numbers and letters.

A *motor starter,* Fig. 23-10, is a device that controls and regulates power to a motor. The starter includes parts that will carry heavy currents. It also includes parts that will carry a smaller control current in the same assembly. The motor starter contains contacts that connect the power source to the motor and a coil energized by a small amount of electric current to activate and close the power contacts. A comparison of the picture and draw-

489

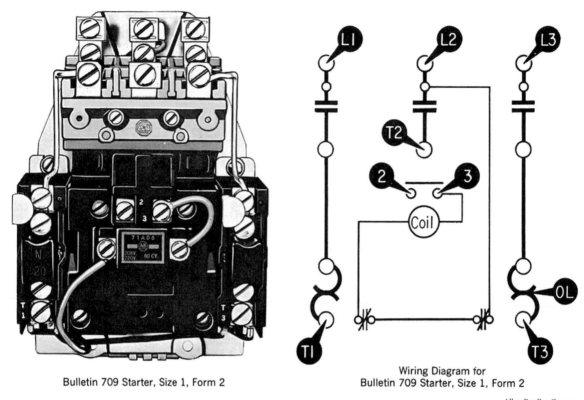

Bulletin 709 Starter, Size 1, Form 2

Wiring Diagram for
Bulletin 709 Starter, Size 1, Form 2

Allen-Bradley Company

Fig. 23-10. **Wiring diagram of a motor starter.**

ing shown in Fig. 23-10 should help you become familiar with the motor starter as it is represented in wiring-diagram form. The principal corresponding parts are labeled so that the wiring diagram can be compared with the actual starter. This should aid in visualizing the starter when studying a wiring diagram. It will help in making connections when it is actually wired up. Note that the wiring diagram shows as many parts as possible in their proper relative positions.

A *list of materials* or parts tabulation should be included with every schematic or wiring diagram. The materials list is necessary for the maintenance of the electrical equipment. It should completely describe the components and their functions. These lists are a valuable aid in understanding the schematic or wiring diagram. They include information that the component symbol alone cannot provide. Manufacturers' catalogs and bulletins are also useful in providing detailed specifications and wiring diagrams of the individual components. An example of such information is shown in Fig. 23-12. This illustrates a wiring diagram for a *reset timer.*

23.9 PRINTED CIRCUITS. Printed circuit boards have largely replaced point-to-point hand wiring in electronic equipment and electrical control assemblies. The printed circuit board is composed of an insulating material. This usually contains glass-epoxy, glass-polyester, or phenolic resin. A thin layer of copper foil is bonded to one or both sides.

The copper is coated with a photosensitive chemical layer and exposed to light through a carefully prepared pattern. This pattern, showing the exact conductive paths between the various electronic parts, is prepared by a skilled drafter. The drafter may use pen-and-ink, preprinted self-adhesive "pads" and tapes, or a CAD program designed for this purpose. Some sophisticated CAD programs will take the information and automatically generate an "autorouted" circuit board layout. A master drawing, Fig 23-11, of the board layout is generally done to a larger scale. It is then photographically reduced to achieve good dimensional accuracy of the final product.

The board is "developed" in a chemical bath. This etches away unwanted portions of the copper, leaving behind the desired conductive paths and "pads." These will have holes drilled through them to carry the wire leads of electronic parts. These leads are then soldered to the copper foil.

23.10 CAD IN ELECTRICAL AND ELECTRONIC DRAFTING.

Electrical and electronic-based software packages were some of the first CAD packages developed in the design industry. The need to work on miniaturized circuits for the development of microchips requiring extremely close tolerances pushed the early use of electronic-based software.

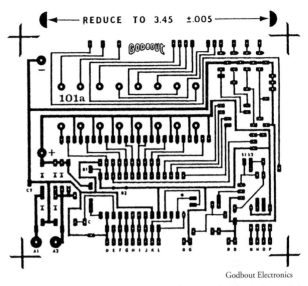

Godbout Electronics

Fig. 23-11. Master drawing of printed circuit board layout.

Electrical CAD software includes a complete set of all the required symbols in a symbol library. Attributes such as capacitance, resistance, inductance, and voltage are attached to the symbols, Fig. 23-13. The electrical symbols are inserted into the drawing in the required layout. A complete bill of materials and a specification sheet are then developed.

Electronics design software is more complex. It performs several additional processes. The use of printed board circuits with multiple layers on miniature chips requires the use of sophisticated software. Such software

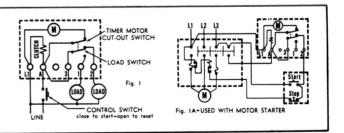

FORM 32

Use with a maintained contact pilot switch which closes and holds the timer energized during its timing interval. At the end of the time delay, contact 3-2 opens and contact 3-1 closes. Opening the pilot switch resets the timer and returns the contacts to their normal position.

Eagle Signal, Division E. W. Bliss Company

Fig. 23-12. Wiring diagram for one form of a reset timer.

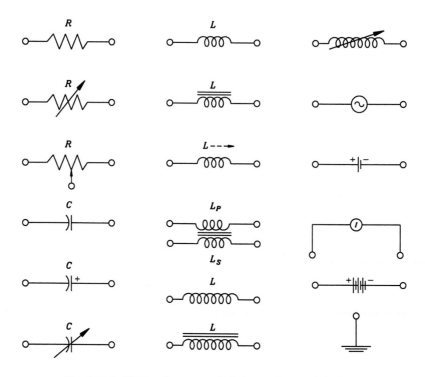

Fig. 23-13. CAD software symbols in an electrical drafting program.

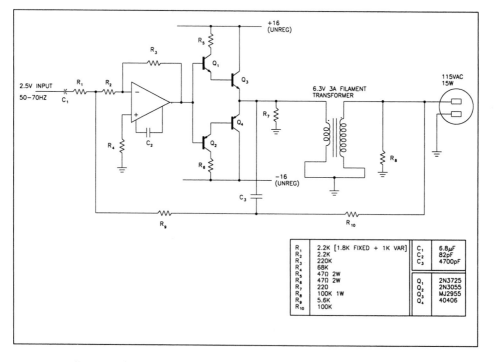

Fig. 23-14. Routing and connections in electronics design software.

is needed to eliminate shorts and to place parts in the exact location required. The software allows the drafter to place parts as designated by the electronics engineer or designer. The drafter then indicates on a data chart how each part is connected to related parts. The software package automatically routes the connecting wires in the most efficient manner. It flags possible shorts and conflicts in connections. The resulting connections are then tested on the computer to determine if all circuits are working properly. Upon completion of the testing, the layout is transferred to a PC board. Production and assembly functions are then performed, Fig. 23-14.

PROBLEMS

PROBLEM 23-A. An electric circuit has three lamps connected in series having the following resistances: 12, 6, and 4 ohms.
  (a) Draw the schematic diagram showing the circuit. Use Layout A in the Appendix. Make the diagram approximately three times as large as the one shown in Fig. 23-5(a).
  (b) What is the total resistance of the circuit?
  (c) What amount of current will flow if the entire circuit is connected to a 24-volt source?

PROBLEM 23-B. The resistances in Problem 23-A are all connected in parallel.
  (a) Draw the schematic diagram showing the circuit. Use Layout A in the Appendix. Make the diagram approximately three times as large as the one shown in Fig. 23-5(b).
  (b) What is the total resistance of the circuit?
  (c) What amount of current will flow through each lamp if the voltage of the source is 9 volts?

PROBLEM 23-C. Using Layout B in the Appendix, make a double-size drawing of the wiring and schematic diagrams of the step-down transformer in the control circuit shown in Fig. 23-10.

PROBLEM 23-D. Using Layout A in the Appendix and a suitable scale, draw an electrical layout wiring diagram of a $9' \times 12'$ room. This room has two wall receptacles and a center ceiling light controlled by a switch at the room entrance.

PROBLEM 23-E. Using Layout C in the Appendix and a suitable scale, draw an electrical layout wiring diagram of your home or apartment similar to the diagram shown in Fig. 24-19.

PROBLEM 23-F. Using Layout A in the Appendix, draw a schematic diagram of two three-way switches controlling one lamp.

Arnold & Brown

CHAPTER 24

IMPORTANT TERMS

specifications   sections

plans   details

elevations   schedule

OBJECTIVES

After studying this chapter, you should be able to:

- Identify the three major divisions of the views of a structure.

- Demonstrate the lettering of notes on an architectural drawing.

- Explain the purpose of elevations.

- Explain the purpose of a schedule.

CHAPTER **24**
Architectural Drafting

Architects deal not only with the design of new buildings but also with the conservation of existing structures. They must consider how best to preserve historic homes and commercial buildings while proposing remodeling that will provide the occupants with modern, comfortable, and safe surroundings.

The architect must be able to visualize, sketch, and make finished drawings that provide all the details needed to show contractors how to proceed with a building or remodeling project. Rigid drawing standards must be adhered to in order to prevent errors in interpretation.

Before a building permit is issued, a set of the plans must be submitted for approval by the local zoning or planning board. The building inspector will check the plans carefully during costruction to make sure that the job is being done properly.

CAREER LINK

To learn more about careers in architecture, contact the American Institute of Architects, 1735 New York Avenue NW, Washington, DC 20006 or visit their web site at *www. aiaonline.com*

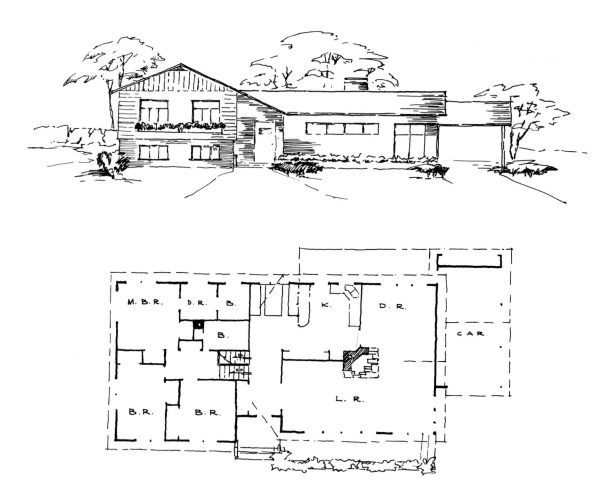

Fig. 24-1. Preliminary sketch.

24.1 ARCHITECTURAL DRAFTING. The story of civilization in its early times is to a large extent the story of the human race's effort to construct permanent "shelters" or buildings for human activities. The buildings, or *architecture,* of any period of history provide an excellent record of that way of life. They also demonstrate the human ability to solve the problems created by a large number of people settled in one place.

When people of the Stone Age started their first settlements, they felt the need for practical, useful, and safe shelters. They soon need-ed to learn new tasks and new skills to construct relatively large buildings. At some point, such buildings were no longer the work of a single person. Others had to contribute their efforts and skills. A new means of communication other than language had to be developed. What one could call early *architectural drawings* were not only instructions to workers describing the structure as to its shape and size. They were also a permanent record of achievement. This dual function of architectural drawing has remained unchanged to the present day.

24.2 ORGANIZATION OF ARCHITECTURAL DRAWINGS. Modern architectural drawings, a collection of plans, inform contractors and workers graphically about the essential parts of the structure and their relationships. They also provide the basis for coordination of the activities of the various trades. They furnish the necessary information for engineering installations such as mechanical and electrical equipment.

As a permanent record architectural drawings permit the checking of the contractor's performance as to completeness and accuracy. They also allow the checking of the quality and quantity of the materials used in construction and the quality of the workmanship. However, a complete and precise description of materials is seldom possible by means of lines and symbols only. For example, a symbol and note on an elevation view may indicate face brick. However, nothing is indicated regarding the texture of the surface or the color. These very special properties of the materials are described in detail in a set of *specifications.* Specifications are written documents accompanying the architectural drawings. They describe in detail the mutual relationships of the owner, the architect, and the builder. They specify the quality and other properties of building materials. They establish the performance standards and quality of workmanship. In general, they describe in words features on the drawings that cannot be adequately described graphically.

The architect, as the creator of the drawings, must be guided by all these requirements. He or she must proceed to describe as completely as possible the site, the shell or structural skeleton, the outside, and the inside of the building. A complete set of such drawings makes up the heart of the collection of all documents connected with the construction. The other parts are the *specifications* and legal forms such as *contracts, performance bonds, liens,* and *titles.* Together they set forth in a legally binding way the individual responsibilities of the architect, the contractor, and the owner.

The organization of architectural drawings is quite different from that of standard machine drawings. However, the objective of both is the same: to describe the object completely as to shape and size. However, the architectural drawing describes an object of comparatively large size — many times larger than the drawing itself. Therefore, it must be drawn at a considerably reduced scale. For example, a house is usually drawn to the scale $\frac{1}{4}$" = 1'-0". At this scale many important details are too small on the drawing. These must be shown by some kind of symbol, or a special kind of "shorthand." Many different views of the structure are needed for a complete description. It is customary to group them into three major divisions: *plans, elevations* and *sections,* and *details.* Each of these divisions may include one or more drawings. Details are customarily drawn to a larger scale than the plans and elevations in order to clarify the elements.

In developing the drawings for a house the architect will first produce a set of *preliminary drawings,* or *preliminary sketches.* These may be freehand sketches of room arrangements. They may include a few sketches of outside elevations, usually not drawn to exact scale, Fig. 24-1. For larger buildings, however, preliminary drawings are always drawn to scale, Fig. 24-2. In some cases, for promotional purposes, the architect will produce a *rendering* of the building, usually in perspective and sometimes in color, Fig. 24-2. From the preliminary drawings a set of working drawings is made. These represent the last stage of the project. Working drawings are very carefully drawn. They are repeatedly checked. In case of a disagreement, they become the legal basis for a settlement. In this respect they must be very carefully coordinated with the specifications so that all problems arising from the construction are properly described.

24.3 NOTES ON ARCHITECTURAL DRAWINGS.

The most carefully drawn and dimensioned set of architectural drawings still would not describe the structure adequately. For example, the sequence of building operations can never be shown on the drawing itself. The drawing shows only the completed construction. Thus, many explanatory *notes* are found on architectural drawings. These are neatly lettered in vertical or inclined letters of the type and size described in Sec. 24.4. Because of the extensive use of notes on architectural drawings, architectural lettering is an essential part of architectural drafting. See Figs. 24-3 to 24-6.

PERSPECTIVE

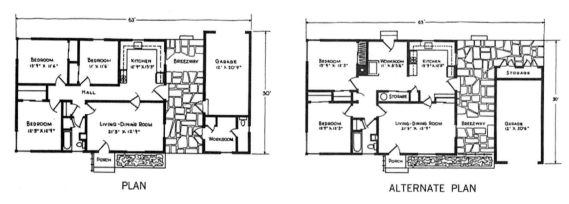

PLAN

ALTERNATE PLAN

U.S. Gypsum Co.

Fig. 24-2. Preliminary drawings.

ABCDEFGHIJKLMNOPQRSTUV
WXYZ 0123456189
ABCDEFGHIJKLMNOPQRSTU
VWXYZ 0123456189
ABCDEFGHIJKLMNOPQRSTUVWXYZ
0123456789

ABCDEFGHIJKLMNOPQRSTUVWXYZ
0123456789

Fig. 24-3. Architectural lettering.

24.4 ARCHITECTURAL LETTERING AND SYMBOLS.

Lettering of notes on the drawing, in the title block, and in the legend must be precise in meaning, brief, legible, and neat. Architectural practice has developed standard wording, standard symbols, a standard alphabet, and many abbreviations. In architectural practice, both vertical and inclined capital letters are used. Good architectural drafters will be proficient in both. However, both styles should not be used on the same drawing or set of drawings. Each individual will in time develop a system of strokes and letter forms best suited to his or her own purposes.

No lowercase letters are used on architectural drawings. The lettering in notes on architectural drawings may be any size from $\frac{1}{16}''$ to $\frac{1}{8}''$ in height. The spacing of horizontal guidelines and the spacing between lines will be done best by use of the Ames Lettering Guide, Fig. 5-13. Experienced drafters can space the guidelines by eye. The general rules

ABCDEFGHIJKLMN
OPQRSTUVWXYZ&
1234567890

Fig. 24-4. Architectural title letters (serif type).

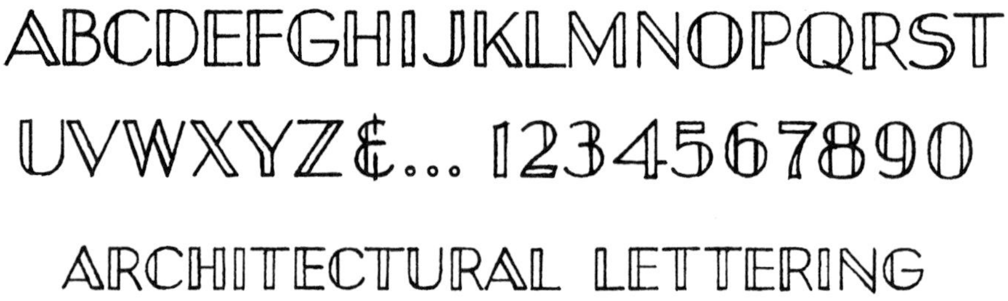

ABCDEFGHIJKLMNOPQRST
UVWXYZ&... 1234567890
ARCHITECTURAL LETTERING

Fig. 24-5. Architectural title letters (Gothic type).

for good engineering lettering are equally applicable to architectural practice. Lettering forms an integral part of architectural drawing. Its proper execution adds not only to legibility but also to appearance. Even an excellent drawing can be ruined by carelessly lettered notes and poorly formed numerals.

Title blocks and titles will sometimes be lettered in serif forms either in solid or outline type, Fig. 24-4. Gothic-type letters are shown in Fig. 24-5. Title blocks on architectural drawings, Fig. 24-6, are somewhat different from title blocks used on machine drawings. Compare this title block with some machine drawing title blocks, for example, that in Fig. 16-12.

Besides notes, a large number of standard abbreviations and graphic symbols are found on any architectural drawing. These abbrevia-

tions and symbols save a great deal of time. Because of their extensive use they are as readable and understandable as full names or precisely drawn views of objects. See Tables 18 and 19 in the Appendix.

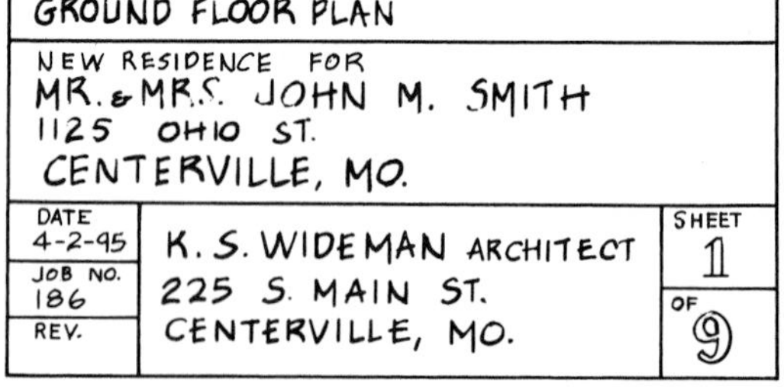

24.5 ELEMENTS OF A HOUSE. A building is composed of many parts or elements. Thus, terms such as *foundations, columns, beams, floors,* and *ceilings* have acquired precisely defined meanings. Some of these are different from the meaning of the word as used in everyday conversation. For example, when we say, "Open the window," we mean in architectural terms, "Open the sash." Then, again, some meanings of architectural terms are rather obscure and generally unknown. For example, the word *astragal* (a small convex molding) is meaningless to most people. Some terms, such as *meeting rail* (the middle horizontal member on a double-hung window) do not describe the part understandably to the average person. For these reasons it will be essential for anyone interested in architectural drawing to become acquainted with the various architectural terms. An immediate identification of symbols on a drawing and the association of such parts with proper names are expected from any architectural drafter.

GROUND FLOOR PLAN		
NEW RESIDENCE FOR MR. & MRS. JOHN M. SMITH 1125 OHIO ST. CENTERVILLE, MO.		
DATE 4-2-95	K. S. WIDEMAN ARCHITECT	SHEET 1
JOB NO. 186	225 S. MAIN ST.	
REV.	CENTERVILLE, MO.	OF 9

Fig. 24-6. Architectural title block.

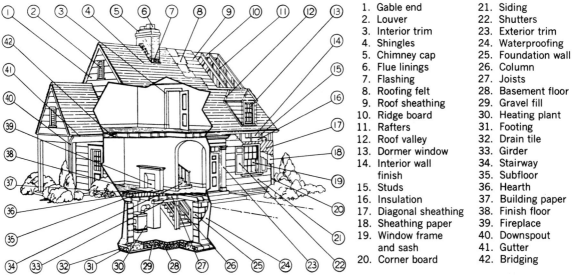

1. Gable end	21. Siding
2. Louver	22. Shutters
3. Interior trim	23. Exterior trim
4. Shingles	24. Waterproofing
5. Chimney cap	25. Foundation wall
6. Flue linings	26. Column
7. Flashing	27. Joists
8. Roofing felt	28. Basement floor
9. Roof sheathing	29. Gravel fill
10. Ridge board	30. Heating plant
11. Rafters	31. Footing
12. Roof valley	32. Drain tile
13. Dormer window	33. Girder
14. Interior wall finish	34. Stairway
15. Studs	35. Subfloor
16. Insulation	36. Hearth
17. Diagonal sheathing	37. Building paper
18. Sheathing paper	38. Finish floor
19. Window frame and sash	39. Fireplace
20. Corner board	40. Downspout
	41. Gutter
	42. Bridging

National Bureau of Standards

Fig. 24-7. **Elements of a house.**

The major elements of a house are *foundations* (including *footings*), *walls, partitions, floors, roof, doorways, windows, stairs, fireplace,* and *chimneys*. The most important structural elements of a house are *columns, beams, lintels, girders,* and *trusses*. The most common parts of a house are identified in Fig. 24-7.

24.6 PLANS. The *plans* contain most of the information and show more of the dimensions than do the elevations. What we call a *plan* is in reality a horizontal section through the house taken above the window-sill height. Thus, all walls appear sectioned. All openings, such as doors and windows, appear as open breaks in walls and partitions, Fig. 24-19.

In general, each floor of the house will be shown on a separate floor plan. Such a plan will contain all information about the sizes, as well as references to the details, cross-sections, and interior views shown elsewhere. It will also contain designations of finished floor materials and the separation of these materials

by edging or moldings. Normally a plan will show the location of walls, doors, windows, electrical and mechanical equipment (such as lighting and plumbing fixtures), heating and air conditioning ductwork and piping, home security, telephone, and television cable wiring.

The dimensions are the most important feature of the plan. There are some significant differences between architectural dimensioning and the dimensioning used in machine drafting. In machine drafting we try to avoid placing dimensions inside the outline of a view. Architectural drawings necessarily abound in these. In machine drawing, all dimensions are related to specific base lines or reference lines, such as center lines. Chain dimensions (continuous lines of dimensions) are usually avoided. On architectural drawings, almost all dimensions are parts of a chain. These differences are clearly a result of the much greater tolerances that building practices allow. See Sec. 10.26. A typical

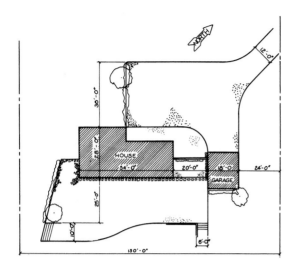

Fig. 24-8. Plot plan.

example of a complete plan is shown in Fig. 24-19. The placing of notes in the open areas of the plan and the arrangement of dimensions on the outside and inside should be especially noted. Besides the general plan, a more complex building will require special plans for floor and roof framing, Fig. 24-23. It will sometimes require them for electrical wiring and heating or air-conditioning systems. These are drawn by drafters specializing in mechanical equipment drafting or electrical drafting.

Most architectural drawings contain a small-scale plan, called a *plot plan*. This shows the location of the building with respect to the property lines. It sometimes shows the location of trees and structures on the property. The building is shown on a plot plan only by its outline or section-lined areas, as in Fig. 24-8.

24.7 ELEVATIONS AND SECTIONS. As the plan shows the horizontal shape and dimensions of the building, the vertical shapes and sizes are defined by *elevations* and *sections*. Elevations could be called outside views of the building.

Usually all such views will be shown. For a rectangular building four regular elevations will suffice. A U-shaped building may require as many as six. The general rule is that every outside surface of the building containing any openings or special features must be shown on some elevation.

Very often the elevation views cannot be placed on the same sheet as the plan. To identify properly each elevation so that it can be coordinated with the respective side of the plan, the architect identifies the sides of the building by the points of a compass, such as NORTH ELEVATION, Fig. 24-20, and EAST ELEVATION, Fig. 24-21. The north direction symbol is usually given on the plot plan or the GROUND FLOOR PLAN. Elevations and sections contain only vertical dimensions arranged in chains. No horizontal dimensions should be placed on elevation views. Sections are usually taken through a window, as shown on Fig. 24-22(a).

Elevation views serve also to identify different materials, or elements, used on the outside of the building. These include *siding, stonework, brick roof shingles, metal flashing, glass* and *glass blocks, cement* and *concrete, downspouts* and *gutters*. In every case where a symbol would not be sufficient to identify the material or element, an abbreviation or word description should be used. Figures 24-20 and 24-21 show many instances of this type, such as *flash* and *counterflash,* and *T.C. Lining (terracotta flue lining)*. Heavy long-and-short-dash lines, indicating floor levels, should also be noted.

24.8 ARCHITECTURAL DETAILS. Next to the plan, the most important drawings in the architectural set are the *details*. Since the plans and elevations are drawn at a scale too small to distinguish the small parts and features, many of these must be shown on the details. The scale must be sufficiently large to identify all component parts. This scale may be any size from 1″ = 1′-0″ to 3″ = 1′-0″ to full size.

Detail drawings may be placed on sheets containing plans, elevations, and sections. In large sets of drawings they are placed on separate sheets. As shown on Figs. 24-19 and 24-22, each detail is identified by the same title on the small-scale drawing and under the detail drawing itself. This cross-identification of small-scale drawings and details is essential. It allows the checker, the specification writer, the estimator, and the contractor to understand the construction, the appearance of the structure, and the materials of which it is constructed.

The details drawn by the architect serve as a basis for the shop drawings prepared by the contractor and the subcontractors. These shop drawings show manufacturing details and fabrication and installation procedures. In this way the architect is able to control not only the exact execution of the design but also the quality of the construction of specific parts of the structure. Architectural practice follows the conventional or standard manufacturing detailing practices for many parts of the house. Such standard details can be readily found in any appropriate handbook.

Because each house or structure is based on an individual design, many special conditions must be covered by details adapted for a particular structure or condition. However, in general it will be found that these special details follow the arrangements and practices shown on standard details. For this reason a careful study of detailing methods and standard detailing practices will be very helpful.

24.9 PRINCIPAL DETAILS. Usually included in this category are all details that repeatedly form a component part of every house or structure. The most important ones are *footings* and *foundations, wall sections, floor* and *ceiling framing, doors, chimneys* and *heating tracts, stairs, roof framing* and *roofs.* Miscellaneous structural details include *lintels, beams, girders,* and *columns,* and details of *built-in installations.*

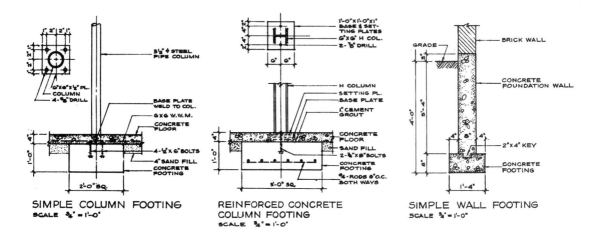

SIMPLE COLUMN FOOTING
SCALE ¾″ = 1′-0″

REINFORCED CONCRETE
COLUMN FOOTING
SCALE ¾″ = 1′-0″

SIMPLE WALL FOOTING
SCALE ¾″ = 1′-0″

Fig. 24-9. Footings.

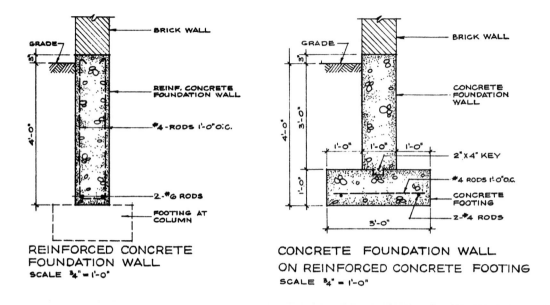

REINFORCED CONCRETE FOUNDATION WALL
SCALE ¾" = 1'-0"

CONCRETE FOUNDATION WALL ON REINFORCED CONCRETE FOOTING
SCALE ¾" = 1'-0"

Fig. 24-10. Foundation walls.

24.10 FOOTINGS AND FOUNDATIONS. The stability of the structure depends on the strength, size, and quality of the footings. Climate variations dictate that the footings be placed at a certain depth below the grade to insure their placement on a frost-free base. The size of the footings depends on the loads that they are to carry. These loads are calculated to include the complete weight of the structure, including the equipment and the loads that will be placed on the floors after the completion of the building. For flat roofs the possible *snow load* must be added. For steep roofs the possible *wind load* must be added. Municipal and county building codes usually require footings of a sufficient size for residential buildings to ensure the safety of the structure under normal loading conditions. Examples of footings of a residential type are shown in Fig. 24-9. Footings support the foundation walls. These in turn support the ground floor framing and the structure above. These walls are usually poured concrete or concrete block. They are waterproofed by different methods from the outside or inside, or both. The continuity of a foundation wall is essential to prevent seepage into the spaces that are placed below the grade. A typical detail of a foundation wall is shown in Fig. 24-10.

24.11 WALL SECTIONS. The most common types of residential construction are the *frame wall,* Fig. 24-11, *brick veneer wall, brick wall, brick* and *concrete block wall,* Fig. 24-12, and *concrete block wall.* Some wood framing details can be found in Fig. 24-15. Brick veneer is a frame structure faced with a 4" thick brick wall on the outside. Different brick and concrete block walls vary in the arrangement of masonry materials. They are frequently combined with stone facing.

24.12 FLOOR AND CEILING FRAMING. Different materials and their combinations create various types of floor and ceiling framing. The sizes of joists and structural members are usually specified in the building codes for normal residential conditions. Figures 24-13 and 24-14 show the most common types of such framing.

24.13 DOORS AND WINDOWS. If made of wood, doors and windows are classified as *mill work*. They are usually factory-made and delivered to the building site as complete units. Their details will vary from one manufacturer to another and can be found readily in manufacturers' catalogs. Standard terms have been established by usage. These include *double-hung window, casement sash, solid wood*

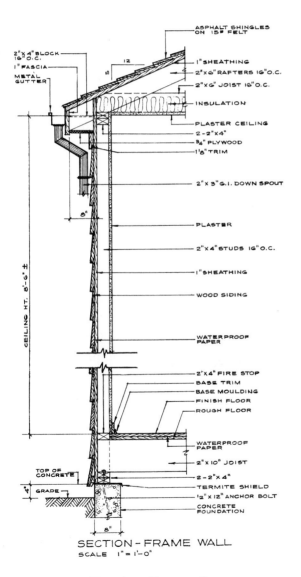

Fig. 24-11. Frame wall.

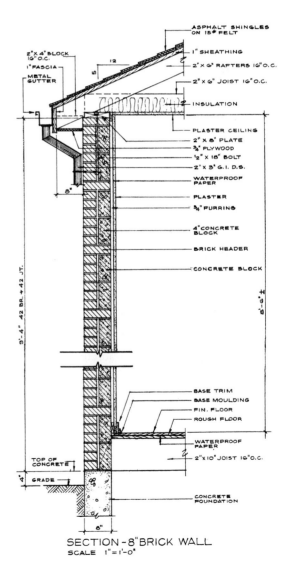

Fig. 24-12. Brick and concrete block wall.

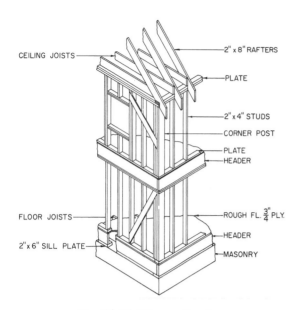

Fig. 24-13. Western framing.

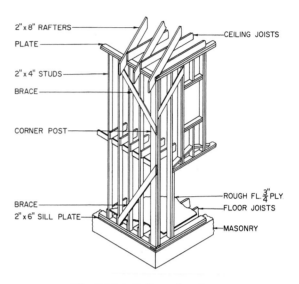

Fig. 24-14. Balloon framing.

door, hollow core door, and *solid core door.* An architectural drafter must be acquainted with these terms and also with the specific manner of their construction. All such details can be found in architectural handbooks and manufacturers' literature. Figures 24-16 and 24-22 show examples of this type of detailing.

24.14 STAIRS. Interior and exterior stairways are made of wood, steel, concrete, or combinations of these materials. Wood stair details are shown in Fig. 24-17.

24.15 MISCELLANEOUS STRUCTURAL DETAILS. A modern residential structure, with its large windows and open plan, requires somewhat more structural detailing than the conventional old-type brick residential building. In many cases it requires a steel skeleton and steel roof framing. Other structural details are details of lintels, girders, and beams. See Fig. 24-18.

24.16 CHIMNEYS AND HEATING DUCTS. Included in this category of details are chimneys, stacks for ventilation, fireplaces, and outdoor ovens and grills. The detailing of these features will be subject to various building code requirements. Typical details can be found in handbooks. Figure 24-22 shows a detail of a small fireplace. A chimney with three flues, one serving for the fireplace, the second for the central heating plant, and the third as a gas range vent, can be seen in Figs. 24-19 to 24-21.

24.17 BUILT-IN INSTALLATIONS. Almost every residential structure will contain some built-in installations, such as kitchen cabinets, library shelves, and access panels to mechanical equipment. These must be carefully detailed and coordinated with the dimensions and design shown on the plan and the elevation drawings.

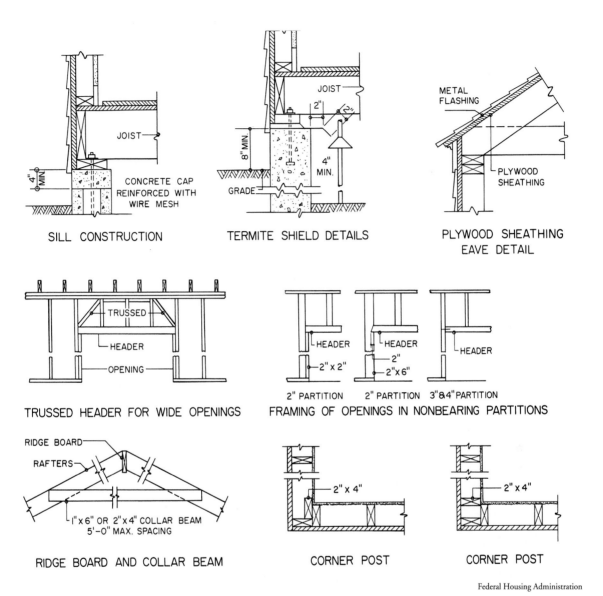

SILL CONSTRUCTION

TERMITE SHIELD DETAILS

PLYWOOD SHEATHING
EAVE DETAIL

TRUSSED HEADER FOR WIDE OPENINGS

FRAMING OF OPENINGS IN NONBEARING PARTITIONS

2" PARTITION   2" PARTITION   3"&4" PARTITION

RIDGE BOARD AND COLLAR BEAM

CORNER POST

CORNER POST

Federal Housing Administration

Fig. 24-15. Miscellaneous framing details.

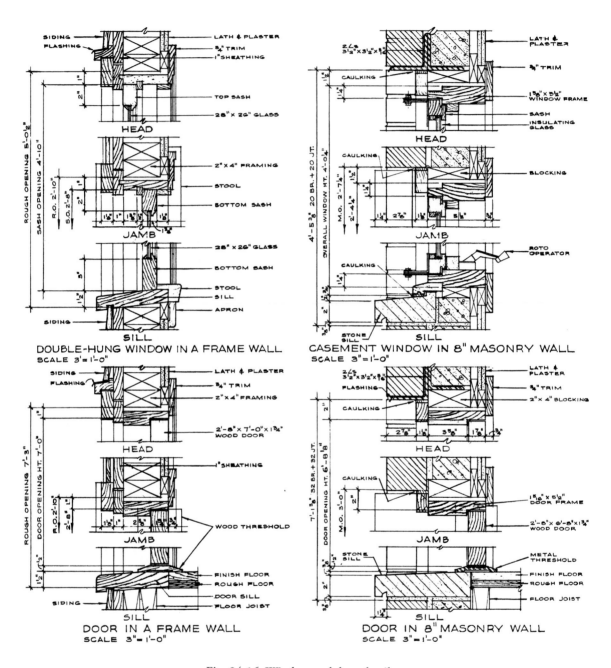

Fig. 24-16. Window and door details.

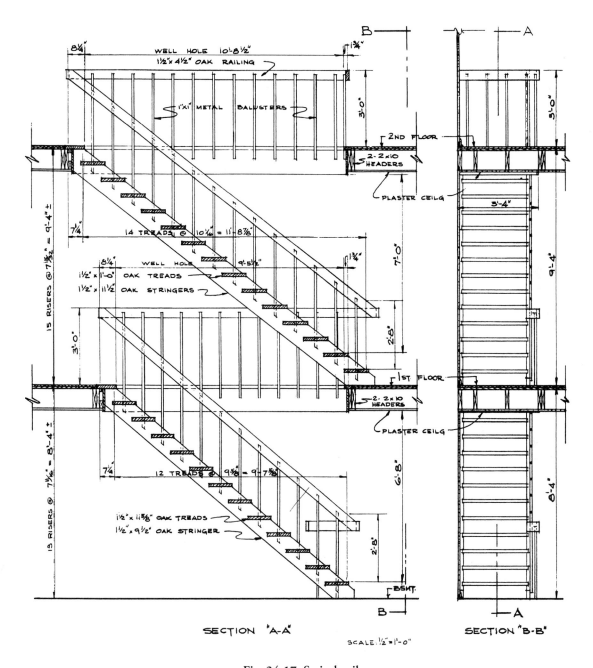

Fig. 24-17. Stair details.

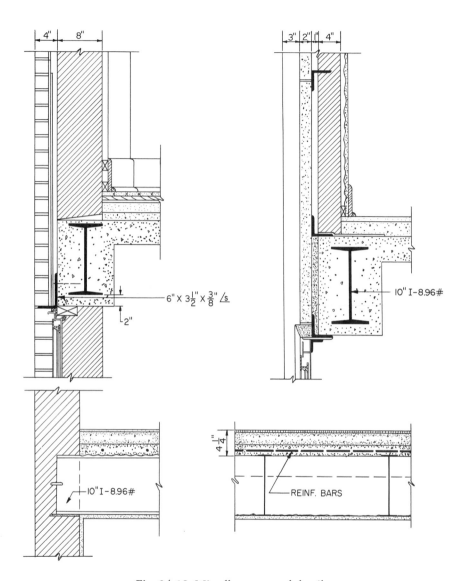

4" 8"

6" X 3½" X ⅜" ⁄s

2"

10" I-8.96#

10" I-8.96#

3" 2" 1" 4"

10" I-8.96#

4¼"

REINF. BARS

Fig. 24-18. Miscellaneous steel details.

24.18 WORKING DRAWINGS. Working drawings represent the final stage of the architect's work. They include the plans, elevations, sections, and details. Before these drawings are completed, many basic decisions will have been made both by the architect and by the owner. The preliminary sketches or drawings will thus have served as the basis for discussions about the size of the building and the location of partitions, stairways, and exits. Now the drafter is ready to proceed on the basis of these decisions and draw the final plans. These plans must be exact and comprehensive. They must give the contractor all the necessary information for the construction of the building.

The drawings in Figs. 24-19 to 24-23 represent such a set of working drawings. This

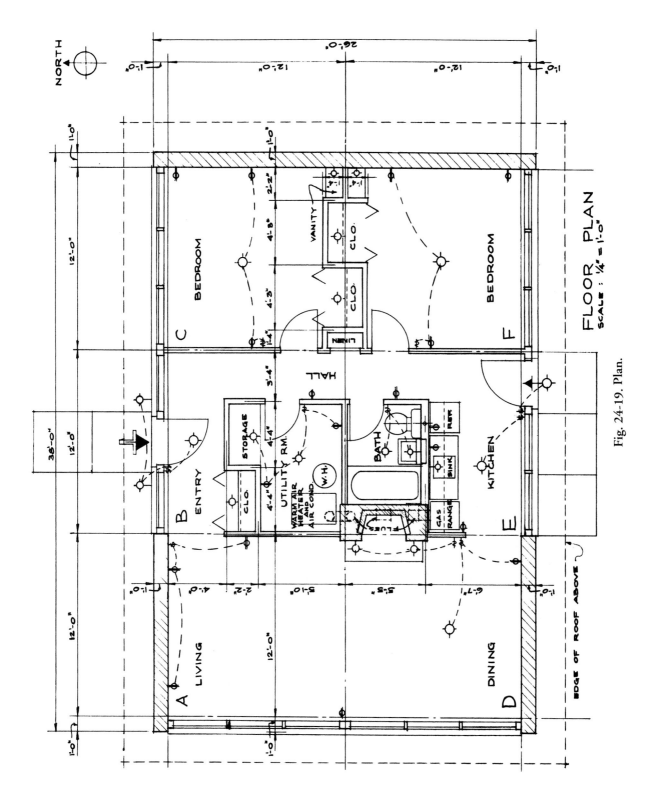

FLOOR PLAN
SCALE : 1/4" = 1'-0"

Fig. 24-19. Plan.

511

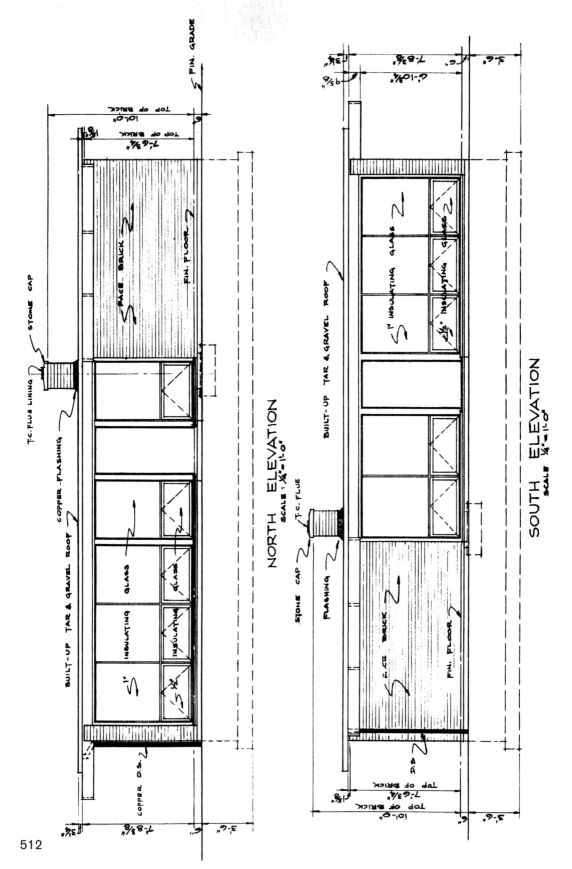

NORTH ELEVATION
SCALE: ¼"=1'-0"

SOUTH ELEVATION
SCALE ¼"=1'-0"

Fig. 24-20. Elevations.

set includes the plans, elevations, and sections, plus the important details drawn at a sufficiently large scale to show the construction clearly and completely. Also, the set may contain different *schedules,* such as a ventilation schedule, lintel schedule, and door and

window schedule. A schedule is a table or chart specifying in detail the types and sizes of a certain feature on the plans. For example, a door schedule will show in tabular form the type of door (flush panel, glass, metal, etc.), its size, and the location where it is to be

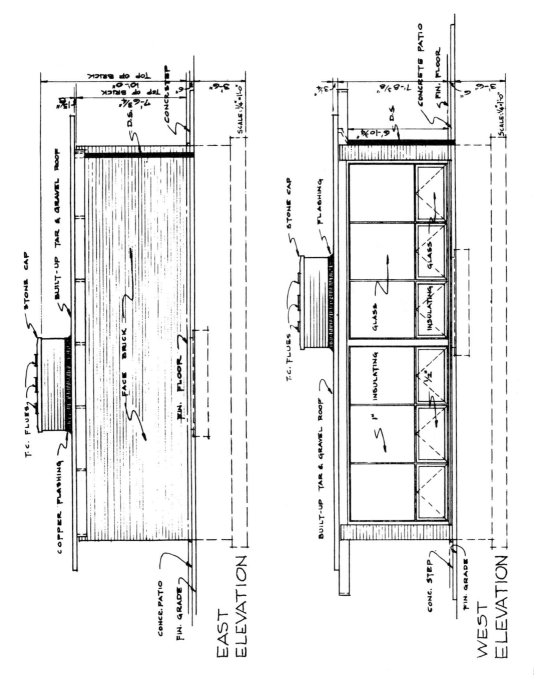

Fig. 24-21. Elevations.

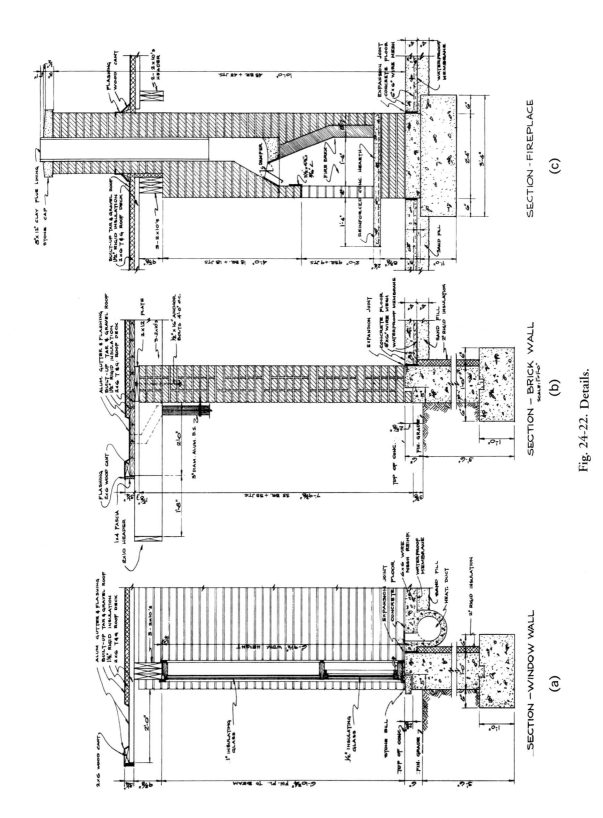

Fig. 24-22. Details.

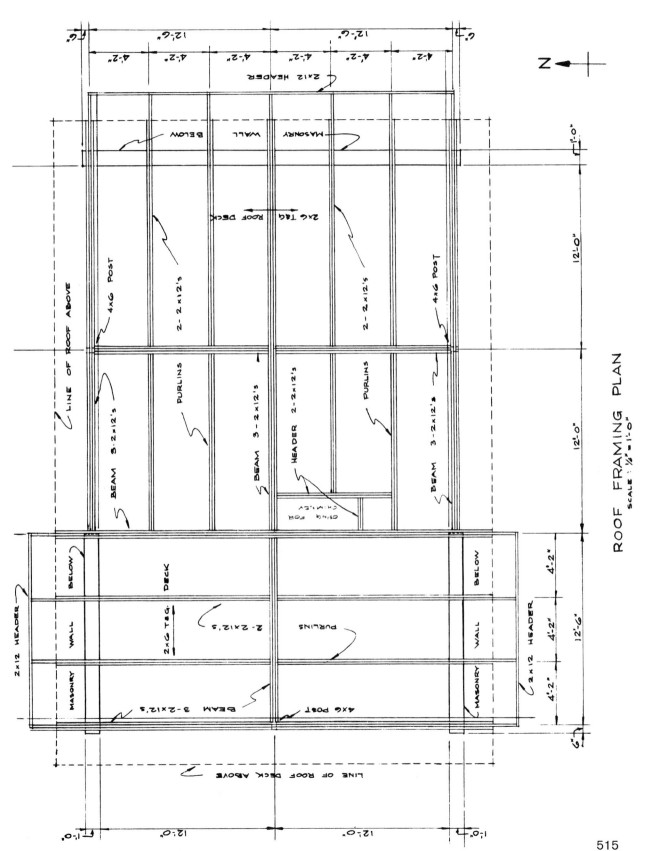

ROOF FRAMING PLAN
SCALE ½" = 1'-0"

Fig. 24-23. Framing plan.

used. A room finish schedule similarly will show in tabular form the finish materials for every room such as the tile floor, acoustical plaster ceiling, or plaster walls. This information is required for construction purposes or by city or state building codes.

Working drawings are started by drawing and completing the ground floor plan, usually at the scale of $\frac{1}{4}'' = 1'-0''$. From this plan the other floor plans, such as the basement floor plan and second floor plan, are traced in outline. They are modified with all necessary features for the respective floor. Thus, a drafter is able to coordinate the basement columns and girders with the load of the first floor partitions.

The elevations are usually drawn after the locations of all openings on the plan have been fixed and dimensioned. Ordinarily they are drawn at the scale of $\frac{1}{4}'' = 1'-0''$. They should indicate materials that are to be used on the outside of the building. These materials are sometimes indicated by symbols and sometimes by abbreviations. For examples, see Figs. 24-20 and 24-21. The elevations usually include the outline of the foundations in hidden lines. Thus the total height of the structure and the relationship of the floor level to the grade level can be determined from the elevations.

In a sense, working drawings represent an elaborate assembly of many parts and many materials. They should be studied from this point of view. They indicate how the different parts of the structure form the whole, how different materials join together, and how the different phases of construction should be coordinated. The student should note particularly the system by which the details, shown in Fig. 24-22, are drawn and related to the plans and elevations.

24.19 THE USE OF CAD IN ARCHITECTURAL DRAFTING.

Architectural CAD software is designed to speed the process of developing a completed plan. Such software has an extensive library of architectural symbols. These symbols are placed in the design. Fig. 24-24.

All walls are drawn at the desired thickness. Corners and intersections are cleaned. Doors and windows are automatically trimmed to the proper sizes. The CAD drafter has the choice of developing the design as a two-dimensional plan (not often done) or as a three-dimensional model. The three-dimensional model performs many functions as the walls are drawn and the symbols are inserted. Wall height, basements, foundation footings and walls, insulation, door and window frames, headers and trimmers, and an attached database of information are some of the items included in the wall development.

Roof styles are specified. The roof is generated as a three-dimensional model. Upon completion of the model, elevations are chosen for the required views. They are placed in the proper positions. Dimensions and notes are then added. The database is used to generate schedules and bills of materials. The three-dimensional model may be used to develop a walk-through of the plan as a means of evaluating the design, Fig. 24-25.

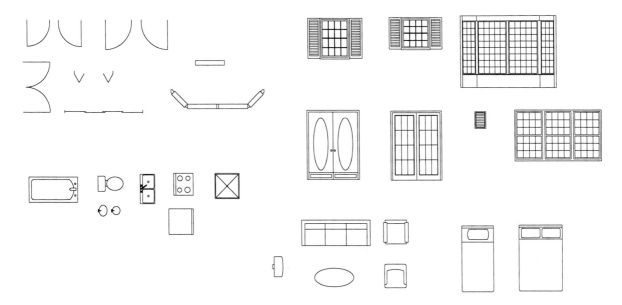

Fig. 24-24. The symbols that are available in architectural CAD software save the drafter time and insure consistency.

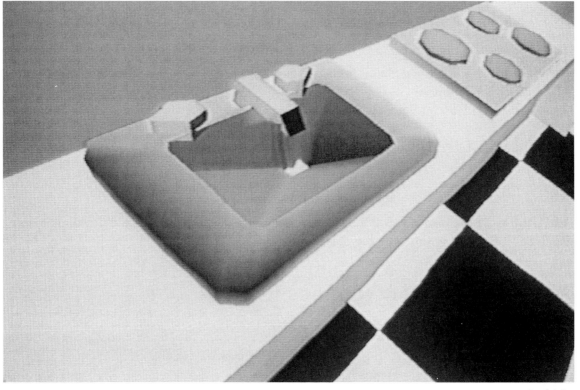

Autodesk, Inc.

Fig. 24-25. Images created on the computer make it possible to take a virtual walk through a building as it is being designed.

PROBLEMS

PROBLEM 24-A. Using architectural letters as shown in Fig. 24-3 (vertical or inclined, as assigned), letter five lines of lettering $\frac{1}{8}''$ high, making lines 6″ long with approximately $\frac{1}{16}''$ spaces between. For text, use the beginning paragraph of this chapter.

PROBLEM 24-B. Using architectural letters as shown in Fig. 24-3 (vertical or inclined, as assigned), letter three lines of lettering $\frac{1}{4}''$ high. Repeat the alphabet. Make the lines 6″ long and spaced approximately $\frac{1}{8}''$ apart.

PROBLEM 24-C. Using architectural title letters shown in Fig. 24-5, letter the following title in letters $\frac{1}{2}''$ high: Architectural Drafting.

PROBLEM 24-D. Using Fig. 24-8 as an example, draw a plot plan of the building where you reside. Scale $\frac{1}{32}''$ = 1′-0″.

PROBLEM 24-E. Study Fig. 24-7. Then identify and locate the following elements of the house shown on the working drawings in Figs. 24-19 to 24-23: flue lining, flashing, insulation, columns, subfloor, fireplace, joists, gravel fill, foundation wall, hearth, and heating plant. Sketch each detail freehand approximately to the scale shown on the working drawing. (Do not trace from the book.)

PROBLEM 24-F. Draw a detail of foundations 1′-0″ thick and 2′-4″ wide on which stands a 1′-4″ concrete foundation wall 4′-6″ high. The top of the wall is to be 8″ above the grade. Scale: 1″ = 1′-0″. (Refer to Figs. 24-9 and 24-10.)

PROBLEM 24-G. Figure 24-12 shows a section through a brick and concrete block wall. Using the data given in Fig. 24-11, convert the section into a brick veneer wall 10″ thick. Scale: $1\frac{1}{2}''$ = 1′-0″.

PROBLEM 24-H. Convert section through the brick wall shown in Fig. 24-22 by including a casement-sash detail as shown in Fig. 24-16. Scale: $1\frac{1}{2}''$ = 1′-0″.

PROBLEM 24-I. The plan of the house shown in Fig. 24-19 consists of six areas between column center lines indicated by the letters A to F. Redesign the plan in a manner such that the "Kitchen" and "Bath," presently located in area E, are relocated properly in area B. The "Entry" and "Utility Room," presently located in area B, should be relocated properly in area E. If necessary, rearrange windows and doors to suit the new plan.

PROBLEM 24-J. Add a 10'-0" × 20'-0" carport adjoining areas C and F on the new plan of Problem 24-I. Scale: $\frac{1}{4}$" = 1'-0".

PROBLEM 24-K. Draw front and rear elevations for completed Problem 24-I, using wall sections of completed Problem 24-H in areas A and D. Scale: $\frac{1}{4}$" = 1'-0".

PROBLEM 24-L. Draw cross-section and framing details of the carport required in Problem 24-J. Scale: $\frac{1}{4}$" = 1'-0" for section and 3" = 1'-0" for details.

PROBLEM 24-M. Draw elevation of the fireplace shown on plan in Fig. 24-J. Using your own design, include an open built-in bookcase on either side of the hearth. Scale: 1" = 1'-0".

PROBLEM 24-N. Draw detail elevation of the kitchen wall indicated in plan, Fig. 24-19, showing all appliances and cabinets. Scale: 1" = 1'-0".

PROBLEM 24-O. Draw plan, two elevations, and a section of a single-car garage (12'-0" × 20'-0") of frame construction.

CHAPTER **25**

descriptive
geometry

slope

edge view

piercing point

O B J E C T I V E

After studying this chapter, you should be able to:

- Describe the ways in which descriptive geometry relates to technical drawing.

CHAPTER 25
Descriptive Geometry

Drafters, designers, and engineers use descriptive geometry to determine how points, lines, and planes relate to each other in space. With descriptive geometry techniques, they can solve problems. For example, the civil engineer can determine the percent grade of a road, as described on pages 522–524.

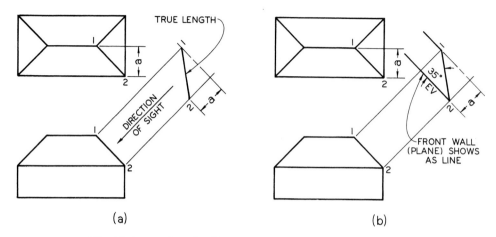

Fig. 25-1. True length of line and angle between line and plane.

25.1 DEFINITION.

A consolidation of several dictionary definitions helps give us a definition of descriptive geometry: *"Descriptive Geometry* is that branch of geometry that provides a graphical solution of a three-dimensional problem by means of projections upon mutually perpendicular planes." In Chapter 7 (and succeeding chapters) we followed that definition. The "three-dimensional problem" we solved was that of representing an object, such as the house shown in Fig. 7-2, by *views* ("projections") on "mutually perpendicular planes." These planes were the front, top, and right-side planes. Actually, people in the technical drawing field have come to mean problem-solving rather than representation when they use the term "descriptive geometry." You have seen in Chapter 13 how auxiliary views may be used to find certain angles, Sec. 13.9, or the true length of an oblique line, Sec. 13.11. These are descriptive geometry solutions.

25.2 ANGLE BETWEEN LINE AND PLANE.

At Fig. 25-1(a) we see a repetition of Fig. 13-14. By assuming a line of sight perpendicular to any view (in this case, the front view), we obtain an auxiliary view showing the true length of hip rafter 1-2.

At (b) a portion of the front wall has been added in the auxiliary view. Since this *frontal plane* appears as a line ("EV," or edge view), the true angle of 35° between line 1-2 and the plane appears. It may be dimensioned as shown.

Of course, the angle between a hip rafter and a wall of a house may not be very useful information for a carpenter. However, the general problem of finding the angle between a line and a plane is important in other fields. For example, Fig. 25-2(a) shows the top (map) and front elevation views of a portion of a roadway. For the purpose at hand, we need be concerned only with the center line 1-2 of the roadway, as at (b).

If we introduce a horizontal reference plane (horizontal in space) through point 1, it appears as a horizontal line X-X in the front view. Now, the true angle between a line and horizontal is sometimes called the *slope* of the line. To find the slope of line 1-2 in Fig. 25-2 (b) we must construct an auxiliary view show-

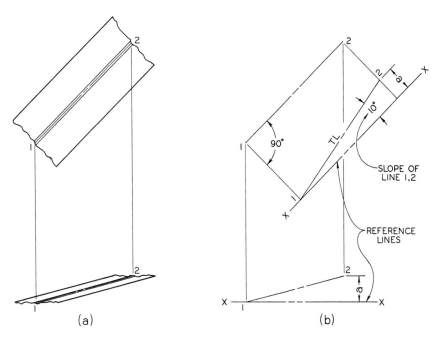

(a)   (b)

Fig. 25-2. Slope of a line.

ing 1-2 in true length (TL) and the horizontal plane in edge view (EV). To do this we simply use our horizontal plane as our reference plane X-X and project at right angles to the top view. Thus X-X appears perpendicular to the projection lines (and parallel to the top view). As in Fig. 25-1(b), we now have a line true length (1-2) and a plane in edge view (X-X). The angle of 10° between them is thus true size and is the slope of line 1-2.

The civil engineer usually expresses map directions as compass bearings, Sec. 22.1, and inclination with horizontal as *percent grade* rather than slope. Thus the previous center line 1-2 would be described, Fig. 25-3, as having a bearing of N 45° E and an upgrade from point 1 of 18 percent. Note the way the grade is measured. Any convenient length (100 units here) is set off along the horizontal plane from point 1. The corresponding distance to line 1-2 at right angles to the horizontal plane, 18 units, is measured. Since the 18 units are measured perpendicular to the horizontal plane, it is vertical *in space*

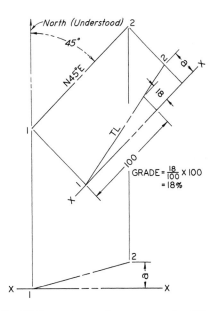

Fig. 25-3. Bearing and percent grade of line.

523

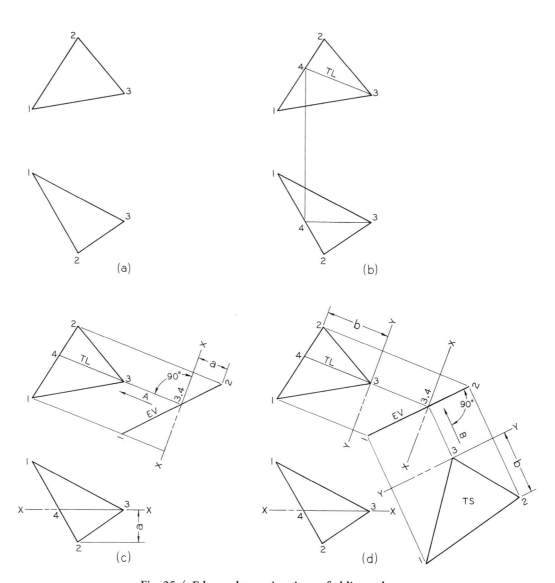

Fig. 25-4. Edge and true-size views of oblique plane.

(although not vertical on the paper). Thus the definition of *grade* is:

$$\text{Percent grade} = \frac{\text{Vertical distance}}{\text{Horizontal distance}} \times 100$$

The two distances are measured between any convenient two points on the line. (In trigonometry, this proportion would be the tangent of the slope angle. Thus tan 10° = .18. From a table of natural tangents, tan 10° = .1763.)

25.3 EDGE AND TRUE SIZE VIEWS OF OBLIQUE PLANE. In Sec. 8.13 we noted how a plane surface appears when viewed from various directions. In Fig. 8-14(a) and (b) we observed that under some circumstances a plane appears as a line — in *edge view* (EV). In Fig. 13-12 we observed further that when two intersecting surfaces appear as lines we can measure the true angle between them. This situation results when we look parallel to their line of intersection, obtaining a "point view" of this line. Finally, in Fig. 13-15 we discovered that we could obtain an edge view of a plane surface by looking parallel to a true-length line in the surface.

Now, how can we apply what we have learned to obtain an edge view of plane 1-2-3 of Fig. 25-4(a)? First, we observe that none of the three given edges is true length. We could construct an auxiliary view showing any one of these lines in true length, but it is simpler to add a line, as at (b), that will be true length in one of the given views. In this case we have chosen to add a horizontal line through point 3 in the front view. This line intersects line 1-2 at point 4. Point 4 is projected to the top view of 1-2. Line 3-4 is true length in the top view as indicated.

As shown at (c), we now select a line of sight A parallel to 3-4. We construct the corresponding auxiliary view. The result is a point view of line 3-4 and, accordingly, the desired edge view of plane 1-2-3.

In Sec. 13.12 we obtained a true-size (TS) view of a plane surface by looking perpendicular to its edge view and constructing the corresponding secondary auxiliary view. In our present case we can do the same thing, Fig. 25-4(d). True-size views are, of course, very useful because all lines and angles of the true-size surface may be measured. In fact, any needed plane-geometry constructions may be performed in such a view.

25.4 PIERCING POINT. We may use the foregoing edge-view construction in another way. We can use it to find the point of intersection (*piercing point*) of a line and an oblique plane. Suppose, in Fig. 25-5(a), we are given the front and top views of plane 1-2-3 and line 4-5. Note that again we do not have a true-

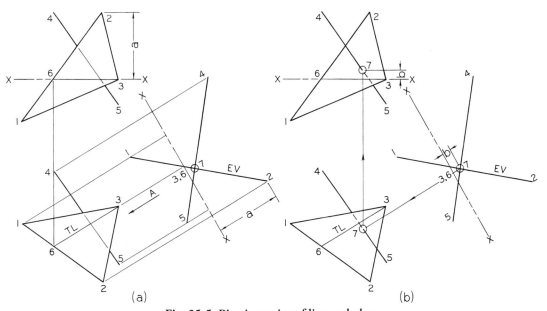

Fig. 25-5. Piercing point of line and plane.

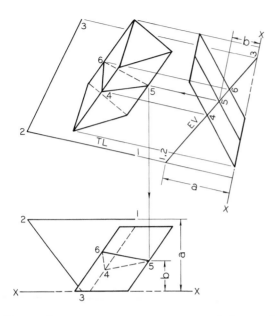

Fig. 25-6. Piercing points—intersection of plane and prism.

length line. This time we have chosen to add line 3-6 in the top view, which, when projected to the front view, appears true length as indicated. This gives us the direction of line

of sight A, and we obtain the edge view of plane 1-2-3. Line 4-5 intersects this edge view at point 7, the desired piercing point. If necessary, we can project point 7 back to the given front and top views as at (b). Note the use of measurement b to check the accuracy of location of point 7 in the top view.

An application of the piercing-point construction is shown in Fig. 25-6. When the edge view of plane 1-2-3 is obtained in the auxiliary view, the piercing points 4, 5, and 6 of the edges of the prism with plane 1-2-3 are evident. Triangle 4-5-6 is the intersection of plane 1-2-3 and the prism.

25.5 COMPUTER GRAPHICS. Descriptive geometry involves the analysis of objects represented by their basic elements – points, lines, and planes. Thus the experienced drafter can use a CAD program to solve problems involving visibility, edge views or true size of planes, piercing points, and intersections.

P R O B L E M S

Problem layouts are given in Figs. 25-7 to 25-9, each designed to occupy one-fourth or one-half of Layout B in the Appendix. Thus, combinations of two, three, or four problems may be assigned for one sheet. For practice, cross-section paper may be used. Rough solutions may be sketched freehand. For accurate answers, of course, the problems must be laid out and solved carefully with instruments. Spacing or layout dimensions should be omitted. Dimensions for given or required measured quantities should be shown. The symbols TL, EV, and TS should also be used where they will help clarify the method of solution.

If assigned by the instructor, construct drawings using decimal-inch or metric scales by converting given fractional dimensions to decimal-inch or metric equivalents. Refer to Table 20 in the Appendix.

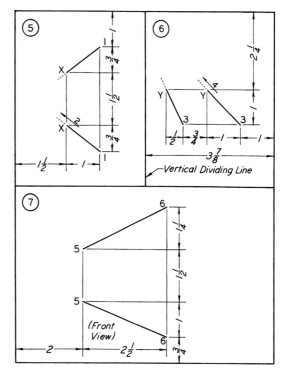

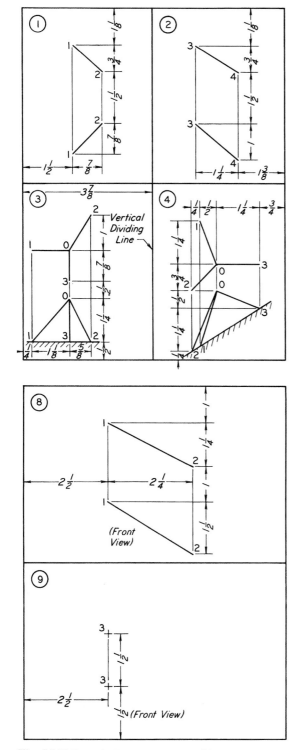

Problem Statements

1. Find and measure the true length of line 1–2.

2. Find and measure the true length of line 3–4.

3. Find and dimension the true lengths of tripod legs 0–1, 0–2, and 0–3. Scale: $1'' = 1'-0$.

4. Find and dimension the true lengths of tripod legs 0–1, 0–2, and 0–3. Scale: $\frac{3}{4}'' = 1'-0$.

5. Line 1–2 has a true length of 2″. Complete the given views.

6. Line 3–4 has a true length of 9′–6. Complete the given views. Scale: $\frac{1}{4}'' = 1'-0$.

7. Find and measure the true length and slope of line 5–6. Scale: $\frac{1}{8}'' = 1'-0$.

8. Find and measure the angle between line 1–2 and a frontal plane. What are the bearing and percent grade of line 1–2?

9. If a line 3–4 has a bearing of N 60° E, a downgrade from point 3 of 30%, and a true length of 3″, complete the front and top views of 3–4.

Fig. 25-7. Descriptive geometry problems. Use Layout B in the Appendix. Omit spacing dimensions, but show dimensions for required numerical answers. Also show symbols TL, EV, and TS where appropriate for clarifying your solutions.

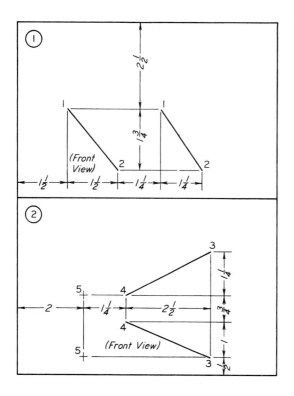

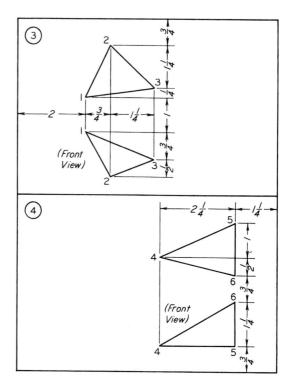

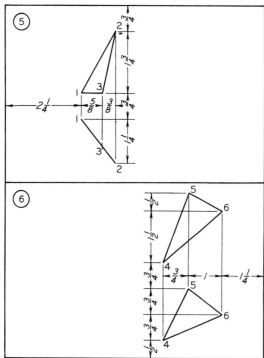

Problem Statements

1. Find the angle between line 1–2 and a profile plane. What is the bearing of line 1–2?

2. Line 3–4 represents a tunnel. Another tunnel is to be constructed starting at point 5 and meeting tunnel 3–4 at a point, 6, which is to be 600′ along 3–4 from point 3. Find the bearing, length, and grade of tunnel 5–6. Scale: $\frac{1}{4}'' = 100'$ (or $1'' = 400'$).

3. Add a horizontal line to plane 1–2–3 through point 3. Find a point view of this horizontal line and the accompanying edge view of plane 1–2–3. This edge view makes it possible to measure the true angle between plane 1–2–3 and a horizontal plane. If assigned, dimension this angle.

4. Find the true angle between plane 4–5–6 and a horizontal plane. (See the statement for problem 3.)

5. Construct a true-size view of plane 1–2–3. If assigned, calculate the area of triangle 1–2–3. Scale: $\frac{1}{4}'' = 10'$ ($1'' = 40'$).

6. Construct a true-size view of triangle 4–5–6. Measure and dimension the true angle at 4.

Fig. 25-8. Descriptive geometry problems. Use Layout B in the Appendix. Omit spacing dimensions, but show dimensions for required numerical answers. Also show symbols TL, EV, and TS where appropriate for clarifying your solutions.

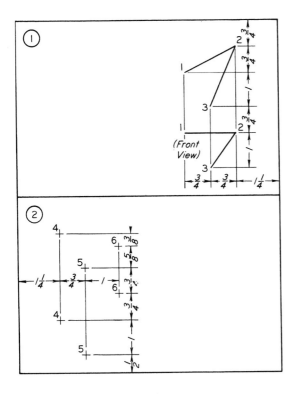

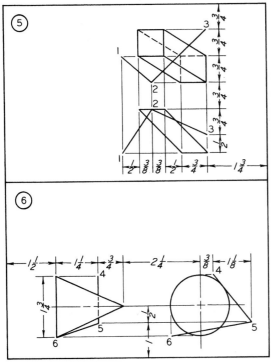

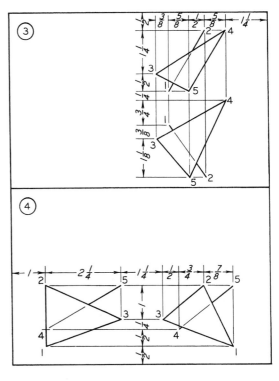

Problem Statements

1. Through the use of a true-size view of plane 1–2–3, find the bisector of angle 1–2–3 and project it back to the given front and top views. If assigned, find and measure the bearing and grade of the bisector.

2. Points 4, 5, and 6 determine a plane. Also three points determine a circle (see Fig. 5–14). Construct a true-size view of plane 4–5–6. Then find the center of the circle through the three points and project this center back to the given views.

3. Find the piercing point of line 1–2 in plane 3–4–5. If assigned, regard triangle 3–4–5 as opaque and show line 1–2 in proper visibility.

4. Find the piercing point of line 4–5 in plane 1–2–3. If assigned, regard triangle 1–2–3 as opaque, and show line 4–5 in proper visibility.

5. Obtain an edge view of plane 1–2–3 and determine in the given views the line of intersection of plane 1–2–3 and the prism.

6. Obtain the line of intersection of plane 4–5–6 and the right-circular cone. If assigned, develop a pattern of the conical surface, including the line of intersection. (See Sec. 18.17.)

Fig. 25-9. Descriptive geometry problems. Use Layout B in the Appendix. Omit spacing dimensions, but show dimensions for required numerical answers. Also show symbols TL, EV, and TS where appropriate for clarifying your solutions.

529

APPENDIX

TABLE OF CONTENTS

1 ABBREVIATIONS FOR USE ON DRAWINGS AND IN TEXT – AMERICAN NATIONAL STANDARD [SELECTED FROM ANSI Y1.1]

A

absolute	ABS
accelerate	ACCEL
accessory	ACCESS.
account	ACCT
accumulate	ACCUM
actual	ACT.
adapter	ADPT
addendum	ADD.
addition	ADD.
adjust	ADJ
advance	ADV
after	AFT.
aggregate	AGGR
air condition	AIR COND
airplane	APL
allowance	ALLOW
alloy	ALY
alteration	ALT
alternate	ALT
alternating current	AC
altitude	ALT
aluminum	AL
American National Standard	AMER NATL STD
Amerian wire gage	AWG
amount	AMT
ampere	AMP
amplifier	AMPL
anneal	ANL
antenna	ANT.
apartment	APT.
apparatus	APP
appendix	APPX
approved	APPD
approximate	APPROX
arc weld	ARC/W
area	A
armature	ARM.
armor plate	ARM-PL
army navy	AN
arrange	ARR.
artificial	ART.
asbestos	ASB
asphalt	ASPH
assemble	ASSEM
assembly	ASSY

assistant	ASST
associate	ASSOC
association	ASSn
atomic	AT
audible	AUD
audio frequency	AF
authorized	AUTH
automatic	AUTO
auto-transformer	AUTO TR
auxiliary	AUX
avenue	AVE
average	AVG
aviation	AVI
azimuth	AZ

B

Babbitt	BAB
back feed	BF
back pressure	BP
back to back	B to B
backface	BF
balance	BAL
ball bearing	BB
barometer	BAR
base line	BL
base plate	BP
bearing	BRG
bench mark	BM
bending moment	M
bent	BT
bessemer	BESS
between	BET.
between centers	BC
between perpendiculars	BP
bevel	BEV
bill of material	B/M
Birmingham wire gage	BWG
blank	BLK
block	BLK
blueprint	BP
board	BD
boiler	BLR
boiler feed	BF
boiler horsepower	BHP
boiling point	BP
bolt circle	BC
both faces	BF

both sides	BS
both ways	BW
bottom	BOT
bottom chord	BC
bottom face	BF
bracket	BRKT
brake	BK
brake horsepower	BHP
brass	BRS
brazing	BRZG
break	BRK
Brinell hardness	BH
British Standard	BR STD
British thermal unit	BTU
broach	BRO
bronze	BRZ
Brown & Sharpe (wire gage, same as AWG)	B&S
building	BLDG
bulkhead	BHD
burnish	BNH
bushing	BUSH.
button	BUT.

C

cabinet	CAB.
calculate	CALC
calibrate	CAL
cap screw	CAP SCR
capacity	CAP
carburetor	CARB
carburize	CARB
carriage	CRG
case harden	CH
cast iron	CI
cast steel	CS
casting	CSTG
castle nut	CAS NUT
catalogue	CAT.
cement	CEM
center	CTR
center line	CL
center of gravity	CG
center of pressure	CP
center to center	C to C
centering	CTR
chamfer	CHAM

change	CHG
channel	CHAN
check	CHK
check valve	CV
chord	CHD
circle	CIR
circular	CIR
circular pitch	CP
cirumference	CIRC
clear	CLR
clearance	CL
clockwise	CW
coated	CTD
cold drawn	CD
cold-drawn steel	CDS
cold finish	CF
cold punched	CP
cold rolled	CR
cold-rolled steel	CRS
combination	COMB.
combustion	COMB
commercial	COML
company	CO
complete	COMPL
compress	COMP
concentric	CONC
concrete	CONC
condition	COND
connect	CONN
constant	CONST
construction	CONST
contact	CONT
continue	CONT
copper	COP.
corner	COR
corporation	CORP
correct	CORR
corrugate	CORR
cotter	COT
counter	CTR
counterbore	CBORE
counter clockwise	CCW
counterdrill	CDRILL
counterpunch	CPUNCH
countersink	CSK
coupling	CPLG
cover	COV
cross section	XSECT
cubic	CU
cubic foot	CU FT
cubic inch	CU IN.
current	CUR
customer	CUST
cyanide	CYN

D

decimal	DEC
dedendum	DED
deflect	DEFL
degree	(°) DEG
density	D
department	DEPT
design	DSGN
detail	DET
develop	DEV
diagonal	DIAG
diagram	DIAG
diameter	DIA
diametral pitch	DP
dimension	DIM
discharge	DISCH
distance	DIST
division	DIV
double	DBL
dovetail	DVTL
dowel	DWL
down	DN
dozen	DOZ
drafting	DFTG
drawing	DWG
drill or drill rod	DR
drive	DR
drive fit	DF
drop	D
drop forge	DF
duplicate	DUP

E

each	EA
east	E
eccentric	ECC
effective	EFF
elbow	ELL
electric	ELEC
elementary	ELEM
elevate	ELEV
elevation	EL
engine	ENG
engineer	ENGR
engineering	ENGRG
entrance	ENT
equal	EQ
equation	EQ
equipment	EQUIP
equivalent	EQUIV
estimate	EST
exchange	EXCH
exhaust	EXH
existing	EXIST.
exterior	EXT
extra heavy	X HVY
extra strong	X STR
extrude	EXTR

F

fabricate	FAB
face to face	F to F
Fahrenheit	F
far side	FS
federal	FED
feed	FD
feet	(') FT
figure	FIG
fillet	FIL
fillister	FIL
finish	FIN
finish all over	FAO
flange	FLG
flat	F
flat head	FH
floor	FL
fluid	FL
focus	FOC
foot	(') FT
force	F
forged steel	FST
forging	FORG
forward	FWD
foundry	FDRY
frequency	FREQ
front	FR
furnish	FURN

G

gage or gauge	GA
gallon	GAL
galvanize	GALV
galvanized iron	GI
galvanized steel	GS
gasket	GSKT
general	GEN
glass	GL
government	GOVT
governor	GOV
grade	GR
graduation	GRAD

graphite	GPH	Keyway	KWY	natural	NAT
grind	GRD			near face	NF
groove	GRV	**L**		near side	NS
ground	GRD	laboratory	LAB	negative	NEG
		laminate	LAM	neutral	NEUT
H		lateral	LAT	nominal	NOM
half-round	$\frac{1}{2}$ RD	left	L	normal	NOR
handle	HDL	left hand	LH	north	N
hanger	HGR	length	LG	not to scale	NTS
hard	H	length over all	LOA	number	NO.
harden	HDN	letter	LTR		
hardware	HDW	light	LT	**O**	
head	HD	line	L	obsolete	OBS
headless	HDLS	locate	LOC	octagon	OCT
heat	HT	logarithm	LOG.	office	OFF.
heat-treat	HT TR	long	LG	on center	OC
heavy	HVY	lubricate	LUB	opposite	OPP
hexagon	HEX	lumber	LBR	optical	OPT
high-pressure	HP			original	ORIG
high-speed	HS	**M**		outlet	OUT.
horizontal	HOR	machine	MACH	outside diameter	OD
horsepower	HP	machine steel	MS	outside face	OF
hot rolled	HR	maintenance	MAINT	outside radius	OR
hot-rolled steel	HRS	malleable	MALL	overall	OA
hour	HR	malleable iron	MI		
housing	HSG	manual	MAN.	**P**	
hydraulic	HYD	manufacture	MFR	pack	PK
		manufactured	MFD	packing	PKG
I		manufacturing	MFG	page	P
illustrate	ILLUS	material	MATL	paragraph	PAR.
inboard	INBD	maximum	MAX	part	PT
inch	(") IN.	mechanical	MECH	patent	PAT.
inches per second	IPS	mechanism	MECH	pattern	PATT
inclosure	INCL	median	MED	permanent	PERM
include	INCL	metal	MET.	perpendicular	PERP
inside diameter	ID	meter	M	piece	PC
instrument	INST	miles	MI	piece mark	PC MK
interior	INT	miles per hour	MPH	pint	PT
internal	INT	millimeter	MM	pitch	P
intersect	INT	minimum	MIN	pitch circle	PC
iron	I	minute	(') MIN	pitch diameter	PD
irregular	IREG	miscellaneous	MISC	plastic	PLSTC
		month	MO	plate	PL
J		Morse taper	MORT	plumbing	PLMB
joint	JT	motor	MOT	point	PT
joint army-navy	JAN	mounted	MTD	point of curve	PC
journal	JNL	mounting	MTG	point of intersection	PI
junction	JCT	multiple	MULT	point of tangent	PT
		music wire gage	MWG	polish	POL
K				position	POS
key	K	**N**		potential	POT.
keyseat	KST	national	NATL	pound	LB

pounds per square inch	PSI
power	PWR
prefabricated	PREFAB
preferred	PFD
prepare	PREP
pressure	PRESS.
process	PROC
production	PROD
profile	PF
propeller	PROP
publication	PUB
push button	PB

Q

quadrant	QUAD
quality	QUAL
quarter	QTR

R

radial	RAD
radius	R
railroad	RR
ream	RM
received	RECD
record	REC
rectangle	RECT
reduce	RED.
reference line	REF L
reinforce	REINF
release	REL
relief	REL
remove	REM
require	REQ
required	REQD
return	RET.
reverse	REV
revolution	REV
revolutions per minute	RPM
right	R
right hand	RH
rivet	RIV
Rockwell hardness	RH
roller bearing	RB
room	RM
root diameter	RD
root mean square	RMS
rough	RGH
round	RD

S

schedule	SCH
schematic	SCHEM
scleroscope hardness	SH
screw	SCR
second	SEC
section	SECT
semi-steel	SS
separate	SEP
set screw	SS
shaft	SFT
sheet	SH
shoulder	SHLD
side	S
single	S
sketch	SK
sleeve	SLV
slide	SL
slotted	SLOT.
small	SM
socket	SOC
space	SP
special	SPL
specific	SP
spot faced	SF
spring	SPG
square	SQ
standard	STD
station	STA
stationary	STA
steel	STL
stock	STK
straight	STR
street	ST
structural	STR
substitute	SUB
summary	SUM.
support	SUP.
surface	SUR
symbol	SYM
system	SYS

T

tangent	TAN.
taper	TPR
technical	TECH
template	TEMP
tension	TENS.

terminal	TERM.
thick	THK
thousand	M
thread	THD
threads per inch	TPI
through	THRU
time	T
tolerance	TOL
tongue & groove	T & G
tool steel	TS
tooth	T
total	TOT
transfer	TRANS
typical	TYP

U

ultimate	ULT
unit	U
universal	UNIV

V

vacuum	VAC
valve	V
variable	VAR
versus	VS
vertical	VERT
volt	V
volume	VOL

W

wall	W
washer	WASH.
watt	W
week	WK
weight	WT
west	W
width	W
wood	WD
Woodruff	WDF
working point	WP
working pressure	WP
wrought	WRT
wrought iron	WI

Y

yard	YD
year	YR

2 SCREW THREADS, AMERICAN NATIONAL, UNIFIED, AND METRIC

AMERICAN NATIONAL STANDARD UNIFIED AND AMERICAN NATIONAL SCREW THREADS[a]

Nominal Diameter	Coarse[b] NC UNC		Fine[b] NF UNF		Extra Fine[c] NEF UNEF		Nominal Diameter	Coarse[b] NC UNC		Fine[b] NF UNF		Extra Fine[c] NEF UNEF	
	Thds. per Inch	Tap Drill[d]	Thds. per Inch	Tap Drill[d]	Thds. per Inch	Tap Drill[d]		Thds. per Inch	Tap Drill[d]	Thds. per Inch	Tap Drill[d]	Thds. per Inch	Tap Drill[d]
0 (.060)			80	3/64			1	8	7/8	12	59/64	20	61/64
1 (.073)	64	No. 53	72	No. 53	....	....	1 1/16	....	....	....	....	18	1
2 (.086)	56	No. 50	64	No. 50	....	....	1 1/8	7	63/64	12	1 3/64	18	1 5/64
3 (.099)	48	No. 47	56	No. 45	....	....	1 3/16	....	....	....	....	18	1 9/64
4 (.112)	40	No. 43	48	No. 42	....	....	1 1/4	7	1 7/64	12	1 11/64	18	1 3/16
5 (.125)	40	No. 38	44	No. 37	....	....	1 5/16	....	....	....	....	18	1 17/64
6 (.138)	32	No. 36	40	No. 33	....	....	1 3/8	6	1 7/32	12	1 19/64	18	1 5/16
8 (.164)	32	No. 29	36	No. 29	....	....	1 7/16	....	....	....	....	18	1 3/8
10 (.190)	24	No. 25	32	No. 21	....	....	1 1/2	6	1 11/32	12	1 27/64	18	1 7/16
12 (.216)	24	No. 16	28	No. 14	32	No. 13	1 9/16	....	....	....	....	18	1 1/2
1/4	20	No. 7	28	No. 3	32	7/32	1 5/8	....	....	....	....	18	1 9/16
5/16	18	F	24	I	32	9/32	1 11/16	....	....	....	....	18	1 5/8
3/8	16	5/16	24	Q	32	11/32	1 3/4	5	1 9/16	....	....	....	....
7/16	14	U	20	25/64	28	13/32	2	4 1/2	1 25/32	....	....	....	....
1/2	13	27/64	20	29/64	28	15/32	2 1/4	4 1/2	2 1/32	....	....	....	....
9/16	12	31/64	18	33/64	24	33/64	2 1/2	4	2 1/4	....	....	....	....
5/8	11	17/32	18	37/64	24	37/64	2 3/4	4	2 1/2	....	....	....	....
11/16	....	....	....	....	24	41/64	3	4	2 3/4	....	....	....	....
3/4	10	21/32	16	11/16	20	45/64	3 1/4	4	....	....	....	....	....
13/16	....	....	....	....	20	49/64	3 1/2	4	....	....	....	....	....
7/8	9	49/64	14	13/16	20	53/64	3 3/4	4	....	....	....	....	....
15/16	....	....	....	....	20	57/64	4	4	....	....	....	....	....

[a] ANSI/ASME B1.1. For 8-, 12-, and 16-pitch thread series, see next page.
[b] Classes 1A, 2A, 3A, 1B, 2B, 3B, 2, and 3.
[c] Classes 2A, 2B, 2, and 3.
[d] For approximate 75% full depth of thread. For decimal sizes of numbered and lettered drills, see Table 3.

2 Screw Threads, American National, Unified, and Metric (continued)

AMERICAN NATIONAL STANDARD UNIFIED AND AMERICAN NATIONAL SCREW THREADS[a] (continued)

Nominal Diameter	8-Pitch[b] Series 8N and 8UN		12-Pitch[b] Series 12N and 12UN		16-Pitch[b] Series 16N and 16UN		Nominal Diameter	8-Pitch[b] Series 8N and 8UN		12-Pitch[b] Series 12N and 12UN		16-Pitch[b] Series 16N and 16UN	
	Thds. per Inch	Tap Drill[c]	Thds. per Inch	Tap Drill[c]	Thds. per Inch	Tap Drill[c]		Thds. per Inch	Tap Drill[c]	Thds. per Inch	Tap Drill[c]	Thds. per Inch	Tap Drill[c]
1/2	....	....	12	27/64	....	....	2 1/16	....	....	....	....	**16**	2
9/16	....	....	12[e]	31/64	....	....	2 1/8	....	....	12	2 3/64	16	2 1/16
5/8	....	....	12	35/64	....	....	2 3/16	....	....	....	....	**16**	2 1/8
11/16	....	....	12	39/64	....	....	2 1/4	8	2 1/8	12	2 11/64	16	2 3/16.
3/4	....	....	12	43/64	16[e]	11/16	2 5/16	....	....	....	....	**16**	2 1/4
13/16	....	....	12	47/64	16	3/4	2 3/8	....	....	12	2 19/64	16	2 5/16
7/8	....	....	12	51/64	16	13/16	2 7/16	....	....	....	....	**16**	2 3/8
5/16	....	....	12	55/64	16	7/8	2 1/2	8	2 3/8	12	2 27/64	16	2 7/16
1	8[e]	7/8	12	59/64	16	15/16	2 5/8	....	....	12	2 35/64	16	2 9/16
1 1/16	....	....	12	63/64	16	1	2 3/4	8	2 5/8	12	2 43/64	16	2 11/16
1 1/8	8	1	12[e]	1 3/64	16	1 1/16	2 7/8	....	....	12	....	16	....
1 3/16	....	....	12	1 7/64	16	1 1/8	3	8	2 7/8	12	....	16	....
1 1/4	8	1 1/8	12	1 11/64	16	1 3/16	3 1/8	....	....	12	....	16	....
1 5/16	....	....	12	1 15/64	16	1 1/4	3 1/4	8	....	12	....	16	....
1 3/8	8	1 1/4	12[e]	1 19/64	16	1 5/16	3 3/8	....	....	12	....	16	....
1 7/16	....	....	12	1 23/64	16	1 3/8	3 1/2	8	....	12	....	16	....
1 1/2	8	1 3/8	12[e]	1 27/64	16	1 7/16	3 5/8	....	....	12	....	16	....
1 9/16	....	....	....	....	16	1 1/2	3 3/4	8	....	12	....	16	....
1 5/8	8	1 1/2	12	1 35/64	16	1 9/16	3 7/8	....	....	12	....	16	....
1 11/16	....	....	....	....	16	1 5/8	4	8	....	12	....	16	....
1 3/4	8	1 5/8	12	1 43/64	16[e]	1 11/16	4 1/4	8	....	12	....	16	....
1 13/16	....	....	....	....	16	1 3/4	4 1/2	8	....	12	....	16	....
1 7/8	8	1 3/4	12	1 51/64	16	1 13/16	4 3/4	8	....	12	....	16	....
1 15/16	....	....	....	....	16	1 7/8	5	8	....	12	....	16	....
2	8	1 7/8	12	1 59/64	16[e]	1 15/16	5 1/4	8	....	12	....	16	....

[a]ANSI/ASME B1.1.
[b]Classes 2A, 3A, 2B, 3B, 2, and 3.
[c]For approximate 75% full depth of thread.
[d]Boldface type indicates American National threads only.
[e]This is a standard size of the Unified or American National threads of the coarse, fine, or extra fine series. See preceding page.

2 SCREW THREADS, AMERICAN NATIONAL, UNIFIED, AND METRIC (CONTINUED)

METRIC SCREW THREADS[a]

Preferred sizes for commercial threads and fasteners are shown in **boldface** type.

Coarse (general purpose)		Fine	
Nominal Size & Thd Pitch	Tap Drill Diameter, mm	Nominal Size & Thd Pitch	Tap Drill Diameter, mm
M1.6 × 0.35	1.25	—	—
M1.8 × 0.35	1.45	—	—
M2 × 0.4	1.6	—	—
M2.2 × 0.45	1.75	—	—
M2.5 × 0.45	2.05	—	—
M3 × 0.5	2.5	—	—
M3.5 × 0.6	2.9	—	—
M4 × 0.7	3.3	—	—
M4.5 × 0.75	3.75	—	—
M5 × 0.8	4.2	—	—
M6 × 1	5.0	—	—
M7 × 1	6.0	—	—
M8 × 1.25	6.8	**M8 × 1**	7.0
M9 × 1.25	7.75	—	—
M10 × 1.5	8.5	**M10 × 1.25**	8.75
M11 × 1.5	9.50	—	—
M12 × 1.75	10.30	**M12 × 1.25**	10.5
M14 × 2	12.00	**M14 × 1.5**	12.5
M16 × 2	14.00	**M16 × 1.5**	14.5
M18 × 2.5	15.50	**M18 × 1.5**	16.5
M20 × 2.5	17.5	**M20 × 1.5**	18.5
M22 × 2.5[b]	19.5	**M22 × 1.5**	20.5
M24 × 3	21.0	**M24 × 2**	22.0
M27 × 3[b]	24.0	**M27 × 2**	25.0
M30 × 3.5	26.5	**M30 × 2**	28.0
M33 × 3.5	29.5	**M30 × 2**	31.0
M36 × 4	32.0	**M36 × 2**	33.0
M39 × 4	35.0	M39 × 2	36.0
M42 × 4.5	37.5	**M42 × 2**	39.0
M45 × 4.5	40.5	M45 × 1.5	42.0
M48 × 5	43.0	**M48 × 2**	45.0
M52 × 5	47.0	M52 × 2	49.0
M56 × 5.5	50.5	**M56 × 2**	52.0
M60 × 5.5	54.5	M60 × 1.5	56.0
M64 × 6	58.0	**M64 × 2**	60.0
M68 × 6	62.0	M68 × 2	64.0
M72 × 6	66.0	**M72 × 2**	68.0
M80 × 6	74.0	**M80 × 2**	76.0
M90 × 6	84.0	**M90 × 2**	86.0
M100 × 6	94.0	**M100 × 2**	96.0

[a]Metric Fasteners Standard, IFI-500 and ANSI/ASME B1.13M.
[b]Only for high strength structural steel fasteners.

3 TWIST DRILL SIZES – AMERICAN NATIONAL STANDARD AND METRIC

AMERICAN NATIONAL STANDARD DRILL SIZES[a]

All dimensions are in inches.
Drills designated in common fractions are available in diameters $\frac{1}{64}$" to $1\frac{3}{4}$" in $\frac{1}{64}$" increments, $1\frac{3}{4}$" to $2\frac{1}{4}$" in $\frac{1}{32}$" increments. $2\frac{1}{4}$" to 3" in $\frac{1}{16}$" increments and 3" to $3\frac{1}{2}$" in $\frac{1}{8}$" increments. Drills larger than $3\frac{1}{2}$" are seldom used, and are regarded as special drills.

Size	Drill Diameter	Size	Drill Diameter	Size	Drill Diameter	Size	Drill Diameter	Size	Drill Diameter	Size	Drill Diameter
1	.2280	17	.1730	33	.1130	49	.0730	65	.0350	81	.0130
2	.2210	18	.1695	34	.1110	50	.0700	66	.0330	82	.0125
3	.2130	19	.1660	35	.1100	51	.0670	67	.0320	83	.0120
4	.2090	20	.1610	36	.1065	52	.0635	68	.0310	84	.0115
5	.2055	21	.1590	37	.1040	53	.0595	69	.0292	85	.0110
6	.2040	22	.1570	38	.1015	54	.0550	70	.0280	86	.0105
7	.2010	23	.1540	39	.0995	55	.0520	71	.0260	87	.0100
8	.1990	24	.1520	40	.0980	56	.0465	72	.0250	88	.0095
9	.1960	25	.1495	41	.0960	57	.0430	73	.0240	89	.0091
10	.1935	26	.1470	42	.0935	58	.0420	74	.0225	90	.0087
11	.1910	27	.1440	43	.0890	59	.0410	75	.0210	91	.0083
12	.1890	28	.1405	44	.0860	60	.0400	76	.0200	92	.0079
13	.1850	29	.1360	45	.0820	61	.0390	77	.0180	93	.0075
14	.1820	30	.1285	46	.0810	62	.0380	78	.0160	94	.0071
15	.1800	31	.1200	47	.0785	63	.0370	79	.0145	95	.0067
16	.1770	32	.1160	48	.0760	64	.0360	80	.0135	96	.0063
										97	.0059

LETTER SIZES

A	.234	G	.261	L	.290	Q	.332	V	.377
B	.238	H	.266	M	.295	R	.339	W	.386
C	.242	I	.272	N	.302	S	.348	X	.397
D	.246	J	.277	O	.316	T	.358	Y	.404
E	.250	K	.281	P	.323	U	.368	Z	.413
F	.257								

[a]ANSI B94.11M.

3 TWIST DRILL SIZES – AMERICAN NATIONAL STANDARD AND METRIC (CONTINUED)

METRIC DRILL SIZES
Decimal-inch equivalents are for rererence only.

Drill Diameter		Drill Diameter		Drill Diameter		Drill Diameter		Drill Diameter		Drill Diameter	
mm	in.	mm	in.	mm	in.	mm	in.	mm	in.	mm	in.
0.40	.0157	1.95	.0768	4.70	.1850	8.00	.3150	13.20	.5197	25.50	1.0039
0.42	.0165	2.00	.0787	4.80	.1890	8.10	.3189	13.50	.5315	26.00	1.0236
0.45	.0177	2.05	.0807	4.90	.1929	8.20	.3228	13.80	.5433	26.50	1.0433
0.48	.0189	2.10	.0827	5.00	.1969	8.30	.3268	14.00	.5512	27.00	1.0630
0.50	.0197	2.15	.0846	5.10	.2008	8.40	.3307	14.25	.5610	27.50	1.0827
0.55	.0217	2.20	.0866	5.20	.2047	8.50	.3346	14.50	.5709	28.00	1.1024
0.60	.0236	2.25	.0886	5.30	.2087	8.60	.3386	14.75	.5807	28.50	1.1220
0.65	.0256	2.30	.0906	5.40	.2126	8.70	.3425	15.00	.5906	29.00	1.1417
0.70	.0276	2.35	.0925	5.50	.2165	8.80	.3465	15.25	.6004	29.50	1.1614
0.75	.0295	2.40	.0945	5.60	.2205	8.90	.3504	15.50	.6102	30.00	1.1811
0.80	.0315	2.45	.0965	5.70	.2244	9.00	.3543	15.75	.6201	30.50	1.2008
0.85	.0335	2.50	.0984	5.80	.2283	9.10	.3583	16.00	.6299	31.00	1.2205
0.90	.0354	2.60	.1024	5.90	.2323	9.20	.3622	16.25	.6398	31.50	1.2402
0.95	.0374	2.70	.1063	6.00	.2362	9.30	.3661	16.50	.6496	32.00	1.2598
1.00	.0394	2.80	.1102	6.10	.2402	9.40	.3701	16.75	.6594	32.50	1.2795
1.05	.0413	2.90	.1142	6.20	.2441	9.50	.3740	17.00	.6693	33.00	1.2992
1.10	.0433	3.00	.1181	6.30	.2480	9.60	.3780	17.25	.6791	33.50	1.3189
1.15	.0453	3.10	.1220	6.40	.2520	9.70	.3819	17.50	.6890	34.00	1.3386
1.20	.0472	3.20	.1260	6.50	.2559	9.80	.3858	18.00	.7087	34.50	1.3583
1.25	.0492	3.30	.1299	6.60	.2598	9.90	.3898	18.50	.7283	35.00	1.3780
1.30	.0512	3.40	.1339	6.70	.2638	10.00	.3937	19.00	.7480	35.50	1.3976
1.35	.0531	3.50	.1378	6.80	.2677	10.20	.4016	19.50	.7677	36.00	1.4173
1.40	.0551	3.60	.1417	6.90	.2717	10.50	.4134	20.00	.7874	36.50	1.4370
1.45	.0571	3.70	.1457	7.00	.2756	10.80	.4252	20.50	.8071	37.00	1.4567
1.50	.0591	3.80	.1496	7.10	.2795	11.00	.4331	21.00	.8268	37.50	1.4764
1.55	.0610	3.90	.1535	7.20	.2835	11.20	.4409	21.50	.8465	38.00	1.4961
1.60	.0630	4.00	.1575	7.30	.2874	11.50	.4528	22.00	.8661	40.00	1.5748
1.65	.0650	4.10	.1614	7.40	.2913	11.80	.4646	22.50	.8858	42.00	1.6535
1.70	.0669	4.20	.1654	7.50	.2953	12.00	.4724	23.00	.9055	44.00	1.7323
1.75	.0689	4.30	.1693	7.60	.2992	12.20	.4803	23.50	.9252	46.00	1.8110
1.80	.0709	4.40	.1732	7.70	.3031	12.50	.4921	24.00	.9449	48.00	1.8898
1.85	.0728	4.50	.1772	7.80	.3071	12.50	.5039	24.50	.9646	50.00	1.9685
1.90	.0748	4.60	.1811	7.90	.3110	13.00	.5118	25.00	.9843		

4 ACME THREADS, GENERAL-PURPOSE[a]

Size	Threads per Inch	Size	Threads per Inch	Size	Threads per Inch	Size	Threads per Inch
¼	16	¾	6	1½	4	3	2
5/16	14	⅞	6	1¾	4	3½	2
⅜	12	1	5	2	4	4	2
7/16	12	1⅛	5	2¼	3	4½	2
½	10	1¼	5	2½	3	5	2
⅝	8	1⅜	4	2¾	3	. . .	. .

[a]ANSI/ASME B1.5.

5 BOLTS, NUTS, AND CAP SCREWS – SQUARE AND HEXAGON – AMERICAN NATIONAL STANDARD AND METRIC

AMERICAN NATIONAL STANDARD SQUARE AND HEXAGON BOLTS[a] AND NUTS[b] AND HEXAGON CAP SCREWS[c]

Boldface type indicates product features unified dimensionally with British and Canadian standards. All dimensions are in inches.

Nominal Size D Body Diameter of Bolt	Regular Bolts					Heavy Bolts		
	Width Across Flats W		Height H			Width Across Flats W	Height H	
	Sq.	Hex.	Sq. (Unfin.)	Hex. (Unfin.)	Hex. Cap Scr.[c] (Fin.)		Hex. (Unfin.)	Hex. Screw (Fin.)
¼ **0.2500**	⅜	⁷⁄₁₆	¹¹⁄₆₄	¹¹⁄₆₄	⁵⁄₃₂	....	....	....
⁵⁄₁₆ **0.3125**	½	½	¹³⁄₆₄	⁷⁄₃₂	¹³⁄₆₄	....	....	....
⅜ **0.3750**	⁹⁄₁₆	⁹⁄₁₆	¼	¼	¹⁵⁄₆₄	....	....	....
⁷⁄₁₆ **0.4375**	⅝	⅝	¹⁹⁄₆₄	¹⁹⁄₆₄	⁹⁄₃₂	....	....	....
½ **0.5000**	¾	¾	²¹⁄₆₄	¹¹⁄₃₂	⁵⁄₁₆	⅞	¹¹⁄₃₂	⁵⁄₁₆
⁹⁄₁₆ **0.5625**	....	¹³⁄₁₆	....	....	²³⁄₆₄	....	....	....
⅝ **0.6250**	¹⁵⁄₁₆	¹⁵⁄₁₆	²⁷⁄₆₄	²⁷⁄₆₄	²⁵⁄₆₄	1¹⁄₁₆	²⁷⁄₆₄	²⁵⁄₆₄
¾ **0.7500**	1⅛	1⅛	½	½	¹⁵⁄₃₂	1¼	½	¹⁵⁄₃₂
⅞ **0.8750**	1⁵⁄₁₆	1⁵⁄₁₆	¹⁹⁄₃₂	³⁷⁄₆₄	³⁵⁄₆₄	1⁷⁄₁₆	³⁷⁄₆₄	³⁵⁄₆₄
1 **1.000**	1½	1½	²¹⁄₃₂	⁴³⁄₆₄	³⁹⁄₆₄	1⅝	⁴³⁄₆₄	³⁹⁄₆₄
1⅛ **1.1250**	1¹¹⁄₁₆	1¹¹⁄₁₆	¾	¾	¹¹⁄₁₆	1¹³⁄₁₆	¾	¹¹⁄₁₆
1¼ **1.2500**	1⅞	1⅞	²⁷⁄₃₂	²⁷⁄₃₂	²⁵⁄₃₂	2	²⁷⁄₃₂	²⁵⁄₃₂
1⅜ **1.3750**	2¹⁄₁₆	2¹⁄₁₆	²⁹⁄₃₂	²⁹⁄₃₂	²⁷⁄₃₂	2³⁄₁₆	²⁹⁄₃₂	²⁷⁄₃₂
1½ **1.5000**	2¼	2¼	1	1	¹⁵⁄₁₆	2⅜	1	¹⁵⁄₁₆
1¾ **1.7500**	....	2⅝	....	1⁵⁄₃₂	1³⁄₃₂	2¾	1⁵⁄₃₂	1³⁄₃₂
2 **2.0000**	....	3	....	1¹¹⁄₃₂	1⁷⁄₃₂	3⅛	1¹¹⁄₃₂	1⁷⁄₃₂
2¼ 2.2500	....	3⅜	....	1½	1⅜	3½	1½	1⅜
2½ 2.5000	....	3¾	....	1²¹⁄₃₂	1¹⁷⁄₃₂	3⅞	1²¹⁄₃₂	1¹⁷⁄₃₂
2¾ 2.7500	....	4⅛	....	1¹³⁄₁₆	1¹¹⁄₁₆	4¼	1¹³⁄₁₆	1¹¹⁄₁₆
3 3.0000	....	4½	....	2	1⅞	4⅝	2	1⅞
3¼ 3.2500	....	4⅞	....	2³⁄₁₆	....	....	....	....
3½ 3.5000	....	5¼	....	2⁵⁄₁₆	....	....	....	....
3¾ 3.7500	....	5⅝	....	2½	....	....	....	....
4 4.0000	....	6	....	2¹¹⁄₁₆	....	....	....	....

[a]ANSI B18.2.1.
[b]ANSI B18.2.2.
[c]Hexagon cap screws and finished hexagon bolts are combined as a single product.

5 BOLTS, NUTS, AND CAP SCREWS – SQUARE AND HEXAGON – AMERICAN NATIONAL STANDARD AND METRIC (CONTINUED)

AMERICAN NATIONAL STANDARD SQUARE AND HEXAGON BOLTS AND NUTS
AND HEXAGON CAP SCREWS (continued)

See ANSI B18.2.2 for jam nuts, slotted nuts, thick nuts, thick slotted nuts, and castle nuts.

Nominal Size D Body Diameter of Bolt		Regular Nuts					Heavy Nuts			
		Width Across Flats W		Thickness T			Width Across Flats W	Thickness T		
		Sq.	Hex.	Sq. (Unfin.)	Hex. Flat (Unfin.)	Hex. (Fin.)		Sq. (Unfin.)	Hex. Flat (Unfin.)	Hex. (Fin.)
¼	0.2500	7/16	7/16	7/32	7/32	7/32	½	¼	15/64	15/64
5/16	0.3125	9/16	½	17/64	17/64	17/64	9/16	5/16	19/64	19/64
3/8	0.3750	5/8	9/16	21/64	21/64	21/64	11/16	3/8	23/64	23/64
7/16	0.4375	3/4	11/16	3/8	3/8	3/8	3/4	7/16	27/64	27/64
½	0.5000	13/16	3/4	7/16	7/16	7/16	7/8 [a]	½	31/64	31/64
9/16	0.5625	....	7/8	....	31/64	31/64	15/16	....	35/64	35/64
5/8	0.6250	1	15/16	35/64	35/64	35/64	1 1/16 [a]	5/8	39/64	39/64
3/4	0.7500	1 1/8	1 1/8	21/32	41/64	41/64	1 1/4 [a]	3/4	47/64	47/64
7/8	0.8750	1 5/16	1 5/16	49/64	3/4	3/4	1 7/16 [a]	7/8	55/64	55/64
1	1.0000	1 1/2	1 1/2	7/8	55/64	55/64	1 5/8 [a]	1	63/64	63/64
1 1/8	1.1250	1 11/16	1 11/16	1	1	31/32	1 13/16 [a]	1 1/8	1 1/8	1 7/64
1 1/4	1.2500	1 7/8	1 7/8	1 3/32	1 3/32	1 1/16	2 [a]	1 1/4	1 1/4	1 7/32
1 3/8	1.3750	2 1/16	2 1/16	1 13/64	1 13/64	1 11/64	2 3/16 [a]	1 3/8	1 3/8	1 11/32
1 1/2	1.5000	2 1/4	2 1/4	1 5/16	1 5/16	1 9/32	2 3/8 [a]	1 1/2	1 1/2	1 15/32
1 5/8	1.6250	....	....	....	....	....	2 9/16	....	....	1 19/32
1 3/4	1.7500	....	....	....	....	....	2 3/4	....	1 3/4	1 23/32
1 7/8	1.8750	....	....	....	....	....	2 15/16	....	....	1 27/32
2	2.0000	....	....	....	....	....	3 1/8	....	2	1 31/32
2 1/4	2.2500	....	....	....	....	....	3 1/2	....	2 1/4	2 13/64
2 1/2	2.5000	....	....	....	....	....	3 7/8	....	2 1/2	2 29/64
2 3/4	2.7500	....	....	....	....	....	4 1/4	....	2 3/4	2 45/64
3	3.0000	....	....	....	....	....	4 5/8	....	3	2 61/64
3 1/4	3.2500	....	....	....	....	....	5	....	3 1/4	3 3/16
3 1/2	3.5000	....	....	....	....	....	5 3/8	....	3 1/2	3 7/16
3 3/4	3.7500	....	....	....	....	....	5 3/4	....	3 3/4	3 11/16
4	4.0000	....	....	....	....	....	6 1/8	....	4	3 15/16

[a] Product feature not unified for heavy square nut.

5 BOLTS, NUTS, AND CAP SCREWS – SQUARE AND HEXAGON – AMERICAN NATIONAL STANDARD AND METRIC (CONTINUED)

METRIC HEXAGON BOLTS, HEXAGON CAP SCREWS, HEXAGON STRUCTURAL BOLTS, AND HEXAGON NUTS

Nominal Size D, mm	Width Across Flats W (max)		Thickness T (max)			
Body Dia and Thd Pitch	Bolts,[a] Cap Screws,[b] and Nuts[c]	Heavy Hex & Hex Structural Bolts[a] & Nuts[c]	Bolts (Unfin.)	Cap Screw (Fin.)	Nut (Fin. or Unfin.)	
					Style 1	Style 2
M5 × 0.8	8.0		3.88	3.65	4.7	5.1
M6 × 1	10.0		4.38	4.47	5.2	5.7
M8 × 1.25	13.0		5.68	5.50	6.8	7.5
M10 × 1.5	16.0		6.85	6.63	8.4	9.3
M12 × 1.75	18.0	21.0	7.95	7.76	10.8	12.0
M14 × 2	21.0	24.0	9.25	9.09	12.8	14.1
M16 × 2	24.0	27.0	10.75	10.32	14.8	16.4
M20 × 2.5	30.0	34.0	13.40	12.88	18.0	20.3
M24 × 3	36.0	41.0	15.90	15.44	21.5	23.9
M30 × 3.5	46.0	50.0	19.75	19.48	25.6	28.6
M36 × 4	55.0	60.0	23.55	23.38	31.0	34.7
M42 × 4.5	65.0		27.05	26.97	....	....
M48 × 5	75.0		31.07	31.07	....	....
M56 × 5.5	85.0		36.20	36.20	....	....
M64 × 6	95.0		41.32	41.32	....	....
M72 × 6	105.0		46.45	46.45	....	....
M80 × 6	115.0		51.58	51.58	....	....
M90 × 6	130.0		57.74	57.74	....	....
M100 × 6	145.0		63.90	63.90	....	....

HIGH STRENGTH STRUCTURAL HEXAGON BOLTS[a] (Fin.) AND HEXAGON NUTS[c]

M16 × 2	27.0	....	10.75	....	....	17.1
M20 × 2.5	34.0	....	13.40	....	....	20.7
M22 × 2.5	36.0	....	14.9	....	....	23.6
M24 × 3	41.0	....	15.9	....	....	24.2
M27 × 3	46.0	....	17.9	....	....	27.6
M30 × 3.5	50.0	....	19.75	....	....	31.7
M36 × 4	60.0	....	23.55	....	....	36.6

[a] ANSI B18.2.3.5M, B18.2.3.6M, B18.2.3.7M.
[b] ANSI B18.2.3.1M.
[c] ANSI B18.2.4.1M, B18.2.4.2M.

6 Cap Screws, Slotted[a] and Socket Head[b] – American National Standard and Metric

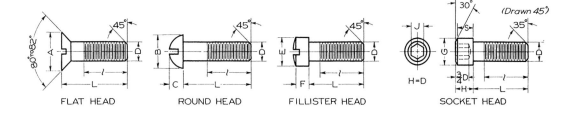

FLAT HEAD   ROUND HEAD   FILLISTER HEAD   SOCKET HEAD

Nominal Size D	Flat Head[a]	Round Head[a]		Fillister Head[a]		Socket Head[b]		
	A	B	C	E	F	G	J	S
0 (.060)	....	....	....	....	....	.096	.05	.054
1 (.073)	....	....	....	....	....	.118	1/16	.066
2 (.086)	....	....	....	....	....	.140	5/64	.077
3 (.099)	....	....	....	....	....	.161	5/64	.089
4 (.112)	....	....	....	....	....	.183	3/32	.101
5 (.125)	....	....	....	....	....	.205	3/32	.112
6 (.138)	....	....	....	....	....	.226	7/64	.124
8 (.164)	....	....	....	....	....	.270	9/64	.148
10 (.190)	....	....	....	....	....	.312	5/32	.171
1/4	.500	.437	.191	.375	.172	.375	3/16	.225
5/16	.625	.562	.245	.437	.203	.469	1/4	.281
3/8	.750	.675	.273	.562	.250	.562	5/16	.337
7/16	.812	.750	.328	.625	.297	.656	3/8	.394
1/2	.875	.812	.354	.750	.328	.750	3/8	.450
9/16	1.000	.937	.409	.812	.375	....	....	....
5/8	1.125	1.000	.437	.875	.422	.938	1/2	.562
3/4	1.375	1.250	.546	1.000	.500	1.125	5/8	.675
7/8	1.625	....	....	1.125	.594	1.312	3/4	.787
1	1.875	....	....	1.312	.656	1.500	3/4	.900
1 1/8	2.062	....	....	....	....	1.688	7/8	1.012
1 1/4	2.312	....	....	....	....	1.875	7/8	1.125
1 3/8	2.562	....	....	....	....	2.062	1	1.237
1 1/2	2.812	....	....	....	....	2.250	1	1.350

[a] ANSI B18.6.2.
[b] ANSI/ASME B18.3. For hexagon-head screws, see Table 5.

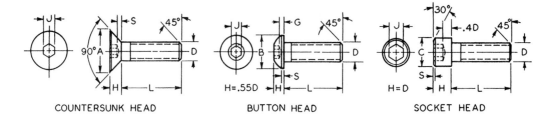

COUNTERSUNK HEAD   BUTTON HEAD   SOCKET HEAD

METRIC SOCKET HEAD CAP SCREWS									
Nominal Size D	Countersunk Head[a]			Button Head[b]			Socket Head[c]		Hex Socket Size
	A (max)	H	S	B	S	G	C	S	J
M1.6 × 0.35	....	....	....	....	....	....	3.0	0.16	1.5
M2 × 0.4	....	....	....	....	....	....	3.8	0.2	1.5
M2.5 × 0.45	....	....	....	....	....	....	4.5	0.25	2.0
M3 × 0.5	6.72	1.86	0.25	5.70	0.38	0.2	5.5	0.3	2.5
M4 × 0.7	8.96	2.48	0.45	7.6	0.38	0.3	7.0	0.4	3.0
M5 × 0.8	11.2	3.1	0.66	9.5	0.5	0.38	8.5	0.5	4.0
M6 × 1	13.44	3.72	0.7	10.5	0.8	0.74	10.0	0.6	5.0
M8 × 1.25	17.92	4.96	1.16	14.0	0.8	1.05	13.0	0.8	6.0
M10 × 1.5	22.4	6.2	1.62	17.5	0.8	1.45	16.0	1.0	8.0
M12 × 1.75	26.88	7.44	1.8	21.0	0.8	1.63	18.0	1.2	10.0
M14 × 2	30.24	8.12	2.0	....	....	....	21.0	1.4	12.0
M16 × 2	33.6	8.8	2.2	28.0	1.5	2.25	24.0	1.6	14.0
M20 × 2.5	19.67	10.16	2.2	....	....	....	30.0	2.0	17.0
M24 × 3	....	....	....	....	....	....	36.0	2.4	19.0
M30 × 3.5	....	....	....	....	....	....	45.0	3.0	22.0
M36 × 4	....	....	....	....	....	....	54.0	3.6	27.0
M42 × 4.5	....	....	....	....	....	....	63.0	4.2	32.0
M48 × 5	....	....	....	....	....	....	72.0	4.8	36.0

[a] IFI 535.
[b] ANSI/ASME B18.3.4M.
[c] ANSI/ASME B18.3.1M.

7 MACHINE SCREWS – AMERICAN NATIONAL STANDARD AND METRIC

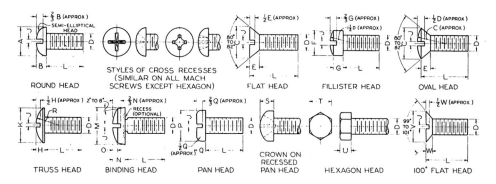

ROUND HEAD STYLES OF CROSS RECESSES (SIMILAR ON ALL MACH SCREWS EXCEPT HEXAGON) FLAT HEAD FILLISTER HEAD OVAL HEAD

TRUSS HEAD BINDING HEAD PAN HEAD CROWN ON RECESSED PAN HEAD HEXAGON HEAD 100° FLAT HEAD

AMERICAN NATIONAL STANDARD MACHINE SCREWS[a]

Length of Thread: On screws 2″ long and shorter, the threads extend to within two threads of the head and closer if practicable; longer screws have minimum thread length of 1¾″.
Points: Machine screws are regularly made with plain sheared ends, not chamfered.
Threads: Either Coarse or Fine Thread Series, Class 2 fit.
Recessed Heads: Two styles of cross recesses are available on all screws except hexagon head.

Nominal Size	Max. Diameter D	Round Head		Flat Heads & Oval Head		Fillister Head		Truss Head			Slot Width
		A	B	C	E	F	G	K	H	R	J
0	0.060	0.113	0.053	0.119	0.035	0.096	0.045	0.131	0.037	0.087	0.023
1	0.073	0.138	0.061	0.146	0.043	0.118	0.053	0.164	0.045	0.107	0.026
2	0.086	0.162	0.069	0.172	0.051	0.140	0.062	0.194	0.053	0.129	0.031
3	0.099	0.187	0.078	0.199	0.059	0.161	0.070	0.226	0.061	0.151	0.035
4	0.112	0.211	0.086	0.225	0.067	0.183	0.079	0.257	0.069	0.169	0.039
5	0.125	0.236	0.095	0.252	0.075	0.205	0.088	0.289	0.078	0.191	0.043
6	0.138	0.260	0.103	0.279	0.083	0.226	0.096	0.321	0.086	0.211	0.048
8	0.164	0.309	0.120	0.332	0.100	0.270	0.113	0.384	0.102	0.254	0.054
10	0.190	0.359	0.137	0.385	0.116	0.313	0.130	0.448	0.118	0.283	0.060
12	0.216	0.408	0.153	0.438	0.132	0.357	0.148	0.511	0.134	0.336	0.067
¼	0.250	0.472	0.175	0.507	0.153	0.414	0.170	0.573	0.150	0.375	0.075
5⁄16	0.3125	0.590	0.216	0.635	0.191	0.518	0.211	0.698	0.183	0.457	0.084
3⁄8	0.375	0.708	0.256	0.762	0.230	0.622	0.253	0.823	0.215	0.538	0.094
7⁄16	0.4375	0.750	0.328	0.812	0.223	0.625	0.265	0.948	0.248	0.619	0.094
½	0.500	0.813	0.355	0.875	0.223	0.750	0.297	1.073	0.280	0.701	0.106
9⁄16	0.5625	0.938	0.410	1.000	0.260	0.812	0.336	1.198	0.312	0.783	0.118
5⁄8	0.625	1.000	0.438	1.125	0.298	0.875	0.375	1.323	0.345	0.863	0.133
¾	0.750	1.250	0.547	1.375	0.372	1.000	0.441	1.573	0.410	1.024	0.149

Nominal Size	Max. Diameter D	Binding Head			Pan Head			Hexagon Head		100° Flat Head		Slot Width
		M	N	O	P	Q	S	T	U	V	W	J
2	0.086	0.181	0.050	0.018	0.167	0.053	0.062	0.125	0.050	....	....	0.031
3	0.099	0.208	0.059	0.022	0.193	0.060	0.071	0.187	0.055	....	....	0.035
4	0.112	0.235	0.068	0.025	0.219	0.068	0.080	0.187	0.060	0.225	0.049	0.039
5	0.125	0.263	0.078	0.029	0.245	0.075	0.089	0.187	0.070	....	....	0.043
6	0.138	0.290	0.087	0.032	0.270	0.082	0.097	0.250	0.080	0.279	0.060	0.048
8	0.164	0.344	0.105	0.039	0.322	0.096	0.115	0.250	0.110	0.332	0.072	0.054
10	0.190	0.399	0.123	0.045	0.373	0.110	0.133	0.312	0.120	0.385	0.083	0.060
12	0.216	0.454	0.141	0.052	0.425	0.125	0.151	0.312	0.155	....	....	0.067
¼	0.250	0.513	0.165	0.061	0.492	0.144	0.175	0.375	0.190	0.507	0.110	0.075
5⁄16	0.3125	0.641	0.209	0.077	0.615	0.178	0.218	0.500	0.230	0.635	0.138	0.084
3⁄8	0.375	0.769	0.253	0.094	0.740	0.212	0.261	0.562	0.295	0.762	0.165	0.094
7⁄16	.4375	....	....	....	.865	.247	.305	....	....	....	....	.094
½	.500	....	....	....	.987	.281	.348	....	....	....	....	.106
9⁄16	.5625	....	....	....	1.041	.315	.391	....	....	....	....	.118
5⁄8	.625	....	....	....	1.172	.350	.434	....	....	....	....	.133
¾	.750	....	....	....	1.435	.419	.521	....	....	....	....	.149

METRIC MACHINE SCREWS[b]

Length of Thread: On screws 36 mm long or shorter, the threads extend to within one thread of the head: on longer screws the thread extends to within two threads of the head.

Points: Machine screws are regularly made with sheared ends, not chamfered.

Threads: Coarse (general-purpose) threads series are given.

Recessed Heads: Two styles of cross-recesses are available on all screws except hexagon head.

Nominal Size & Thd Pitch	Max. Dia. D, mm	Flat Heads & Oval Head		Pan Heads			Hex Head		Slot Width
		C	E	P	Q	S	T	U	J
M2 × 0.4	2.0	3.5	1.2	4.0	1.3	1.6	3.2	1.6	0.7
M2.5 × 0.45	2.5	4.4	1.5	5.0	1.5	2.1	4.0	2.1	0.8
M3 × 0.5	3.0	5.2	1.7	5.6	1.8	2.4	5.0	2.3	1.0
M3.5 × 0.6	3.5	6.9	2.3	7.0	2.1	2.6	5.5	2.6	1.2
M4 × 0.7	4.0	8.0	2.7	8.0	2.4	3.1	7.0	3.0	1.5
M5 × 0.8	5.0	8.9	2.7	9.5	3.0	3.7	8.0	3.8	1.5
M6 × 1	6.0	10.9	3.3	12.0	3.6	4.6	10.0	4.7	1.9
M8 × 1.25	8.0	15.14	4.6	16.0	4.8	6.0	13.0	6.0	2.3
M10 × 1.5	10.0	17.8	5.0	20.0	6.0	7.5	15.0	7.5	2.8
M12 × 1.75	12.0	....	....	....	....	....	18.0	9.0	....

Nominal Size	Metric Machine Screw Lengths—L[c]																					
	2.5	3	4	5	6	8	10	13	16	20	25	30	35	40	45	50	55	60	65	70	80	90
M2 × 0.4	PH	A	A	A	A	A	A	A	A	A												
M2.5 × 0.45		PH	A	A	A	A	A	A	A	A	A											
M3 × 0.5			PH	A	A	A	A	A	A	A	A	A										
M3.5 × 0.6				PH	A	A	A	A	A	A	A	A	A									
M4 × 0.7				PH	A	A	A	A	A	A	A	A	A	A								
M5 × 0.8					PH	A	A	A	A	A	A	A	A	A	A	A						
M6 × 1						A	A	A	A	A	A	A	A	A	A	A	A					
M8 × 1.25						A	A	A	A	A	A	A	A	A	A	A	A	A	A	A	A	
M10 × 1.5							A	A	A	A	A	A	A	A	A	A	A	A	A	A	A	A
M12 × 1.75							A	A	A	A	A	A	A	A	A	A	A	A	A	A	A	A

Min. Thd Length—28 mm

Min. Thd Length—38 mm

[b]Metric Fasteners Standard. IFI-513.

[c]PH = recommended lengths for only pan and hex head metric screws;

A = recommended lengths for all metric screw head-styles.

8 Keys – Square, Flat, Plain Taper,[a] and Gib Head

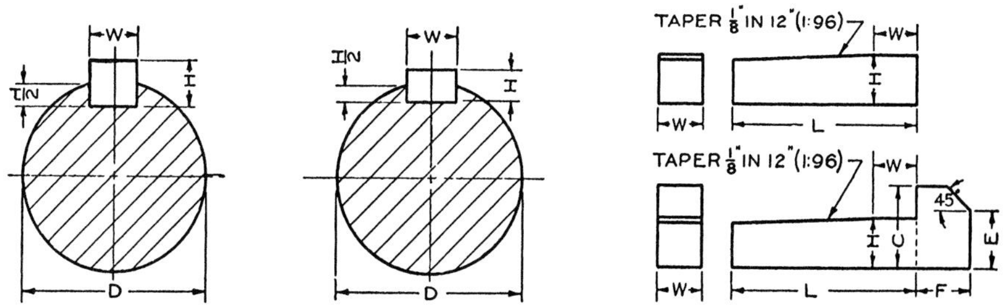

Shaft Diameters	Square Stock Key	Flat Stock Key	Gib Head Taper Stock Key					
			Square			*Flat*		
			Height	Length	Height to Chamfer	Height	Length	Height to Chamfer
D	W = H	W × H	C	F	E	C	F	E
1/2 to 9/16	1/8	1/8 × 3/32	1/4	7/32	5/32	3/16	1/8	1/8
5/8 to 7/8	3/16	3/16 × 1/8	5/16	9/32	7/32	1/4	3/16	5/32
15/16 to 1 1/4	1/4	1/4 × 3/16	7/16	11/32	11/32	5/16	1/4	3/16
1 5/16 to 1 3/8	5/16	5/16 × 1/4	9/16	13/32	13/32	3/8	5/16	1/4
1 7/16 to 1 3/4	3/8	3/8 × 1/4	11/16	15/32	15/32	7/16	3/8	5/16
1 13/16 to 2 1/4	1/2	1/2 × 3/8	7/8	19/32	5/8	5/8	1/2	7/16
2 5/16 to 2 3/4	5/8	5/8 × 7/16	1 1/16	23/32	3/4	3/4	5/8	1/2
2 7/8 to 3 1/4	3/4	3/4 × 1/2	1 1/4	7/8	7/8	7/8	3/4	5/8
3 3/8 to 3 3/4	7/8	7/8 × 5/8	1 1/2	1	1	1 1/16	7/8	3/4
3 7/8 to 4 1/2	1	1 × 3/4	1 3/4	1 3/16	1 3/16	1 1/4	1	13/16
4 3/4 to 5 1/2	1 1/4	1 1/4 × 7/8	2	1 7/16	1 7/16	1 1/2	1 1/4	1
5 3/4 to 6	1 1/2	1 1/2 × 1	2 1/2	1 3/4	1 3/4	1 3/4	1 1/2	1 1/4

[a] Plain taper square and flat keys have the same dimensions as the plain parallel stock keys, with the addition of the taper on top. Gib head taper square and flat keys have the same dimensions as the plain taper keys, with the addition of the gib head.

Stock lengths for plain taper and gib head taper keys: The minimum stock length equals 4W, and the maximum equals 16W. The increments of increase of length equal 2W.

9 Screw Threads[a] – Square and Acme

Size	Threads per Inch	Size	Threads per Inch	Size	Threads per Inch	Size	Threads per Inch
3/8	12	7/8	5	2	2 1/2	3 1/2	1 1/3
7/16	10	1	5	2 1/4	2	3 3/4	1 1/3
1/2	10	1 1/8	4	2 1/2	2	4	1 1/3
9/16	8	1 1/4	4	2 3/4	2	4 1/4	1 1/3
5/8	8	1 1/2	3	3	1 1/2	4 1/2	1
3/4	6	1 3/4	2 1/2	3 1/4	1 1/2	over 4 1/2	1

[a] See Table 4 for General Purpose Acme Threads.

10 Woodruff Keys[a] – American National Standard

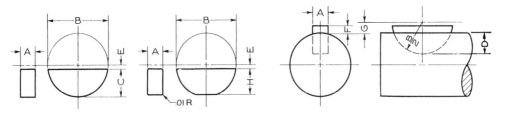

Key No.[b]	Nominal Sizes					Maximum Sizes			Key No.[b]	Nominal Sizes					Maximum Sizes		
	A × B	E	F	G		H	D	C		A × B	E	F	G		H	D	C
204	1/16 × 1/2	3/64	1/32	5/64		.194	.1718	.203	808	1/4 × 1	1/16	1/8	3/16		.428	.3130	.438
304	3/32 × 1/2	3/64	3/64	3/32		.194	.1561	.203	809	1/4 × 1 1/8	5/64	1/8	13/64		.475	.3590	.484
305	3/32 × 5/8	1/16	3/64	7/64		.240	.2031	.250	810	1/4 × 1 1/4	5/64	1/8	13/64		.537	.4220	.547
404	1/8 × 1/2	3/64	1/16	7/64		.194	.1405	.203	811	1/4 × 1 3/8	3/32	1/8	7/32		.584	.4690	.594
405	1/8 × 5/8	1/16	1/16	1/8		.240	.1875	.250	812	1/4 × 1 1/2	7/64	1/8	15/64		.631	.5160	.641
406	1/8 × 3/4	1/16	1/16	1/8		.303	.2505	.313	1008	5/16 × 1	1/16	5/32	7/32		.428	.2818	.438
505	5/32 × 5/8	1/16	5/64	9/64		.240	.1719	.250	1009	5/16 × 1 1/8	5/64	5/32	15/64		.475	.3278	.484
506	5/32 × 3/4	1/16	5/64	9/64		.303	.2349	.313	1010	5/16 × 1 1/4	5/64	5/32	15/64		.537	.3908	.547
507	5/32 × 7/8	1/16	5/64	9/64		.365	.2969	.375	1011	5/16 × 1 3/8	3/32	5/32	8/32		.584	.4378	.594
606	3/16 × 3/4	1/16	3/32	5/32		.303	.2193	.313	1012	5/16 × 1 1/2	7/64	5/32	17/64		.631	.4848	.641
607	3/16 × 7/8	1/16	3/32	5/32		.365	.2813	.375	1210	3/8 × 1 1/4	5/64	3/16	17/64		.537	.3595	.547
608	3/16 × 1	1/16	3/32	5/32		.428	.3443	.438	1211	3/8 × 1 3/8	3/32	3/16	9/32		.584	.4065	.594
609	3/16 × 1 1/8	5/64	3/32	11/64		.475	.3903	.484	1212	3/8 × 1 1/2	7/64	3/16	19/64		.631	.4535	.641
807	1/4 × 7/8	1/16	1/8	3/16		.365	.2500	.375	. . . .	. . . .	. .	. .	. .		. . . .	. . . .	. . . .

[a]ANSI B17.2.
[b]Key numbers indicate nominal key dimensions. The last two digits give the nominal diameter B in eighths of an inch, and the digits before the last two give the nominal width A in thirty-seconds of an inch.

11 Woodruff Key Sizes for Different Shaft Diameters[a]

Shaft Diameter	5/16 to 3/8	7/16 to 1/2	9/16 to 3/4	13/16 to 15/16	1 to 1 3/16	1 1/4 to 1 7/16	1 1/2 to 1 3/4	1 13/16 to 2 1/8	2 3/16 to 2 1/2
Key Numbers	204	304 305	404 405 406	505 506 507	606 607 608 609	807 808 809	810 811 812	1011 1012	1211 1212

[a]Suggested sizes; not standard.

12 WASHERS,[a] PLAIN – AMERICAN NATIONAL STANDARD

For parts lists, etc., give inside diameter, outside diameter, and the thickness; for example, .344 × .688 × .065 TYPE A PLAIN WASHER.

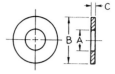

PREFERRED SIZES OF TYPE A PLAIN WASHERS[b]

Nominal Washer Size[c]			Inside Diameter A	Outside Diameter B	Nominal Thickness C
. . . .	. . . .		0.078	0.188	0.020
. . . .	. . . .		0.094	0.250	0.020
. . . .	. . . .		0.125	0.312	0.032
No. 6	0.138		0.156	0.375	0.049
No. 8	0.164		0.188	0.438	0.049
No. 10	0.190		0.219	0.500	0.049
3/16	0.188		0.250	0.562	0.049
No. 12	0.216		0.250	0.562	0.065
1/4	0.250	N	0.281	0.625	0.065
1/4	0.250	W	0.312	0.734	0.065
5/16	0.312	N	0.344	0.688	0.065
5/16	0.312	W	0.375	0.875	0.083
3/8	0.375	N	0.406	0.812	0.065
3/8	0.375	W	0.438	1.000	0.083
7/16	0.438	N	0.469	0.922	0.065
7/16	0.438	W	0.500	1.250	0.083
1/2	0.500	N	0.531	1.062	0.095
1/2	0.500	W	0.562	1.375	0.109
9/16	0.562	N	0.594	1.156	0.095
9/16	0.562	W	0.625	1.469	0.109
5/8	0.625	N	0.656	1.312	0.095
5/8	0.625	W	0.688	1.750	0.134
3/4	0.750	N	0.812	1.469	0.134
3/4	0.750	W	0.812	2.000	0.148
7/8	0.875	N	0.938	1.750	0.134
7/8	0.875	W	0.938	2.250	0.165
1	1.000	N	1.062	2.000	0.134
1	1.000	W	1.062	2.500	0.165
1 1/8	1.125	N	1.250	2.250	0.134
1 1/8	1.125	W	1.250	2.750	0.165
1 1/4	1.250	N	1.375	2.500	0.165
1 1/4	1.250	W	1.375	3.000	0.165
1 3/8	1.375	N	1.500	2.750	0.165
1 3/8	1.375	W	1.500	3.250	0.180
1 1/2	1.500	N	1.625	3.000	0.165
1 1/2	1.500	W	1.625	3.500	0.180
1 5/8	1.625		1.750	3.750	0.180
1 3/4	1.750		1.875	4.000	0.180
1 7/8	1.875		2.000	4.250	0.180
2	2.000		2.125	4.500	0.180
2 1/4	2.250		2.375	4.750	0.220
2 1/2	2.500		2.625	5.000	0.238
2 3/4	2.750		2.875	5.250	0.259
3	3.000		3.125	5.500	0.284

[a]From ANSI B18.22.1. For complete listings, see the standard.
[b]Preferred sizes are for the most part from series previously designated "Standard Plate" and "SAE." Where common sizes existed in the two series, the SAE size is designated "N" (narrow) and the Standard Plate "W" (wide).
[c]Nominal washer sizes are intended for use with comparable nominal screw or bolt sizes.

13 Washers,[a] Lock – American National Standard

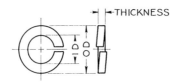

For parts lists, etc., give nominal size and series; for example, ¼ REGULAR LOCK WASHER

PREFERRED SERIES

Nominal Washer Size[b]		Inside Diameter, Min.	Regular		Extra Duty		Hi-Collar	
			Outside Diameter, Max.	Thick-ness, Min.	Outside Diameter, Max.	Thick-ness, Min.	Outside Diameter, Max.	Thick-ness, Min.
No. 2	0.086	0.088	0.172	0.020	0.208	0.027	. . . .	. . . .
No. 3	0.099	0.101	0.195	0.025	0.239	0.034	. . . .	. . . .
No. 4	0.112	0.115	0.209	0.025	0.253	0.034	0.173	0.022
No. 5	0.125	0.128	0.236	0.031	0.300	0.045	0.202	0.030
No. 6	0.138	0.141	0.250	0.031	0.314	0.045	0.216	0.030
No. 8	0.164	0.168	0.293	0.040	0.375	0.057	0.267	0.047
No. 10	0.190	0.194	0.334	0.047	0.434	0.068	0.294	0.047
No. 12	0.216	0.221	0.377	0.056	0.497	0.080	. . . .	. . . .
¼	0.250	0.255	0.489	0.062	0.535	0.084	0.365	0.078
⁵⁄₁₆	0.312	0.318	0.586	0.078	0.622	0.108	0.460	0.093
³⁄₈	0.375	0.382	0.683	0.094	0.741	0.123	0.553	0.125
⁷⁄₁₆	0.438	0.446	0.779	0.109	0.839	0.143	0.647	0.140
½	0.500	0.509	0.873	0.125	0.939	0.162	0.737	0.172
⁹⁄₁₆	0.562	0.572	0.971	0.141	1.041	0.182	. . . .	. . . .
⅝	0.625	0.636	1.079	0.156	1.157	0.202	0.923	0.203
¹¹⁄₁₆	0.688	0.700	1.176	0.172	1.258	0.221	. . .	. . . .
¾	0.750	0.763	1.271	0.188	1.361	0.241	1.111	0.218
¹³⁄₁₆	0.812	0.826	1.367	0.203	1.463	0.261	. . . .	. . . .
⅞	0.875	0.890	1.464	0.219	1.576	0.285	1.296	0.234
¹⁵⁄₁₆	0.938	0.954	1.560	0.234	1.688	0.308	. . . .	. . . .
1	1.000	1.017	1.661	0.250	1.799	0.330	1.483	0.250
1¹⁄₁₆	1.062	1.080	1.756	0.266	1.910	0.352	. . . .	. . . .
1⅛	1.125	1.144	1.853	0.281	2.019	0.375	1.669	0.313
1³⁄₁₆	1.188	1.208	1.950	0.297	2.124	0.396	. . . .	. . . .
1¼	1.250	1.271	2.045	0.312	2.231	0.417	1.799	0.313
1⁵⁄₁₆	1.312	1.334	2.141	0.328	2.335	0.438	. . . .	. . . .
1⅜	1.375	1.398	2.239	0.344	2.439	0.458	2.041	0.375
1⁷⁄₁₆	1.438	1.462	2.334	0.359	2.540	0.478	. . . .	. . . .
1½	1.500	1.525	2.430	0.375	2.638	0.496	2.170	0.375

[a]From ANSI B18.21.1. For complete listing, see the standard.
[b]Nominal washer sizes are intended for use with comparable nominal screw or bolt sizes.

14 WELDING SYMBOLS AND PROCESSES – AMERICAN WELDING SOCIETY STANDARD[a]

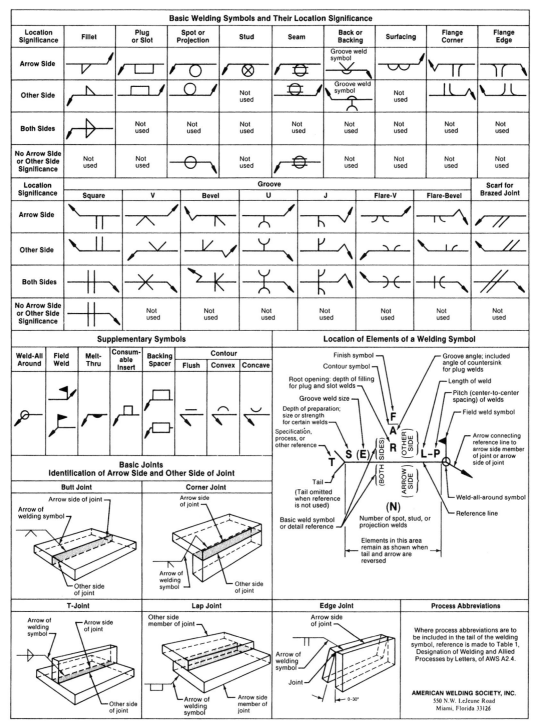

AMERICAN WELDING SOCIETY, INC.
550 N.W. LeJeune Road
Miami, Florida 33126

[a]ANSI/AWS A2.4. It should be understood that these charts are intended only as shop aids. The only complete and official presentation of the standard welding symbols is in A2.4.

15 TOPOGRAPHIC SYMBOLS

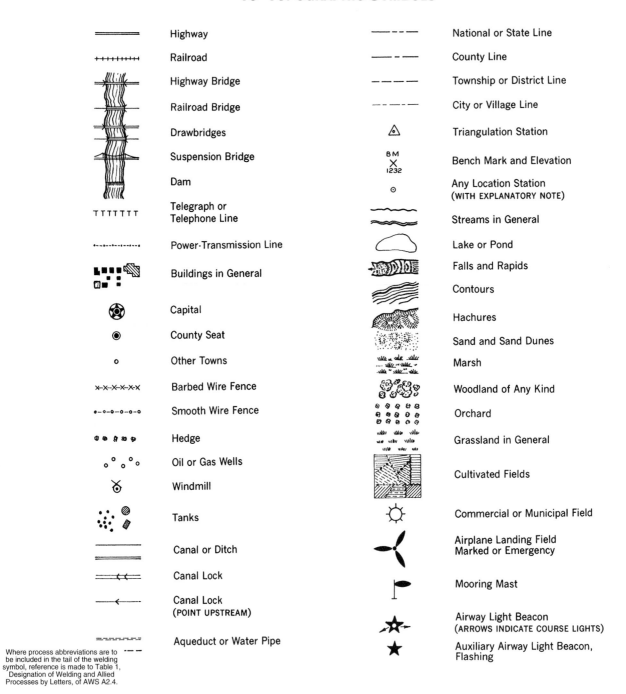

Highway		National or State Line	
Railroad		County Line	
Highway Bridge		Township or District Line	
Railroad Bridge		City or Village Line	
Drawbridges		Triangulation Station	
Suspension Bridge		Bench Mark and Elevation	
Dam		Any Location Station (WITH EXPLANATORY NOTE)	
Telegraph or Telephone Line		Streams in General	
Power-Transmission Line		Lake or Pond	
Buildings in General		Falls and Rapids	
Capital		Contours	
County Seat		Hachures	
Other Towns		Sand and Sand Dunes	
Barbed Wire Fence		Marsh	
Smooth Wire Fence		Woodland of Any Kind	
Hedge		Orchard	
Oil or Gas Wells		Grassland in General	
Windmill		Cultivated Fields	
Tanks		Commercial or Municipal Field	
Canal or Ditch		Airplane Landing Field Marked or Emergency	
Canal Lock		Mooring Mast	
Canal Lock (POINT UPSTREAM)		Airway Light Beacon (ARROWS INDICATE COURSE LIGHTS)	
Aqueduct or Water Pipe		Auxiliary Airway Light Beacon, Flashing	

Where process abbreviations are to be included in the tail of the welding symbol, reference is made to Table 1, Designation of Welding and Allied Processes by Letters, of AWS A2.4.

AMERICAN WELDING SOCIETY, INC.
550 N.W. LeJeune Road
Miami, Florida 33126

16 Piping Symbols[a] – American National Standard

	FLANGED	SCREWED	BELL & SPIGOT	WELDED	SOLDERED
1. Joint					
2. Elbow—90°					
3. Elbow—45°					
4. Elbow—Turned Up					
5. Elbow—Turned Down					
6. Elbow—Long Radius					
7. Reducing Elbow					
8. Tee					
9. Tee—Outlet Up					
10. Tee—Outlet Down					
11. Side Outlet Tee—Outlet Up					
12. Cross					
13. Reducer—Concentric					
14. Reducer—Eccentric					
15. Lateral					
16. Gate Valve—Elev.					
17. Globe Valve—Elev.					
18. Check Valve					
19. Stop Cock					
20. Safety Valve					
21. Expansion Joint					
22. Union					
23. Sleeve					
24. Bushing					

[a] ANSI/ASME Y32.2.3.

553

17 Standard Graphical Symbols for Electronic Diagrams

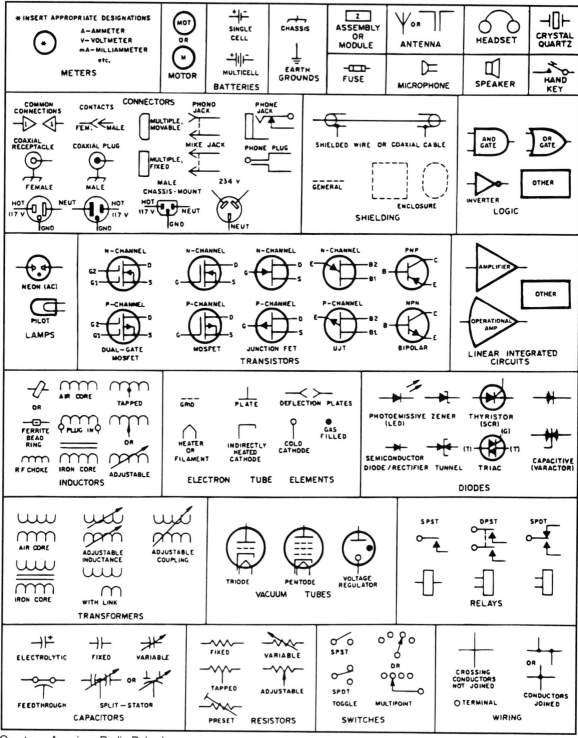

Courtesy: American Radio Relay League

554

18 GRAPHICAL ELECTRICAL WIRING SYMBOLS FOR ARCHITECTURAL AND ELECTRICAL LAYOUT DRAWINGS[a]

LIGHTING OUTLETS

Ceiling	Wall	
○	—○	Surface or Pendant Incandescent Mercury Vapor or Similar Lamp Fixture
ⓡ	—ⓡ	Recessed Incandescent Mercury Vapor or Similar Lamp Fixture
▭○▭		Surface or Pendant Individual Fluorescent Fixture
▭○ᴿ▭		Recessed Individual Fluorescent Fixture
▭○▭▭		Surface or Pendant Continuous-Row Fluorescent Fixture
▭○ᴿ▭▭		* Recessed Continuous-Row Fluorescent Fixture
├——┤——┤		** Bare-Lamp Fluorescent Strip
Ⓧ	—Ⓧ	Surface or Pendant Exit Light
Ⓧᴿ	—Ⓧᴿ	Recessed Exit Light
Ⓑ	—Ⓑ	Blanked Outlet
Ⓙ	—Ⓙ	Junction Box
Ⓛ	—Ⓛ	Outlet Controlled by Low-Voltage Switching When Relay is Installed in Outlet Box

* In the case of combination continuous-row fluorescent and incandescent spotlights, use combinations of the above standard symbols.

** In the case of continuous-row bare-lamp fluorescent strip above an area-wide diffusing means, show each fixture run, using the standard symbol; indicate area of diffusing means and type of light shading and/or drawing notation.

RECEPTACLE OUTLETS

Ungrounded	Grounding	
—⊖	—⊖ᴳ	Single Receptacle Outlet
—⊜	—⊜ᴳ	Duplex Receptacle Outlet
—⊛	—⊛ᴳ	Triplex Receptacle Outlet
—⊕	—⊕ᴳ	Quadruplex Receptacle Outlet
—⊜	—⊜ᴳ	Duplex Receptacle Outlet—Split Wired
—⊛	—⊛ᴳ	Triplex Receptacle Outlet—Split Wired
—⊿*	—⊿ᴳ*	* Single Special-Purpose Receptacle Outlet
—⊿*	—⊿ᴳ*	* Duplex Special-Purpose Receptacle Outlet
—⊜ᴿ	—⊜ᴿᴳ	Range Outlet
—◖ᴅᵂ	—◖ᴳᴅᵂ	Special-Purpose Connection or Provision For Connection. Use subscript Letters To Indicate Function (DW—Dishwasher; CD—Clothes Dryer, etc.)
⊖ x″	⊖ᴳ x″	Multi-Outlet Assembly. (Extend arrows to limit of installation. Use appropriate symbol to indicate type of outlet. Also indicate spacing of outlets as x inches.)
Ⓒ	—Ⓒᴳ	Clock Hanger Receptacle
Ⓕ	—Ⓕᴳ	Fan Hanger Receptacle
⊟	⊟ᴳ	Floor Single Receptacle Outlet
⊟	⊟ᴳ	Floor Duplex Receptacle Outlet
◭*	◭ᴳ*	* Floor Special-Purpose Outlet
◀		Floor Telephone Outlet—Public
◁		Floor Telephone Outlet—Private

* Use numeral or letter either within the symbol or as a subscript alongside the symbol keyed to explanation in the drawing list of symbols to indicate type of receptacle or usage.

SWITCH OUTLETS

S	Single-Pole Switch
S₂	Double-Pole Switch
S₃	Three-Way Switch
S₄	Four-Way Switch
Sₖ	Key-Operated Switch
Sₚ	Switch and Pilot Lamp
Sₗ	Switch for Low-Voltage Switching System
Sₗₘ	Master Switch for Low-Voltage Switching System
—⊖S	Switch and Single Receptacle
⊜S	Switch and Double Receptacle
Sᴅ	Door Switch
Sₜ	Time Switch
S꜀ʙ	Circuit Breaker Switch
Sₘ꜀	Momentary Contact Switch or Pushbutton For Other Than Signalling System
Ⓢ	Ceiling Pull Switch

SIGNALLING SYSTEM OUTLETS
RESIDENTIAL OCCUPANCIES

▣	Pushbutton
⊽	Buzzer
◫	Bell
⊽◫	Combination Bell-Buzzer
CH	Chime
◇	Annunciator
D	Electric Door Opener
M	Maid's Signal Plug
▢	Interconnection Box
BT	Bell-Ringing Transformer
▶	Outside Telephone
▷	Interconnecting Telephone
R	Radio Outlet
TV	Television Outlet

CIRCUITING

Wiring Method Identification By Notation On Drawing Or In Specifications

———————— Wiring Concealed in Ceiling or Wall

— — — — Wiring Concealed in Floor

------------ Wiring Exposed

Note: Use heavy-weight line to identify service and feeders. Indicate empty conduit by notation CO (Conduit only)

———⤻²⁄¹ Branch Circuit Home Run to Panel Board. Number of arrows indicates number of circuits. (A numeral at each arrow may be used to identify circuit number.) Note: Any circuit without further identification indicates two-wire circuit. For a greater number of wires, indicate with cross lines, e.g.:

——⫴— 3 wires; ——⫼— 4 wires, etc.

Unless indicated otherwise, the wire size of the circuit is the minimum size required by the specification. Identify different functions of wiring system, e.g., signalling system by notation or other means.

———○ Wiring Turned Up

———● Wiring Turned Down

[a] ANSI Y32.9.

19 ARCHITECTURAL SYMBOLS

Symbol	Name	Symbol	Name
	BRICK		WINDOW IN A BRICK VENEER WALL
	CONCRETE BLOCK		WINDOW IN A BRICK WALL WITH PLASTER
	CLAY TILE		DOOR IN A FRAME WALL
	CONCRETE		DOOR IN A BRICK WALL
	STONE		DOOR IN A BRICK VENEER WALL
	ROUGH WOOD		DOOR IN A BRICK WALL WITH PLASTER
	FINISHED WOOD		FIREPLACE
	STEEL		CHIMNEY
	FILL		STAIRS
	SAND		BUILT-IN-TUB
	GRAVEL		WATER CLOSET
	INSULATION		LAVATORY
	WINDOW IN A FRAME WALL		COUNTER SINK
	WINDOW IN A BRICK WALL		RANGE

556

20 FRACTIONAL AND DECIMAL-INCH AND MILLIMETER EQUIVALENTS

Decimal measurements may be set off directly on drawings with the aid of an engineers scale.
Metric measurements may be set off directly on drawings with a metric scale.

4ths	8ths	16ths	32nds	64ths	To 4 Places	To 3 Places	To 2 Places	Milli-meters	4ths	8ths	16ths	32nds	64ths	To 4 Places	To 3 Places	To 2 Places	Milli-meters
				1/64	.0156	.016	.02	.397					33/64	.5156	.516	.52	13.097
			1/32		.0312	.031	.03	.794				17/32		.5312	.531	.53	13.494
				3/64	.0469	.047	.05	1.191					35/64	.5469	.547	.55	13.891
		1/16			.0625	.062	.06	1.588			9/16			.5625	.562	.56	14.288
				5/64	.0781	.078	.08	1.984					37/64	.5781	.578	.58	14.684
			3/32		.0938	.094	.09	2.381				19/32		.5938	.594	.59	15.081
				7/64	.1094	.109	.11	2.778					39/64	.6094	.609	.61	15.478
	1/8				.1250	.125	.12	3.175		5/8				.6250	.625	.62	15.875
				9/64	.1406	.141	.14	3.572					41/64	.6406	.641	.64	16.272
			5/32		.1562	.156	.16	3.969				21/32		.6562	.656	.66	16.669
				11/64	.1719	.172	.17	4.366					43/64	.6719	.672	.67	17.066
		3/16			.1875	.188	.19	4.762			11/16			.6875	.688	.69	17.462
				13/64	.2031	.203	.20	5.159					45/64	.7031	.703	.70	17.859
			7/32		.2188	.219	.22	5.556				23/32		.7188	.719	.72	18.256
				15/64	.2344	.234	.23	5.953					47/64	.7344	.734	.73	18.653
1/4					.2500	.250	.25	6.350	3/4					.7500	.750	.75	19.050
				17/64	.2656	.266	.27	6.747					49/64	.7656	.766	.77	19.447
			9/32		.2812	.281	.28	7.144				25/32		.7812	.781	.78	19.844
				19/64	.2969	.297	.30	7.541					51/64	.7969	.797	.80	20.241
		5/16			.3125	.312	.31	7.938			13/16			.8125	.812	.81	20.638
				21/64	.3281	.328	.33	8.334					53/64	.8281	.828	.83	21.034
			11/32		.3438	.344	.34	8.731				27/32		.8438	.844	.84	21.431
				23/64	.3594	.359	.36	9.128					55/64	.8594	.859	.86	21.828
	3/8				.3750	.375	.38	9.525		7/8				.8750	.875	.88	22.225
				25/64	.3906	.391	.39	9.922					57/64	.8906	.891	.89	22.622
			13/32		.4062	.406	.41	10.319				29/32		.9062	.906	.91	23.019
				27/64	.4219	.422	.42	10.716					59/64	.9219	.922	.92	23.416
		7/16			.4375	.438	.44	11.112			15/16			.9375	.938	.94	23.812
				29/64	.4531	.453	.45	11.509					61/64	.9531	.953	.95	24.209
			15/32		.4688	.469	.47	11.906				31/32		.9688	.969	.97	24.606
				31/64	.4844	.484	.48	12.303					63/64	.9844	.984	.98	25.003
					.5000	.500	.50	12.700						1.0000	1.000	1.00	25.400

Sheet Layouts

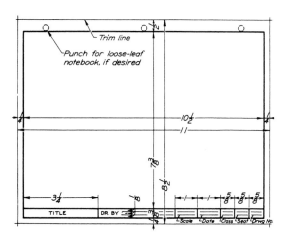

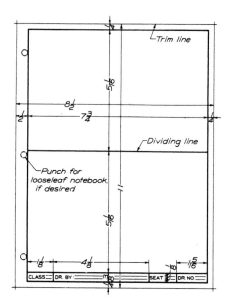

Layout A
See Fig. 4-43 showing steps in drawing this layout.

Layout B

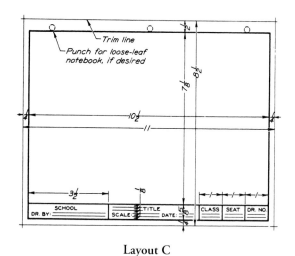

Layout C

Sheet Sizes

American National Standard

A – 8.50″ × 11.00″
B – 11.00″ × 17.00″
C – 17.00″ × 22.00″
D – 22.00″ × 34.00″
E – 34.00″ × 44.00″

International Standard

A4 – 210 mm × 297 mm
A3 – 297 mm × 420 mm
A2 – 420 mm × 594 mm
A1 – 594 mm × 841 mm
A0 – 841 mm × 1189 mm

(25.4 mm = 1.00″)

Note: Sheet sizes are also available from suppliers in decimal-inch or metric equivalents.

SHEET LAYOUTS (CONTINUED)

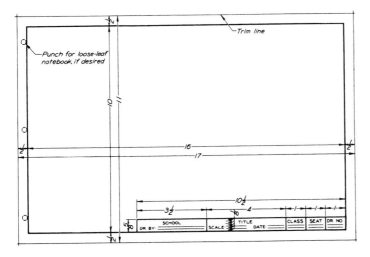

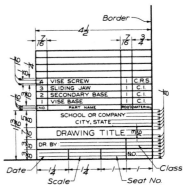

Layout D

Block title and parts list for Layout E

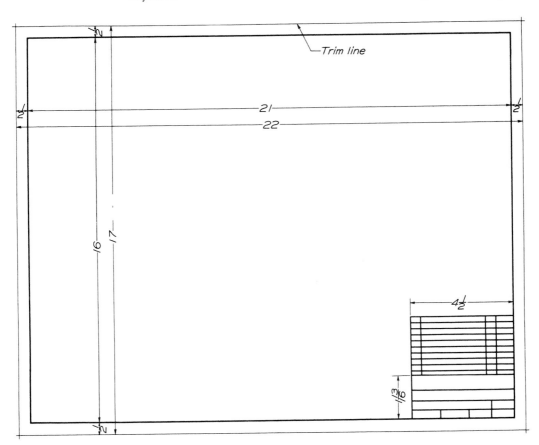

Layout E
See block title and parts list above.

GLOSSARY

This glossary provides definitions of key terms in technical drawing and manufacturing processes. Many of these definitions are followed by the number of the text section where the term is discussed.

Each term is designated as being either a noun (*n*) or a verb (*v*).

"The beginning of wisdom is to call things by their right names."

— CHINESE PROVERB

A

Acme *(n).* Screw thread form, Secs. 15.4, 15.11.

addendum *(n).* Radial distance from pitch circle to top of gear tooth. Sec. 20.5.

aeronautical drafting *(n).* Technical drawing used in aircraft manufacturing. Sec. 1.5.

aligned system *(n).* A dimensioning system in which the figures are lettered so as to be read from the bottom of the sheet, from the right side, or between these positions. Sec. 10.6.

allowance *(n).* Minimum clearance between mating parts. Secs. 10.26, 10.27.

alphabet of lines *(n).* The lines used in drafting. Sec. 4.10.

American National thread *(v).* An older standard thread form that is now being replaced. Sec. 15.16.

ampere *(n).* The unit in which current is measured. Sec. 23.3.

anneal *(v).* To heat and cool gradually, to reduce brittleness and increase ductility. Sec. 11.17.

architects scale *(n).* An all-round scale with a full-size scale of inches divided into sixteenths. It also has a number of reduced-size scales in which inches or fractions of an inch represent feet. Sec. 4.20.

architectural drawing *(n).* A drawing that shows the shape and size of a structure. Sec. 1.5.

arrowhead *(n).* A pointer drawn at the end of a line. Sec. 10.3.

assembly drawing *(n).* A drawing that shows how part or all of a machine or structure is put together. Sec. 16.8.

atlas *(n).* Maps bound together in book form. Sec. 22.1.

auxiliary view *(n).* A true-size view of an inclined face obtained by viewing the object at right angles to that face through a special inclined auxiliary plane parallel to it. Sec. 13.2.

axis of revolution *(n).* The axis about which an object is revolved. Sec. 14.2.

B

bar chart *(n).* A chart that shows the percentage ratio of various parts to a given whole, using bars of various heights. The total length of the bar represents 100 percent. This distance is divided into segments that are proportional parts of the whole. The bars may be drawn horizontally or vertically. Appropriate shading or crosshatching is used to distinguish between the segments. Sec. 19.2.

basic size *(n)*. The size from which the limits of size are derived by the application of allowances and tolerances. Sec. 10.27.

beams *(n)*. Horizontal structural members that help distribute loads and thus support the structure. Sec. 21.1.

bevel *(n)*. An inclined edge, not at a right angle to the adjoining surface.

bevel gears *(n)*. Gears that connect shafts whose axes intersect. Sec. 20.8.

blind hole *(n)*. A hole that does not go all the way through a piece.

block *(n)*. The name given to several entities assembled in one group. Also called a *pattern*. Sec. 9.8.

blueprint paper *(n)*. Ordinary paper coated on one side with a chemical preparation sensitive to light. The coated surface is a pale green color. Sec. 8.4.

blueprinting *(n)*. A method of reproducing drawings that uses sensitized paper and light. Sec. 8.4.

bolt circle *(n)*. A circular center line on a drawing, containing the centers of holes about a common center. Sec. 10.17.

bore *(v)*. To enlarge a hole with a boring bar or tool in a lathe, drill press, or boring mill. Secs. 11.11, 11.12.

broach *(n)*. A long cutting tool with a series of teeth that gradually increase in size. The tool is forced through a hole or over a surface to produce a desired shape. Sec. 11.16.

broken-out section *(n)*. A small part of a view that is sectioned to show some detail of inside construction. Sec. 12.5.

C

CAD/CAM (computer-aided drafting and computer-aided manufacturing) *(n)*. Software programs that integrate design and manufacturing. Sec. 3.11.

CAD system (computer-aided drafting system) *(n)*. A drafting system that consists of a software program, which performs the commands, and computer hardware, which runs the software program. Sec. 3.2.

calipers *(n)*. Instrument (of several types) for measuring diameters. Sec. 11.10.

cam *(n)*. A rotating member for changing circular motion to reciprocating motion. Sec. 20.1.

carburize *(v)*. To heat a low-carbon steel to approximately 2000°F. in contact with material that adds carbon to the surface of the steel, and to cool slowly in preparation for heat treatment. Sec. 11.17.

caseharden *(v)*. To harden the outer surface of carburized steel by heating and then quenching. Sec. 11.17.

casting *(n)*. A metal object produced by pouring molten metal into a mold. Sec. 11.2.

cast iron *(n)*. Iron melted and poured into molds. Sec. 11.2.

CD-ROM (compact-disk read-only memory) *(n)*. A disk that holds digital computer information. Also spelled compact disc. Sec. 3.9.

center lines *(n)*. Lines used to indicate axes of symmetry. They are also used in place of extension lines for locating holes and other features. Sec. 10.2.

central processing unit (CPU) *(n)*. The main chip inside a computer. It receives and analyzes instructions, performs the required operations, and sends information to storage. Sec. 3.4.

chamfer *(n).* A narrow inclined surface along the intersection of two surfaces.

CHAMFER

change strip *(n).* A block on the drawing that briefly describes changes to the drawing. Sec. 16.4.

chart *(n).* (1) A graphical presentation of numerical data. Sec. 19.1. (2) A map prepared primarily for navigation. Sec. 22.1.

cheek *(n).* The middle portion of a three-piece flask used in molding, Sec. 11.2.

chuck *(n).* A mechanism for holding a rotating tool or workpiece.

colorharden *(v).* Same as caseharden, except that it is done to a shallower depth, usually for appearance only.

command line *(n).* The area on the computer screen where you can type commands from the keyboard. This area also displays error messages and other system prompts. Sec. 9.4.

compact disk *(n).* See **CD-ROM.**

computer-aided drafting system *(n).* See **CAD system.**

computer graphics *(n).* A term applied to the use of computers to produce drawings, designs, graphs, charts, and other graphical data. Sec. 1.5.

concurrent engineering *(n).* A process in which the design and production teams work together to ensure that a design is workable and can be manufactured in an efficient, cost-effective way. Sec. 3.12.

cone *(n).* A single-curved surface that has straight-line elements. Sec. 6.2.

connection diagram *(n).* See **wiring diagram.**

contour *(n).* In map drafting, an imaginary line on the earth's surface, passing through points of equal elevation, or height above sea level. Sec. 22.1.

contour interval *(n).* In map drafting, the distance between the contours. Sec. 22.1.

contour lines *(n).* Lines on a map that represent the elevations of the various portions. Sec. 22.1.

contour pen *(n).* A pen used to draw contours on a map. Sec. 22.1.

conventional break *(n).* In drawings, a break used to remove a considerable length from a long object. Sec. 12.12.

cope *(n).* The upper portion of a flask used in molding. Sec. 11.2.

core *(v).* To form a hollow portion in a casting by using a dry-sand core or a green-sand core in a mold. Sec. 11.2.

core box *(n).* In sand casting, the box that holds a prepared mixture of sand and a binding substance. Sec. 11.2.

cotter pin *(n)* A split pin used as a fastener, usually to prevent a nut from unscrewing.

counterbore *(v).* To enlarge an end of a hole cylindrically with a counterbore tool., Sec. 11.12.

COUNTERBORE

countersink *(v).* To enlarge an end of a hole conically, usually with a countersink tool. Sec. 11.12.

COUNTERSINK

current *(n)*. The time rate of flow of electricity. Sec. 23.3.

cursor location coordinates *(n)*. In CAD, the numbers that indicate the location of the cursor. The cursor coordinates follow the standard Cartesian coordinate system. The lower left-hand corner of the drawing window is usually the origin, coordinate (0,0). Sec. 9.5.

cutting-plane line *(n)*. In a top view, the line that represents the edge view of a cutting plane. Sec. 12.1.

cylinder *(n)*. A single-curved surface with straight line elements.

D

dedendum *(n)*. The depth of a gear tooth below the pitch circle. Sec. 20.5.

depth auxiliary view *(n)*. An auxiliary view named for the principle dimensions of the object shown in the auxiliary view. Sec. 13.3.

descriptive geometry *(n)*. That branch of geometry that provides a graphical solution of a three-dimensional problem by means of projections upon mutually perpendicular planes. Sec. 25.1.

design drawing *(n)*. A drawing that indicates the general arrangement of a structure or product and gives specifics of the different members. Sec. 21.1.

design size *(n)*. The size from which the limits of size are derived by the application of tolerances. When there is no allowance, the design size and basic size are the same. Sec. 10.27.

detail drawings or **details** *(n)*. Drawings that give all of the information needed to make a part (Sec. 16.1) or construct a structural member (Sec. 21.1). Details in architectural drawings are drawn in a scale large enough to distinguish the small parts and features. Detail drawings may be placed on sheets containing plans, elevations, and sections. Sec. 24.8.

development *(n)* Drawing of the surface of an object unfolded or rolled out on a plane.

diametral pitch *(n)* Number of gear teeth per inch of pitch diameter. Sec. 20.5.

diazo-dry process *(n)*. A reproduction process that uses paper coated with special chemicals. The image is developed by contact with ammonia vapors. Sec. 8.5.

die *(n)*. A tool used to cut small external threads by hand. Sec. 15.12.

die casting *(n)*. Process of forcing molten metal under pressure into metal dies or molds, producing a very accurate and smooth casting, Sec. 11.7.

die stamping *(n)*. Process of cutting or forming a piece of sheet metal with a die.

digital versatile disk. See **DVD.**

digitizer pad (or **tablet**) *(n)*. A computer input device. It consists of (1) a tablet that is a little larger than a notebook. (2) a plastic sheet that fits over the tablet, and (3) a hand-held puck with several buttons. The tablet contains invisible spots that sense the position of the puck. Each spot is programmed with a different CAD command. Sec. 3.7.

dimension line *(n)*. A line that has an arrowhead at each end indicating the extent of the dimension. A gap is left (except in architectural and structural drawing) near the middle for the dimension figure. On small drawings, dimension lines are spaced at least 3/8″ (.375″) from the object and at least 1/4″ (.250″) apart. The spacing must be uniform throughout the drawing. Sec. 10.2.

direction of revolution *(n)*. The direction in which an object is revolved about the axis of revolution. The direction of revolution is clockwise or counterclockwise. Sec. 14.2.

dividers *(n)*. An instrument used for subdividing distances into equal spaces or for transferring distances in which the spacing between the points is approximately 1″ or over. Sec. 4.28.

dog *(n)*. A small auxiliary clamp used to prevent work from rotating in relation to the face plate of a lathe. Sec. 11.11.

dowel *(n)*. A cylindrical pin, commonly used between two contacting flat surfaces to prevent sliding.

DOWEL

draft *(n)*. The tapered shape of the parts of a pattern that permit it to be easily withdrawn from the sand or, on a forging, that permit it to be easily withdrawn from the dies. Sec. 11.2.

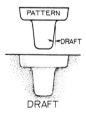

PATTERN

DRAFT

DRAFT

drafting machine *(n)*. A drafting tool that combines all the functions of the T-square, triangles, scales, and a protractor. Sec. 4.32.

drag *(n)*. Lower portion of a flask used in molding. Sec. 11. 2.

drawing window *(n)*. In CAD, the large blank area in the middle of the computer screen. This is the area where drawings are created, viewed, and edited. Sec. 9.3.

drill *(v)*. To cut a cylindrical hole with a drill. Sec. 11.12.

drill press *(n)*. A machine for drilling and other hole-forming operations. Sec. 11.12.

drop forge *(v)*. To form a piece while hot between dies in a drop hammer or with great pressure. Sec. 11.6.

DVD (digital versatile disk) *(n)*. An optical disk similar to a compact disk but able to store much more data. Sec. 3.9.

E

edge view *(n)*. A view of a plane surface obtained by looking parallel to a true-length line in the surface. Sec. 25.3.

electrical drawing *(n)*. Technical drawing used in the electronics industries. Sec. 1.5.

electrical or **electronic drafting** *(n)*. The drawing of electrical wiring diagrams, interconnection diagrams, block diagrams, layout diagrams, and schematics. Sec. 23.1.

electric circuit *(n)*. The path through which a current flows. Sec. 23.3.

element *(n)*. An imaginary straight line. For example, an element of a cylinder is an imaginary straight line on the surface parallel to the axis. Sec. 18.8.

elevation *(n)*. An outside view of a building. Sec. 24.7.

ellipse *(n)*. A geometric form resembling a circle viewed at an angle. Sec. 2.9.

engineers scale *(n)*. A scale with a series of scales in which inches are divided into 10, 20, 30, 40, 50, 60, or 80 parts. Sec. 4.20.

entities *(n)*. Basic geometric shapes such as lines, circles, and arcs. Also called objects. Sec. 9.6.

entrepreneur *(n)*. Someone who starts a business.

erection diagram *(n)*. A diagram that shows piece markings and indicates the sequence to be followed in the final assembly of the structure. Sec. 21.1.

extension line *(n)*. In drafting, a line that "extends" from the object, with a gap of about 1/16″ (.062″) next to the object. It continues to about 1/8″ (.125″) beyond the outermost arrowhead. Sec. 10.2.

F

face *(v)*. To finish a surface at right angles, or nearly so, to the center line of rotation on a lathe. Sec. 11.11.

FAO *(n)*. Finish all over. Sec. 10.7.

figure of intersection *(n)*. The complete intersection between two solids. Sec. 18.22.

fillet *(n)*. An interior rounded intersection between two surfaces. Sec. 11.3.

fin *(n)*. See **flash.**

finish mark *(n)*. A symbol indicating that a surface is to be finished or machined. Sec. 10.7.

fit *(n)*. Degree of tightness or looseness between two mating parts, as a loose fit, a snug fit, or a tight fit, Secs. 10.26, 10.28.

fixture *(n)*. A special device for holding the work in a machine tool, but not for guiding the cutting tool.

flange *(n)*. A relatively thin rim around a piece.

FLANGE

flash *(n)*. A thin extrusion of metal at the intersection of dies or sand molds. Sec. 11.6.

flask *(n)*. A box made of two or more parts for holding the sand in sand molding, Sec. 11.2.

forging *(n)*. A metal shape produced by hammering heated bars or billets of metal between dies. Sec. 11.6.

foundation plans *(n)*. Plans that show the footings or the location of piles to be driven. Sec. 21.1.

friction gear *(n)*. A gear that, if placed in contact with another friction gear and rotated, will transmit motion to the other gear. Sec. 20.5.

front view or **front elevation** *(n)*. A view that shows the true width and height of an object, but not the depth. The view is obtained by extending perpendicular projections from all points on the object to the plane. Sec. 7.1.

full section *(n)*. A sectional view obtained by passing a cutting plane fully through the object. Sec. 12.1.

function keys *(n)*. On a computer keyboard, the keys labeled F1 through F12. They are shortcut keys for issuing commands. Sec. 3.7.

G

gate *(n)*. The opening in a sand mold at the bottom of the sprue through which the molten metal passes to enter the cavity or mold. Sec. 11.2.

general assembly *(n)*. A drawing that shows how the parts fit together and how the assembly functions. Sec. 16.11.

giant bow compass *(n)*. A large bow compass with a maximum radius about equal to that of the conventional large compass, but with the rigidity of the small bow instrument. Sec. 4.25.

graduate *(v)*. To set off accurate divisions on a scale or dial.

grid *(n)*. In CAD, a horizontal and vertical pattern of dots spaced at regular, preset intervals. Sec. 9.5.

grind *(v)*. To remove metal by means of an abrasive wheel, often made of carborundum. Used chiefly where accuracy is required. Sec. 11.15.

H

half section *(n)*. The section obtained if the cutting plane is passed only halfway through an object and then the quarter of the object in front of the cutting plane is removed. Sec. 12.4.

half view *(n)*. A view sometimes used for simple symmetrical objects. Center lines instead of break lines are used to limit the half views. Sec. 13.8.

hard disk *(n)*. The primary storage medium in a computer. Hard disks are measured by their capacity in bytes and by their speed in milliseconds. The speed of a drive is a measure of how fast you can read or write information to the drive. Sec. 3.9.

harden *(v)*. To heat steel above a critical temperature and then quench in water or oil. Sec. 11.17.

heat-treat *(v)*. To change the properties of metals by heating and then cooling. Sec. 11.17.

height-auxiliary view *(n)*. An auxiliary view that shows the principal dimension, height. Sec. 13.3.

helix (pronounced *he' liks*) *(n)*. The design made by screw threads as they wind around a shaft in a curve. Sec. 15.1.

hidden lines *(n)*. Dashed lines that show the hidden parts of an object. Secs. 7.4, 7.15.

I

icon *(n)*. A graphic symbol. Sec. 9.2.

inclined letters *(n)*. On a drawing, letters that lean to the right. Sec. 5.2.

integrated manufacturing *(n)*. The type of manufacturing that integrates CAD with manufacturing processes. Sec. 3.11.

interchangeable part *(n)*. Refers to a part made to limit dimensions so that it will fit any mating part similarly manufactured. Sec. 10.26.

involute system *(n)*. The most widely used gear tooth form. Sec. 20.6.

irregular, or **French, curves** *(n)*. Curves made of amber or clear plastic used to draw irregular curves. Sec. 4.30.

isometric drawing *(n)*. A drawing prepared so that the front edge (and those edges parallel to it) will appear vertical. The two lower edges (and those edges parallel to it) will appear vertical. The two lower edges (and those parallel to them) will appear about 30° to the horizontal. Sec. 17.3.

isometric lines *(n)*. Lines that are parallel to the isometric axes. Sec. 17.7.

J

jig *(n)* A device for guiding a tool in cutting a piece. Usually it holds the work in position.

joints *(n)*. In truss construction, the points at which the parts of a truss connect. Sec. 21.1.

K

key *(n)*. A small piece of metal sunk partly into both shaft and hub to prevent rotation, Sec. 15.28.

keyboard *(n)*. The most common computer input device. It consists of typewriter keys, navigation keys, function keys, and a number keypad. Sec. 3.7.

keyseat *(n)*. A slot or recess in a shaft to hold a key. Sec. 15.28.

KEYSEAT

keyway *(n)*. A slot in a hub or portion surrounding a shaft to receive a key, Sec. 15.28.

KEYWAY

knurl *(v)*. To impress a pattern of dents in a turned surface with a knurling tool to produce a better hand grip. Sec. 11.11.

L

lathe *(n)* A machine used to shape metal or other materials by rotating against a tool. Sec. 11.11.

layer *(n)*. In CAD, an electronic "sheet of paper" on which a drawing is created. Different parts of the drawing can be drawn on different layers. Sec. 9.5.

layout drawing *(n)*. An assembly drawing made to scale and including the views necessary to show the size and shape of each part. Dimensions are omitted. Sec. 16.1.

leader *(n)*. A thin solid line that "leads" from a note or dimension. It is terminated by an arrowhead touching the part to which attention is directed. Sec. 10.2.

lettering devices *(n)*. Instruments used with a special guide or template to form letters. Sec. 5.16.

line chart *(n)*. A chart on which numbers plotted on a grid are connected by straight lines. Sec. 19.4.

line of intersection *(n)*. The point at which two surfaces intersect. Sec. 18.22.

lowercase letters *(n)*. Letters that are not capital letters. They are called lowercase because printers used to keep such letters in the lower case of type. Sec. 5.2.

M

machine drawing *(n)*. Technical drawing used in the manufacturing industries. Sec. 1.5.

major axis *(n)*. The long axis of a shape (as an ellipse). Sec. 2.9.

malleable casting *(n)*. A casting that has been made less brittle and tougher by annealing.

map *(n)*. A drawing of the earth's surface or a part of it. Sec. 22.1.

marine drawing *(n)*. Technical drawing used in ship construction. Sec. 1.5.

mating dimensions *(n)*. Certain dimensions that must correspond to make parts fit together. Sec. 10.22.

mating parts *(n)*. Two or more parts that fit together. Sec. 10.22.

mechanical drawing *(n)*. A drawing made with precision drawing instruments.

members *(n)*. In truss construction, the various parts of a truss. Sec. 21.1.

menu *(n)*. In computer software, a list of items displayed on the monitor. The user selects from this list. Sec. 9.4.

metric system *(n)*. A decimal system of measurement. Sec. 10.30.

microfilming *(v)*. A photographic process that reduces drawings and records to 1/15-1/16 of their original size. Sec. 8.6.

micrometer caliper *(n)*. A precision measuring instrument that allows to four decimal places, or 1/10,000″. Sec. 11.10.

mill *(v)*. To remove material by means of a rotating cutter on a milling machine. Sec. 11.14.

milling machine *(n)*. A machine with rotating cutters. The workpiece (usually metal) is fastened to the table and fed against the cutters. Sec. 11.14.

minor axis *(n)*. The short axis of a shape (as an ellipse). Sec. 2.9.

mold *(n)*. The mass of sand or other material that forms the cavity into which molten metal is poured. Sec. 11.2.

mouse *(n)*. A computer input device. It has a rubber ball on the bottom that rolls on the desktop. When you move the mouse up or down, the cursor on the screen moves up or down. Sec. 3.7.

N

navigation keys *(n)*. On a computer keyboard, the keys that move the cursor around the monitor screen. The main navigation keys are the four cursor keys: left, right, up, and down. Sec. 3.7.

nominal size *(n)*. The size designation that is used for the purpose of general identification. Sec. 10.27.

non-isometric lines *(n)*. Lines that are not parallel to the isometric axes. Sec. 17.7.

normalize *(v)*. To heat steel above its critical temperature and then to cool it in air. Sec. 11.17.

O

objects *(n)*. See **entities.**

object snap *(n)*. In CAD, a feature that causes the crosshairs to jump to specified parts of an object, such as the midpoint or endpoint. Sec. 9.5.

offset section *(n)*. A section in which the cutting plane is bent or "offset" to pass through several features of an object that are not in a straight line. Sec. 12.8.

ohm *(n)*. The unit in which resistance is measured. Sec. 23.3.

Ohm's law *(n)*. The law that expresses the relationship among current, voltage, and resistance. Current (in amperes) equals potential difference (in volts) divided by resistance (in ohms). Sec. 23.3.

one-point perspective *(n)*. A perspective drawing with one vanishing point. Sec. 17.32.

operating system *(n)*. The system that translates the commands from the software program into language the computer hardware can understand. Sec. 3.6.

ortho *(n)*. In CAD, a feature that restricts crosshair movement to either vertical or horizontal directions. Sec. 9.5.

outline assembly (*n*). A drawing that shows one or more views of an assembly "in outline." Little or no sectioning is generally needed. Unimportant details are omitted. Sec. 16.9.

outside spring calipers (*n*). Calipers used to check the nominal size of outside diameters. Sec. 11.10.

P

parallelepiped (*n*). A prism having bases that are parallelograms. Sec. 6.2.

parallel-line development (*n*). A rectangular pattern such as the type produced when a prism or cylinder is rolled out. Sec. 18.1.

parallel ruling straightedge (*n*). A straightedge controlled by a system of cords and pulleys. These permit it to be moved up or down on the board, always maintaining a true horizontal position. Sec. 4.31.

partial view (*n*). A view that gives a complete shape description, while also omitting the drawing of difficult curves. Sec. 13.8.

pattern (*n*). (1) A model, usually of wood, used in forming a mold for a casting. Sec. 11.2. (2) In CAD, a group of entities. Sec. 9.8.

pattern drawing (*n*). A drawing that gives only the information needed in the pattern shop. Sec. 11.1.

perspective (*n*). The technique of representing objects so that they appear progressively smaller as they are farther away. Sec. 17.28.

pictograph (*n*). A graph or chart in which pictures are used as symbols to represent units or quantities. Sec. 19.6. Also, picture writing. Sec. 5.1.

pictorial drawing (*n*). One in which the object is viewed in such a position that several faces appear in a single view.

pie chart (*n*). A chart that shows how the whole is split up into portions. Sec. 19.5.

piercing point (*n*). The point of intersection of a line and an oblique plane. Sec. 25.4.

pinion (*n*). The smaller of two mating gears. Sec. 20.5.

pitch (of a thread) (*n*). The distance, parallel to the axis, from a point on one thread to the corresponding point on the next adjoining thread. Sec. 15.5.

pitch circle (*n*). An imaginary circle corresponding to the circumference of the friction gear from which the spur gear was derived. Sec. 20.6.

pixels (*n*). Dots displayed on the screen of a monitor. Sec. 3.8.

plan (*n*). A top view. In architectural drawing, a plan is a horizontal section through the house taken above the window-sill height.

plat (or plot) (*n*). A map of a small "parcel" of land. Sec. 22.1.

plotter (*n*). A computer output device. Some plotters have ink pens or pencil points that draw lines, letters, etc., on paper. Other plotters use ink jet or electrostatic mechanisms to transfer ink to paper. Sec. 3.10.

primary auxiliary views (*n*). The three types of ordinary auxiliary view—the depth auxiliary, height auxiliary, and width auxiliary views. Sec. 13.3.

printed circuit (*n*). An electronic circuit made by placing conductive material in paths from terminal to terminal on an insulating surface. Sec. 23.9.

prism *(n)*. A geometric shape having plane faces parallel to an imaginary axis. Sec. 6.2.

production illustration *(n)*. An illustration that shows the way parts fit together and the sequence of operations to be performed. Sec. 17.34.

profile *(n)*. (1) An outline. (2) A sectional view through the earth produced by a vertical sectioning plane. Sec. 22.1.

proportion *(n)*. The relation in size between the various parts of a drawing. A drawing is said to be in proportion if all its parts are the correct sizes when compared to all other parts. Sec. 2.7.

pyramid *(n)*. A geometric shape having plane triangular faces that intersect at a common point. Sec. 6.2.

Q, R

quadrilateral *(n)*. A geometric shape having four sides. Sec. 6.2.

rack *(n)*. A flat bar with gear teeth in a straight line to engage with teeth in a gear.

radial-line development *(n)*. A development in which the edges of elements "radiate" like spokes in a wheel from a point, instead of being parallel. These lines usually do not show true length in the regular views. Sec. 18.13.

random access memory (RAM) *(n)*. Temporary memory that is stored on integrated circuit chips inside a computer. Sec. 3.4.

rapid prototyping *(n)*. Any one of several processes by which a 3D CAD drawing is used to directly generate a physical model of an object. One such method is **stereolithography.**

ream *(v)*. To enlarge a finished hole slightly to give it greater accuracy. Sec. 11.11.

regular polygon *(n)* A geometric shape with four equal sides. Sec. 6.2.

regular solid *(n)*. A three-dimensional geometric shape with faces that are regular polygons. Sec. 6.2.

removed section *(n)*. In drafting, a section that is moved from its normal position to some more convenient position on the sheet. In such cases, the removed section may be drawn to a larger scale if desired. Sec. 12.7.

resistance *(n)*. The opposition to the flow of electricity. Sec. 23.3.

resolution *(n)*. The quality of display of a monitor screen. It is determined by the number of dots, or pixels, displayed on the screen. Sec. 3.8.

revolved section *(n)*. A section obtained by passing a cutting plane at right angles through the central portion of the object and then revolving the cutting plane. Sec. 12.6.

rib or **web** *(n)*. A thin flat part of an object used for bracing or adding strength. Sec. 12.9.

right-side view or **elevation** *(n)*. The view that shows the true height and depth of an object, but not the width. Sec. 7.1.

rivet *(v)*. To connect with rivets, or to clench over the end of a pin by hammering.

round *(n)*. A rounded outside corner on a casting. Sec. 11.3.

round *(n)*. An exterior rounded intersection of two surfaces. Sec. 11.3.

S

SAE *(n)*. Society of Automotive Engineers.

sand casting *(n)*. A manufacturing process in which a casting is made by pouring molten metal (iron, steel, aluminum, brass, or some other metal) into a cavity in damp sand. Sec. 11.2.

sandblast *(v)*. To blow sand at high velocity with compressed air against castings or forgings to clean them.

schedule *(n)*. A table or chart specifying in detail the types and sizes of certain features on the plans. Sec. 24.18.

schematic or **elementary diagram** *(n)*. An electrical diagram that shows the electrical connections and functions of a circuit by means of standard graphical symbols. It shows them without any regard to the actual physical size, shape, or location of the electrical devices or parts that make up the circuit. Sec. 23.4.

scleroscope *(n)*. An instrument for measuring hardness of metals.

scrape *(v)*. To remove metal by scraping with a hand scraper, usually to fit a bearing.

secondary auxiliary view *(n)*. Any auxiliary view that is projected form a primary auxiliary view. Sec. 13.12.

sections *(n)*. A cutaway view of an object. Sec. 12.1 and 24.7.

shape *(v)*. to remove metal from a piece with a shaper. Sec. 11.13.

shaper *(n)*. A tool in which the stock is held in a vise. A single-pointed, non-rotating cutting tool similar to that used in the lathe cuts as it is forced forward in a straight line past the stationary work. The tool then returns to the starting position to take another cut. Sec. 11.13.

sheet-metal drawing *(n)*. Technical drawing used in the heating, ventalating, and air-conditioning industries. Sec. 1.5.

shrink rule *(n)*. A rule used by a patternmaker. The units on a shrink rule are slightly oversize. Sec. 11.2.

single-line or **one-line diagram** *(n)*. An electrical diagram that presents the various circuit information simply by means of single lines and standard graphical symbols. Sec. 23.4.

slope *(n)*. The true angle between a line and the horizontal. Sec. 25.2.

snap *(n)*. In CAD, a feature that causes the crosshairs to jump to preset points. Sec. 9.5.

specifications *(n)*. Written documents accompanying architectural drawings. Specifications describe the special properties of materials. Sec. 24.2.

sphere *(n)*. A solid geometric shape whose surface is at all points equally distant from its center.

spline *(n)*. A keyway, usually one of a series cut around a shaft or hole.

SPLINED HOLE

spotface *(v)*. To produce a round spot or bearing surface around a hole, usually with a spotfacer. Sec. 11.12.

SPOTFACE

sprue *(n)*. In casting, a hole in the sand leading to the *gate* that leads to the mold, through which the metal enters. Sec. 11.2.

spur gears *(n)*. Gears that connect parallel shafts. Sec. 20.5.

status display *(n)*. In CAD, an area on the computer screen that shows which options are turned on and off. It also shows the current coordinates of the crosshairs. Sec. 9.5.

stereolithography *(n)*. A rapid prototyping process in which the object is built up from a series of thin cross sections made of a light-sensitive plastic resin.

storage media *(n)*. Items, such as a hard disk, diskettes, CD-ROM, DVD's, and streaming tape, that are used to store computer data.

streaming tape *(n)*. A magnetic tape used to store computer data. Sec. 3.9.

stretchout line *(n)*. The perimeter of an item's base, or the total distance around the base, laid out in a straight line. Sec. 18.4.

structural drawing *(n)*. Technical drawing used in the construction industries where structural steel is used for large buildings and bridges.

subassembly *(n)*. A drawing that shows a group of related parts that form a unit of the whole machine. Sec. 16.11.

super video graphics array (SVGA) *(n)*. A type of computer monitor capable of displaying 1280 or more pixels horizontally and 1024 or more pixels vertically. Sec. 3.8.

surface broach *(n)*. A tool that machines a surface by being forced through or across the work. Each succeeding tooth bites deeper and deeper until the final teeth form the required hole or surface. Sec. 11.16.

symbolic thread symbols *(n)*. Symbols used for small diameters, under approximately 1″ on the drawing, to represent all forms of threads. Sec. 15.13.

T

tap *(n)*. A tool used to cut small internal threads. Sec. 15.12.

tap *(v)*, To cut relatively small internal threads with a tap. Sec. 15.12.

taper *(n)* Conical form given to a shaft or a hole. Also refers to the slope of a plane surface, Sec. 10.16.

tapped holes *(n)*. Small internal threads, usually cut with a tap. Sec. 15.12.

tapping *(n)*. The threading of a small hole with one or more taps. Sec. 11.12.

temper *(v)*. To reheat hardened steel to bring it to a desired degree of hardness. Sec. 11. 17.

template *(n)*. A guide or pattern used to mark out the work, guide the tool in cutting it, or check the finished product.

tin *(n)*. A silvery metal used in alloys and for coating other metals, such as tin plate.

title strip *(n)*. A block on a drawing that presents in an organized way all the necessary information not shown on the drawing itself. Sec. 16.3.

tolerance *(n)*. Total amount of variation permitted in limit dimensions of a part. Secs. 10.26, 10.27.

topographic map *(n)*. A map that shows physical features (such as lakes and mountains) and structures (like bridges and buildings). Sec. 22.1.

torus *(n)*. A geometric shape resembling a doughnut. Sec. 6.2.

triangle *(n)*. A plane figure bounded by three sides. Sec. 6.2.

triangulation *(n)*. The process of dividing a surface into a number of triangles and then transferring each of them in turn to the pattern. Sec. 18.18.

trim *(n)*. The CAD command that erases those parts of lines that overlap. Sec. 9.6.

truss *(n)*. A support formed of members connected to each other at joints. All of the truss members are in the same vertical plane. Sec. 21.1.

T-square *(n)*. A drawing instrument with a head and blade. Sec. 4.4.

turn *(v)*. To produce, on a lathe, an external cylindrical surface. Sec. 11.11.

two-point perspective *(n)*. A perspective drawing with two vanishing points. Sec. 17.31.

U, V

unidirectional system *(n)*. A dimensioning system in which the figures are lettered so as to be read from the bottom of the sheet. Sec. 10.6.

Unified National thread series *(n)*. One of the basic thread forms. Sec. 15.4.

vernier caliper *(n)*. A precision measuring instrument that measures to four places, or 1/10,000″. Sec. 11.10.

vertical letters *(n)*. On a drawing, letters that stand upright. Sec. 5.2.

view *(n)*. A drawing that shows an object from one position. There are different views, such as top, front, and right-side views.

voice recognition *(n)*. A computer input device. It consists of a microphone and special software that recognizes a person's unique voice patterns. Sec. 3.7.

volt *(n)*. The unit in which voltage is measured. Sec. 23.3.

voltage *(n)*. The difference in electrical pressure that causes the flow of electricity. Sec. 23.3.

W, X, Y, Z

web *(n)*. See **rib**.

weld *(v)*. To join two pieces of metal by means of heat, with or without the application of pressure. Sec. 11.9.

width auxiliary view *(n)*. An auxiliary view that, as its principal dimension, shows the width of the object. Sec. 13.3.

wiring diagram or **connection diagram** *(n)*. An electrical diagram that shows the connections of an installation, or the electrical devices or parts that comprise the circuit. Sec. 23.4.

Woodruff key *(n)*. A semicircular flat key. Sec. 15.28.

WOODRUFF KEYS

working drawing *(n)*. A complete drawing or set of drawings prepared so that the object represented can be built without additional information. Sec. 1.4.

working drawing assembly *(n)*. A combined detail and assembly drawing giving complete dimensions and notes for all parts. Sec. 16.10.

INDEX